LIBRARY

# Astronomy and Astrophysics Library

More information about this series at http://www.springer.com/series/848

Nicolas Thomas

# An Introduction to Comets

## Post-Rosetta Perspectives

Nicolas Thomas
Space Research and Planetary Sciences Division
Physics Institute
University of Bern
Bern, Switzerland

ISSN 0941-7834 ISSN 2196-9698 (electronic)
Astronomy and Astrophysics Library
ISBN 978-3-030-50576-9 ISBN 978-3-030-50574-5 (eBook)
https://doi.org/10.1007/978-3-030-50574-5

Cover illustration: Collage of an artist's impression of a comet seen in the sky, superposed on a precise 3D shape model of the nucleus of comet 67P/Churyumov-Gerasimenko with the Rosetta spacecraft (not to scale) orbiting (ESA, Rhiannon Thomas). The graphic is intended to symbolize the gain in knowledge (from naked-eye impressions to precise measurements) acquired using modern technology.

This Springer imprint is published by the registered company Springer Nature Switzerland AG.
The registered company address is: Gewerbestrasse 11, 6330 Cham, Switzerland

*For BLT, . . . . even if it's not what you originally intended!*

# Preface (Motivation and Scope)

The historical significance of the irregular appearance and motion of bright comets across the sky has often been referred to in scientific literature, and there can be no doubt that our ancestors would have been mightily impressed by celestial objects similar to the naked-eye comets such as C/1996 B2 (Hyakutake) and C/1995 O1 (Hale–Bopp) that have been seen in recent times. While this might be sufficient on its own to justify detailed scientific investigation of comets, the possibility that they might be relics from the Solar System formation process presents a more scientifically exciting reason for studying these objects.

Figure 1 shows the remarkable image of HL Tauri (HL Tau), a young T Tauri star, acquired with the Atacama Large Millimeter Array (ALMA). The image shows material in the form of a disc surrounding the parent star. Within the disc there are rings or gaps that have almost certainly been produced by the formation of proto-planets (e.g. Clery 2018; Pérez et al. 2019). The gravitational field of a proto-planet within the disc attracts material from its vicinity clearing out a ring around the star. What is noticeable, however, is that even though proto-planets have already formed in the disc, there is still a large amount of material in the disc which has not yet been accreted onto the proto-planets. Furthermore, this material may be masking smaller objects which are growing but are not yet large enough to clear a ring.

Our Solar System shows that once the system has fully evolved the regions between the major planets are essentially void. In our Solar System, sometime between the stage illustrated by the HL Tau image and today, the material between the planets was removed. Much of that material must have impacted other objects in our system, but some of it almost certainly did not. Close, rather than impacting, encounters with the planets and proto-planets would have resulted in significant orbit modification placing the objects on more eccentric orbits with larger aphelion distances. This process implies that objects that had yet to reach planetary size escaped the vicinity of the larger proto-planets and could potentially have survived through to the present day. Planetary system evolution codes are now able to explore this in a little more detail (e.g. Fig. 2). They show that for systems with a low initial amount of solid mass, only 30% of the material ends up in planets larger than one

**Fig. 1** The protoplanetary disc of HL Tauri (HL Tau) observed at sub-millimetre wavelengths by the Atacama Large Millimeter Array (ALMA). Image Credit: ALMA (ESO/NAOJ/NRAO), NSF

Earth mass, and hence, there is considerably more material residing in other reservoirs. Furthermore, as first Edgeworth and then Kuiper deduced (Edgeworth 1949; Kuiper 1951), planetary formation processes did not end at the orbit of Neptune. The orbital period of a small object around a central star is

$$P = 2\pi\sqrt{\frac{a_s{}^3}{GM}} \tag{1}$$

where $a_s$ is the semi-major axis, and $GM$ is the geopotential of the star which takes a value of $1.32712 \times 10^{20}$ m$^3$ s$^{-2}$ for our Sun. Assuming a circular orbit, the mean orbital speed is then

$$\bar{v} = \sqrt{\frac{GM}{a_s}} \tag{2}$$

where the proportionality to $\sqrt{(1/a_s)}$ shows that relative speeds in the outer Solar System decrease with heliocentric distance, $r_h$. This led to a reduction in the number

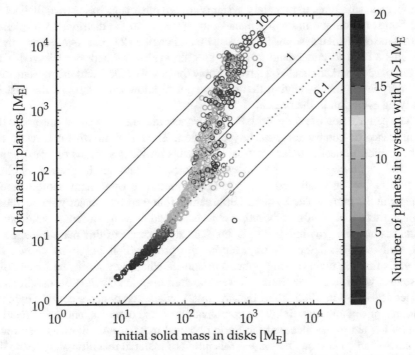

**Fig. 2** The total mass of material in formed planetary systems as a function of the initial solid mass in the discs computed from a planetary formation model by C. Mordasini (pers. comm.). The resulting amount of mass in the formed planetary systems is about 30% of total solid mass in the original disc for low initial amounts. This increases to over 100% (the planets can also accumulate light gases such as H and He) with increasing initial solid mass

of interactions between outer Solar System objects leaving a population of bodies beyond the orbit of Neptune that may have experienced relatively little change over the past 4.6 billion years.

It is the fact that comets are active that leads to the suspicion that they were involved in this process. This activity is produced by the sublimation of ice (mostly water ice) from the surfaces of small, irregular-shaped, solid nuclei. The probable presence of ices (including many that are far more volatile than water ice such as carbon monoxide) suggests that these objects have not been significantly thermally processed since their formation. Other objects in our Solar System have been heated through gravitational or collisional processes and have either lost their surface ice or the ice has been structurally modified (e.g. through melting). In the case of some objects, notably asteroids, water ice might still be present at depth, but this is, at least partially, the subject of speculation.

The possibility that the Solar System still contains remnants that have hardly been altered since the completion of the planetary formation process is a tantalizing prospect. Understanding the conditions under which Solar System formation began would clearly be a major step in trying to establish the frequency with which planetary systems like our own form. The current drive to determine the number and structures of planetary systems around other stars has shown that, while our Sun is not unique in having planets, the diversity in the distribution of planets within the other solar systems is far greater than imagined 25 years ago. Hence, there remains a need to understand how our specific system formed and evolved. Carl Sagan once remarked that "you have to know the past to understand the present" and one may invert this by saying that one can get to know something of the past by studying evolution in the present.

The perturbation of comet orbits by the planets raises further questions about the significance of comets for planetary evolution. Their motion with respect to the rather uniform quasi-circular orbits of the planets provides a means of transporting material over a large range of heliocentric distances. Although the present numbers of comets may be small, Fig. 1 also indicates that there were many more of these objects in the early Solar System. Perturbation followed by impact with accreting planets was a means of incorporating objects formed at many different heliocentric distances into the growing proto-planets. The significance of this mass transport is not well established, but its implications are profound. It has been known for 35 years that the main driving volatile in comets when they reach the inner Solar System is water ice. Furthermore, it was shown during the detailed observations of comet 1P/Halley in 1985–1986 that the less volatile components contain copious amounts of organic material. This combination of ice, organics, and large relative motion has led to the idea that the Earth obtained most, if not all, of its water and organics from comets and that it was this influx of material that ultimately led to the development of life. There is no doubt that the surfaces of the terrestrial planets have been impacted by comets many times over the lifetime of the Solar System. It should also be noted that meteor showers, which are the products of dust ejected from comets, enter the Earth's atmosphere on a regular basis. But the full significance remains unclear and provides further grounds for detailed investigation of cometary material.

Given the importance of cometary research, there have been rather few books on the subject and one of the motivations for this work is the absence of good introductory texts. Probably the closest in nature to the concept of this book is the 2010 third edition of *Physics of Comets* by K.S. Krishna Swamy. This has a number of excellent introductions to various aspects of cometary physics. It is quite focused on spectroscopy and gas emission, whereas here I have weighted the text more towards the nucleus and the innermost coma as a direct result of the observations of comet 67P/Churyumov–Gerasimenko by the European Space Agency's (ESA) Rosetta spacecraft. Similarly Brandt and Chapman's book *Introduction to Comets* from 2004 (2nd edition) emphasizes the plasma aspects of comets and is somewhat out of date with respect to the nucleus.

Although *Physics and Chemistry of Comets* edited by W.F. Huebner is now nearly 30 years old, much of the text remains relevant. Huebner collected eight chapters from experts on all aspects of cometary research and produced an excellent summary roughly 4 years after the Giotto encounter with 1P/Halley. On the other hand, it was written prior to the discovery of the first Kuiper–Edgeworth Belt Object in 1992 and well before several major space missions to comets (including not only

ESA's Rosetta but also NASA's Deep Space 1, Stardust, and Deep Impact) that have significantly enhanced our knowledge.

*Comets II*, the second book on comets in the University of Arizona Space Science series after the original book, *Comets*, edited by Wilkening in 1982, was edited by Festou, Keller, and Weaver, and is a collection of 37 20–30-page papers covering the state of cometary research prior to the launch of Rosetta in 2004. While each paper has a review nature, it does not really provide a simple introduction to the topic and is rather inhomogeneous in approach. But it is nonetheless extremely useful for looking for specific details.

The proceedings of the 35th Saas Fee Conference on Trans-Neptunian Objects and Comets contains a large chapter by Heike Rauer (2010) covering many aspects of cometary research and can also be considered as a good introduction to the subject. The chapters by Dave Jewitt ("The Kuiper Belt and comets") and by Alessandro Morbidelli ("Comets and their reservoirs") in this book are also very informative (Jewitt 2010; Morbidelli 2010).

Apart from the above, there are a number of books on specific aspects that will be referred to in the rest of the text. I draw attention here to some notable examples. The ISSI Scientific Report, "Heat and Gas Diffusion in Comet Nuclei", edited by W.F. Huebner et al. from 2006, provides a thorough analytical discussion of the physics of sublimation from ice–dirt mixtures. This work arose from Huebner's efforts to compare several models which were producing markedly different results for what were thought to be fundamental quantities in comet nucleus evolution. It is therefore almost a required starting point for students seeking to understand active processes on cometary nuclei.

A little known book edited by N. Kömle et al., *Theoretical Modelling of Comet Simulation Experiments*, provides insights into a set of laboratory experiments designed to look at cometary processes. These were carried out in the late 1980s in Germany under the title "KOSI" ("KOmeten-SImulation"). The individual papers here drew attention to the need for laboratory experiments but also detailed some of the difficulties in modelling the complex processes expected on cometary nuclei.

The textbook on *Gaskinetic Theory* by Tamas Gombosi is an excellent text on the approaches needed to understand and appreciate cometary gas outflow. The plasma distribution arising from a comet's interaction was studied in detail at the time of the Halley encounters. The multiple encounters at many positions with respect to the nucleus were of significant advantage in studying the phenomena. Tom Cravens's book on the *Physics of Solar System Plasmas* gives a brief review of these results but also includes the theoretical background and is an important reference text for space plasma physics. It is in some ways complementary to the book by Gombosi. Several chapters in *Introduction to Space Physics* edited by Kivelson and Russell are also of interest to students because of the simplicity with which fundamental concepts are brought forward.

Three useful books on the topic of celestial mechanics are *Solar System Dynamics* by Murray and Dermott, *Methods of Celestial Mechanics* by my former colleague, Gerhard Beutler, and *Celestial Mechanics: A Computational Guide for the Practitioner* by L.G. Taff (1985). The latter is slightly sneered upon by some of the professionals for its jokey style but, for amateurs (including myself), it is very clear.

Up until recently there has not been a really good treatment of photometry. Fortunately, Shepard has rectified this with his *Introduction to Planetary Photometry* in 2017. I have retained a section on this subject for practical reasons, but Shepard's book should be consulted for detailed definitions and derivations. In the field of spectroscopy and non-local thermal equilibrium processes, I have found *Non-LTE Radiative Transfer in the Atmosphere* by Lopez-Puertas and Taylor to be very useful.

Hans Rickman completed a book on *The Origin and Evolution of Comets Ten Years after the Nice Model and One Year after Rosetta* at the end of 2017 where the emphasis is on the origin and dynamics of comets with respect to Solar System evolution models. We will touch briefly on this subject here, but readers are referred to Rickman's text for greater detail.

It is with the relative absence of good introductory texts on comets in mind that the concept for this book has been developed while simultaneously discussing new information acquired at the Rosetta target comet. It is intended as a companion for an introductory course on cometary physics at master's level. Obviously, this drives certain decisions about content, and inevitably, I cannot reference every work about comets. This is all the more clear when you look in the ADS abstracting service and discover that between 2014 and 2019 alone, there were more than 4800 entries with the work "comet" in the title with over 1400 of them being refereed. As a consequence of this, I have tended to refer only briefly to certain types of object (e.g. sun-grazing comets for which Jones et al. (2018) have written a review) and limited discussions of the plasma environment (which has been covered recently in a review by Gombosi (2014)). Several authors have been rigorous in deriving important equations within their texts. (Both Cravens and Gombosi have done this extremely well, for example.) Here, derivations are limited often simply to reduce the time and space necessary to develop them. But it should be emphasized that without study of the mathematical detail, understanding can only be limited. Hence, students are encouraged to search out the more mathematical treatments to flesh out what is written here and references have been given where appropriate. Furthermore, this work is not intended as a comprehensive review. Reviewing 5000+ papers would be ludicrous and the reference list would end up being around 150 pages alone. However, I have tried to work with papers that are informative on the key topics. I apologize in advance if your particular favourite has not been tagged.

The text is divided into five main sections. We begin by looking at orbital dynamics and classification. In the next sections, we look at the nucleus, the gas distribution and composition, the dust emission, and the plasma interactions. This seems to be a natural way to divide the subject to provide structure, but the reader should be aware, even at this point, that treating one aspect in isolation is not a viable approach to the study of these complex objects, and it is the purpose of this book to provide sufficient background in all aspects of cometary research such that links between phenomena can be appreciated. I have included three smaller chapters at the end on related topics.

Since the first far spacecraft encounter with 21P/Giacobini-Zinner by NASA's ISEE-3 spacecraft in 1985, there have been numerous fly-bys of comets that have brought us a vast amount of new information (Table 1). Here, though, we do draw

**Table 1** Spacecraft that have encountered comets

| Spacecraft | Agency | Target | Type of encounter | Date of fly-by/ rendezvous | Closest approach | General Reference | Comment |
|---|---|---|---|---|---|---|---|
| ICE (originally ISEE-3) | NASA | 21P/Giacobini-Zinner | Fly-by | 11 Sept. 1985 | 7800 km | Farquhar (1983); Cowley (1987) | First fly-by of a comet |
|  |  | 1P/Halley | Far fly-by | End March 1986 | 28 Mkm |  |  |
| Vega 1 | Roscosmos | 1P/Halley | Fly-by | 6 March 1986 | 8889 km | Sagdeev et al. (1986) |  |
| Vega 2 | Roscosmos | 1P/Halley | Fly-by | 9 March 1986 | 8030 km |  |  |
| Sakigake | JAXA | 1P/Halley | Far fly-by | 11 March 1986 | 6.99 Mkm | Hirao (1986) |  |
| Suisei | JAXA | 1P/Halley | Fly-by | 8 March 1986 | 151,000 km |  |  |
| Giotto | ESA | 1P/Halley | Fly-by | 13 March 1986 | 596 km | Reinhard (1986) | Closest fly-by of 1P and highest resolution data |
| Deep Space 1 | NASA | 19P/Borrelly | Fly-by | 22 Sept. 2001 | 2171 km | Boice et al. (2002) |  |
| Stardust | NASA | 81P/Wild 2 | Fly-by | 2 Jan. 2004 | 237 km | Brownlee et al. (2003) |  |
| Deep Impact | NASA | 9P/Tempel 1 | Fly-by and impact | 4 Jul. 2005 | 575 km | A'Hearn et al. (2005) | First impactor on a comet |
| EPOXI (Deep Impact) | NASA | 103P/Hartley 2 | Fly-by | 4 Nov. 2010 | 700 km | A'Hearn et al. (2011) | Re-use of Deep Impact spacecraft |
| NExT (Stardust) | NASA | 9P/Tempel 1 | Fly-by | 15 Feb. 2011 | 181 km | Veverka et al. (2013) | Re-use of the Stardust spacecraft |
| Rosetta | ESA | 67P/Churyumov-Gerasimenko | Rendezvous | 6 Aug. 2014 to 30 Sept. 2016 | 0 km (impact) | Glassmeier et al. (2007) | First rendezvous |

heavily on results from the recently completed Rosetta mission to comet 67P/ Churyumov-Gerasimenko (hereafter 67P—it is the only object which I abbreviate purely to this type of designation). Rosetta was a mission designed to study the nucleus and its immediate environment. The data are available from the Planetary Science Archive (PSA) of ESA, and only Rosetta data available through this source have been used herein. The spacecraft, even when driven away from the nucleus by the activity as the comet approached perihelion, was never more than 2000 km from the nucleus during its prime mission. This is in stark contrast to earlier missions to comets which were fly-bys. Unlike Rosetta, these missions were also equipped with more extensive instrumentation for investigating the plasma produced by the ionization of cometary gases and the subsequent interaction of that plasma with the solar wind. While this plasma interaction is an interesting physical problem and its historical importance (e.g. the deduction of the existence of the solar wind through studies of cometary plasma tails by Biermann in 1951) cannot be underestimated, I have decided to tip the balance of this work towards the source of this material because, for the reasons discussed above, it is the wish to access this "unprocessed" nucleus that forms the driving goal in cometary physics today.

Bern, Switzerland                                                    Nicolas Thomas

# Acknowledgements

I would first like to thank the members of our group in Bern who have worked with me, over the past 15 years or so, on many of the aspects discussed. They are Ivano Bertini, Yann Brouet, Jo Ann Egger, Ramy El-Maarry, Clement Feller, Susanne Finklenburg, Selina-Barbara Gerig, Antonio Gracia Berná, Clemence Herny, Tra-Mi Ho, Bernhard Jost, Ying (Tracy) Liao, Raphael Marschall, Olga Pinzon, Olivier Poch, Antoine Pommerol, Michael von Gunten, and Zuriñe Yoldi. It will be apparent that there are many original diagrams in the text. This group has made a major contribution to these diagrams. Most have been financed through the grants from the Swiss National Science Foundation and National Centre for Competence in Research, PlanetS. Many others in Bern have also supported this work in some way, but I would like to mention Martin Rubin, Kathrin Altwegg, Andre Bieler, Helen Tzou, Peter Wurz, Andre Galli, Axel Murk, Martin Jutzi, Willy Benz, Christoph Mordasini, Yann Alibert, Simon Grimm, and Tina Rothenbühler for specific help.

We received a Horizon 2020 grant from the European Commission for a project called MiARD. The interaction with the groups included in MiARD has been extremely helpful. There were 40 or so persons involved in the project, but I would like to thank in particular the institutional leadership of Ekkehard Kührt, Laurent Jorda, Ian Wright, Paul Hartogh, Kokou Dadzie, James Whitby, and Rafael Rodrigo. It is perhaps a little unfair to single out others in the MiARD team, but I must express my appreciation for contributions from Chariton Christou, Olivier Groussin, Stubbe Hviid, David Kappel, Joerg Knollenberg, David Marshall, Stefano Mottola, Frank Preusker, Ladi Rezac, Frank Scholten, Jean-Baptiste Vincent, and John Zarnecki.

I have had many discussions (sometimes only brief within a workshop context) which have helped formulate approaches to the subjects addressed herein. I would therefore like to thank, in no particular order, Gerhard Schwehm, Matt Taylor, Horst Uwe Keller, Jong-Shinn Wu, Björn Davidsson, Vladimir Zakharov, Valentin Bickel, Colin Snodgrass, Carsten Güttler, Bastian Gundlach, Jürgen Blum, Yuri Skorov,

Sonia Fornasier, Yves Langevin, Jessica Sunshine, Casey Lisse, Geronimo Villanueva, Ludmilla Kolokolova, and Dominique Bockelée-Morvan.

Finally, I would like to thank the w's for their patience—especially at rugby, cricket, and gymnastics tournaments!

# Formal Acknowledgements

This research and document has made use of data and/or services provided by the International Astronomical Union's Minor Planet Center. The work also includes plots of database material from the JPL Horizons web site.

I have used data from the Rosetta instruments for many plots and images. I acknowledge the experiment teams for the provision of data to ESA's Planetary Science Archive and have used that archive for almost all original diagrams based on Rosetta data. In the following table (Table 2), I provide a list of the instruments and the principal investigators. This table also serves as an acronym list for the instruments in the rest of the text.

**Table 2** Rosetta instruments and principal investigators contributing data to the Planetary Science Archive used herein

| Acronym | Instrument type | Subsystem | Subsystem type | Principal investigator |
|---|---|---|---|---|
| **Orbiter** | | | | |
| OSIRIS | Imaging system | NAC | Narrow angle, high-resolution camera | H.U. Keller (original) H. Sierks (comet operations phase) (MPS, Göttingen) |
| | | WAC | Wide angle, low-resolution camera | |
| VIRTIS | Infrared spectrometer | M | Imaging low-resolution spectrometer | A. Coradini (original) F. Capaccioni (IAPS-INAF, Rome) |
| | | H | High-resolution spot spectrometer | |
| MIRO | Microwave spectrometer | | | S. Gulkis (original) M. Hofstetter (operations phase) (NASA/JPL, Pasadena) |

(continued)

**Table 2** (continued)

| Acronym | Instrument type | Subsystem | Subsystem type | Principal investigator |
|---|---|---|---|---|
| ROSINA | Ion and neutral mass spectrometer | DFMS | Dual focusing mass spectrometer | H. Balsiger (original) K. Altwegg (post-launch) (University of Bern) |
| | | RTOF | High-resolution reflection time-of-flight mass spectrometer | |
| | | COPS | Pressure (density) sensor | |
| COSIMA | Dust mass spectrometer and particle imager | COSISCOPE | Microscopic camera | J. Kissel (original) M. Hilchenbach (operations phase) (MPS, Göttingen) |
| MIDAS | Atomic force microscope | | | W. Riedler, K. Torkar, M. Bentley, and M. Thurid (IWF, Graz) |
| ALICE | UV imaging spectrometer | | | S.A. Stern (SWRI, Boulder) |
| RSI | Radio science investigation | | | M. Pätzold (University of Cologne) |
| CONSERT | Radio sounding experiment | | | W. Kofman (LPG, Grenoble) |
| GIADA | Dust counter | | | L. Colangeli (original) A. Rotundi (Università degli studi di Napoli "Parthenope") |
| RPC | Plasma sciences package | ICA | Ion composition analyzer | H. Nilsson (SISP, Kiruna) |
| | | IES | Ion and electron sensor | J. Burch (SwRI, San Antonio) |
| | | LAP | Langmuir probe | A. Eriksson (SISP, Uppsala) |
| | | MAG | Magnetometer | K.-H. Glassmeier (TU Braunschweig) |
| | | MIP | Mutual Impedance Probe (electron density and temperature) | J.G. Trotignon (original) P. Henri (LPC2E/CNRS, Orleans) |
| **Lander (Philae) PI—H. Rosenbauer (original), J.-P. Bibring, H. Boehnhardt (post-launch)** | | | | |
| APXS | Alpha-proton-X-ray spectrometer | | | G. Klingelhöfer (Gutenburg-University, Mainz) |

(continued)

**Table 2** (continued)

| Acronym | Instrument type | Subsystem | Subsystem type | Principal investigator |
|---|---|---|---|---|
| CIVA | Imaging system | P | Panoramic camera | J.-P. Bibring (IAS, Orsay) |
| | | M | Optical microscope/ IR imager | |
| COSAC | Gas chromatograph and mass spectrometer | | | H. Rosenbauer (original) F. Goesmann (operations) (MPS, Göttingen) |
| MUPUS | Heat probe | | | T. Spohn (Muenster/ DLR Berlin) |
| PTOLEMY | Gas chromatograph and mass spectrometer | | | I. Wright (Open University) |
| ROLIS | Imaging system | | | S. Mottola (DLR, Berlin) |
| ROMAP | Magnetometer | | | U. Auster (TU Braunschweig) |
| SD2 | Sampler, drill, and distribution system | | | A. Ercoli-Finzi (Politecnico di Milano, Milano) |
| SESAME | Mechanical and electrical properties suite | | | K.J. Seidensticker, (original) M. Knapmeyer (DLR, Köln, Berlin) |

# Acronyms[1]

| | |
|---|---|
| ALMA | Atacama Large Millimeter Array |
| BCCA | Ballistic cluster–cluster agglomeration |
| BPCA | Ballistic particle-cluster aggregation |
| BRDF | Bidirectional reflectance distribution function |
| CBE | Collisionless Boltzmann equation |
| CBOE | Coherent backscatter opposition effect |
| CME | Coronal mass ejection |
| DDA | Discrete dipole approximation |
| DSMC | Direct Simulation Monte Carlo |
| GMC | Giant molecular cloud |
| HIFI | Heterodyne instrument for the far infrared |
| HMC | Halley Multicolour Camera |
| HTC | Halley-type comet |
| iSALE | Impact simplified arbitrary Lagrangian-Eulerian (computer code) |
| JFC | Jupiter family comet |
| KBO | (Edgeworth-)Kuiper Belt Object |
| KHI | Kelvin–Helmholtz Instability |
| LAM | Long axis mode |
| LTE | Local thermodynamic equilibrium |
| MHD | Magnetohydrodynamics |
| MSPCD | Multi-resolution Stereo-photoclinometry by Deformation |
| NGF | Non-gravitational force |
| NPA | Non-principal axis |
| PSF | Point spread function |
| PSG | Planetary Spectrum Generator |
| RVR | Refractory to volatile ratio |

(continued)

---

[1]Acronyms associated with the names of Rosetta instruments (e.g., ALICE) and the names of spacecraft (e.g., ROSAT) have been excluded.

| SAM | Short axis mode |
|------|------|
| SDO | Scattered disc object |
| SFD | Size-frequency distribution |
| SHOE | Shadow-hiding opposition effect |
| SPC | Short-period comet/Stereo-photoclinometry |
| SPG | Stereo-photogrammetry |
| SPH | Smooth particle hydrodynamics |
| SRP | Solar radiation pressure |
| STP | Standard temperature and pressure |
| SWAN | Solar wind anisotropy instrument |
| TNO | Trans-Neptunian object |
| TNT | Trinitrotoluene |
| VCDT | Vienna-Canyon Diablo Troilite |
| VDF | Velocity distribution function |
| VSMOW | Vienna Standard Mean Ocean Water |

# Symbols Used

| Symbol | Meaning | Typical unit | Introducing equation |
|---|---|---|---|
| $a$ | Particle radius | [m] | 2.133 |
| $A\ (J', J_l)$ | Transition probability | | 3.119 |
| $A, B, C, D$ | Fitting constants for equilibrium vapour pressure curves | | 2.93, 2.94 |
| $\overline{A}$ | Average reflectance value over a circle around the nucleus | | 4.60 |
| $A_l$ | Fractional area of the surface at $T_l$ | $[m^2]$ | 2.124 |
| $A_{1,2,3}$ | Coefficients of non-gravitational forces | $[AU\ day^{-2}]$ | 1.21 |
| $A_{fric}$ | Threshold friction velocity | Unitless | 2.133 |
| $Af\rho$ | Measure of the dust production rate | [m] | 4.58 |
| $a_g$ | Surface gravitational acceleration | $[m\ s^{-2}]$ | 2.28 |
| $A_G$ | A geometric factor supporting determination of the critical period | Unitless | 2.37 |
| $A_H$ | Directional-hemispherical albedo | Unitless | 2.4 |
| $a_{imp}$ | Impactor radius | [m] | 2.129 |
| $a_m$ | Largest liftable dust radius | [m] | 4.91 |
| $A_M$ | Minnaert albedo | Unitless | 2.76 |
| $a_n$ | Acceleration of a neutral resonance scattering | $[m\ s^{-2}]$ | 3.123 |
| $A_N$ | Normal albedo | Unitless | 2.79 |
| $a_{NGF}$ | Non-gravitational acceleration of the nucleus | $[m\ s^{-2}]$ | 1.26 |
| $a_p$ | Pore radius | [m] | 3.103 |
| $a_s$ | Semi-major axis | [AU] | 1 |
| $a_s(\theta)$ | Slope distribution function | $[rad^{-1}]$ | 2.74 |
| $a_{tu}$ | Tube radius | [m] | 3.105 |
| $A_V$ | Modified dimensionless velocity threshold for dust motion | Unitless | 2.135 |

(continued)

| Symbol | Meaning | Typical unit | Introducing equation |
|---|---|---|---|
| $A_{var}$ | Temporary variable used in the definition of the Pareto distribution | | 4.85 |
| $b$ | Impact parameter | [m] | 3.2 |
| $\mathbf{B}$ | Magnetic field vector | | 3.92 |
| $b_d$, $b_c$ | Power-law index for dust particle size calculations ($b_c$ is used to differentiate between the coma and the surface) | Unitless | 4.75, 4.102 |
| $B_{CB}$ | Model for coherent backscatter opposition effect (CBOE) | Unitless | 2.69 |
| $b_{far}$ | Impact parameter where optical depth effects are negligible | [m] | 4.107 |
| $B_{SH}$ | Model for shadow-hiding opposition effect (SHOE) | Unitless | 2.69 |
| $B_{\vartheta,\lambda}$ | Spectral distribution of radiated flux (in frequency and wavelength) | [W m$^{-2}$ sr$^{-1}$ Hz$^{-1}$], [W m$^{-2}$ sr$^{-1}$ m$^{-1}$] | 2.8, 2.9 |
| $c$ | Speed of light | [m s$^{-1}$] | 3.24 |
| $C_D$ | Drag coefficient | Unitless | 4.88 |
| $C_{ex}$ | Collisional excitation rate | [m$^{-3}$ s$^{-1}$] | 3.39 |
| $c_{mix}$ | Specific heat capacity of an ice–silicate mixture | [J kg$^{-1}$ K$^{-1}$] | 2.120 |
| $c_{NGF}$, $m_{NGF}$, $k_{NGF}$, $n_{NGF}$ | Constants for the determination of $g(r_h)$ | Unitless | 1.22 |
| $c_s$ | Speed of sound | [m s$^{-1}$] | 3.53 |
| $c_{SHC}$ | Specific heat capacity of a material | [J K$^{-1}$ kg$^{-1}$] | 2.104 |
| $C_x$ | Jacobi constant with respect to planet x | | 1.18 |
| $D$ | A size distribution where $D_{min}$ and $D_{max}$ are the minimum and maximum sizes possible in the size-frequency distribution (Pareto distribution) | | 4.82 |
| $d_{el}$ | Electrical skin depth of a layer | [m] | 2.110 |
| $D_f$ | Fractal mass dimension | Unitless | 4.37 |
| $D_g$ | Gas diffusion coefficient | [m$^2$ s$^{-1}$] | 2.136 |
| $D_K{}^{eff}$ | Effective diffusivity | [m$^2$ s$^{-1}$] | 3.104 |
| $d_l$ | Lever arm with respect to rotation axis | [m] | 4.114 |
| $d_m$ | Molecular diameter | [m] | 3.100 |
| $dQ_d/da$ | Differential particle size production rate for particle radius $a$ | | 4.58 |
| $D_p$ | Pore area fractal dimension | Unitless | 2.119 |
| $d_t$ | Thermal diffusivity | [m$^2$ s$^{-1}$] | 2.103 |
| $D_t$ | Tortuosity fractal dimension | | 2.119 |
| $\mathbf{E}$ | Electric field vector | [V m$^{-1}$] | 2.131 |
| $E$ | Quantized potential energy | [J] | 3.23 |

(continued)

| Symbol | Meaning | Typical unit | Introducing equation |
|---|---|---|---|
| $E$ | Source–sink term in the energy conservation equation | [J] | 3.58 |
| $e, i$ | Angle of emission (incidence) | [rad] | 2.3 |
| $E_a$ | Eccentric anomaly | [rad] | 1.7 |
| $E_a$ | Maximum distance from the nucleus in sunward direction | [m] | 4.74 |
| $E_{in, out}$ | Energy input, output for energy conservation | [J] | 2.99 |
| $E_J$ | Energy of a molecule in a rotational level with total angular momentum quantum number J | [J] | 3.33 |
| $E_l$ | Energy of the level | [J] | 3.36 |
| $E_n$ | Energy transfer to the neutral by collision | [J] | 3.117 |
| $E_{o,p}, I_{o,p}$ | Energy level (spin angular momentum) for ortho-para calculations | [J] | 3.127 |
| $E_{rad}$ | Power that is thermally re-radiated by a particle | [W] | 4.118 |
| $E_{rot}$ | Rotational kinetic energy | [J] | 2.38 |
| $e_s$ | Total specific energy | | 3.67 |
| $E_{sol}$ | Absorbed solar energy per second by a particle | [W] | 4.117 |
| $e_t$ | A transmission angle for thermal radiation through a layer | [rad] | 2.110 |
| $e_x$ | Orbital eccentricity | Unitless | 1.4 |
| $f$ | True anomaly | [rad] | 1.6 |
| $f$ | Filling factor | Unitless | 4.58 |
| $F$ | Flux from a surface | [W m$^{-2}$] | 2.59 |
| $F$ | External force (e.g., gravity) in the Euler equations | [N] | 3.55 |
| $f(v)$ | Velocity distribution function of gas molecules | | 3.45, 3.84 |
| $F_\odot(z)$ | Insolation at a certain depth, z | [W m$^{-2}$] | 2.113 |
| $F_\odot$ | Solar flux at 1 AU | [W m$^{-2}$ nm$^{-1}$] | 2.1 |
| $f_A$ | Axial ratio of the spheroid | Unitless | 2.37 |
| $F_A$ | Applied force | [N] | 4.114 |
| $F_b$ | Fraction of backscattered molecules | Table 3.7 | |
| $F_C$ | Cohesive forces | [J] | 4.96 |
| $F_C, F_{ig}$ | Plasma drag force, intergrain Coulomb force | [N] | 4.124 |
| $f_{col}$ | Collision frequency | [s$^{-1}$] | 3.116 |
| $F_D$ | Drag force | [J] | 4.87 |
| $f_e$ | Electron impact ionization frequency | | 5.34 |
| $F_E$ | Electric force on dust particle | [N] | 2.131 |

(continued)

| Symbol | Meaning | Typical unit | Introducing equation |
|---|---|---|---|
| $F_{gss}$ | Perturbations arising from other massive objects in the Solar System | [N] | 4.124 |
| $f_i$ | Ionization frequencies for electron impacts with water molecules | [s$^{-1}$] | 5.11 |
| $F_i$ | Recoil force acting on the i$^{th}$ facet of a body | [N] | 1.24 |
| $f_{ice}$ | Mass fraction of ice | | 2.120 |
| $F_N$ | Emitted thermal flux from a resolved surface | [W m$^{-2}$] | 2.124 |
| $F_N$ | Reflected flux from unresolved nucleus in visible wavelength | [W m$^{-2}$] or [W m$^{-2}$ nm$^{-1}$] | 2.1 |
| $f_{osc}$ | Oscillator strength for a particular transition | | 3.119 |
| $f_{ph}$ | Photo-ionization frequency | | 5.34 |
| $F_{pr}$ | Radiation pressure force | [N] | 4.67 |
| $F_r, F_t$ | Radial and transverse forces acting on the nucleus | [N] | 1.23 |
| $F_{ref}$ | The reflected flux from an unresolved nucleus seen by an observer | [W m$^{-2}$] | 2.6 |
| $f_{k_T,ph}$ | Photo-ionization frequency for species k$_T$ | | 5.6 |
| $f_{tan}$ | Effective tangential fraction of the theoretical volatile flux at zero solar phase angle that contributes to a net torque | Unitless | 2.35 |
| $F_{th}$ | Thermal flux from an unresolved isothermal object | [W m$^{-2}$] | 2.13 |
| $F_{vdW}$ | Van der Waals force | [J] | 4.95 |
| $f_{vol}$ | Factor that defines escaping volatile content of dust | | 4.127 |
| $g$ | Gravitational acceleration | [m s$^{-2}$] | 2.129 |
| $g(r_h)$ | Function intended to account for variation of outgassing strength with heliocentric distance | Unitless | 1.22 |
| $G_d$ | Integral of the dust column density on a circle about a point source | [kg m$^{-1}$] | 4.3 |
| $g_f$ | Band emission rate | [photon s$^{-1}$ molecule$^{-1}$] | 3.38 |
| $G_{g,c,s}$ | Products of the impact parameter and the integral of the column density on a circle multiplied by variables to produce invariant values in the free radial expansion approximation | | 3.5–3.7 |
| $GM$ | Geopotential of an object | [m$^3$ s$^{-2}$] | (1) |
| $h$ | Planck's constant | [J s] | 2.8 |
| $h_e$ | Specific enthalpy | | 3.58 |
| $H$ | Hamaker constant | [J] | 4.95 |
| $H(\mu, \mu_0)$ | Functions within the Hapke formalism for reflectance | | 2.66 |

(continued)

| Symbol | Meaning | Typical unit | Introducing equation |
|---|---|---|---|
| $H_D$ | Surface heat content | [J] | 2.106 |
| $I$ | Radiance (intensity) from reflection, scattering, or emission | [W m$^{-2}$ sr$^{-1}$] | 2.3 |
| $I$ | Emission intensity from a vibrational band | [W m$^{-2}$ sr$^{-1}$] | 3.38 |
| $I, Q, U, V$ | Stokes parameters | | 4.6 |
| $I_\perp, I_\parallel$ | Intensity of electromagnetic field perpendicular (parallel) to scattering plane as scattered by a particle | | 4.122 |
| $I_{a,b,c}, I_N$ | Moment of inertia (axes) and total for the nucleus | [kg m$^2$] | 2.33 |
| $I_i$ | Incident irradiance | | 4.20 |
| $I_L$ | Radiance from a perfectly diffusing Lambertian surface | [W m$^{-2}$ sr$^{-1}$] | 2.80 |
| $I_M$ | Moment of inertia of a molecule | | 3.34 |
| $I_{ref}$ | Reflected spectral intensity from a particle | [W m$^{-2}$ sr$^{-1}$ nm$^{-1}$] | 4.119 |
| $I_s, I_l$ | Moment of inertia about the short (long) axis of an irregular object | [kg m$^2$] | 2.38 |
| $I_{th}$ | Thermal spectral intensity | | 4.121 |
| **Iv, Fu, Ro** | Coefficients for the efficiency of the entrainment of a particle within the gas flow, the gravitational interaction, and radiation pressure effects in the Zakharov formulation | | 4.99 |
| $i_x$ | Orbital inclination | [rad] | 1.17 |
| $I_\vartheta$ | Radiance in a frequency interval | [W m$^{-2}$ sr$^{-1}$ Hz$^{-1}$] | 3.10 |
| $J$ | Total angular momentum quantum number | Unitless | 3.33 |
| $J', J_i$ | Energy levels in a resonance scattering process | | 3.119 |
| $j_{flu}$ | Flux from a reservoir into vacuum | [molecule m$^{-2}$ s$^{-1}$] | 3.101 |
| $j_x$ | Gas flux | [m$^{-2}$ s$^{-1}$] | 2.136 |
| $j_\upsilon$ | Emission coefficient | [W m$^{-3}$ sr$^{-1}$ Hz$^{-1}$] | 3.15 |
| $J_\upsilon$ | Source function | [W m$^{-2}$ sr$^{-1}$ Hz$^{-1}$] | 3.17 |
| $k$ | Boltzmann's constant | [J K$^{-1}$] | 2.8 |
| $K$ | Shape factor | Unitless | 2.54 |
| $k_E$ | Parameter of the Lunar-Lambert function | Unitless | 2.79 |
| $k_{in}$ | Ion-neutral momentum transfer collision rate | [cm$^3$ s$^{-1}$] | 5.33 |
| $k_M$ | Parameter of Minnaert function | Unitless | 2.76 |
| $k_N$ | Normalization constant for B$_\lambda$ | | 2.10 |
| $k_N$ | Wavenumber of incident radiation | [m$^{-1}$] | 4.17 |
| $Kn, Kn_0$ | Knudsen number and Knudsen number of the source | Unitless | 3.47, 3.49 |

(continued)

| Symbol | Meaning | Typical unit | Introducing equation |
|---|---|---|---|
| $Kn_p$ | Knudsen penetration number | Unitless | 3.99 |
| $k_p, k_p"$ | Proportionality constants connected to dust size distributions | | 4.75, 4.77 |
| $k_s$ | Sublimation coefficient | Unitless | 2.96 |
| $k_T$ | Species type used to indicate different species | | 4.43 |
| $L$ | Angular momentum | [kg m$^2$ s$^{-1}$] | 2.32 |
| $L_0$ | Characteristic scale length for conductive path | [m] | 2.119 |
| $L_{Kn}, L_0$ | Characteristic dimension of a gas source | [m] | 3.47 |
| $L_S, L_{H2O}$ | Enthalpy (latent heat) of phase change (and specifically for water ice sublimation) | [J kg$^{-1}$] | 2.88 |
| $L_{th}$ | Thermal luminosity | [W] | 2.12 |
| $L_{th}$ | Thickness of a layer | [m] | 3.104 |
| $L_{tu}$ | Length of the tube | [m] | 3.105 |
| $M$ | Reactive torque caused by activity distribution | [kg m$^2$ s$^{-2}$] | 2.31 |
| $M(\mu_0,\mu)$ | Model for multiple scattering within the Hapke formalism | | 2.69 |
| $M_\odot$ | Solar mass | [kg] | 1.12 |
| $m_\odot$ | Astronomical magnitude of the Sun | Unitless | 2.2 |
| $m_0$ | Mass of the individual monomer | [kg] | 4.38 |
| $m_1$ | Observed visual magnitude of a comet | Unitless | 1.1 |
| $m_{1,2}$ | Atomic masses of two species | [kg] | 3.34 |
| $M_a$ | Mean anomaly | [rad] | 1.8 |
| $Ma$ | Mach number | Unitless | 3.53 |
| $m_e, m_n$ | Mass of electron (neutral) | [kg] | 3.123 |
| $m_g$ | Mass of a molecule | [kg] | 3.54 |
| $m_{H2O}$ | Molecular mass of water (or another species) | [kg] | 2.35 |
| $m_i, m_n$ | Mass of an ion (neutral) | [kg] | 3.117 |
| $m_{imp}$ | Impactor mass | [kg] | 2.128 |
| $M_M$ | Molar mass | [kg mol$^{-1}$] | 3.24 |
| $m_N$ | Magnitude of the nucleus | Unitless | 2.2 |
| $M_N$ | Mass of the nucleus | [kg] | 1.23 |
| $m_p$ | Mass of a projectile | [kg] | 7.1 |
| $M_p$ | Mass of a planet | | 2.87 |
| $m_{ref}$ | Refractive index | Unitless | 4.27 |
| $m_{ref,r}, m_{ref,i}$ | Real and imaginary parts of the refractive index | | 4.27 |
| $M_s$ | Sonic Mach number for the solar wind | Unitless | 5.22 |
| $\tilde{n}$ | Mean motion of masses around barycentre | [rad s$^{-1}$] | 1.8 |
| $N$ | A number of observations or objects in a bin | | 4.83 |
| $n_0$ | Number density of an imaginary sub-surface reservoir of gas molecules | [m$^{-3}$] | 3.80 |

(continued)

| Symbol | Meaning | Typical unit | Introducing equation |
|---|---|---|---|
| $N_A$ | Avogadro number | Unitless | 2.90 |
| $N_d$ | Dust column density | [molecule m$^{-2}$] | 4.2 |
| $n_{dau}$ | Number density of the daughter product | | 5.4 |
| $n_e$ | Electron density | [m$^{-3}$] | 3.39 |
| $n_{g,d}$ | Local gas (dust) density | [molecule m$^{-3}$], [kg m$^{-3}$] | 2.98, 2.136, 4.1 |
| $n_{gp}$ | Local parent density | [m$^{-3}$] | 3.111 |
| $n_h$ | Power-law exponent describing the drop in brightness with $r_h$ | Unitless | 1.3 |
| $n_i$, $n_n$ | Ion (neutral) density | [m$^{-3}$] | 3.116 |
| $N_m$ | Number of monomers in a particle structure | Unitless | 4.41 |
| $n_N$ | Local density of the neutral (similar to $n_g$) | [m$^{-3}$] | 3.39 |
| $N_p$ | Number of particles per unit volume | # | 2.116 |
| $n_{par}$ | Number density of a parent molecule | [molecule m$^{-3}$] | 5.4 |
| $N_s$ | Column density from the Sun to a point in the coma sunward of the nucleus | [m$^{-2}$] | 5.7 |
| OPR | Ratio of ortho-$H_2$ to para-$H_2$ | | 3.127 |
| $p$ | Geometric albedo | Unitless | 2.1 |
| $P$ | Orbital period | [s] | 1 |
| $\boldsymbol{P}$ | Position in space | [m] | 2.30 |
| $\mathbb{P}$ | Degree of polarization | Unitless | 4.11 |
| $p(\alpha)$ | Phase function for single scattering at a phase angle, $\alpha$ by atoms or molecules | | 3.118 |
| $p_{a,b}$, $T_{a,b}$ | Pressure and temperature on either side of a layer or tube | [Pa] | 3.104 |
| $P_{crit}$ | Critical period either of an oblate spheroid for the splitting of the nucleus or of the balance between centrifugal and gravitational acceleration | [s] | 2.36, 2.57 |
| $p_d$ | Geometric albedo of dust particles | Unitless | 4.35 |
| $p_{dyn}$ | Dynamic pressure | [Pa] | 5.1 |
| $p_L$ | Semi-latus rectum of an orbit | [AU] | 1.23 |
| $P_l$ | Population in a quantized energy level | | 3.36 |
| $\mathbb{P}_L$, $\mathbb{P}_C$ | Degree of linear (circular) polarization | Unitless | 4.13, 4.14 |
| $p_{mag}$ | Magnetic pressure | [Pa] | 5.2 |
| $p_s$ | Equilibrium vapour pressure | [Pa] | 2.89 |
| $p_s$ | Albedo normalized to geometric albedo in backscattering geometry but follows the behaviour of the scattering function | Unitless | 4.36 |
| $p_{STP}$ | Pressure under STP conditions | [Pa] | 3.31 |
| $q$ | Periapsis distance for parabolic motion | [m] | 1.12 |
| $\boldsymbol{q}$ | Vectorial heat flux | [W m$^{-2}$] | 3.65 |
| $q_c$ | Particle charge | [C] | 2.131 |

(continued)

| Symbol | Meaning | Typical unit | Introducing equation |
|---|---|---|---|
| $Q_{col}$ | Specific impact energy of a collision | | 7.1 |
| $Q_{d\_direct}$ | Production rate of dust particles removed from activity source and escaping | [molecule s$^{-1}$], [kg s$^{-1}$] | 4.126 |
| $Q_{d\_fallback}$ | Number of particles removed from activity source and falling back | [molecule s$^{-1}$], [kg s$^{-1}$] | 4.127 |
| $Q_{d\_outgassed}$ | Final refractory mass loss from the comet | [molecule s$^{-1}$], [kg s$^{-1}$] | 4.125 |
| $Q_{damp}$ | Quality factor of the material | Unitless | 2.54 |
| $q_f$ | Gas flux through a porous layer or through a tube | [m$^{-2}$ s$^{-1}$] | 3.104 |
| $Q_{g,d}$ | Gas (dust) production rate | [molecule s$^{-1}$], [kg s$^{-1}$] | 1.27, 2.112, 3.10, 4.1 |
| $Q_{g\_direct}$ | Direct gas production through activity | [molecule s$^{-1}$], [kg s$^{-1}$] | 4.125 |
| $Q_{g\_extended}$ | Gas production caused by volatiles being lost by escaping particles | [molecule s$^{-1}$], [kg s$^{-1}$] | 4.125 |
| $Q_{g\_indirect}$ | Gas lost from the comet through outgassing of dust that falls back | [molecule s$^{-1}$], [kg s$^{-1}$] | 4.125 |
| $Q_{gp}$ | Parent production at source in the Haser model | [molecule s$^{-1}$] | 3.107 |
| $q_p, q_d$ | Total number of parent (daughter) molecules crossing a spherical surface from a point source in the Haser model | [molecule s$^{-1}$] | 3.107, 3.112 |
| $Q_{pr}$ | Radiation pressure efficiency factor | Unitless | 4.68 |
| $Q_s$ | Source term in the mass conservation equation for fluids | | 3.50 |
| $Q_{sca}, Q_{ext}, Q_{abs}$ | Scattering, extinction, absorption efficiencies | Unitless | 4.33 |
| $Q_T$ | Total production rate from a defined source | [molecule s$^{-1}$] | 3.97 |
| $r$ | Cometocentric distance | [m] | 3.1 |
| $r_0$ | Heliocentric distance inside which bulk of solar insolation goes into subliming ice at the surface of the nucleus | [AU] | 1.22 |
| $r_a, r_p$ | Apoapsis (periapsis) distance | [AU] | 1.4 |
| $R_{BB}$ | Radiant exitance of a black body | [W m$^{-2}$] | 2.23 |
| $r_c$ | Gyroradius | [m] | 5.18 |
| $R_c$ | Characteristic radius of a particle | [m] | 4.40 |
| $r_{ce}$ | Radius of the electron collisionopause | [m] | 5.34 |
| $R_{CO_2}$ | Ratio of different isotopic compositions of $CO_2$ | Unitless | 3.124 |
| $r_{ej}$ | Distance from rotation axis in normal direction | [m] | 2.34 |
| $r_{eq}$ | Radius of the outer edge of the equlibrium regime in the coma | [m] | Fig. 3.2 |

(continued)

| Symbol | Meaning | Typical unit | Introducing equation |
|--------|---------|--------------|----------------------|
| $R_g$ | Ideal gas constant | $[\text{J mol}^{-1} \text{ K}^{-1}]$ | 3.24 |
| $R_{gyr}$ | Radius of gyration of a dust particle | [m] | 4.39 |
| $r_h$ | Heliocentric distance | [AU] | 1.1 |
| $r_{HD}$ | Radius of the hydrodynamic limit | [m] | 3.46 |
| $r_{Hill}$ | Hill radius | [m] | 4.108 |
| $r_i$ | Interatomic separation | [m] | 3.34 |
| $r_i$ | Distance of the individual monomer from the centre of mass of the whole particle | [m] | 4.38 |
| $R_i$ | Ionization frequency of cometary species | $[\text{s}^{-1}]$ | 5.25 |
| $r_{Knp}$ | Distance to the interaction axis from the source in the definition of the Knudsen penetration number | [m] | 3.97 |
| $r_m$ | Molecular radius introduced by expansion of the collision cross section | [m] | 3.98 |
| $R_m$ | Radius of an individual monomer within a particle | [m] | 4.39 |
| $r_{ml}$ | Substitution variable | | 5.28 |
| $r_N, r_{sn}$ | Nucleus radius (sub-nucleus radius) | [m] | 2.2 |
| $\boldsymbol{r_N}$, Italic | Radius vector of the $i^{th}$ facet in the body frame of the nucleus | [m] | 2.31 |
| $R_{obs}$ | Observed radiant exitance | $[\text{W m}^{-2}]$ | 2.24 |
| $r_{op}$ | Observer–particle distance | [m] | 4.18 |
| $\boldsymbol{r_p}$ | Vector from point $\mathbf{P}$ to a volume element d$V$ | [m] | 2.30 |
| $R_p$ | Planetary disturbing function | | 1.21 |
| $r_x$ | Distances of massless objects from a mass $x$ | [m] | 1.18 |
| $s$ | Distance along the line of sight | [m] | 3.2 |
| $S$ | Entropy | $[\text{J K}^{-1}]$ | 2.88 |
| $\mathbf{S}$ | Stokes vector | | 4.6 |
| $S'$ | Spectral gradient | $[\% \ (100 \text{ nm})^{-1}]$ | 2.85 |
| $S_\odot$ | Solar flux at 1 AU integrated over the full wavelength range | $[\text{W m}^{-2}]$ | 2.7 |
| $S_c$ | Numerical constant in a formulation for cohesive forces | Unitless | 4.96 |
| $s_d$ | Eroded depth through sublimation | [m] | 2.130 |
| $S_i$ | Ion source term | | 5.23 |
| $S_{jk}$ | Components of the Müller matrix | | 4.17 |
| $S_s$ | Spectral line intensity | $[\text{cm molecule}^{-1}]$ | 3.23 |
| $\mathbf{S}_{unp}, \mathbf{S}_{pol}$ | Completely unpolarized (completely polarized) component of the Stokes vector | | 4.9 |
| $T$ | Temperature (gas or surface) | [K] | 1.25 |
| $\boldsymbol{T, n}$ | Transverse and radial components to NGFs | | 1.21 |
| $T_\odot$ | Effective solar temperature | [K] | 4.119 |
| $T_\parallel$ | Temperature in direction parallel to flow | [K] | 3.82 |

(continued)

| Symbol | Meaning | Typical unit | Introducing equation |
|---|---|---|---|
| $T_\perp$ | Temperature in direction orthogonal to flow | [K] | 3.83 |
| $T_0$ | Temperature of an imaginary sub-surface reservoir of gas molecules | [K] | 3.82 |
| $T_b$ | Brightness temperature | [K] | 2.21 |
| $T_i$ | Temperature of the ith facet on a model nucleus | [K] | 1.25 |
| $T_i, T_n, T_e$ | Ion (neutral, electron) temperature | [K] | 3.117, 5.38 |
| $T_{PERI}$ | Time of periapsis | [s] | 1.8 |
| $t_R$ | Characteristic timescale for radiation | [s] | 2.108 |
| $T_S$ | Tensile strength | [N m$^{-2}$] | 2.36 |
| $t_{STP}$ | Mean time between collisions at standard temperature and pressure (STP) | [s] | 3.29 |
| $T_{STP}$ | Temperature under STP conditions | [K] | 3.31 |
| $T_x$ | Tisserand parameter with respect to planet x | Unitless | 1.17 |
| $u$ | Bulk or drift velocity of a gas or plasma in a flow | [m s$^{-1}$] | 3.45 |
| $\boldsymbol{u}$ | Bulk fluid velocity vector in the conservation equations for the Boltzmann equation | [m s$^{-1}$] | 3.95 |
| $U$ | Impactor velocity | [m s$^{-1}$] | 2.128 |
| $u_\infty$ | Upstream solar wind speed | [m s$^{-1}$] | 5.27 |
| $u_{sw}$ | Flow speed of solar wind | [m s$^{-1}$] | 4.124 |
| $u_{th}$ | Fluid threshold velocity | [m s$^{-1}$] | 2.133 |
| $\vartheta$ | Frequency | [Hz] | |
| $V$ | Volume | [m$^3$] | 2.89 |
| $v_\perp, v_\parallel$ | Velocity perpendicular (parallel) to the field | [m s$^{-1}$] | 5.14 |
| $\bar{v}$ | Mean orbital speed | [m s$^{-1}$] | 2 |
| $v_\infty$ | Gas terminal velocity | [m s$^{-1}$] | 3.44 |
| $v_0$ | Gas velocity at the surface | [m s$^{-1}$] | 4.91 |
| $v_a$ | Velocity of a spherical particle | [m s$^{-1}$] | 4.89, 4.124 |
| $v_{col}$ | Collision velocity | [m s$^{-1}$] | 7.1 |
| $v_{ej}$ | Velocity of ejected mass in direction normal to the surface | [m s$^{-1}$] | 2.35 |
| $v_{esc}$ | Escape velocity | [m s$^{-1}$] | 2.29 |
| $v_g, v_d$ | Gas (dust) velocity | [m s$^{-1}$] | 2.98, 4.1 |
| $v_{gp}$ | Velocity of the parent in the Haser model | [m s$^{-1}$] | 3.108 |
| $v_h$ | Heliocentric velocity | [m s$^{-1}$] | 3.120 |
| $v_{H2O}$ | Effective velocity with which water molecule leaves a particle | [m s$^{-1}$] | 4.112 |
| $v_i$ | Effective velocity of ejected material at the source from the i$^{th}$ facet | [m s$^{-1}$] | 1.24 |
| $V_{imp}$ | Volume of an impact crater | [m$^3$] | 2.97 |
| $v_q$ | Vibrational quantum number | Unitless | 3.35 |

(continued)

| Symbol | Meaning | Typical unit | Introducing equation |
|---|---|---|---|
| $v_r$ | Rotational velocity | [m s$^{-1}$] | 2.56 |
| $v_{rel}$ | Relative velocity | [m s$^{-1}$] | 3.39 |
| $v_{th}$ | Most probable velocity in an equilibrated gas | [m s$^{-1}$] | 3.28 |
| $v_{th}$ | Thermal expansion velocity of the gas at the surface multiplied by a correction factor to allow for expansion effects | [m s$^{-1}$] | 4.92 |
| $v_v$ | Harmonic wavenumber | | 3.35 |
| $V_V$ | Volume of the voids | [m$^3$] | 2.115 |
| $V_x$ | Unperturbed relative velocity at which a comet encounters planet x | Units of the planet's orbital velocity | 1.19 |
| $w_l$ | Statistical weight | | 3.36 |
| $W_s$ | Scattered power | [W] | 4.20 |
| $x$ | Size parameter | Unitless | 4.25 |
| $x$ | Distance along the line of sight towards the Sun from the plane orthogonal to the Sun-comet line containing the nucleus | [m] | 5.7 |
| $X$ | Anharmonicity constant | | 3.35 |
| $X*$ | Quantity X (e.g. gas velocity) at the sonic point | | 3.71 |
| $x_l$ | Thermal skin depth | [m] | 2.105 |
| $x_i$ | Effective active fraction of the surface emitting at the given production rate from the i$^{th}$ facet | Unitless | 1.24 |
| $X_{Kn}$ | Any macroscopic variable (e.g. density) | | 3.48 |
| $Y$ | Material strength parameter | | 2.128 |
| $z$ | Depth below surface or depth of a bed of material | [m] | 2.102, 2.132 |
| $z$ | Distance between the sources | [m] | 3.100 |
| $Z(T)$ | Gas emission (sublimation) rate | [molecule m$^{-2}$ s$^{-1}$] | 2.35 |
| $z_0$ | Particle to surface distance | [m] | 4.95 |
| $Z_i$ | Mass loss rate per unit area from the i$^{th}$ facet | [molecule m$^{-2}$ s$^{-1}$] | 1.24 |
| $Z_T$ | Partition function | | 3.37 |
| $\alpha$ | Phase angle | [rad] | 2.1 |
| $\alpha_B, \beta_B$ | Coefficients in Barker's equation | [rad] | 1.13, 1.14 |
| $\alpha_D$ | Doppler line width | [Hz] | 3.23 |
| $\alpha_L$ | Pressure broadening line width | [Hz] | 3.29 |
| $\alpha_{pa}$ | Pitch angle | [rad] | 5.19 |
| $\beta$ | Ratio of radiation pressure force to gravitational force | Unitless | 4.72 |

(continued)

| Symbol | Meaning | Typical unit | Introducing equation |
|---|---|---|---|
| $\beta_1, \beta_2$ | Reciprocal scale length of the parent and daughter species in the Haser model | $[m^{-1}]$ | 3.107, 3.112 |
| $\beta_A, \chi$ | Angles in the definition of polarization | [rad] | 4.11 |
| $\beta_{mag}$ | Relative importance of magnetic pressure with respect to thermal pressure | Unitless | 5.3 |
| $\gamma$ | Ratio of the specific heats | Unitless | 3.60 |
| $\Gamma$ | Thermal inertia of a surface | $[J\ m^{-2}\ K^{-1}\ s^{-\frac{1}{2}}]$, [TIU] | 2.107 |
| $\gamma_c$ | Empirical constant describing cohesive forces in saltation models | | 2.135 |
| $\gamma_H$ | Variable related to the single-scattering albedo used in the Hapke formalism for reflectance | Unitless | 2.67 |
| $\Delta$ | Comet-Earth distance | [AU] | 1.1 |
| $\delta^{13}C$ | Isotopic enrichment relative to a standard | "per mil" | 3.125 |
| $\delta_{imp}$ | Mass density of the impactor | $[kg\ m^{-3}]$ | 2.128 |
| $\Delta\delta$ | Isotopic fractionation | | 3.126 |
| $\Delta\omega$ | Precession of the argument of perihelion from relativistic effects | Arcseconds per comet orbit about the Sun | 1.20 |
| $\varepsilon$ | Thermal emissivity | Unitless | 2.7 |
| $\varepsilon_{MG}$ | Average dielectric function derived by Maxwell Garnett | | 4.49 |
| $\varepsilon_p, \varepsilon_h, \varepsilon_i$ | Permittivity of the material, a host material, and an inclusion | Unitless | 4.49, 4.50 |
| $\varepsilon_r, \varepsilon_i$ | Real and imaginary parts of the dielectric constant | | 4.27 |
| $\varepsilon_s$ | Parameter accounting for specular reflection from the surface of a sphere | | 4.88 |
| $\eta$ | Momentum transfer coefficient | Unitless | 1.24 |
| $\eta_{th}$ | Infrared beaming parameter | Unitless | |
| $\theta$ | Scattering angle ($\pi$-$\alpha$) | [rad] | 4.5 |
| $\bar{\theta}$ | Mean slope angle | [rad] | 2.75 |
| $\Theta$ | Thermal parameter | Unitless | 2.109 |
| $\nu$ | Viscosity | | 3.104 |
| $\kappa$ | Thermal conductivity | $[W\ m^{-1}\ K^{-1}]$ | 2.102 |
| $\kappa_m$ | Effective conductivity | $[J\ m^{-1}\ K^{-1}]$ | 2.117 |
| $\kappa_p$ | Radiative conductivity across pores | $[J\ m^{-1}\ K^{-1}]$ | 2.117 |
| $\kappa_p$ | Permeability | Unitless | 3.104 |
| $\kappa_s, \kappa_g, \kappa_e$ | Conductivities (solid, gas, effective) supporting empirical determination of heat transfer in porous media | $[J\ m^{-1}\ K^{-1}]$ | 2.119 |
| $\kappa_v$ | Absorption coefficient | $[m^{-1}]$ | 3.10 |
| $\lambda$ | Wavelength | [m] | 2.9 |

(continued)

| Symbol | Meaning | Typical unit | Introducing equation |
|---|---|---|---|
| $\lambda_{ce}$ | Mean free path for electrons | [m] | 5.36 |
| $\lambda_M$ | Photometric longitude | [rad] | 2.77 |
| $\lambda_{max}$ | Maximum pore size | [m] | 2.119 |
| $\lambda_{MFP}$ | Mean free path | [m] | 3.42 |
| $\lambda_{ref}$ | Reference wavelength | [m] | 2.86 |
| $\lambda_{s,th}$ | Photoionization threshold for species s | | 5.6 |
| $\mu$, $GM$ $(GM_\odot)$ | Geopotential of a star (the Sun) | $[m^3\ s^{-2}]$ | 1 |
| $\mu$ | Cosine of the emission angle, e | Unitless | 2.65 |
| $\mu_0$ | Cosine of the incidence angle, i | Unitless | 2.60 |
| $\mu_0$ | Permeability of free space | | 5.2 |
| $\mu_{imp}, \nu_{imp}$ | Coefficients in the impact cratering equations | | 2.129 |
| $\mu_r$ | Reduced mass of the nucleus and the projectile | | 7.2 |
| $\mu_{rig}$ | Rigidity | [Pa] | 2.54 |
| $\mu_v$ | Coefficient of dynamic viscosity | | 3.63, 4.94 |
| $\vartheta$ | Frequency | [Hz] | 2.8 |
| $\nu_0$ | Rest frequency of emission line | [Hz] | 3.23 |
| $\nu_k$ | Kinematic viscosity | | 3.62 |
| $\xi$ | Asymmetry parameter | Unitless | 2.71 |
| $\xi_P$ | Light penetration scale length | [m] | 2.114 |
| $\xi_T$ | Tortuosity | Unitless | 3.103 |
| $\rho$ | Radial size of a circular aperture on the sky | [km] | 4.58 |
| $\rho_\infty$ | Upstream solar wind mass density | $[kg\ m^{-3}]$ | 5.28 |
| $\rho_a$ | Dust particle bulk density | $[kg\ m^{-3}]$ | 4.66 |
| $\rho_C$ | Reflectance coefficient | Unitless | 2.60 |
| $\rho_{dl}$ | Reflectance of a large dust particle | Unitless | 4.109 |
| $\rho_F$ | Reflectance factor, REFF | Unitless | 2.62 |
| $\rho_g$ | Local gas mass density | $[kg\ m^{-3}]$ | 2.132 |
| $\rho_N, \rho_d$ | Bulk mass density of a body (dust grains) | $[kg\ m^{-3}]$ | 2.128 |
| $\rho_{ref}$ | Reflectance factor at a reference wavelength | Unitless | 2.85 |
| $\rho_s$ | Particle mass density | $[kg\ m^{-3}]$ | 2.132 |
| $\rho_{s-d}$ | Reflectance of a single particle at a specific phase angle | Unitless | 4.110 |
| $\rho_{sw}$ | Mass density of the solar wind | | 5.1 |
| $\rho_V$ | Density of a volume element dV | $[kg\ m^{-3}]$ | 2.30 |
| $\sigma$ | Stefan–Boltzmann constant | $[W\ m^{-2}\ K^{-4}]$ | 2.7 |
| $\sigma_{abs,sca,ext}$ | Absorption, scattering, extinction cross section | $[m^2]$ | 3.12 |
| $\sigma_{col}$ | Collision cross section | $[m^2]$ | 3.39 |
| $\sigma_{cx,H}$ | Proton–hydrogen atom charge exchange cross section | $[m^2]$ | 5.9 |

(continued)

| Symbol | Meaning | Typical unit | Introducing equation |
|---|---|---|---|
| $\sigma_{DR}$ | Electron impact dissociative recombination cross section | [m$^2$] | 5.38 |
| $\sigma_{en}$ | Electron-neutral collisional cross section | [m$^2$] | 5.36 |
| $\sigma_{H2O}$ | Spectral photodissociation cross section of H$_2$O | [m$^2$] | 5.8 |
| $\sigma_i$ | Surface area of the i$^{th}$ facet | [m$^2$] | 1.24 |
| $\sigma_{in}$ | Ion-neutral collisional cross section | [m$^2$] | 3.116 |
| $\sigma_p$ | Projected cross-sectional area of a spheroid | [m$^2$] | 2.26 |
| $\sigma_{s,ph}$ | Ionization cross section | | 5.5 |
| $\sigma_T$ | Tidally induced stress | [N m$^{-2}$] | 2.87 |
| $\sigma_x$ | Geometric cross section of the particles | [m$^2$] | 4.30 |
| $\tau_d$ | Dust optical depth | Unitless | 4.46 |
| $\tau_{damp}$ | Damping timescale | [s] | 2.54 |
| $\tau_{diff}$ | Thermal diffusive timescale | [s] | 2.122 |
| $\tau_p$ | Lifetime of the parent species | [s] | 3.108 |
| $\tau_v$ | Viscous stress tensor | | 3.61 |
| $\tau_\vartheta$ | Optical depth | Unitless | 3.14 |
| $\varphi$ | Rotation angle about the vector defining the direction of the incident light | [rad] | 4.19 |
| $\Phi$ | Scattering angle weighted with the scattering function | [rad] | 4.68 |
| $\varphi(\alpha)$ | Reflected phase function | Unitless | 2.1 |
| $\phi, \varphi, \theta$ | Euler angles | [rad] | 2.48, 2.50, 2.51 |
| $\phi_A$ | Angle from the source to the interaction point | [rad] | 3.97 |
| $\phi_M$ | Photometric latitude | [rad] | 2.77 |
| $\Phi_S$ | Single particle angular scattering function | [sr$^{-1}$] | 2.66, 4.22 |
| $\phi_{th}(\alpha)$ | Thermal phase function | Unitless | 2.18 |
| $\chi$ | Dust to gas mass loss ratio | | 2.121 |
| $\Psi$ | Porosity | Unitless | 2.115 |
| $\Psi_v$ | Conversion of the bulk motion of the fluid into internal energy via viscous dissipation | | 3.65 |
| $\omega$ | Single-scattering albedo of the individual particles (also that contributing to surface scattering) | Unitless | 2.65, 4.34 |
| $\Omega$ | Collision operator | | 3.87 |
| $\Omega_d$ | Angular velocity of a dust particle | [rad s$^{-1}$] | 4.114 |
| $\Omega_f$ | Gyrofrequency | [Hz] | 5.17 |
| $\Omega_N$ | Rotational angular velocity of the nucleus | [$^\circ$ day$^{-1}$], [rad s$^{-1}$] | 2.34, 2.105 |
| $\Omega_s$ | Solid angle (usually for radiative transfer) | [sr] | 2.58 |
| $\epsilon$ | Specific internal energy | [J] | 3.58 |

# Contents

| 1 | **Light Curves, Orbits, and Reservoirs** | | 1 |
|---|---|---|---|
| | 1.1 | Light Curves | 1 |
| | 1.2 | Orbits and Origins | 6 |
| | 1.3 | Non-gravitational Forces | 23 |
| 2 | **The Nucleus** | | 27 |
| | 2.1 | Sizes and Shapes of Unresolved Nuclei | 27 |
| | 2.2 | Sizes and Shapes of Resolved Objects | 34 |
| | 2.3 | Mass and Density | 38 |
| | 2.4 | Rotational Properties | 41 |
| | 2.5 | Centripetal Accelerations | 54 |
| | 2.6 | Surface Reflectance | 54 |
| | 2.7 | Visible Colour | 66 |
| | 2.8 | Interior Structure | 69 |
| | | 2.8.1 Large-Scale Structure | 69 |
| | | 2.8.2 The Surface Layer and the Strength of Cometary Material | 72 |
| | 2.9 | Surface Processes | 75 |
| | | 2.9.1 Introduction | 75 |
| | | 2.9.2 Sublimation of Ices | 76 |
| | | 2.9.3 Energy Balance | 80 |
| | |     2.9.3.1 Simple Surface Energy Balance with Sublimation and Conduction | 80 |
| | |     2.9.3.2 Volume Absorption, Solid-State Greenhouse Effect, and Porosity | 95 |
| | |     2.9.3.3 Multi-volatile Models and Ice Fractionation | 101 |
| | |     2.9.3.4 The Amorphous-Crystalline Transition of Water Ice | 104 |
| | |     2.9.3.5 Thermal Emission from Resolved Surfaces | 107 |

2.9.3.6  Surface Roughness, Infrared Beaming,
                            Self-Shadowing and Self-Heating . . . . . . . .  109
        2.10  Surface Appearance and Cometary "Geology" . . . . . . . . . . . . . .  111
                2.10.1  Regional Classification . . . . . . . . . . . . . . . . . . . . . . . .  113
                2.10.2  Textural Differences . . . . . . . . . . . . . . . . . . . . . . . . . .  118
                2.10.3  Impact Cratering . . . . . . . . . . . . . . . . . . . . . . . . . . . .  119
                2.10.4  Depressions, Pits, and Other Quasi-Circular
                            Structures . . . . . . . . . . . . . . . . . . . . . . . . . . . . . . . .  125
                2.10.5  Fracturing . . . . . . . . . . . . . . . . . . . . . . . . . . . . . . . . .  131
                        2.10.5.1  Torque-Induced and Other Possible
                                    Tectonic Fractures . . . . . . . . . . . . . . . . . . .  131
                        2.10.5.2  Thermal Fractures . . . . . . . . . . . . . . . . . . .  133
                        2.10.5.3  Polygonal Networks . . . . . . . . . . . . . . . . .  136
                2.10.6  Heat Trapping . . . . . . . . . . . . . . . . . . . . . . . . . . . . . .  137
                2.10.7  Activity Induced Mass Wasting . . . . . . . . . . . . . . . . . .  138
                2.10.8  Sedimentary Processes . . . . . . . . . . . . . . . . . . . . . . . .  144
                2.10.9  Activity in Dust-Covered Areas . . . . . . . . . . . . . . . . . .  150
                2.10.10  Dust Ponding . . . . . . . . . . . . . . . . . . . . . . . . . . . . . . .  151
                        2.10.10.1  The Surface Fluidization Mechanism . . . . .  154
                2.10.11  Surface Changes in Smooth Terrains . . . . . . . . . . . . . .  156
                2.10.12  Other Circular Structures . . . . . . . . . . . . . . . . . . . . . .  159
                2.10.13  Surface Dust Transport . . . . . . . . . . . . . . . . . . . . . . .  161
                2.10.14  Dune Pits . . . . . . . . . . . . . . . . . . . . . . . . . . . . . . . . .  168
                2.10.15  Ice Exposures . . . . . . . . . . . . . . . . . . . . . . . . . . . . . .  169
                2.10.16  Other Surface Changes . . . . . . . . . . . . . . . . . . . . . . .  173
                2.10.17  Evidence for Large-Scale Mass Loss . . . . . . . . . . . . . .  176

3  Gas Emissions Near the Nucleus . . . . . . . . . . . . . . . . . . . . . . . . . . . . .  179
        3.1  Fundamentals . . . . . . . . . . . . . . . . . . . . . . . . . . . . . . . . . . . . .  179
        3.2  Major Species and Their Emissions . . . . . . . . . . . . . . . . . . . . . .  181
        3.3  Minor Species . . . . . . . . . . . . . . . . . . . . . . . . . . . . . . . . . . . . .  206
        3.4  Gas Expansion . . . . . . . . . . . . . . . . . . . . . . . . . . . . . . . . . . . . .  210
                3.4.1  The Initial Conditions . . . . . . . . . . . . . . . . . . . . . . . . . .  210
                3.4.2  The Knudsen Number . . . . . . . . . . . . . . . . . . . . . . . . . .  214
                3.4.3  Fluid Expansion . . . . . . . . . . . . . . . . . . . . . . . . . . . . . .  216
                        3.4.3.1  Fluid Equations for Equilibrium Flow . . . .  216
                        3.4.3.2  The Initial Evolution of the Velocity
                                    Distribution Function . . . . . . . . . . . . . . . .  220
                        3.4.3.3  Analytical Solutions for the Fluid
                                    Equations . . . . . . . . . . . . . . . . . . . . . . . . . .  221
                3.4.4  The Knudsen Layer and Low Density Flow . . . . . . . . .  224
                        3.4.4.1  The Boltzmann Equation and the Direct
                                    Simulation Monte Carlo Method . . . . . . . .  224
                        3.4.4.2  The Knudsen Penetration Number . . . . . . .  228
                        3.4.4.3  The Influence of Composition . . . . . . . . . .  232

|  |  | 3.4.5 | Examples of Near-Nucleus Gas Flow . . . . . . . . . . . . . . | 234 |
|  |  |  | 3.4.5.1 | Cases with Spherical Sources . . . . . . . . . . | 234 |
|  |  |  | 3.4.5.2 | Cases with Realistic Shapes . . . . . . . . . . . | 238 |
|  |  |  | 3.4.5.3 | Deviations from Insolation-Driven Activity . . . . . . . . . . . . . . . . . . . . . . . . . | 238 |
|  |  |  | 3.4.5.4 | Degeneracy in Surface Activity Distributions . . . . . . . . . . . . . . . . . . . . . | 240 |
|  |  | 3.4.6 | Re-Condensation and Surface Reflection of Gas Molecules . . . . . . . . . . . . . . . . . . . . . . . . . . . . | 241 |
|  |  | 3.4.7 | The Initial Gas Temperature . . . . . . . . . . . . . . . . . . . | 243 |
|  |  | 3.4.8 | Effects of Porosity on Small Scales . . . . . . . . . . . . . | 246 |
|  |  | 3.4.9 | Nightside Outgassing . . . . . . . . . . . . . . . . . . . . . . . | 247 |
|  | 3.5 | Reaction Chemistry and the Extended Coma . . . . . . . . . . . . . . | 251 |
|  |  | 3.5.1 | Daughter Products and the Haser Model . . . . . . . . . . | 251 |
|  |  | 3.5.2 | Detailed Reaction Kinetics . . . . . . . . . . . . . . . . . . . | 257 |
|  |  | 3.5.3 | The Swings and Greenstein Effects . . . . . . . . . . . . . | 263 |
|  |  | 3.5.4 | Prompt Emission . . . . . . . . . . . . . . . . . . . . . . . . . . | 265 |
|  |  | 3.5.5 | Other Notable UV Line Emissions . . . . . . . . . . . . . . | 267 |
|  |  | 3.5.6 | Spatial and Temporal Variations of Parent and Daughter Species . . . . . . . . . . . . . . . . . . . . . . | 269 |
|  | 3.6 | Compositional Variation with Heliocentric Distance . . . . . . . . . | 272 |
|  | 3.7 | Radiation Pressure on Gas Molecules and Radicals: The Neutral Tail(s) . . . . . . . . . . . . . . . . . . . . . . . . . . . . . . . | 274 |
|  | 3.8 | Isotopic Ratios . . . . . . . . . . . . . . . . . . . . . . . . . . . . . . . . . | 276 |
|  | 3.9 | Ortho to Para Ratios . . . . . . . . . . . . . . . . . . . . . . . . . . . . . . | 279 |
| **4** | **Dust Emission from the Surface** . . . . . . . . . . . . . . . . . . . . . . . . . . | 281 |
|  | 4.1 | The Point Source Approximation . . . . . . . . . . . . . . . . . . . . . . | 281 |
|  | 4.2 | Scattering of Light by Dust . . . . . . . . . . . . . . . . . . . . . . . . . | 282 |
|  |  | 4.2.1 | Introduction and Rayleigh Scattering . . . . . . . . . . . . | 282 |
|  |  | 4.2.2 | Scattering by Particles Close to the Wavelength . . . . . | 283 |
|  |  |  | 4.2.2.1 | Preliminaries . . . . . . . . . . . . . . . . . . . . . | 283 |
|  |  | 4.2.3 | Mie Theory . . . . . . . . . . . . . . . . . . . . . . . . . . . . . . | 288 |
|  |  | 4.2.4 | The T-Matrix Method . . . . . . . . . . . . . . . . . . . . . . . | 295 |
|  |  | 4.2.5 | Computer Simulated Particles . . . . . . . . . . . . . . . . . | 296 |
|  |  | 4.2.6 | Observed Particle Structures . . . . . . . . . . . . . . . . . . | 298 |
|  |  | 4.2.7 | The Discrete Dipole Approximation . . . . . . . . . . . . . | 299 |
|  |  | 4.2.8 | The Observed Phase Function at 67P . . . . . . . . . . . . | 300 |
|  |  | 4.2.9 | The Observed Radiance . . . . . . . . . . . . . . . . . . . . . | 302 |
|  |  |  | 4.2.9.1 | The Optical Thin Case . . . . . . . . . . . . . . . | 302 |
|  |  |  | 4.2.9.2 | Optical Thickness Effects . . . . . . . . . . . . . | 303 |
|  |  | 4.2.10 | Inhomogeneous Particles and Maxwell Garnett Theory . . . . . . . . . . . . . . . . . . . . . . . . . . . . . . . . | 304 |
|  | 4.3 | Af$\rho$ ("Afrho") . . . . . . . . . . . . . . . . . . . . . . . . . . . . . . . . . . | 306 |
|  | 4.4 | Radiation Pressure . . . . . . . . . . . . . . . . . . . . . . . . . . . . . . . | 310 |

       4.4.1    Radiation Pressure Efficiency . . . . . . . . . . . . . . . . . . . 311

       4.4.2    The Fountain Model . . . . . . . . . . . . . . . . . . . . . . . . . 312

       4.4.3    The Dynamics of Fluffy Particles . . . . . . . . . . . . . . . . 315

  4.5    Dust Size Distributions . . . . . . . . . . . . . . . . . . . . . . . . . . . . . 316

  4.6    The Lifting Dust Ejection Process . . . . . . . . . . . . . . . . . . . . . . 322

       4.6.1    Drag Force at the Nucleus Surface . . . . . . . . . . . . . . . . 322

       4.6.2    Cohesive Forces . . . . . . . . . . . . . . . . . . . . . . . . . . . 325

       4.6.3    Advanced Dust Ejection Concepts . . . . . . . . . . . . . . . . 327

  4.7    The Influence of Drag on the Equations of Motion for the Dust . . . 329

       4.7.1    Analytical Solutions . . . . . . . . . . . . . . . . . . . . . . . . . 329

       4.7.2    Numerical Solutions . . . . . . . . . . . . . . . . . . . . . . . . . 331

       4.7.3    Gas-Dust Energy Exchange within the Coma . . . . . . . . 334

  4.8    Converting A$f\rho$ to a Dust Loss Rate . . . . . . . . . . . . . . . . . . . . . 336

  4.9    Observation of Non-uniform Dust Emission . . . . . . . . . . . . . . . 338

       4.9.1    Large-Scale Structures . . . . . . . . . . . . . . . . . . . . . . . 338

       4.9.2    Small-Scale Structures . . . . . . . . . . . . . . . . . . . . . . . 342

       4.9.3    Transient Jet/Filament Structures . . . . . . . . . . . . . . . . 347

  4.10  Processes in the Innermost Coma . . . . . . . . . . . . . . . . . . . . . . 350

       4.10.1  Comparisons of Models of 67P with Data . . . . . . . . . . 350

       4.10.2  Deviations from Force-Free Radial Outflow . . . . . . . . . 350

       4.10.3  Dust Above the Nightside Hemisphere . . . . . . . . . . . . 357

  4.11  Slow (Large) Moving Particles in the Coma . . . . . . . . . . . . . . . 359

       4.11.1  Observations and Significance of Slow-Moving

                 Particles . . . . . . . . . . . . . . . . . . . . . . . . . . . . . . . . . 359

       4.11.2  Individual Particle Dynamics . . . . . . . . . . . . . . . . . . . 368

       4.11.3  Neck-Lines and Dust Trails . . . . . . . . . . . . . . . . . . . . 370

  4.12  Radiometric Properties of Dust . . . . . . . . . . . . . . . . . . . . . . . . 375

       4.12.1  The Colour of Dust . . . . . . . . . . . . . . . . . . . . . . . . . 375

       4.12.2  The Thermal Properties of Dust and Sublimation . . . . . 376

       4.12.3  The Polarization of the Scattered Light . . . . . . . . . . . . 380

  4.13  Equation of Motion of Charged Dust . . . . . . . . . . . . . . . . . . . . 383

  4.14  The Non-volatile Composition of Dust and the Nucleus . . . . . . . 384

  4.15  Refractory to Volatile Ratios . . . . . . . . . . . . . . . . . . . . . . . . . 392

**5   The Plasma Environment** . . . . . . . . . . . . . . . . . . . . . . . . . . . . . 399

  5.1    Initial Considerations and the Solar Wind . . . . . . . . . . . . . . . . 399

  5.2    The Production of Cometary Ions . . . . . . . . . . . . . . . . . . . . . . 402

       5.2.1    Photoionization . . . . . . . . . . . . . . . . . . . . . . . . . . . . 402

       5.2.2    Charge-Exchange . . . . . . . . . . . . . . . . . . . . . . . . . . 406

       5.2.3    Electron Impact Ionization . . . . . . . . . . . . . . . . . . . . 407

  5.3    Dynamics of the Interaction . . . . . . . . . . . . . . . . . . . . . . . . . 409

  5.4    Processes within the Diamagnetic Cavity . . . . . . . . . . . . . . . . . 418

  5.5    Spectroscopy and Imaging of Plasma Tails . . . . . . . . . . . . . . . . 422

       5.5.1    Main Ion Species . . . . . . . . . . . . . . . . . . . . . . . . . . . 422

       5.5.2    Ion Velocities . . . . . . . . . . . . . . . . . . . . . . . . . . . . . 423

       5.5.3    Disconnection Events . . . . . . . . . . . . . . . . . . . . . . . . 423

**6 Comet-Like Activity in Related Objects** . . . . . . . . . . . . . . . . . . . . . . 427
   6.1    Active Asteroids . . . . . . . . . . . . . . . . . . . . . . . . . . . . . . . . . . 427
   6.2    Activity of Centaurs . . . . . . . . . . . . . . . . . . . . . . . . . . . . . . . 430
   6.3    Interstellar Visitors . . . . . . . . . . . . . . . . . . . . . . . . . . . . . . . 430
   6.4    Comets in Close Proximity to the Sun . . . . . . . . . . . . . . . . . . . 431

**7 The Loss of Comets** . . . . . . . . . . . . . . . . . . . . . . . . . . . . . . . . . . 433

**8 Future Investigations of Comets** . . . . . . . . . . . . . . . . . . . . . . . . . 439

**9 Final Remarks** . . . . . . . . . . . . . . . . . . . . . . . . . . . . . . . . . . . . . 447

**References** . . . . . . . . . . . . . . . . . . . . . . . . . . . . . . . . . . . . . . . . . . 449

**Index** . . . . . . . . . . . . . . . . . . . . . . . . . . . . . . . . . . . . . . . . . . . . . . 497

# Chapter 1
# Light Curves, Orbits, and Reservoirs

## 1.1 Light Curves

The reason for the ancient fascination with comets was almost certainly because of their rapid brightening and their apparent motion against the background stars combined with their distinctly non-star-like appearance. The first aspect to be understood scientifically was the orbital motion which was explained by Edmund Halley. He correctly deduced that there were similarities between the comets of 1531, 1607, and 1682 and predicted that it was the same comet. He thus established that this comet was not on a hyperbolic orbit but on a very eccentric orbit (Lancaster-Brown 1985) and predicted its return. He also supported the publication of Newton's work on celestial mechanics from which it could be determined that such a comet would be on a (roughly) elliptical orbit with the Sun at one of the foci of the ellipse and an axis of symmetry along the long axis of the ellipse passing through the Sun and the pericentre position. However, the concept of why brightening and fading occurs and its relationship to the sublimation of an ice-dirt mixture was only clearly understood in the middle of the twentieth century.

It is now known that the cause of the non-stellar appearance is the sublimation of a gas/dust mixture from the surface of a small (typically 10 km diameter) nucleus. It should be recalled that the existence of a single solid nucleus at the centre of cometary activity was still the subject of serious scientific debate in the 1950s (e.g. Lyttleton 1951) and not conclusively proven until the spacecraft encounters with comet 1P/Halley in 1986 (Fig. 1.1; Keller et al. 1986). But long before confirmation of this concept, observers of a bright comet in the sky could identify three main elements.

The coma (or "head") of the comet is nebulous in appearance and often roundish (Fig. 1.2). The brightness arises from a combination of reflected sunlight from dust particles and resonant fluorescence from neutral gas radicals. The dust tail is seen in reflected sunlight and its tail-like nature results from solar radiation pressure (SRP) acting on emitted dust particles. It extends in the anti-sunward direction away from

© Springer Nature Switzerland AG 2020
N. Thomas, *An Introduction to Comets*, Astronomy and Astrophysics Library,
https://doi.org/10.1007/978-3-030-50574-5_1

**Fig. 1.1** The nucleus of comet 1P (Credit: ESA, MPAe). Halley as imaged by the Halley Multicolour Camera from onboard Giotto on 14 March 1986

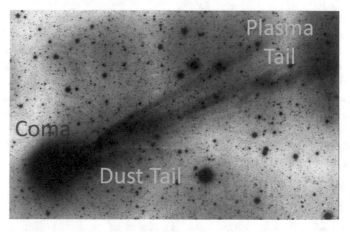

**Fig. 1.2** Negative image of C/1996 B2 (Hyakutake) taken with ESO 1-m Schmidt telescope at La Silla on February 28.36 UT 1996 by ESO night assistant Oscar Pizarro. ©ESO (CC BY 4.0)

the head. As can be seen in Fig. 1.2, the dust tail can be structured. It can also have a curved appearance as a result of the motion of the comet about the Sun combined with the observing geometry. The third visible component is the plasma tail that arises from the ionization of gas molecules and radicals followed by pick-up and acceleration by the solar wind. In a dark environment, all three of these components can become visible to the naked eye for a very bright comet, with C/1995 O1 (Hale-

Bopp) being a good example. The appearance of a similar comet thousands of years ago would have undoubtedly impressed our forebears. Similarly, the brightening and fading again within a relatively short period of time (of the order of a few months) would also have led to considerable amazement and, with the absence of a visible source, it is not surprising that our ancestors were sufficiently perplexed by these objects that they tried to illustrate what they saw in works such as the Bayeux Tapistry and the Augsburg Book of Miracles (Borchert and Waterman 2017).

A final element that is not apparent to the naked-eye of even bright comets is the presence of a neutral tail. Gas species are also subject to radiation pressure before dissociation and/or ionization. Some species (atomic sodium being a prime example) can receive significant anti-sunward acceleration through the resonant fluorescence process that results in the production of a neutral tail. A neutral sodium tail was first seen in observations of C/1995 O1 (Hale-Bopp) and was subsequently found in observations of C/1996 B2 (Hyakutake) (Cremonese and the European Hale-Bopp Team 1997; Cremonese et al. 1997). For most neutral species, lifetimes are relatively short and hence gas tails produced in this way are not of major significance.

Making standardized photometric orbital light curves of comets from multiple independent observations is not straightforward but attempts were driven forward by the need to place constraints on the activity of 1P/Halley for its return in 1986 (e.g. Newburn and Yeomans 1982). A modern example of an orbital light curve is shown in Fig. 1.3 and is also for 1P/Halley (Ferrin 2010).

Cometary light curves are usually expressed using the magnitude system. Though its use is ubiquitous in Earth-based optical astronomy it does have three drawbacks

– it is an inverse scale, with fainter stars having larger magnitudes,
– it is a logarithmic scale and,
– the base of the logarithm is 2.512.

The observed visual magnitude of a comet is denoted by $m_1(\Delta, r_h)$ where $\Delta$ is the comet-Earth distance and $r_h$ is the Sun-comet distance. The visual magnitude is estimated by making differential photometry with nearby objects of known brightness. This has a level of uncertainty. However, accuracy at the 0.1 magnitudes level is achievable under good conditions.

Combining results from several observers and reducing the data into a standard system (e.g. the Johnson-Morgan V magnitude system) is not straightforward as the photometric responses of equipment used to measure the target can differ significantly and atmospheric opacity (extinction) will vary with time and the observing site (see e.g. Sterken and Manfroid 1992). On the other hand, as Fig. 1.3 shows, the change in cometary visual magnitude from its first telescopic detection to perihelion passage can be more than 15 magnitudes (a change in brightness of a factor of $10^6$) with occasional comets (such as 1P/Halley) appearing to brighten by factors of 100 (5 magnitudes) or more to naked-eye observers and then fading again before being lost.

Visual magnitudes are usually expressed as reduced magnitudes. Here the magnitude is converted to a unit distance from the observer—normally a comet-Earth distance of 1 AU—by dividing by $\Delta^2$. This can be expressed in equation form as

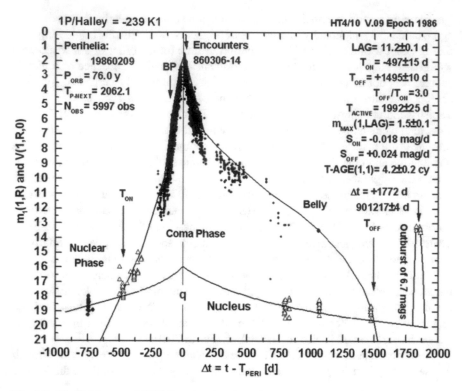

**Fig. 1.3** The light curve of 1P/Halley showing the reduced magnitude ($m(1, R_h)$) as a function of time with respect to perihelion. Some of the properties of the curves are discussed in the text (from Ferrin 2010)

$$m_1(1, r_h) = m_1(\Delta, r_h) - 5 \, log_{10}\Delta \qquad (1.1)$$

This has been performed in Fig. 1.3 which also shows several other points of interest. Firstly, the plot shows the brightness relative to the time to pericentre, $T_{PERI}$. However, this brightness is not symmetric about $T_{PERI}$. The value, LAG (which refers to the lag in the maximum brightness with respect to perihelion) indicates that the maximum in brightness occurred 11.2 days after perihelion. The maximum brightness is not at the time of maximum solar input. Work by Ferrin suggests that for 67P, using pre-Rosetta apparitions, the lag was 33 days after perihelion. Hansen et al. (2016) used ROSINA/COPS measurements of the gas density at 67P and combined this with a simple free-radial expansion model of the gas to show that gas production at 67P peaked between 18 and 22 days after perihelion. They also showed that the gas production was highly correlated with ground-based dust brightness measurements. Consequently, although there may be small differences in the exact result, the existence of a pre- and post-perihelion brightness asymmetries with respect to perihelion is demonstrable and is related to activity and nucleus gas and dust production.

While a positive value of the lag might be explicable through some form of thermal inertia on orbital timescales, there are exceptions that challenge this hypothesis. For example, two other comets visited by spacecraft, 81P/Wild 2 and 9P/Tempel 1, both have negative lags—the peaks of their brightnesses are pre-perihelion by 13 and 10 days respectively. On the other hand, the shape of the nucleus may play a role because the surface area exposed to sunlight may change significantly through perihelion if the rotation axis is not orthogonal to the orbital plane. However, if this were the sole reason for the lag then one would expect lags for a sample of comets to be equally distributed about perihelion if the rotation axes are randomly distributed with respect to the orbital plane and, as Ferrin shows, they are clearly not with far more positive lags than negative ones. We can see from this that the pre-/post-perihelion asymmetries of cometary brightnesses measured from ground remain inadequately explained.

Another aspect of Fig. 1.3 is that the brightness is greater post-perihelion than at the same heliocentric distance pre-perihelion and that the curve is appreciably steeper pre-perihelion than post-perihelion. Once again, this is not universally the case and there are exceptions including comets such as C/1995 O1 Hale-Bopp and 2P/Encke. Simplistically, one might expect the brightness, $m_1(1,r_h)$, to be proportional to $1/r_h^4$. This would arise from the decreasing solar flux with heliocentric distance combined with the reduction in the energy being available for sublimation. However, cometary light curves generally indicate steeper dependencies on $r_h$, particularly pre-perihelion. These features would persist if further linearization of the orbital light curves would be performed by removing the $1/r_h^2$ dependence of the illumination, viz.,

$$m_1(1,1) = m_1(\Delta, r_h) - 5 \, log_{10}\Delta - 5 log_{10} r_h \tag{1.2}$$

and indeed it is common to re-write the above equation as

$$m_1(1,1) = m_1(\Delta, r_h) - 5 \, log_{10}\Delta - 2.5 \, n_h \, log_{10} r_h \tag{1.3}$$

where $n_h$ is a power-law exponent describing the decrease in brightness with $r_h$. The dependencies can be fit with separate power laws for pre- and post-perihelion giving values that are useful to model non-gravitational forces on the nucleus (see below).

Finally in Fig. 1.3, at least one "outburst" or anomalous brightening can be seen where the comet's brightness increased by nearly 7 magnitudes (a factor of 500) 5 years after perihelion. These features are commonly seen in the orbital light curves of comets. It is also noticeable that the short-term variability can be quite large with a spread of measurement of the order of 2 magnitudes. While the physics of outbursts remains a subject of considerable discussion, the spread in measured brightnesses seen in the light curves is attributed to variations in production rate with rotational phase and dynamical changes arising from the outgassing process itself. For further investigation of the reasons behind these phenomena, we will need to look at the source and the production rates of gas and dust.

## 1.2  Orbits and Origins

Edmund Halley's successful prediction showed that his comet is gravitationally bound to the Sun. In the simple assumption of the restricted two-body problem, the orbit of an inactive test particle around the Sun can be described as in Table 1.1. For closed orbits (where the orbital eccentricity, $e_x$, is less than 1), the Sun is at one of the foci of an ellipse with the orbital period, $P$, being given by Kepler's third law, as seen in Eq. (1).

The eccentricity can be calculated from the periapsis and apoapsis distances, $r_p$ and $r_a$ respectively, through

$$e_x = \frac{r_a - r_p}{r_a + r_p} \tag{1.4}$$

that are in turn related to the semi-major axis through

$$r_p = a_s(1 - e_x) \qquad \text{and} \qquad r_a = a_s(1 + e_x) \tag{1.5}$$

The radial distance from the Sun, $r_h$, at a given time can be computed from

$$r_h = \frac{a_s(1 - e_x^2)}{1 + e_x \cos f} \tag{1.6}$$

where $f$ is the true anomaly. Even in the simple case of the two-body problem, the introduction of the time dependence leads to an equation that can only be solved numerically. The simplest way is to use Newton's false-root method to iterate on the value of the eccentric anomaly, $E_a$, which arises from the definition

$$r_h = a_s(1 - e_x \cos E_a) \tag{1.7}$$

and is related to the mean anomaly, $M_a$, through the equation

$$M_a = ñ(t - T_{PERI}) = E_a - e_x \sin E_a \tag{1.8}$$

where ñ is the mean motion of the object about the fixed mass in units of, for example, [rad s$^{-1}$] and time is relative to the last time of periapsis. The true anomaly is related to $E_a$ through the equation

**Table 1.1** Orbit types described through their eccentricity

| Orbit type | Eccentricity, $e_x$ |
|---|---|
| Circular | 0.0 |
| Elliptical | $0.0 < e_x < 1.0$ |
| Parabolic | 1.0 |
| Hyperbolic | $e_x > 1.0$ |

$$\cos f = \frac{\cos E_a - e_x}{1 - e_x \cos E_a} \tag{1.9}$$

The iteration is then performed using the step

$$E_n = E_o - \frac{(E_o - e_x \sin E_o - M)}{1 - e_x \cos E_o} \tag{1.10}$$

where $E_o$ is an initial estimate for $E_a$ and $E_n$ is the new estimate. Iteration occurs until the difference between $E_o$ and $E_n$ is negligible for your application. A simpler relationship between $f$ and $E$ can be derived (see Murray and Dermott 1999) as

$$\tan \frac{f}{2} = \sqrt{\frac{1 + e_x}{1 - e_x}} \tan \frac{E_a}{2} \tag{1.11}$$

so that knowing $E_a$, one can derive $r_{\rm h}$ and $f$ uniquely.

In the special case of a parabolic orbit about the Sun, Kepler's equation for parabolic motion is

$$\tan \frac{f}{2} + \frac{1}{3} \left( \tan \frac{f}{2} \right)^3 = \sqrt{\frac{GM_\odot}{2q^3}} (t - T_{PERI}) \tag{1.12}$$

where $q$ is the periapsis distance, $M_\odot$ is the solar mass and $G$ is Newtonian gravitational constant. (The product, $GM_\odot$ is often referred to as the standard gravitational parameter and given a symbol, $\mu$.) This is a form of Barker's equation which relates the time of flight to the true anomaly of a parabolic trajectory and can be solved analytically for the time dependence by

$$\cot \beta_B = 3 \sqrt{\frac{GM_\odot}{2q^3}} (t - T_{PERI}) \tag{1.13}$$

where

$$\sqrt[3]{\cot \frac{\beta_B}{2}} = \cot \alpha_B \tag{1.14}$$

and

$$2 \cot 2\alpha_B = \tan \frac{f}{2} \tag{1.15}$$

The angles $\alpha_B$ and $\beta_B$ are useful angles supporting the derivation. The speed of the object at any point on the orbit is given by

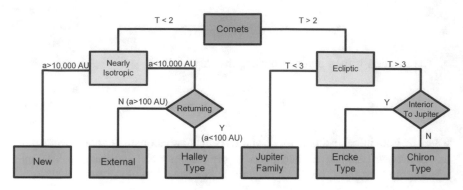

**Fig. 1.4** The Levison classification of comets

$$|v| = \sqrt{\frac{2GM_\odot}{r_h}} \ . \tag{1.16}$$

This can be used to calculate the heliocentric speed of a comet on a parabolic orbit at the Earth's distance from the Sun. This is around 42 km s$^{-1}$ which, when combined with the Earth's orbital speed (~30 km s$^{-1}$), gives the worst case impact velocity at the Earth for a comet entering the Solar System on a parabolic, retrograde, orbit and illustrates the potential danger from such an object. In order to imagine the consequences, a 67P-sized comet hitting the Earth at this velocity would be equivalent to around $6 \times 10^{12}$ tons of TNT. For comparison, the "Fat Man" atomic bomb that was exploded over the Japanese city of Nagasaki at the end of the Second World War was equivalent to $2 \times 10^4$ tons of TNT. This also gives an estimate of the maximum fly-by velocity that a spacecraft might encounter a comet with and is a consideration in the design of ESA's forthcoming Comet Interceptor mission. For comparison, Giotto completed a fly-by of 1P/Halley at a relative velocity of 68.373 km s$^{-1}$.

Currently, comets are classified according to their orbital characteristics. Figure 1.4 illustrates Levison's classification from 1996 with some minor modifications.

The orbital period is used to separate comets into two main categories. Short-period comets (SPCs) are those with P < 200 years whereas long-period comets (LPCs) have P > 200 years. The naming convention for comets is related to this division.[1] Comets that are defined to be SPCs or when there are confirmed observations at more than one perihelion passage are referred to as periodic and their names are prefixed with a number and the designation P/ (e.g. comet 19P/Borrelly). Some periodic, numbered, comets no longer exist or are deemed to have disappeared. The prefix D/ is used in this case (e.g. 20D/Westphal). There are currently eight comets in this category. Reappearance or re-discovery cannot be excluded and has indeed

---

[1]http://www.minorplanetcenter.net/iau/lists/CometResolution.html

**Fig. 1.5** The aphelion distance of numbered comets plotted against their inclination. The positions of 1P/Halley and 153P/Ikeya-Zhang on the plot are marked. The grey circles indicate the positions of the giant planets. The open squares indicate comets with P < 20 years and the diamonds are comets with P > 20 years (Data source: JPL Horizons)

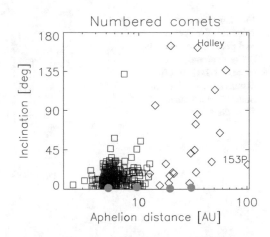

occurred in the specific cases of 226P/Pigott-LINEAR-Kowalski (formally D/1783 W1 (Pigott)) and 289P/Blanpain (formally D/1819 W1 (Blanpain)) (Rickman 2017). The use of 200 years is arbitrary but roughly corresponds to the time since systematic observations with instruments could be made. The prefix C/ is used for comets that are not periodic according to the above definition. This is a little misleading in that many comets have now been identified as being on closed, elliptical orbits and will thus return to the inner solar system at some point although their orbital periods far exceed 200 years. In spite of this, the C/ prefix is also used for comets on near-hyperbolic orbits that, in principle should never return. This has been recently augmented by the prefix I/ for objects that are clearly interstellar in origin. The prefix X/ is used for comets for which an orbit cannot be reasonably established.

As pointed out by Rickman (2017), there are numerous inconsistencies in the definitions and use of designations but there is rarely confusion.

The distribution of numbered comets is shown in Fig. 1.5. Here the inclination of the orbit with respect to the ecliptic plane has been plotted against the aphelion distance in astronomical units. The positions of the giant planets are also shown. Of the numbered comets only 153P/Ikeya-Zhang has a period of >200 years (Fig. 1.5). The plot gives the impression that SPCs can be separated into two categories. Comets with an aphelion distance of around 10 AU or less seem to be more concentrated towards the ecliptic plane with a large population having aphelia close to the heliocentric distance of Jupiter. Comets with larger aphelia are more distributed in inclination with some (including 1P/Halley which is indicated) being retrograde (inclination >90°). This has led to SPCs being divided into Halley-type comets (HTCs) and Jupiter-family comets (JFCs) with HTCs having P > 20 years and JFCs having P < 20 years. The reference to a Jupiter-family is even more clear when unnumbered periodic comets are included and a histogram made of the aphelion distance (Fig. 1.6). The peak at Jupiter's heliocentric distance is striking and it should be noted that there is also the suggestion of a further maximum at the distance of Saturn.

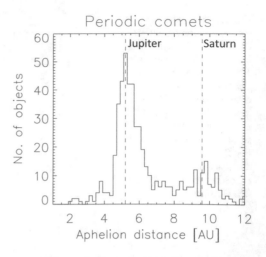

**Fig. 1.6** The distribution of aphelion distances of 543 periodic comets. The broken vertical lines represent the heliocentric distances of Jupiter and Saturn (Data source: JPL Horizons)

In Fig. 1.5, we have used the period alone to differentiate between two classes of object. For many simple calculations, for example the computation of heat fluxes on a cometary nucleus over a period of several months, the restricted two-body problem is adequate and the orbital period is assumed not to change. However, the precise computation of the positions of nuclei over several decades or centuries requires detailed numerical integration taking into account the gravitational influence of the planets. The major planets are continuously perturbing the orbits of SPCs and quite significant changes in the period can occur on relatively short timescales. This is especially apparent when a comet's orbit makes passes close to one of the giant planets. Interactions with Jupiter in particular are highly significant in cometary research. Rickman (2017) lists objects that have undergone recent major perturbations. These include 67P that made a close approach to Jupiter (at a minimum distance of 0.052 AU) in 1959 which resulted in a substantial reduction of its perihelion distance (from 2.76 AU to 1.29 AU).

Levison (1996) introduced the use of the Tisserand parameter to analyse this issue. The Tisserand parameter is a dynamical quantity that arises from a circular restricted three-body system and is approximately conserved (at the 1% level—see Murray and Dermott 1999) during an encounter between a planet orbiting a central star and a small massless object. It therefore provides a way to connect the post-encounter dynamical properties with the pre-encounter properties. The Tisserand parameter also provides a measure of the relative speed of an object when it crosses the orbit of a planet. Therefore, for a given object, different Tisserand parameters exist for different planets.

The Tisserand parameter with respect to Jupiter, $T_J$, is given by

$$T_J = \frac{a_J}{a_s} + 2\sqrt{(1 - e_x^2)\frac{a_s}{a_J}} \cos i_x \qquad (1.17)$$

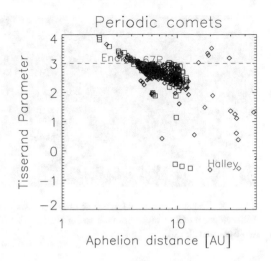

**Fig. 1.7** The Tisserand parameter for Jupiter ($T_J$) for all periodic comets (per 31 Dec 2019). The horizontal broken line represents $T_J = 3$. Numbered comets are shown as diamonds, unnumbered as squares. Comets 1P/Halley, 2P/Encke and 67P are marked

where $a_J$ is Jupiter's semi-major axis, and $i_x$ is the object's inclination. To derive the Tisserand parameter, one must express the Jacobi constant (the only known conserved quantity in the circular restricted three-body problem) in terms of orbital parameters, i.e.

$$C_J = ñ^2 \left(x^2 + y^2\right) + 2\left(\frac{\mu_1}{r_1} + \frac{\mu_2}{r_2}\right) - \left(\dot{x}^2 + \dot{y}^2 + \dot{z}^2\right). \qquad (1.18)$$

Here $r_1$ and $r_2$ are the distances of the massless object from the other two masses and $\mu_x = GM_x$ for the two masses. ñ is the mean motion of the masses about the common barycentre.

If a comet intersects the orbit of the planet, then the Tisserand parameter with respect to a planet $x$, $T_x$, is related to the unperturbed relative velocity, $V_x$, at which it encounters the planet through the equation

$$V_x = \sqrt{3 - T_x} \qquad (1.19)$$

where $V_x$ is in units of the planet's orbital velocity (see Duncan et al. 2004). Inspection of this equation shows that $T_x$ cannot be greater than 3 and thus an object with $T_x > 3$ cannot intersect the planet's orbit in the circular restricted case.

As $T_x$ becomes closer to 3, the influence of the planet on the comet's orbit becomes stronger and, as can be seen in Fig. 1.7, many of the periodic comets (both numbered and unnumbered) are in the region between $2 < T_J < 3$ which leads to a more physical definition of JFCs. It can also be seen that a number of comets with aphelion distances within our planetary system (and therefore short periods) have $T_J \ll 2$. The HTC category is therefore no longer through the period alone but through a $T_J$ value of $<2$. The Tisserand criterion is to some extent imperfect (Jewitt 2012) because Eq. (1.17) assumes that Jupiter's orbit is circular and that no other

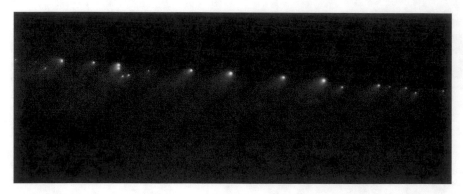

**Fig. 1.8** Comet Shoemaker-Levy 9 as observed using the Hubble Space Telescope in May 1994 following a previous close apprach to Jupiter in 1992 (Credit: Dr. Hal Weaver and T. Ed Smith (STScI), and NASA)

objects influence the motion of the comet. However, it is clearly very useful for describing most cases.

There are in addition a number of active objects well inside the orbit of Jupiter with $T_J > 3$. This has led to the description of a new class of object sometimes referred to as Main Belt Comets (MBCs) although they are perhaps more correctly referred to as "active asteroids" (Jewitt 2012). We will address these objects in a later section.

Objects with $T_J$ slightly greater 3 are themselves of considerable interest. Duncan et al. (2004) have pointed out that these objects can experience temporary and possibly repeated capture by Jupiter. This can result in tidal disruption and break-up through passages within the Roche limit and/or low velocity impact with Jupiter. The prototype for this behaviour was D/Shoemaker-Levy 9 which had a very close encounter with Jupiter in 1992 that led to its break-up (Fig. 1.8) and the impact of more than 20 sub-nuclei with Jupiter 2 years later. These types of events may be fairly common as indicated by the presence of catenae (chains of impacts) on the surfaces of the outer Galilean moons (Melosh and Schenk 1993). Figure 1.9 shows an example (Enki Catana) on Ganymede. It is inferred here that a close approach to Jupiter led to the break-up of an object which then, almost immediately, impacted Ganymede as the object fragments were on their outbound path following the Jupiter encounter.

Of the numbered comets, less than 6% are HTCs (after 1P/Halley itself, 55P/ Tempel–Tuttle is probably the best known) but there is clearly an observational selection effect caused by their longer periods compared to JFCs. On the other hand, well over 100 comets fitting the basic criterion ($20 < P < 200$ years) to be HTCs have been detected but observed only over one perihelion passage. Several of these (e.g. more recently discovered objects such as P/2004 A1 and P/2011 S1) have values of the Tisserand parameter with respect to Saturn, $T_S$, very close to 3 combined with low orbital inclination. This indicates that interactions with Saturn are likely. As of the end of 2019, 89 comets have been classified unambiguously as

**Fig. 1.9** Enki Catena on the Galilean satellite, Ganymede (Credit: Galileo Project, Brown University, JPL, NASA)

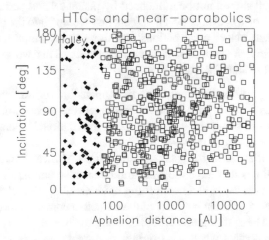

**Fig. 1.10** HTCs as classified in the Horizons database generated by JPL plotted as cosine of orbital inclination vs. aphelion distance (diamonds). The orbital inclinations appear to be randomly distributed. The squares show other comets with P > 200 years but $e_x < 1$ (Data source: JPL Horizons)

HTCs (with one now missing) and are plotted in Fig. 1.10. The inclinations of the orbits appear to be well distributed.

There is also a population of comets with near-parabolic orbits ($0.985 < e_x < 1.0$) with periods greater than 200 years. These comets also appear to show a random distribution in inclination. In addition, over 320 comets have been computed to have hyperbolic orbits. However, these orbits are the osculating orbits—the gravitational Kepler orbits (i.e. the conics) that the objects would have about the Sun if perturbations were not present computed from observations close to their perihelion passage. Oort, in his quite remarkable paper from 1950, used the value $1/a_s$ as a proxy for the orbital energy. If the orbits are integrated backwards taking the perturbations into account, then the distribution in orbital energy changes significantly to produce a spike in the distribution corresponding to semi-major axes between 27,000 AU and 36,000 AU (or roughly ½ light year) depending on the selection criteria used (Dones et al. 2004). The source of these comets is assumed to be an isotropic cloud of comets surrounding our Sun now referred to as the Oort cloud. In defining the distance to, and breadth of, the Oort cloud, there remain numerous issues. In particular, gravitational perturbations may be dominated by non-gravitational forces resulting from outgassing from the surface near perihelion which in turn affects our knowledge of the original orbit. Furthermore, the gravitational forces arising from the galactic mass distribution (so-called galactic tides) and passing giant molecular clouds (GMCs) and stars are also significant as has been recently demonstrated in a series of papers by Fouchard et al. (see, for example, Fouchard et al. 2017a, b). Given the uncertainties in the properties of the Oort cloud, definitions of its characteristics such as that given by Brasser and Schwamb (2015) in which $a_s > 250$ AU and a perihelion distance of >45 AU, are necessarily broad and generally reflect the idea that the Oort cloud should not be influenced by planet-related perturbing forces.

Fouchard et al. distinguish between objects that "creep" into the planetary domain with multiple giant planet perturbations and those that "jump" into the planetary system with a rapid reduction in perihelion distance. This work has also shown several important results, including

- that planetary perturbations are primarily responsible for removal of Oort cloud objects but that Galactic tides or stellar encounters are needed initially to bring the perihelia into the planetary domain so that gravitational interactions can then "kick" the objects out of the system,
- that, over the lifetime of the solar system, comet "showers" may have occurred one or more times that have radically modified the cloud and resulted in major mass loss from the cloud and re-shuffling of its members (Fouchard et al. 2014b),
- that a slight but significant majority of the modelled comets entering the planetary domain should be in retrograde orbits, contrary to observation. A clear preference for retrograde comets when the original semi-major axis is <25,000 AU was found. However, this will depend strongly on the initial distribution of objects in the cloud and the fundamental assumption that the cloud is isotropic (Fouchard et al. 2017b),
- that, in general, the assumed initial properties of the Oort cloud have been shown in models to have a strong effect on both its long term dynamics and final structure.

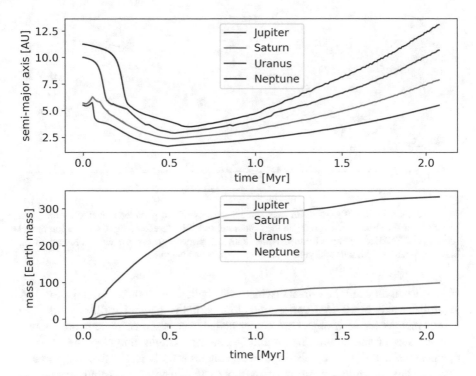

**Fig. 1.11** One possible scenario of mass growth and semi-major axis development for the giant planets within the first 2 million years (courtesy of Y. Alibert)

The Oort cloud is often considered to be the source of the HTCs with $T_J < 2$ and it appears that perturbation of returning LPCs by Jupiter and Saturn can result in reducing the semi-major axes and "leakage" of Oort cloud comets into the planetary domain. Inactive long-period asteroids with $T_J < 2$ have also been found and are now being referred to as Damocloids following the prototype inactive asteroid (5335) Damocles in this HTC-like orbit (Jewitt 2005). It is now assumed that Damocloids are dormant, inactive, HTCs. Similarly, there are over 20 inactive asteroids in JFC type orbits ($r_p < 1.5$ AU and $r_a > 4.9$ AU) including (3552) Don Quixote and (5370) Taranis that suggest that some JFCs may ultimately become totally inactive or dormant. (We shall refer to these as "Taranoids" in a subsequent diagram for convenience but this is not a term that has been used by other authors.)

The difficulty in deducing the source regions in the primordial solar nebula for Oort cloud comets is illustrated by considering Figs. 1.11 and 1.12. Figure 1.11 shows a model of how the giant planets grew and their orbits changed within the first 2 million years of Solar System. The plot shows the effects of Solar System evolution. If we now assume that cometary material also originated at this time and place test particles representing comets within this system, we can use a Monte Carlo approach to determine their fate.

Figure 1.12 was produced using an N-body code (GENGA; Grimm and Stadel 2014) and shows the semi-major axes and eccentricities (expressed as $1-e_x^2$) of

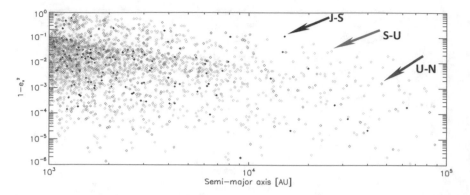

**Fig. 1.12** Distribution of semi-major axis and eccentricity of test particles scattered by the giant planets when the latter were evolving in the primordial solar nebula according to one current model (courtesy of Y. Alibert and S. Grimm). White diamonds started between Jupiter and Saturn, green between Saturn and Uranus and yellow between Uranus and Neptune

objects when they left the domain (which was limited to 1000 AU from the Sun). A semi-major axis of $10^4$ AU corresponds to a period of about 1 million years. The source regions for each object are colour-coded. Three source regions are shown, between Jupiter and Saturn, Saturn and Uranus, and Uranus and Neptune. What is apparent is that the resulting distribution of particles is fairly insensitive to the source region. Thus, the initial position of the object within the solar nebula cannot be determined from its orbit once ejected from the vicinity of the giant planets. Although this is possibly an extreme example, it illustrates that linking individual Oort cloud comets to a specific region in the original disc requires much more knowledge than we currently have and as Fouchard et al. (2013) suggested may only ever be indicative on the basis of statistical arguments.

A further issue is the structure of the solar nebula in its early stages and the effect of gas drag on the growing cometesimals. Brasser et al. (2007) investigated the influence of gas drag on the formation of the Oort cloud. Using a minimum mass solar nebula and assumptions about its radial extent and decay time, they showed that if the primordial solar nebula extended into the Saturn-Neptune region, then the Oort cloud could not be populated in the initial phase. The primordial solar nebula could also act as a size-sorting mechanism whereby smaller objects, most influenced by drag, would be less likely to be ejected into the cloud but remain within the planetary system and subject to collisions with the giant planets. These objects (typically $\leq 100$ m in size) would support planetary growth (e.g. Fortier et al. 2013). Larger objects might still be ejected, especially out of the plane of the nebula in this early phase, depending upon the exact properties of the nebula at the time. This remains the subject of some controversy.

The perihelion distances of objects ejected from the region of the giant planets remain close to the giant planet orbits and their inclinations should also be distributed about the invariable plane so that, in the absence of additional forces, further interactions will occur. However, once these objects are beyond about 10,000 AU, galactic tides and stellar encounters can influence the motion raising the perihelion

**Fig. 1.13** The semi-major axes of 2478 Trans-Neptunian Objects (marked as diamonds), Scattered Disc objects and Centaurs (marked as squares) in the Minor Planet Center database (per 31 Dec. 2016) expressed against their orbital eccentricity

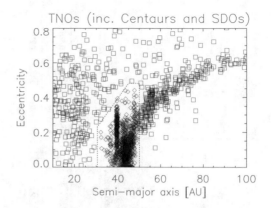

distance and eliminating further planetary encounters. The stellar encounters, which could number as many as 50,000 within 1 parsec of the Sun over the lifetime of the Solar System (Dones et al. 2004; Morbidelli 2010), and the tides thus serve to both populate and de-populate the Oort cloud.

For the JFCs, the Tisserand parameter already suggests that different pathways are involved. While, the study of their origin is a little more complicated, significant progress has been made in the past 25 years and there is a vast amount of literature related (to a greater or lesser extent) to this topic. The existence of objects crossing the orbits of the giant planets (a class of object known as Centaurs) was already known in 1977 following the discovery of the prototype, the unusual asteroid (2060) Chiron, by Kowal et al. (1977). This object occasionally shows comet-like activity despite its perihelion distance being more than 8.4 AU (e.g. Luu and Jewitt 1990). However, the key observation was the confirmation of the predictions by Edgeworth and Kuiper of the existence of a class of objects beyond the orbit of Neptune by Jewitt and Luu (1993) in 1992 that were generically referred to as Kuiper Belt Objects (KBOs). When combined with the Centaurs, nearly 2500 objects are now known and division into sub-categories is now possible although the nomenclature associated with these categories is sometimes confusing (Fig. 1.13).

The Centaurs are now defined as orbiting between the orbits of Jupiter and Neptune and having $T_J > 3.05$. They are a sub-category of the cis-Neptunian objects (Table 1.2) which includes, in addition, the Neptune Trojans—the objects in a 1:1 resonance with Neptune (of which 22 were known per 31 Dec 2019). Trans-Neptunian Objects (TNOs) are any minor planet with a semi-major axis greater than the orbit of Neptune but excluding objects associated with the Oort cloud (Table 1.2). The TNO category can itself be sub-divided into two main categories. Objects, with semi-major axes between that of Neptune and about 50 AU, and with low eccentricities and low inclinations, form a group that have relatively stable orbits. In Fig. 1.13, these objects (as defined by the Minor Planet Center and marked as diamonds) all fall within the area delineated by the broken line on the eccentricity versus semi-major axis graph. Some of these objects, Pluto being the prototype, have perihelion distances closer than the orbit of Neptune. Pluto is not perturbed by

**Table 1.2** Classes of potential cometary pre-cursors in the Outer Solar System

| Category | Sub-category | Class | Resonance and/or Tisserand | Form | Inclination and eccentricity | Inner boundary | Outer boundary | Examples | Comment |
|---|---|---|---|---|---|---|---|---|---|
| Cis-Neptunian objects | | Centaurs | None but $T_J > 3.05$ | Disc-like | Low | | Perihelion <30 AU | (2060) Chiron | |
| | | Neptune Trojans | 1:1 with Neptune | Clumped at Lagrange points | Low | | | | |
| Trans-Neptunian objects | Kuiper Belt objects | Classical KBO ("cubewano") | None | Toroidal | Low inc., low ecc. | | | (50000) Quaoar, (136472) Makemake | |
| | | Resonant KBOs | Various with Neptune | | | | | (15809) 1994 JS (3:5 resonance) | |
| | | Plutinos | 2:3 with Neptune | | | Around Neptune's orbit | | (90482) Orcus | Sub-set of the resonant KBOs |
| | | Twotinos | 1:2 with Neptune | | | Near Neptune's orbit | | (119979) 2002 $WC_{19}$ | Sub-set of the resonant KBOs |
| | Scattered disk objects | Resonant SDOs | Many 1:$\geq$3 | Disc-like | Intermediate inc., moderate ecc. | | | 2000 $CR_{105}$ (Gomes et al. 2008) | |
| | | Detached objects, extreme TNOs and Sednoids | None | | High eccentricity | ~50 AU | >150 AU | (136199) Eris (90377) Sedna, 2013 $FT_{28}$ | Defined by some as also being inner Oort cloud objects. Also called extended scattered disc objects (E-SDO), distant detached objects (DDO), or scattered–extended objects (Elliot et al. 2005) |
| Oort cloud objects | | Inner Oort cloud objects | None | Disc-like | Low | ~250 AU | ~30,000 AU | 2012 VP113 | |
| | | Outer Oort cloud objects | None | Isotropic | Isotropic | | | | |

**Fig. 1.14** Histogram of the number of TNOs found to date (per 31 Dec. 2019) expressed as their period ratioed to that Neptune. The positions of the 3:4, 2:3 and 2:5 resonances are also marked (Data from the Minor Planet Center)

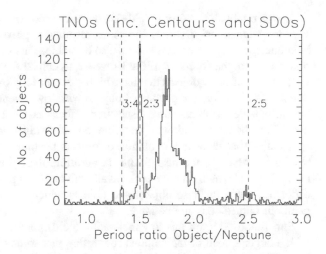

**Fig. 1.15** Histogram of the inclination of objects (resonant and non-resonant) in the classical Kuiper belt

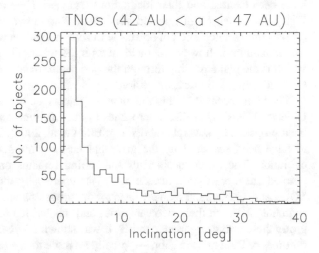

Neptune because of the well known 2:3 mean motion resonance combined with its inclination. In Fig. 1.13, the large number of objects with $a_s = 39.48$ AU over a range of eccentricities corresponds to this resonance. Other resonances are evident as can be seen in Fig. 1.14 which shows a histogram of the ratio of the periods of the objects to that of Neptune. Several other resonances with smaller numbers of members (e.g. 4:7) have also been found. Note the existence of a 2:5 resonance implying that objects with $a_s \sim 55.5$ AU also show interaction with Neptune.

Other objects (roughly 2/3rd) between 42 and 48 AU are not in resonance. These are sometimes referred to as "cubewanos" after the first detected TNO, 1992 QB$_1$ or "classical Kuiper belt objects". (50000) Quaoar is a particular interesting example because of the detection of crystalline water ice on its surface (Jewitt and Luu 2004). It is interesting to note that the distribution in inclination of these objects indicates two populations (Fig. 1.15).

It was established fairly quickly after their discovery that the cubewanos have very long dynamical lifetimes and as the resonances prevent close encounters with Neptune, the resonant objects should also be stable. This stability suggests that these objects are not the source of JFCs. However, as can be seen in Fig. 1.13, there are a vast number of other objects that have higher eccentricity over a wide range of semimajor axis. Note that the paucity of objects with low eccentricity above 50 AU may be attributable to observational selection effects as these objects will have large perihelion distances and will therefore be relatively faint. Nonetheless there is speculation that there is a depletion of objects beyond 50 AU—a hypothesis that should be solved in the near future. In addition, there are probably many objects beyond 60 AU which do not interact at all with our giant planet system. The physical difference between these objects and objects which we occasionally see as near-parabolic comets may not be large.

Superimposed upon this low inclination (cold) population, there is a population with a larger spread of inclinations suggesting that some of the original population have been excited and their inclinations changed. Note also that the histogram in Fig. 1.15, which is an updated version of that shown in Morbidelli and Brown (2004), shows a depletion of objects with zero inclination with respect to the ecliptic. The maximum is however coincident with the invariable plane of the solar system which is the plane passing through the barycentre (centre of mass) perpendicular to the total angular momentum vector.

The high eccentricity, high inclination objects are now classed as Scattered Disc Objects (SDOs). (Detailed approaches to classification and sub-classification have been proposed by several authors, e.g. Elliot et al. 2005). These objects are thought to have been scattered by the giant planets at some stage during Solar System evolution. The exact mechanism and timing remains uncertain. It was however pointed out more than a decade ago that dynamically there is little to distinguish SDOs and Centaurs. Frequent changes in orbital parameters can lead to objects switching between these two categories and this has led the Minor Planet Center to group these two categories together. It was shown 20 years ago that the dynamical lifetimes of Centaurs are short—typically no more than a few million years although the distribution in lifetime is very broad. This suggests replenishment of Centaurs on relatively short timescales and the SDOs appear to be the only viable source. However, getting SDOs into Centaur orbits and from there to JFC orbits seems to be a sophisticated interplay between the objects and the giant planets. Morbidelli (2010) describes the probable process succinctly.

An SDO is essentially in a region dominated by gravitational interaction with Neptune. When interacting, the Tisserand parameter is conserved but the perihelion distance is reduced. From Eq. (1.17), one can see that a range of solutions exist fulfilling this requirement (Fig. 1.16). The conservation of the Tisserand parameter limits the reduction in the periapsis but this would still be sufficient for the object to come under the influence of Uranus in many cases. Levison and Duncan (1997) gave the example of an object close to the 2:3 resonance with an eccentricity of 0.25 which, following interaction ended with a perihelion distance inside the orbit of Uranus. An interaction can then take place with Uranus reducing the perihelion

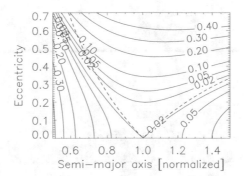

**Fig. 1.16** Contour plot of the absolute difference between a given Tisserand parameter value (in this case 2.98) and values calculated for other eccentricities and semi-major axes. A broad minimum is evident showing the range of possibilities for orbit change as a result of a close encounter with a giant planet. The dashed line gives values of $1-(1/a_s)$ and $(1/a_s)-1$

distance further while conserving the $T_U$ parameter. This process of transfer from an outer giant planet to an inner one continues until the object comes under the control of Jupiter and moves towards a JFC-type orbit. Hence, as had been speculated since their discovery, Centaurs are objects in transit between SDOs and JFCs. However, calculations by Tiscareno and Malhotra (2003) suggest that the process has additional complexity. In their model, ~2/3rds of Centaurs will be ejected from the Solar System by gravitational interactions with only 1/3rd finally contributing to the JFC population. About 20% of Centaurs have lifetimes within Centaur-like orbits shorter than 1 Myr. Further investigation of this process through numerical simulation has been performed by, for example, Grazier et al. (2019).

Unfortunately, this link between SDOs and JFCs does not guarantee that one is sampling an object from the scattered disk when investigating one of these objects. The work of Fouchard et al. (2017a, b) has shown that Oort cloud comets can achieve Centaur-like orbits directly. There are also unusual objects in the planetary system that do not obviously fall into any category unambiguously. As a result, only a statistical argument can be made to claim that a specific object is from a specific source. This should be borne in mind when inferences are made for one specific object.

Sarid et al. (2019) have performed numerical simulations of a Centaur population and showed that objects reaching non-Jupiter crossing, low eccentricity, orbits, but with aphelia well separated from Saturn, will frequently reach JFC-type orbits on rather short (<1000 year) timescales. This "Gateway" into the inner Solar System is currently populated by the active objects 29P/Schwassmann-Wachmann 1, P/2010 TO20 Linear-Grauer, and P/2008 CL94 Lemmon and the asteroid 2016 LN8 (Sarid et al. 2019). It is also noted that JFCs can pass back through this gateway to become Centaurs again.

These considerations leave us with an evolutionary picture that looks like Fig. 1.17 although it must be stressed that there may be objects that can follow other pathways through unusual circumstances.

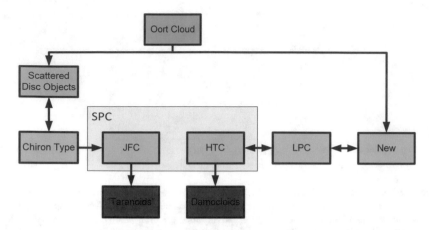

**Fig. 1.17** Evolutionary pathways for comets. Observed active objects are shown in the green boxes. The red boxes indicate asteroidal object types that are inactive counterparts. This diagram is general and some specific objects may have followed other (unusual) paths

**Table 1.3** Selected albedos of Centaurs reported by Duffard et al. (2014)

| Centaur | Albedo | Diameter [km] |
|---|---|---|
| (95626) 2002 GZ32 | $0.037 \pm 0.004$ | $237 \pm 8$ |
| (281371) 2008 FC76 | $0.067^{+0.017}_{-0.011}$ | $68^{+6}_{-7}$ |
| (145486) 2005 UJ438 | $0.256^{+0.097}_{-0.076}$ | $16^{+1}_{-2}$ |
| (248835) 2006 SX368 | $0.052^{+0.007}_{-0.006}$ | $76 \pm 2$ |
| (5145) Pholus | $0.155^{+0.076}_{-0.049}$ | $99^{+15}_{-14}$ |
| (8405) Asbolus | $0.056^{+0.019}_{-0.015}$ | $85^{+8}_{-9}$ |
| (10370) Hylonome | $0.051^{+0.030}_{-0.017}$ | $74 \pm 16$ |
| (10199) Chariklo | $0.035 \pm 0.01$ | $248 \pm 18$ |

Note that Centaurs have a range of albedo

While the dynamical relationship between JFCs and Centaurs seems to have been established, there may still be some issues with the physical relationship. As we shall see, comets have low albedos. Thermal infrared studies of Centaurs (Stansberry et al. 2008) in general point to comet-like albedos (Table 1.3). However, there are objects with appreciable higher albedos (e.g. (5145) Pholus) and all objects are much larger than any known JFC (cf. Lamy et al. 2011, for example). The higher albedos may be an observational issue associated with activity (Duffard et al. 2014). There may also be a relationship between colour and albedo (Peixinho et al. 2020). But if these types of objects really are from a family of pre-cursors of JFCs, then either there are many more, much smaller, Centaurs that we simply have not detected yet or else the mass loss from these objects must be enormous at some stage during the transition.

## 1.3   Non-gravitational Forces

Comets can be considered massless for Solar System dynamics calculations and their motion, incorporating perturbation by the planets, can be calculated directly through numerical integration. However, additional complexity arises from two other sources.

First, comets on eccentric orbits with small semi-major axes can experience non-negligible relativistic effects resulting in radial acceleration towards the Sun (Shahid-Saless and Yeomans 1994). The perihelion precession of the orbit is probably the most measurable effect and can be determined through the equation

$$\Delta\omega = \frac{0.0384}{a_s(1 - e_x{}^2)} \tag{1.20}$$

where $\Delta\omega$ is the precession of the argument of perihelion is arcseconds per comet orbit about the Sun and $a_s$ is again the semi-major axis in units of [AU].

The second, and more physically interesting, complexity arises from the outgassing and mass loss from the cometary surface which provide additional forces (non-gravitational forces or NGFs) on the nucleus that are sufficient to modify both its orbit and its rotational characteristics. The equation of motion, ignoring the relativistic effects (see Beutler (2005) for how these can be included) and assuming the Sun is at the barycentre of Solar System, can be written as

$$\frac{d^2\mathbf{r_h}}{dt^2} = -GM_\odot \frac{\mathbf{r_h}}{r_h^3} + \frac{dR_p}{d\mathbf{r_h}} + A_1 g(r_h)\widehat{\mathbf{r_h}} + A_2 g(r_h)\widehat{\mathbf{T}} + A_3 g(r_h)\widehat{\mathbf{n}} \tag{1.21}$$

where $\mathbf{r_h}$ is the heliocentric distance with the bold face indicating that it is being used vectorially in this equation rather than the usual scalar magnitude. Here,

$$g(r_h) = c_{NGF}\left(\frac{r_h}{r_0}\right)^{-m_{NGF}}\left(1 + \left(\frac{r_h}{r_0}\right)^{n_{NGF}}\right)^{-k_{NGF}}. \tag{1.22}$$

In the second term on the right hand side, $R_p$ is a planetary disturbing function describing the influence of the other Solar System bodies on the comet's motion. The NGFs are represented by the last three terms and the coefficients, $A_{1,2,3}$. This semi-empirical approach was introduced in the 1970s by Marsden and is conceptually straightforward. The outgassing from the nucleus produces accelerations. A radial acceleration arises because outgassing is predominately from the illuminated surface which leads to a reaction force in the anti-Sun direction. Terms arise in the other two directions (transverse, $\widehat{\mathbf{T}}$, and normal, $\widehat{\mathbf{n}}$, to the orbital plane) as a result of the rotation of the nucleus and the obliquity (the angle between the rotation axis and the normal to the orbital plane) combined with the thermal inertia of the nucleus and latitudinal variations. The form of $g(r_h)$ is intended to account for the variation in outgassing strength with heliocentric distance. $r_0$ is the heliocentric distance inside

**Table 1.4** Constants for the determination of g(r) in the standard description of non-gravitational forces (from Yeomans and Chodas 1989)

| Constant | Value |
|---|---|
| $r_0$ | 2.808 AU (for water ice) |
| $c_{NGF}$ | 0.111262 |
| $m_{NGF}$ | 2.15 |
| $k_{NGF}$ | 4.6142 |
| $n_{NGF}$ | 5.093 |

which the bulk of the solar insolation goes into subliming ices at the surface of the nucleus. The constants, $c_{NGF}$, $m_{NGF}$, $k_{NGF}$, and $n_{NGF}$ have been determined through fitting as in Table 1.4. Typical values for $A_1$, $A_2$, and $A_3$ are $0.3 \times 10^{-8}$, $\pm 0.05 \times 10^{-8}$, and $\pm 0.07 \times 10^{-8}$ in units of [AU day$^{-2}$] although order of magnitude variations between individual comets are evident. The fact that $A_1$ is usually larger than $A_2$ and $A_3$ is indicative of the dominance of outgassing from the dayside hemisphere.

It should be noted here that the formulation is symmetric about perihelion which from inspection of Fig. 1.3 is clearly not the case. On the other hand, the formulation can also allow for time-dependence of the $A$ coefficients which removes this issue (see for example Yeomans and Chodas 1989; Aksnes and Mysen 2011). The power law form of $g(r_h)$ should also be noted as this can be related to the power laws used to describe the change in cometary brightness with heliocentric distance described earlier.

The change in the orbital period resulting from the radial and transverse forces ($F_r$ and $F_t$, respectively) can be computed from

$$\Delta P = \frac{6\pi\sqrt{1 - e_x^2}}{M_N \tilde{n}^2} \left[ \frac{e_x}{p_L} \int_0^P \left( F_r \, \sin f + \frac{F_t}{r_h} \right) dt \right] \tag{1.23}$$

where $M_N$ is the mass of the nucleus, $t$ is the time, $\tilde{n}$ is the mean motion, $p_L$ is the semi-latus rectum, and, as before, $e_x$ is the eccentricity and $f$ is the true anomaly (Rickman 2017).

Although this approach to NGF modelling has been useful in defining the orbits of short-period comets, one question that arises is whether something can be learnt about the comet from the observed NGFs. One can assume that the outgassing rate is proportional to the surface area as will be discussed below. If the total outgassing rate can be computed in this way then the non-gravitational accelerations depend upon the nucleus mass and if the volume can be established through observations of the bare nucleus then the density can be derived. This idea has prompted some authors to look at more physical approaches to determining the NGF terms in Eq. (1.23).

The basic idea is to partially invert the problem by modelling the outgassing from the surface accurately using heat balance equations (Sect. 2.9.3) applied to facets describing the shape of the nucleus. Where the shape is essentially unknown, spheres or ellipsoids can be used as approximations (e.g. Maquet et al. 2012). The water production rate is then compared to observation (for example through ground-based observation of OH emission from its 18 cm lines; Sect. 3.2) to provide a first constraint. The recoil force is then calculated at facet level and integrated to provide the radial, traverse, and normal forces at each timestep which, after division by the mass, can be included in the equation of motion. We can express this at one point in time as

$$F_i = -\eta x_i Z_i v_i \boldsymbol{\sigma}_i \qquad (1.24)$$

where $\boldsymbol{\sigma}_i$ is the surface area of each facet, i. Note that $\boldsymbol{\sigma}$ is here a vector to define the normal of the facet because it is assumed that outflow is orthogonal to the surface. $Z_i$ is the mass loss (or sublimation) rate per unit area at the facet level, $v_i$ is the velocity of the ejected material at the source, and $x_i$ is an effective active fraction of the surface that is emitting at the given production rate.

Although the activity distribution and its time dependence, $Z_i(t)$, might be difficult to assess, $v_i$ is also not trivial. Both the gas and the dust contribute to the mass loss at a roughly equal levels but their effective velocities differ. The terminal velocities of the two are also not representative of the value of $v_i$ needed to compute the reactive torque on the nucleus because the gas transfers rotational energy to translational degrees of freedom through collisions as the gas expands and the translational energy of the gas is used to accelerate the dust above the nucleus. $v_i$ may also be a function of $Z_i$ if the gas density at the source is collisionally thick. However, it is reasonable to assume for the purposes of this equation that the mass loss rate is dominated by the sublimation of the icy constituent and that the velocity of the ejected material can be approximated by the gas velocity at the source. The latter is usually assumed to be given by the thermal velocity

$$v_i = \sqrt{\frac{8kT_i}{\pi m}} \qquad (1.25)$$

which, in turn, assumes thermal equilibrium between the evolved gas and the surface temperature of facet, i. $m$ here is the mean molecular mass of the gas molecules (which can usually be assumed to be that of the water molecule) and $k$ is Boltzmann's constant. The use of the surface temperature in this equation is not a trivial assumption as will be seen in Sect. 3.4.7.

The parameter, $\eta$, is a momentum transfer coefficient that describes how well the back thrust resulting from the acceleration of the gas into the vacuum of space is coupled to the nucleus. It should be apparent that errors resulting from using the gas thermal velocity rather than a more complex description will end up affecting the momentum transfer coefficient. $\eta$ is poorly constrained but values of the order of 0.7 are considered to be reasonable. The resulting force on each facet, $F_i$, can then be used to compute a non-gravitational acceleration through the equation

$$a_{NGF} = \frac{\sum_i F_i}{M_N} \qquad (1.26)$$

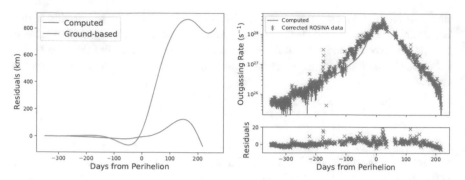

**Fig. 1.18** Left: In blue residuals from a fit to the orbital position of the nucleus using a model of the outgassing. The residuals in dark yellow (called ground-based) are based upon the NGF model of Marsden et al. (1973) with A parameters obtained from the NASA/JPL Horizons system. Right: The required total gas production rate to produce the fit (yellow line) is compared to ROSINA estimates of the total gas production rate (blue crosses). From Attree et al. (2019) A&A, reproduced with permission © ESO

where $M_N$ is the mass of the nucleus. The equations must be consistent with the measured total gas production rates so that, at any one time,

$$Q_g(t) = \sum_i x_i Z_i(t)\sigma_i(t) \tag{1.27}$$

where $Q_g(t)$ is the gas production rate at time, $t$, on the comet's orbit determined through observation.

There are a large number of free parameters in this approach but it is a clear improvement over the Marsden approach because the $g(r_h)$ term is replaced by a physical description rather than empirical constants. Such a physical description requires outgassing rates proportional to the solar insolation. An example of a fit to the measured orbit of 67P using this approach is shown in Fig. 1.18. Here, Attree et al. (2019) reduced the number of free parameters by splitting the surface of the nucleus into six large regions where facets within each region responded to solar insolation in the same way. They also introduced a time-varying solution with stronger outgassing relative to the insolation at perihelion. This need to introduce such behaviour can be linked to the steepness of the brightness variation with heliocentric distance that we saw in, for example, Fig. 1.3. While it is clear that the physical explanations for the choice of this specific model are lacking, the fits to the observables are reasonable and significantly improve upon the NGF models used to date. Furthermore, the gas production rate required for the fit shows reasonable agreement with rates estimated from data acquired by the ROSINA instrument onboard Rosetta. Hence, this approach appears to have promise. To go further we need to look at the source of the activity in more detail.

# Chapter 2
# The Nucleus

## 2.1 Sizes and Shapes of Unresolved Nuclei

For ground-based observations and observations using telescopes in Earth-orbit (e.g. the Hubble Space Telescope), cometary nuclei are usually much smaller than the resolution element (e.g. Lamy et al. 2011). Nonetheless, unresolved observations remain very important because they can place constraints on the diversity of nuclei. Of particular importance are size, albedo, colour and surface composition.

The flux from an unresolved nucleus in the visible can be computed from

$$F_N(\lambda) = F_\odot(\lambda)\, p\, \varphi(\alpha) \frac{r_N^2}{\Delta^2 r_h^2} \qquad (2.1)$$

where $p$ is the geometric albedo, $\varphi$ is the value of the phase function at the phase angle, $\alpha$, and $F_\odot$ is the solar spectral flux at 1 AU within a wavelength interval reduced to a unit of [W m$^{-2}$ nm$^{-1}$] or similar (Tomasko 1976). The geometric albedo is the ratio of its actual brightness at zero phase angle ($\alpha = 0$) to that of a diffusively scattering, perfectly reflecting, disc with the same cross-section. Fits to observations of $\varphi(\alpha)$ have been made in the past using simple linear functions resulting in values of $d\varphi(\alpha)/d\alpha$ of between 0.03 and 0.06 magnitudes per degree (e.g. Jewitt et al. (2003) for 143P/Kowal-Mrkos and Jewitt and Sheppard (2004) for 48P/Johnson). The solar spectral flux (which we will often refer to as just "the solar flux") at 1 AU can be obtained from numerous sources (e.g. Kurucz et al. 1984). An evaluation of the solar flux over an extended wavelength range has been provided by Claire et al. (2012).

An alternative way of expressing this used by ground-based observers is to take an approach developed by Russell (1916) (see e.g. Meech et al. 2004 or Lamy et al. 2011) by writing

© Springer Nature Switzerland AG 2020

N. Thomas, *An Introduction to Comets*, Astronomy and Astrophysics Library,
https://doi.org/10.1007/978-3-030-50574-5_2

$$pr_N^2 = 2.24 \; 10^{22} \; \Delta^2 r_h^2 \; 10^{0.4\left(m_\odot - m_N + \alpha\frac{d\varphi}{d\alpha}\right)} \tag{2.2}$$

where $m_\odot$ is the solar magnitude ($= -26.74$ in the V filter of the Johnson-Morgan photometric system) and $m_N$ is the magnitude of the nucleus thereby providing a result relative to the standard astronomical unit of brightness.

A surface is said to be Lambertian if light falling on it is scattered such that the reflected intensity of the surface to an observer is independent of the observer's angle of view. In other words, a Lambertian surface is one that appears uniformly bright from all directions of view. Such a surface is sometimes referred to as being perfectly diffusing. Again rigorously, if a surface is Lambertian then it has the property of being an ideal, perfectly diffusing, reflecting surface and reflects the entire incident light. A surface that is somewhat absorbing but still perfectly diffusing is often referred to as being Lambertian. Although this is not strictly correct, we will also use this less restrictive definition herein.

Definitions of a Lambertian surface can be confusing and can appear contradictory. There are two things to keep in mind. Firstly, the reflected intensity from a point on the Lambertian surface is proportional to the cosine of the emission angle (cos $e$). But secondly, the observer at a specific emission angle sees a surface area that is proportional to 1/cos $e$. Thus the two cosines cancel out and the reflected intensity can be written as

$$I(\alpha, e, i) = const. \tag{2.3}$$

where $\alpha$, $e$, and $i$ are the phase, emission and incidence angles, respectively. For a Lambertian surface, $p$ is related to the directional-hemispherical albedo, $A_H$ through the relation

$$A_H = \frac{3}{2}p. \tag{2.4}$$

which can be derived easily for a sphere by integration. $A_H$ is an important quantity in cometary physics. It describes the ratio of the total light reflected from a cometary surface, integrated over all emission directions to the incoming flux. The term $(1 - A_H)$ therefore describes the total light absorbed by the surface and hence is of significant importance in defining the energy balance at the surface.

We use the symbol, $S_\odot$ for the solar spectral flux integrated over the full wavelength range at 1 AU, i.e.,

$$S_\odot = \int_0^\infty F_\odot \, d\lambda \tag{2.5}$$

so that the reflected flux from the nucleus at zero phase angle becomes

$$F_{ref} = \frac{2}{3} S_\odot A_H \frac{r_N^2}{\Delta^2 r_h^2} \tag{2.6}$$

While Eq. (2.4) is useful for illustration purpose, in real cases, it does not hold because $I(\alpha, e, i)$ is not constant with the photometric angles. We shall discuss this in more detail later. However, this is sufficient at this stage because Eq. (2.1) shows that there is degeneracy between the geometric albedo, $p$, and the nucleus radius, $r_N$. This can be resolved by making additional observations in the thermal infrared.

In the isothermal approximation and in the absence of thermal inertia, a simple energy balance between the illumination of the nucleus and the thermal re-radiation can be used to determine the surface temperature of a spherical surface, viz.,

$$T^4 = \frac{S_\odot(1 - A_H)}{4\sigma r_h^2} \tag{2.7}$$

where $\sigma$ is the Stefan-Boltzmann constant.

The radiated flux has a spectral distribution defined by the Planck function as

$$B_\vartheta(\vartheta, T) = \frac{2h\vartheta^3}{c^2} \frac{1}{e^{h\vartheta/kT} - 1} \tag{2.8}$$

in units of [W m$^{-2}$ sr$^{-1}$ Hz$^{-1}$] where $\vartheta$ is the frequency. It is often convenient to express the Planck function in terms of wavelength, $\lambda$, as

$$B_\lambda(\lambda, T) = \frac{2hc^2}{\lambda^5} \frac{1}{e^{hc/\lambda kT} - 1} \tag{2.9}$$

in units of [W m$^{-2}$ sr$^{-1}$ m$^{-1}$]. It is also convenient to normalize this function by dividing the integral over all wavelengths (Tucker 1975). The normalization constant, $k_N$ is

$$k_N = \frac{15}{2} \frac{h^3 c^2}{\pi^4 k^4 T^4} \tag{2.10}$$

which can be seen to be related to the definition of Stefan-Boltzmann's constant, $\sigma$,

$$\sigma = \frac{2\pi^5 k^4}{15\, h^3 c^2}. \tag{2.11}$$

The flux (integrated over all wavelengths) seen by an observer can be calculated simply from the thermal luminosity, $L_{th}$, in this case which is

$$L_{th} = 4\pi r_N^2 \sigma T^4 \tag{2.12}$$

and leads to the thermal flux from an unresolved isothermal object being

$$F_{th} = \sigma T^4 \frac{r_N^2}{\Delta^2} \tag{2.13}$$

A slightly more realistic case can also be studied analytically. If we assume that a spherical object is observed at zero phase angle and that the thermal emission is locally in instantaneous equilibrium with the solar insolation then the thermal flux can also be calculated. Here the local surface temperature is obtained from the energy balance equation

$$\sigma T(i)^4 = \frac{S_\odot(1 - A_H)}{r_h^2} \cos i \tag{2.14}$$

where $i$ is again the solar incidence angle. Integration over all wavelengths followed by substitution for $T(i)$ leads to

$$F_{th} = \frac{2}{3} S_\odot(1 - A_H) \frac{r_N^2}{\Delta^2 r_h^2} \tag{2.15}$$

This equation can also be derived using

$$F_{th} = \int_0^{r_N} \frac{B_\lambda(T)}{\Delta^2} \, 2\pi r \, dr \tag{2.16}$$

as a starting point where $B_\lambda(T)$ is the black-body radiance at surface temperature, $T$, integrated over all wavelengths.

Comparing Eq. (2.15) with Eq. (2.1) shows that summing the reflected and radiated fluxes, we have energy conservation with the input solar flux. However, the point to note from this discussion is that the ratio of the thermal flux, $F_{th}$, to the reflected flux, $F$, is related to the hemispherical albedo through the proportionality

$$\frac{F_{th}}{F_{ref}} \propto \frac{(1 - A_H)}{A_H} \tag{2.17}$$

so that $A_H$ can be separated from $r_N$.

There are several further complexities to account for when applying this approach to observations. Firstly, the equations apply to observations at zero phase angle (opposition geometry). As with the reflected flux, a thermal phase function, $\varphi_{th}$, can be introduced to account for non-opposition geometry by using the form

$$F_{th} = \frac{2}{3} S_\odot(1 - A_H) \frac{r_N^2}{\Delta^2 r_h^2} \varphi_{th}(\alpha) \tag{2.18}$$

Secondly, we need to introduce the infrared emissivity, $\varepsilon$, which is a material property. The emissivity is a single value and is quantitatively the ratio of the thermal radiation from a surface to the radiation from an ideal black-body at the same temperature.

The integrated thermal flux is not a quantity that is easily measured. Extrapolation from a measurement within a frequency or wavelength range is necessary and Planck's radiation law (Eq. 2.8) used. At a specific frequency, the thermal flux is then

$$F_{th}(\vartheta) = 2\pi\varepsilon \frac{1}{\Delta^2} \int_0^{r_N} B_\vartheta(\vartheta, T) r\, dr \qquad (2.19)$$

which again assumes instantaneous surface energy balance at zero phase angle. We note here an alternate form of this equation giving the observable thermal flux at a single wavelength as (Delbo and Harris 2002)

$$F_{th}(\lambda) = \frac{2\pi r_N^2}{\Delta^2} \varepsilon \int_0^{\frac{\pi}{2}} B(\lambda, T) \cos i \sin i\, di \qquad (2.20)$$

Numerical models of the surface temperature distribution will be introduced in Sect. 2.9.3.1 to account for the effects of thermal inertia.

One concept of relevance here is that of brightness temperature. Brightness temperature is the temperature that a black body in thermal equilibrium with its surroundings would have to be in order to duplicate the observed intensity of a grey body object within a defined frequency range. The brightness temperature is therefore not a "temperature" in the way that it is used for, for example, a pot on a heating plate. It characterizes the radiated power from a surface and, depending on the mechanism of the emitted radiation, can differ considerably from the physical temperature of the radiating body. The brightness temperatures of strong line emissions, for example, can be very large indeed if the bandwidth is small. On the other hand, the brightness temperature of a surface is usually an expression of a measurement and therefore does not contain assumptions that are needed to get from a radiated power to an actual (thermodynamic) temperature. Consequently, it is useful way to express measured constraints on models.

We can invert Planck's radiation law to express a temperature, $T_b$, as a function of an observed intensity, $I_\lambda$, i.e.

$$T_b = \frac{hc}{k\lambda} \frac{1}{\ln\left(1 + \frac{2hc^2}{I_\lambda \lambda^5}\right)} \qquad (2.21)$$

If we assume that the source is a thermal radiator, then the observed intensity depends upon the emissivity, $\varepsilon$, so that the brightness temperature is usually less than the actual temperature of the object. At high temperatures and long wavelengths, the

Rayleigh-Jeans law can be used and this leads to a simple relationship between the brightness temperature and the actual temperature

$$T_b = \varepsilon T \tag{2.22}$$

In the more general case, the emissivity can be thought of as the ratio of the thermal radiation from a surface to the radiation from an ideal black surface at the same temperature as given by the Stefan–Boltzmann law. The emitted flux (radiant exitance), $R_{BB}$, of a black-body is

$$R_{BB} = \sigma T^4 \tag{2.23}$$

in units of [W m$^{-2}$]. So that

$$\varepsilon = \frac{R_{obs}}{R_{BB}} \tag{2.24}$$

where $R_{obs}$ is the observed radiant exitance.

It is here important to note that the brightness temperature is not a constant over all wavelengths because, in the general case, the emitted power from a surface within a wavelength interval is proportional to the emission coefficient at that wavelength and only in the specific case of a black-body is this independent of wavelength and equal to one. This is a consequence of Kirchhoff's law which can be loosely defined as saying that if an arbitrary body in thermodynamic equilibrium, is emitting and absorbing thermal radiation, the emissivity is equal to the absorptivity. However, the absorptivity of a surface is not constant with wavelength—objects have colour and thus the spectral dependence of emissivity is a property of the material and its grain size as has been shown by laboratory investigations.

The integrated emissivity over all wavelengths is required for the energy balance calculation and is therefore an integral over wavelength weighted by a Planck function, $B_\lambda(T)$, at the temperature of the surface such that

$$\varepsilon = \frac{\int_0^\infty B_\lambda(T)(1 - A_H(\lambda))\, d\lambda}{\int_0^\infty B_\lambda(T)\, d\lambda}. \tag{2.25}$$

where the wavelength dependence of emissivity has been replaced by 1-absorptivity using the hemispherical albedo. We can now see that the spectral radiance from a real surface is not necessarily distributed according to a Planck function at the temperature of that surface. It is this property of the brightness temperature concept that can lead to significant confusion.

A further complexity arises from surface roughness. The effect of surface roughness on the sizes of nuclei measured from Earth-based telescopes is probably small, particularly if the measurements are made close to zero phase angle. However, the significance is much greater for resolved surfaces at intermediate phase angles where

**Fig. 2.1** Illustration of how brightness temperature can differ from the actual temperature of a rough surface. Solid line: 10% of the surface is assumed to be at 300 K, the rest is at absolute zero. Dashed line: 50% of the surface is at 300 K. A wavelength independent emissivity of 1 has been assumed. The brightness temperature of the integrated surface has then been calculated at each wavelength

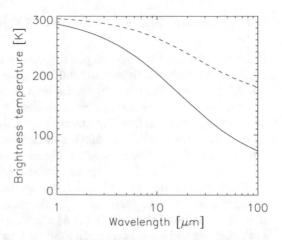

local shadowing becomes of importance or where there are large compositional variations. This leads to non-uniform surface emission. Rough surfaces will have elements oriented towards the Sun that become significantly hotter than an inclined flat surface. Multiple scattering of radiation between rough surface elements increases the total energy absorbed by the surface (Rozitis and Green 2011). Hence, ε is poorly constrained outside the laboratory. Values for planetary surfaces between 0.75 and 0.9 are often assumed.

As an illustration, let us imagine an extremely rough surface such that 10% of the surface is at 300 K and the remaining 90% of the surface is at 0 K and therefore not emitting at all. The emissivity is assumed to be 1. A measure of the spectral intensity distribution of emitted radiation from this surface will show the form of a black-body at 300 K. But the measured intensity will be a factor of 10 lower than from a uniform black-body leading to a brightness temperature that is both lower than the actual temperature and varying with wavelength as shown in Fig. 2.1. A second example with 50% of the surface emitting at 300 K is also shown.

For Earth-based observations, assumptions on several of the unknowns can now be made with some degree of confidence leading to reasonably accurate values for the sizes of nuclei. Lamy et al. (2004) critically reviewed values for 65 comets and also assessed the non-sphericity of several objects. Further results have been presented in Lamy et al. (2009, 2011). They noted that the projected cross-sectional area ($\sigma_p$) of a prolate or oblate spheroid in a simple rotation about an axis is given by

$$\sigma_p = \pi a b^2 \sqrt{\left(\left[\frac{\sin^2\phi}{a^2} + \frac{\cos^2\phi}{b^2}\right]\sin^2\alpha + \frac{\cos^2\alpha}{b^2}\right)} \qquad (2.26)$$

where $\alpha$ is the angle between the spin vector of the nucleus and the direction to the observer and $\phi$ is the rotation angle. This can be used to determine the axial ratio ($a/b$) and give a first, course, assessment of the nucleus shape.

## 2.2    Sizes and Shapes of Resolved Objects

It should be recalled that only 40 years ago, the existence of a single solid nucleus at the heart of cometary activity could still be disputed (although Whipple's concept of a solid "dirty snowball" was by far the most accepted theory). Prior to the Halley fly-bys, the size of the nucleus was estimated by assuming that the nucleus was active over its entire surface. The thermodynamics of sublimation could then be used to determine how large the emitting surface area needed to be to produce the observed water production rate. This minimum surface area could then be used to determine a lower limit for the radius of the nucleus.

There are numerous assumptions in this approach and hence the observations of the nucleus of 1P/Halley were of major significance in challenging these assumptions. The picture taken by the Halley Multicolour Camera onboard Giotto (Fig. 1.1) not merely proved the existence of a solid nucleus but also showed its size to be far larger than the minimum size needed to produce the observed outgassing.

The sizes of nuclei are usually expressed as dimensions along three axes even though it is now quite apparent that nucleus shapes are far from regular. Nonetheless describing nuclei as tri-axial ellipsoids based on these axis dimensions can be useful as first order approximations to their shapes.

A determination of the shape of 1P/Halley from the spacecraft fly-by data was not straightforward. The fast fly-bys ($\sim$70 km s$^{-1}$ relative velocity in all cases) provided only snapshots while the phase angle of the approaches to the nucleus were all such that more than half the visible nucleus was unilluminated. The darkness of the unilluminated limb against a background of illuminated dust did however provide additional constraints. The dimensions finally derived were 15.3 km $\times$ 7.8 km $\times$ 7.4 km (Keller et al. 1995). The uncertainties on each axis length are of the order of 0.5 km.

At the time, thermal IR observations of small bodies were in their infancy and the surface reflectance had to be derived by combining the reflected flux at high heliocentric distance and the estimated cross-sectional area found by the resolved spacecraft observations. This led to a geometric albedo estimate of 0.04—a remarkably low value which was subsequently confirmed by photometric analyses of the Vega and Giotto data (Sagdeev et al. 1986a, b).

The subsequent observations of nuclei have shown that low geometric albedos are usual. Data from spacecraft fly-bys tend to have large errors because the flyby geometry dictates the minimum phase angle of the observation and obtaining the geometric albedo requires extrapolation to zero phase. However, observations of the nucleus of 19P/Borrelly (observed by NASA's Deep Space 1) and that of 9P/Tempel 1 (observed by Deep Impact) are both consistent with geometric albedos of <0.07 and, taking into account coma contributions in the analysis of the 9P/Tempel 1 data, probably <0.05 (Table 2.1). Values obtained by combining thermal emission and photometric observations are fully consistent with this picture.

The irregular shape of 1P/Halley drew considerable attention at the time. The primary reason was that it challenged the idea that comets are uniformly shrinking "dirty snowballs" in the way that Whipple expressed them in his rightly celebrated

**Table 2.1**  Geometric albedos and photometric phase dependencies of nuclei resolved by space-craft fly-bys or rendezvous

| Comet | Geometric albedo | Phase function [mag deg$^{-1}$] | "Bond" albedo | SSA | Approx. centre wavelength | Reference |
|---|---|---|---|---|---|---|
| 1P/Halley | 0.04 | | | | 645 nm | Keller et al. (1987) |
| 9P/ Tempel 1 | 0.059 ± 0.009 | 0.046 | 0.014 | 0.043 | Several (converted to V) | Li et al. (2013a) |
| 19P/ Borrelly | 0.072 ± 0.020 | 0.043 | | 0.057 | 660 nm (converted to V) | Li et al. (2007) |
| 67P | 0.065 ± 0.002 | 0.047 | | 0.046 | 649 nm | Fornasier et al. (2015), Feller et al. (2016) |
| 81P/Wild 2 | 0.059 ± 0.004 | 0.051 | | 0.038 | 700 nm | Li et al. (2009) |
| 103P/ Hartley 2 | 0.045 ± 0.009 | 0.046 | 0.012 | 0.036 | 610 nm (converted to V) | Li et al. (2013b) |

paper in 1950. Although this paper was a landmark in cometary research, the Halley observations showed that the dark material was dominant (leading Keller to describe comets as "icy dirtballs" as a counterpoint to Whipple's phrase) and that they really could not be described as "balls" at all. The original description of comets by Whipple (1950, 1951) referred to an icy conglomerate of ices with meteoritic material. Compared to this concept, the mass of dark, non-volatile, material appeared to be greater although it is important to note that Whipple never actually quantified the non-volatile to ice ratio he envisaged. The idea of uniformly subliming, spherical, totally water-ice dominated, object was no longer tenable. The elongated shape with axes roughly in the ratio 2:1:1 also seemed to be more consistent with a formation process in which larger planetesimals accreted rather than growing uniformly by collection of gas and dust around a core.

The shapes of other nuclei observed by spacecraft (Fig. 2.2) have shown a wide range (Table 2.2).

Even from a distance of 30,000 km, it was clear that the nucleus of 67P was highly irregular in shape and it quickly became apparent that the nucleus had two distinct lobes. The appearance of the nucleus in the early images was referred to as being like a small duck leading to the smaller of the two lobes being referred to as the head with the larger lobe being the body. A neck separated the two. (Quite why cometary scientists feel the need to compare the shapes of nuclei to ducks, avocados, baked potatoes, peanuts, etc. is something of a mystery to me but we will use the terms head, neck and body for 67P purely out of solidarity!) This shape immediately provided a challenge to the Rosetta science team in defining coordinate systems and approaches to defining the 3D shape of the nucleus.

There are two main techniques for determining the shape of a resolved irregular object. Stereo photogrammetry (SPG) uses parallax between images of the same field taken from different viewing geometries to establish the elevation with respect to an image plane. By combining many images of the object, the full 3D shape can be reconstructed in a self-consistent way. The technique is particularly powerful for

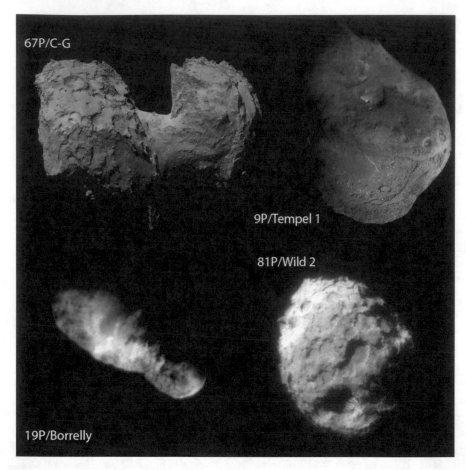

**Fig. 2.2**  Montage of the nuclei of comets 67P, 9P/Tempel 1, 19P/Borrelly and 81P/Wild 2 (Credits: 9P (NASA/JPL/Uni. Maryland, 19P (NASA/JPL), 67P (see Table 1) and 81P (NASA—NSSDC Master Catalog))

object or regions on objects where the topographic relief is large. On the other hand, flat, smooth terrains are not so easily reconstructed and errors on these types of terrain can be significant. An example of an SPG reconstruction is shown in Fig. 2.3 and its application to 67P has been described by Preusker et al. (2017).

An alternative technique is stereo photoclinometry (SPC). Here it is assumed that the reflectance properties of the surface elements are identical. Changes in the brightness then correspond to changes in the photometric angles that in turn constrain the local shape. SPC is very powerful for flat, smooth terrains but is less effective where topographic relief is high. Its application to 67P has been described by Jorda et al. (2016). This technique is complimentary to SPG and combining the results from applying both these techniques to one data set should provide an optimum result. The Multi-resolution Stereo-photoclinometry by Deformation

**Table 2.2**  The sizes and shapes of cometary nuclei observed directly by spacecraft

| Comet | Rough description | Size | |
|---|---|---|---|
| 1P/Halley | Prolate ellipsoid | $15.3 \times 7.8 \times 7.4$ km $\pm 0.5$ km | Keller et al. (1987) |
| 9P/Tempel 1 | Almost pyramidal | Mean radius: 2.81 km $\pm$ 0.1 km<br>Volume: 95.2 km$^3$<br>Principal moment ratios:<br>a/c: 0.688; a/c: 0.930 | Thomas et al. (2013a) |
| 19P/Borrelly | Elongated ellipsoidal | Axes: $4.0 \pm 0.1 \times 1.60 \pm 0.02$ $\times 1.60 \pm 0.02$ km | Buratti et al. (2004) |
| 67P/ Churyumov-Gerasimenko | Axially asymmetric bi-lobate | Axes: $4.34 (\pm 0.02) \times 2.60 (\pm 0.02)$ $\times 2.12 (\pm 0.06)$ km<br>Volume: $18.7 \pm 0.2$ km$^3$ | Jorda et al. (2016), Pätzold et al. (2016) |
| 81P/Wild 2 | Roughly spheroid | Axes: $1.65 \times 2.00$ $\times 2.75$ km $\pm 0.05$ km | Duxbury et al. (2004) |
| 103P/Hartley 2 | Axially symmetric bi-lobate | Mean radius: 0.58 ($\pm 0.018$) km<br>Volume: 0.809 ($\pm 0.077$) km$^3$<br>Principal moment ratios: a/c: $0.166 \pm 0.004$, b/c: $0.979 \pm 0.002$ | Thomas et al. (2013b) |

The descriptions are crude but are intended to give an impression of the shape

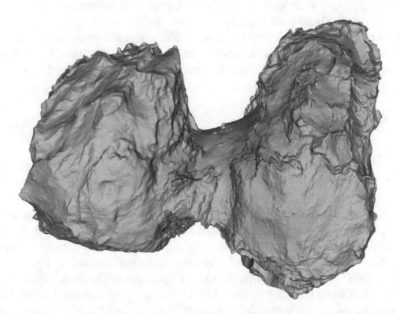

**Fig. 2.3**  A view of the 3D shape model of the nucleus of 67P reconstructed using the SPG (stereo photogrammetry) technique created by Preusker et al. (2017)

(MSPCD) approach (Capanna et al. 2013) was designed to address this issue. This is however technically and computationally challenging for high resolution data.

Cometary shapes have, in the past, been described in terms of the best-fit triaxial ellipsoid (Table 2.2). However, describing 67P as a tri-axial ellipsoid is clearly ridiculous but, for the sake of comparison, Jorda et al. provided the nominal figures. From the same paper, values for the body and the head were determined independently and found to be $4.10 \times 3.52 \times 1.63$ km and $2.50 \times 2.14 \times 1.64$ km, respectively with their volumes being $12.4 \pm 0.6$ km$^3$ and $5.1 \pm 0.3$ km$^3$ representing 66% and 27% of the total volume of the nucleus. The volume of the neck could be derived by subtraction and was found to be $1.4 \pm 0.4$ km$^3$ (about 7% of the total volume) but this carries some uncertainty because of the need to define the interfaces between the three parts of the nucleus. Thomas et al. (2013a, b) adopted an alternate approach of giving a mean radius and values for the principal moments derived from shape models but even this approach has difficulties if there are large surface areas almost radial (or indeed overhanging) with respect to the centre of figure.

It was recognized quickly after the first images of 67P were returned from Rosetta that the neck of the nucleus would be subject to stresses as a result of self-gravity, mechanical strength (assuming a linear elastic material) and the forces coming from cometary activity. Self-gravity stress results from the bilobate shape of the nucleus, while activity stress results from the surface sublimation of water ice and changes diurnally (because of the rotation of the nucleus) and with heliocentric distance. The resulting stresses have been modelled by S.F. Hviid (pers. comm.) using a finite element model and are shown in Fig. 2.4. Similar calculations have been performed by Hirabayashi et al. (2016). The stresses reach around 300 Pa. The evidence that we may also have witnessed the effects of these stresses will be discussed later.

## 2.3   Mass and Density

Prior to the start of development of the instrumentation of the Rosetta mission, the uncertainty in the bulk density of the nucleus was a major issue. Engineers designing the landing system of what was eventually to become the Philae lander required constraints on the density and mass of the nucleus to optimize the approach to landing and anchoring the small spacecraft on the nucleus. At the time, the only way to attack the problem was through analysis of the non-gravitational forces perturbing the orbit. These analyses required numerous assumptions and values between 200 and 2400 kg m$^{-3}$ for the density were discussed at the time. However, the dominance of water ice in the nucleus and the high porosity inferred from studies of dust particles (e.g. arising from the first polarimetric measurements such as those of Dollfus (1989)) led to a consensus that values $<1000$ kg m$^{-3}$ were highly likely and possibly much less. Davidsson (2001) studied the rotation periods of 14 comets and concluded that, for seven objects and assuming zero tensile strength, these nuclei needed to have densities in the range 200 kg m$^{-3} \leq \rho \leq 530$ kg m$^{-3}$ to avoid being

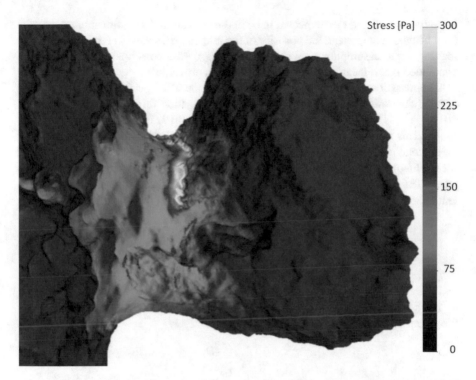

**Fig. 2.4** Tensile stresses in the neck of the nucleus of 67P in units of Pascal (courtesy of S.F. Hviid and O. Groussin)

split by the centrifugal forces. As measurement and modelling capabilities improved further, refinement of the estimates appeared possible. Davidsson and Gutierrez (2004), for example, used NGFs to derive a value of 180–300 kg m$^{-3}$ for 19P/Borrelly. A higher value (490 kg m$^{-3}$) had previously been found by Farnham and Cochran (2002) but it was clear here that much of the difference between the two results could be traced to the assumptions made.

A major breakthrough came with the Deep Impact experiment where modelling of the observed ejecta cloud from the impact of the copper projectile allowed a nucleus mass estimate that could be converted to a bulk density using imaging observations of the volume. This resulted in a density of 400–500 kg m$^{-3}$ (Richardson et al. 2007; Holsapple and Housen 2007) clearly indicating densities around ½ of that of solid water ice.

The close orbit of Rosetta about the nucleus of 67P has, as expected, led to a very precise value of the mass by studying the Doppler shift of the two-way coherent radio link between Earth ground stations and the spacecraft. The standard gravitational parameter was found to be $\mu_N = GM_N = 666.2 \pm 0.2$ m$^3$ s$^{-2}$ giving a mass of $9.982(\pm 0.003) \times 10^{12}$ kg (Pätzold et al. 2016). The nucleus volume is required to obtain the bulk density. This is less precisely known than the mass but the high quality shape models now lead to a bulk density, $\rho_N$, of 537.8($\pm$0.6) kg m$^{-3}$. With a large fraction of water ice expected to be in the nucleus and a significant non-volatile (and heavier) component, the porosity of the nucleus, $\Psi$, seems to be substantial. By making some assumptions about the densities of the constituents, Pätzold et al. concluded that a mean porosity of 72–74% was probable.

The mass lost by the nucleus of 67P during the 2015 perihelion passage could also be determined by the Radio Science Investigation on Rosetta (Pätzold et al. 2019). A value of $10.5 \pm 3.4 \times 10^9$ kg, or about 0.1% of the nucleus mass, was found. This is a very important number as it provides, by far, the tightest constraint on the total mass lost from 67P during its 2015 apparition and also gives a timescale for the lifetime against complete erosion of the nucleus in the current orbit of ~6000 years. The corresponding globally averaged change in radius (assuming an equivalent sphere) is

$$\frac{dr_N}{dM_N} = \frac{1}{\sqrt[3]{36\,\pi M_N^2 \rho_N}} \tag{2.27}$$

and therefore 0.55 m per apparition at the present time.

For a spherical nucleus, the surface gravity and the escape velocity are given by the equations known from school physics, respectively

$$a_g = \frac{GM_N}{r_N^2} \tag{2.28}$$

and

$$v_{esc} = \sqrt{\frac{8}{3}\pi G \rho_N r_N^2} \tag{2.29}$$

where for $v_{esc}$ we have re-written the usual equation so that it is in terms of nucleus radius and density. For typical values of small nuclei, the gravitational acceleration at the surface is of the order of $3 \times 10^{-4}$ m s$^{-2}$, while the escape velocity is of the order of 1 m s$^{-1}$. These are, of course, low values but they are non-negligible especially for large particles in cases where the drag force at the surface arising from sublimation of ices is barely sufficient to lift the particles.

Cometary nuclei are irregular with 67P, 19P/Borrelly and 103P/Hartley 2 being bilobate and highly elongated. This results in a strongly varying surface gravitational potential. In the general case, a brute force method can be applied by dividing up the nucleus into volume elements and integrating so that the acceleration at position, $\boldsymbol{P}$, in space is given by

$$\boldsymbol{a}_g(\boldsymbol{P}) = G\int_V \rho_V \frac{\boldsymbol{r}_P}{|\boldsymbol{r}_P|^3}\,dV \tag{2.30}$$

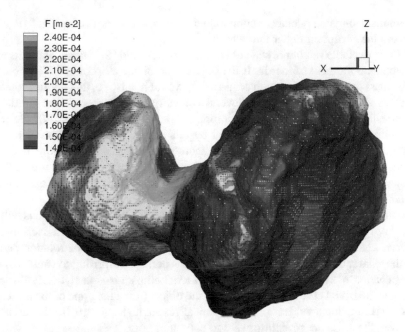

**Fig. 2.5**   The gravitational acceleration at the surface of the nucleus of 67P colour coded in units of [m s$^{-2}$] (Courtesy of R.M. Marschall)

where $\rho_V$ is the density of a volume element, $dV$, and $r_\mathrm{p}$ is the vector from point $P$ to the volume element. This is illustrated in Fig. 2.5 where the gravitational acceleration at the surface of 67P is shown assuming a constant density distribution. A minimum is seen in the neck region whilst a maximum is seen on the flat region on the base of the body (to the right). It can be seen that this area is in a small depression. If a uniform density can indeed be assumed, there are faster computational methods as shown by Werner and Scheeres (1997). The variation over the surface is approaching a factor of 1.7 and this leads to the escape velocity from the surface being strongly dependent upon position.

The gravitational field above the nucleus but within one nucleus radius of the surface is also distorted in the neck region. This has implications for the flow velocities of slow moving particles originating in the neck region.

## 2.4   Rotational Properties

So far, we have treated nuclei analytically as uniform spheres. The observations of 1P/Halley by the Halley Armada clearly established the dubiousness of this assumption and subsequent resolved observations of 19P/Borrelly and 103P/Hartley 2 have shown that elongated nuclei are common with quasi-spherical nuclei (e.g. 81P/Wild 2) being the exception rather than the rule.

The most stable rotational state of an object corresponds to rotation about its axis of maximum moment of inertia. If the object is an oblate ellipsoid with a homogeneous mass distribution and with semi-major axes of $a = b > c$, this implies rotation about its short principal axis, c. Several cometary nuclei appear more similar to prolate ellipsoids which have axes according to the relation $a > b = c$. 1P/Halley and 19P/Borrelly are fairly well represented by this shape. A perfectly biaxial prolate spheroid has two axes of equal maximum moment of inertia and, therefore, does not have a stable rotation pole but this situation is hypothetical for comets unless there are special cases in which erosional effects lead to a transition from one axis having the maximum moment of inertia to another.

In the general case, the reflected light from an ellipsoidal, inactive, nucleus will be modulated at the rotation rate as the illuminated cross-sectional area seen by the observer changes as shown in Eq. (2.26) for a spheroid. Hence, the rotation period and the axial ratio can be derived. However, it is perhaps surprising to some that the determination of cross-sectional areas of tri-axial ellipsoids is analytically difficult except in the trivial cases where one views the object from along one of the principal axes. This remains a subject of mathematical research (Klein 2012). Numerically, however, the areas are straightforward to compute.

In opposition geometry (with the Sun, observer, and object along a straight line) with rotation about the short axis, the amplitude of the modulation is a maximum if the rotation axis is orthogonal to the observer-object line and becomes zero if the rotation axis is aligned along observer-object line (Fig. 2.6). In the absence of applied forces, the rotation axis remains fixed in inertial space. But as the object moves around the Sun, the orientation of the rotation axis with respect to an observer changes and thus the modulation also changes. Hence, multiple observations provide a means of constraining the rotation axis position while the amplitude constrains the axial ratios of the ellipsoid. Both the phase angles of the multiple observations and the different inclinations of the orbits of the object and the observer must also be taken into account. The axial ratios can be converted into absolute dimensions either

**Fig. 2.6** The cross-sectional area of a rotating tri-axial ellipsoid (16 km × 9 km × 7 km). Dashed line: The rotation axis (the short axis) is in the plane of the image (orthogonal to the observer-object line). Solid line: The rotation axis is tilted 30° out of the plane. Note that in the latter case, the cross-sectional area is larger but also that the amplitude of the modulation is smaller

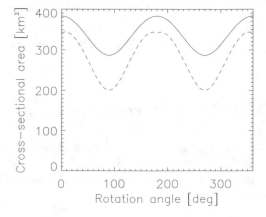

by assuming a surface reflectance or by using observations of the thermal flux as described above.

Size and axis determination of irregular small bodies has reached a high level of sophistication using light curve analysis in combination with other techniques (e.g. stellar occultation and radar delay-Doppler). An example of this was the prediction made for the asteroid (21) Lutetia which was the target of a fly-by by the Rosetta spacecraft 4 years before it reached 67P. The spin axis determined prior to the fly-by was found to be accurate to within 2°, while the size determinations were within 2% of the Rosetta-derived values (Carry et al. 2012).

There are several issues that make achievement of comparable accuracies for cometary nuclei extremely challenging. Firstly, cometary nuclei are, typically, at least a factor of 10 smaller in dimension than (21) Lutetia and hence signal levels are appreciably lower. Second, the activity of comets as they approach the Sun can influence the observed cross-sectional area. To study a "bare" nucleus with no dust coma requires observation at $\gg$2.7 AU thereby further reducing signal levels. Third, we shall see that nuclei are extremely dark and small absolute variations in the surface reflectance can have large effects on the observed fluxes from the object.

Approaches to tackling these problems have been studied for more than two decades. For example, modelling of the dust coma of the comet can be used to remove the signal resulting from activity and thereby obtain the signal from the bare nucleus alone. This has been performed using Hubble Space Telescope observations of a number of comets (Lamy et al. 2004). An accurate knowledge of the point spread function (PSF) of the instrument in combination with PSF correction techniques enhances the accuracy of the determination. Issues concerning surface reflectance can also be addressed by looking at the thermal light curve which will deviate from following the reflected light curve if reflectance variations are present. This has become possible for cometary nuclei using space-borne infrared observatories such as Spitzer. A recent compilation of rotation periods for some of the numbered comets, derived from multiple observational methods, has been provided by Kokotanekova et al. (2017).

Further complexity arises from deviations of objects from regular (spherical to tri-axial ellipsoidal) shapes. Observed light curves can deviate markedly from simple sine curves. This is clearly evident in unresolved observations of 67P taken by the OSIRIS instrument on Rosetta before the spacecraft made its rendezvous (Fig. 2.7). Attempts to determine the shape of the nucleus prior to the start of the Rosetta rendezvous manoeuvres using numerical models to fit the details of the light curves were subsequently shown to be somewhat less successful than expected. The highly irregular shape of the nucleus (with two parts linked by a "neck") was not predicted.

As can be seen in Fig. 2.7, 67P was found to have a rotation period of 12.404 h prior to rendezvous. It had been expected, however, that variations in both the rotation period and the orientation of the rotation axis would occur as a result of torques produced by activity and mass loss (Gutierrez et al. 2005). The sublimation from the surface is anisotropic even in the simplest situation where gas emission is solely related to the cosine of the incidence angle. The irregular geometries of nuclei combined with possible deviations from isolation-driven activity can lead to

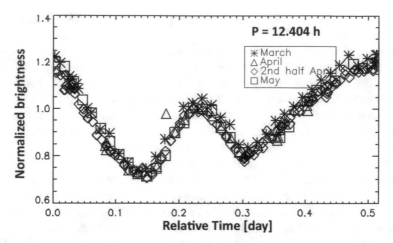

**Fig. 2.7** Light curve of 67P obtained with data from the OSIRIS instrument onboard Rosetta in March–May 2014 (4–6 months before rendezvous)

substantial anisotropy. Hence, numerical solutions must be sought by dividing the surface into facets of varying activity as also discussed in Sect. 1.3.

The reactive torque, $M$, caused by this activity distribution can then be calculated through the equation

$$M(t) = -\sum_{i=1}^{N} Z_i(t) \, (r_{N,i} \times v_i) \qquad (2.31)$$

where $Z_i$ is the mass ejection rate from the i-th facet (Eq. 1.24), $r_{N,i}$ is the radius vector of the facet in the body frame, and $v_i$ is the effective velocity of the ejected matter (Samarasinha and Belton 1995; Neishtadt et al. 2002). Ideally, the quantities, $Z_i$ and $v_i$ as a function of time, should be consistent with the equation for the non-gravitational force (Eq. 1.24). The difficulty in determining these two quantities was discussed earlier.

The change in angular momentum can then be computed through the usual relation to the torque,

$$M(t) = \frac{dL(t)}{dt} \qquad (2.32)$$

and this relates to the change in angular velocity through

$$M(t) = I_N \frac{d\Omega_N(t)}{dt} \qquad (2.33)$$

if the moment of inertia, $I_N$, can be assumed to remain constant (which is not strictly correct of course because of the mass loss producing the torque but is a reasonable

approximation over short timescales). A crude estimate demonstrates quickly that
the effect is significant and needs to be accounted for through detailed modelling.

By simplifying to a spherical object accelerated by a mass ejection, $dM_N/dt$, at a
velocity, $v_{ej}$, acting in the plane orthogonal to the rotation axis at a distance, $r_{ej}$ away
from it, the angular acceleration is given by

$$\frac{d\Omega_N}{dt} = \frac{5}{2} v_{ej} r_{ej} \frac{dM_N}{dt} \frac{1}{M_N \, r_N^2}. \tag{2.34}$$

For reasonable numbers in this highly simplified case, the change in angular
velocity corresponds to changes in the rotation period of the order of seconds per day
for 2 km-sized nuclei at even moderate levels of activity. It should be clear that the
surface area to volume (and therefore mass) ratio increases as the object radius
decreases and hence the effect of outgassing on the rotational properties must
increase as the object gets smaller unless changes in $Z_i$ occur.

Steckloff and Samarasinha (2018) have provided a more rigorous, general,
approach and derived

$$\frac{\overline{d\Omega_N}}{dt} = \frac{3}{4\pi} \overline{Z} m_{H2O} \langle v_{ej} \rangle \frac{f_{tan}}{\rho_N \, r_N^2} \tag{2.35}$$

where $\langle v_{ej} \rangle$ is the average molecular outflow velocity of the sublimating water
molecules in the direction normal to the surface, $\overline{Z}$ is the average production rate in
[molecule m$^{-2}$ s$^{-1}$] of water molecules of mass, $m_{H2O}$, and $f_{tan}$ is the effective
tangential fraction of the theoretical volatile flux at zero solar phase angle that
contributes to a net torque. The bars in Eq. 2.35 indicate use of orbitally averaged
values. In both equations one can see the dependencies upon ejection velocity, mass
loss rate and the mass of the nucleus.

The repeated spacecraft imaging of 9P/Tempel 1 by first the Deep Impact
spacecraft and then the Stardust-NExT mission led to an assessment of the change
in the rotation period of this particular comet when combined with ground-based and
Hubble Space Telescope light curve analysis (Belton et al. 2011). The derived
angular velocities are shown in Table 2.3 and correspond to a total change in the
rotation period of 16.8 ± 0.3 min during the 2000 perihelion passage and
13.7 ± 0.2 min for the 2005 passage. The Deep Impact measurements just prior to
perihelion in 2005 suggest that the torque is mostly applied prior to perihelion in this
case (Belton et al. 2011).

Table 2.3  Derived angular velocities for the rotation of the nucleus of 9P/Tempel 1

| Timeframe | Angular velocity [deg day$^{-1}$] |
|---|---|
| Prior to 2000 perihelion passage | 209.023 ± 0.025 |
| Between 2000 and 2005 perihelion passages | 210.448 ± 0.016 |
| Just prior to 2005 perihelion | 211.856 ± 0.030 |
| Between 2005 and 2010 perihelion passages | 211.625 ± 0.012 |

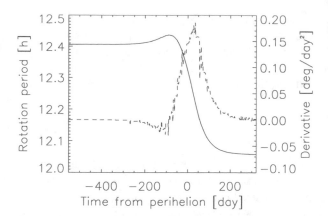

**Fig. 2.8** The rotation period of 67P over the duration of the Rosetta near-comet activities. Solid line: The rotation period itself (left axis). Broken line: The first derivative in units of [deg day$^{-2}$]. The data have been smoothed over using a 51 h boxcar smoother. The maximum of the derivative can be seen around 30 days post-perihelion

For 67P, effects of torques could be easily observed using imaging data over the duration of the near-comet activities of Rosetta to reconstruct the nucleus orientation. This was performed by the European Space Operations Centre in Darmstadt and values provided in the form of reconstructed SPICE kernels (Acton 1996). Figure 2.8, which was produced using these results, shows that the rotation period of the comet changed by nearly 25 min during the perihelion passage initially rising during the approach to the Sun but falling rapidly within the 150 day period about perihelion itself. The first derivative is also plotted in Fig. 2.8 as a rate of change of the angular velocity ($d\Omega_N/dt$) with time computed over a 50 h moving average. This indicates that the maximum torque was around 30 days post perihelion which is consistent with the lag with respect to perihelion of the maximum emission from the nucleus of 67P of $33 \pm 8$ days found from photometric analyses (Ferrin 2007) and measurements of the hydrogen emission that will be discussed later (Fig. 3.48).

A spin-down (a decrease in the rotational angular velocity) of 41P/Tuttle–Giacobini–Kresák has been observed by Schleicher et al. (2019) (see also Bodewits et al. 2018) indicating that an increasing angular velocity due to torques is not universal. On the other hand, 49P/Arend–Rigaux has been shown to have a very small rotation period change of <14 s per apparition (Eisner et al. 2017) over a measurement period of 28 years.

The spin-up observed for 67P and 9P/Tempel 1 can ultimately result in splitting of the nucleus if the tensile strength of the comet is low. The critical periods for spheroids have been derived analytically by Davidsson (2001). For the most probable case of an oblate spheroid, the critical period is given by

$$P_{crit} = \frac{\pi}{\sqrt{\frac{G\rho_N A_G}{4} + \frac{T_S}{\rho a^2}}} \tag{2.36}$$

where $\rho_N$ is the bulk density of the object, $a$ is its semi-major axis, $T_S$ is its tensile strength, and $A_G$ arises from the geometry of the object through

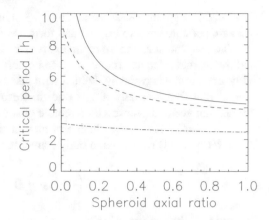

**Fig. 2.9** Critical rotation periods for splitting as a consequence of spin-up following Davidsson (2001). Solid line: Long axis length = 8 km, tensile strength of 20 Pa and a bulk density of 600 kg m$^{-3}$. Dashed line: long axis length reduced to 2 km. Dot-dashed line: long axis length reduced to 2 km and the tensile strength increased to 200 Pa

$$A_G = \left[ \frac{2\pi f_A}{(1 - f_A^2)^{3/2}} \tan^{-1} \sqrt{\frac{1}{f_A^2} - 1} \right] - \frac{2\pi f_A^2}{1 - f_A^2} \qquad (2.37)$$

where $f_A$ is the axial ratio of the spheroid. This equation is plotted for three reasonably realistic cases in Fig. 2.9. For 67P, the present rate of angular accceleration would suggest a lifetime against splitting of only 200 years.

As indicated by Eq. (2.34), the angular acceleration is inversely proportional to the nucleus mass and the square of the radius so that small or unmeasurable rotation period changes observed for comets such as 14P/Wolf, 143P/Kowal-Mrkos, and 162P/Siding Spring (Kokotanekova et al. 2018) may be indicative of larger nucleus sizes and much longer timescales.

In the general case, a cometary nucleus can be in an excited rotational state. We have seen that nuclei can have complex shapes. Dynamically, the most stable rotational state is the one that requires the least amount of rotational kinetic energy for a given total rotational angular momentum. This corresponds to rotation about its axis of maximum moment of inertia. In this state, the moment of inertia is related to the rotational angular momentum, $L$, and the rotational kinetic energy, $E_{rot}$, through

$$I_s = \frac{L^2}{2E_{rot}} \qquad (2.38)$$

where the subscript s indicates that rotation is about the short axis (assuming a uniform density distribution within the object).

However, the external torque resulting from jet activity (Eq. 2.31) is a vector. Thus, it may not merely modify the rotation period but can also push the comet into a rotationally more excited state. It should also not be forgotten that major mass loss events (e.g. splitting of a nucleus), collisions, or tidal effects (e.g. through a close interaction with Jupiter) can produce abrupt changes in the principal moments of inertia leading to more excited, rotational states. These types of rotation are known as non-principal-axis rotational states (NPA rotational states) (Samarasinha and

Mueller 2015; see also Kaasalainen 2001). Rotation about the intermediate and long axes are possible but also excited and form special cases of principal axis rotation.

When the rotation is more complex, two reference frames, the body-fixed frame and the inertial reference frame, need to be defined and kept clearly separated. The body-fixed frame is typically defined with respect to the principal axes of the body. The inertial frame usually aligns one of the axes to the total rotational angular momentum vector as viewed by a remote observer.

In the inertial frame, the change of angular momentum is related to the torque through Eq. (2.34) while, in the body frame, the comparable equation is

$$\frac{d\boldsymbol{L}}{dt} + \boldsymbol{\Omega}_N \times \boldsymbol{L} = \boldsymbol{M} \tag{2.39}$$

where $\boldsymbol{\Omega}_N$ is the angular velocity.

Euler's equations for rigid body rotation can be used to establish changes in angular velocity about each of the three principal axes as a result of torques. By substituting,

$$\boldsymbol{L} = I_N \boldsymbol{\Omega}_N \tag{2.40}$$

the resulting equations are

$$
\begin{aligned}
I_a \dot{\Omega}_a &= (I_b - I_c)\Omega_b \Omega_c + M_a \\
I_b \dot{\Omega}_b &= (I_c - I_a)\Omega_c \Omega_a + M_b \\
I_c \dot{\Omega}_c &= (I_a - I_b)\Omega_a \Omega_b + M_c
\end{aligned}
\tag{2.41}
$$

where $I$ is again the moment of inertia (see e.g. Landau and Lifshitz 1976 or Samarasinha and Belton 1995). The subscripts represent the three (orthogonal) principal axes of the body which we represent as $a$, $b$, and $c$. The three axes form a right-handed coordinate system in the body frame.

Integration of these equations over time requires an assumption that the total mass lost is a negligible fraction of the mass of the nucleus such that the moments of inertia remain constant. This assumption is probably adequate for one perihelion passage. Integration then provides the evolution of the rotational state for non-zero torques. When $\boldsymbol{M} = 0$, however, conservation of energy and momentum are described by (Samarasinha and Mueller 2015),

$$I_a \Omega_a{}^2 + I_b \Omega_b{}^2 + I_c \Omega_c{}^2 = 2E_{rot} \tag{2.42}$$

and

$$I_a{}^2 \Omega_a{}^2 + I_b{}^2 \Omega_b{}^2 + I_c{}^2 \Omega_c{}^2 = L^2. \tag{2.43}$$

**Fig. 2.10** Definition of the short-axis mode (SAM) and long-axis mode (LAM) for the rotation of an irregular object (redrawn after Julian 1987)

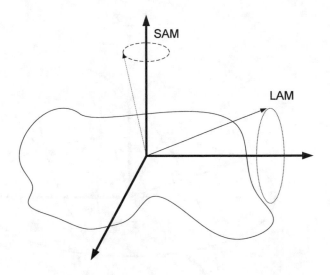

The NPA rotation can be in one of two modes (Fig. 2.10). In the Short Axis Mode (SAM), the short principal axis of the object circulates around the total rotational angular momentum vector. (Samarasinha et al. (2004) described this motion as the short axis "encircling" the angular momentum vector—this is perhaps conceptually easier to envisage.) In the Long Axis Mode (LAM), it is the long principal axis of the object that circulates around the total rotational angular momentum vector.

The rotational energy and angular momentum can be used to distinguish between the short-axis and long axis modes by comparison with Eq. (2.38). Since

$$I_c \leq I_b \leq I_a \tag{2.44}$$

then

$$I_c \leq \frac{L^2}{2E_{rot}} \leq I_a \tag{2.45}$$

and when

$$I_c \leq \frac{L^2}{2E_{rot}} < I_b \tag{2.46}$$

then the motion is a LAM, and if

$$I_b < \frac{L^2}{2E_{rot}} \leq I_a \tag{2.47}$$

then the motion is a SAM. The three Euler angles then describe the rotational motion of the body (Fig. 2.11). They, and their first derivatives, are functions of the

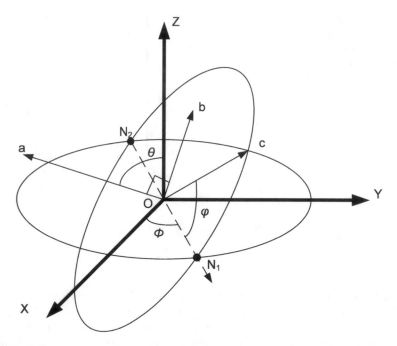

**Fig. 2.11** Euler angle definitions (Following Landau and Lifshitz (1976); Copyright Elsevier 1976. Used with permission)

moments of inertia and the angular velocity as shown by Samarasinha and Mueller (2015). The Euler angle, $\varphi$, is given by

$$\varphi = \tan^{-1}\left(\frac{I_c\Omega_c}{I_b\Omega_b}\right) \tag{2.48}$$

and its first derivative is

$$\dot{\varphi} = -\Omega_a\left[\frac{(I_a - I_c)I_c\Omega_c^2 + (I_a - I_b)I_b\Omega_b^2}{I_c^2\Omega_c^2 + I_b^2\Omega_b^2}\right] \tag{2.49}$$

The other first derivatives are

$$\dot{\phi} = L\left[\frac{I_c\Omega_c^2 + I_b\Omega_b^2}{I_c^2\Omega_c^2 + I_b^2\Omega_b^2}\right] \tag{2.50}$$

and

$$\dot{\theta} = \Omega_c \cos\varphi - \Omega_b \sin\varphi \tag{2.51}$$

The two remaining Euler angles are then defined by

$$\phi = \int \dot{\phi}\, dt \qquad (2.52)$$

and

$$\theta = \cos^{-1}\left(\frac{I_a \Omega_a}{M_N}\right) \qquad (2.53)$$

The detailed investigation of excited rotational states arose because of the observations of comet 1P/Halley. Initially spacecraft observations suggested a rotation period of around 2.2 days (e.g. Wilhelm et al. 1986; see also Belton et al. 1986). Some ground-based observations showed repeatability of coma structures on similar timescales. However, other observations and re-analysis of previous results suggested a period of approximately 7.4 days (e.g. Schulz and Schlosser 1989). The viability of these two possibilities hinged on imaging of the nucleus orientation by the Vega spacecraft. However, the cameras onboard both spacecraft had experienced problems and their observations of the nucleus orientation at the times of their fly-bys (particularly those of Vega 1) are open to interpretation. The more broadly accepted solution is a LAM model with a 7.1 day rotation about the long-axis and a 3.69 day precession of the long axis around the rotational angular momentum vector (Belton et al. 1991).

The case for an excited rotational state of 103P/Hartley 2 appears to be clearer. Observations within the EPOXI mission (the name given to the follow-on mission of NASA's Deep Impact) led to the nucleus being found in a LAM rotational state with changes occurring as a result of the activity-induced torques. At the time of closest approach, the long axis of the nucleus was circulating around the rotational angular momentum vector with a period of 18.40 ± 0.13 h and tilted with respect to the vector by an angle of 81.2 ± 0.6°. Simultaneously the body was rolling around the long axis with a period of 26.72 ± 0.06 h (Belton et al. 2013).

The rotation axis of 67P was found to be directed towards right ascension = 69.54° ± 0.1°, declination = 64.11° ± 0.05° in the J2000 coordinate system. A small NPA component to the rotation appears to be significant (Preusker et al. 2015) and may result from the torques producing the period changes. Jorda et al. (2016) performed a periodogram analysis of the direction of the rotation axis of the comet in celestial coordinates which was obtained as a by-product of the shape reconstruction. This analysis indicated a minimum at 11.5 ± 0.5 day clearly suggesting an excited (SAM) rotational state with an amplitude of 0.15° ± 0.03°. Interpretations were discussed by Gutierrez et al. (2016).

The results from the spacecraft observations show that both changes in the rotation period and the excited rotational states can occur and they are most probably as a result of the activity-induced torques on the nucleus. One curiosity is that despite a fairly sizeable change in the main rotation period, the rotational state of 67P is close to a pure spin.

Internal friction within the nucleus can lead to a reduction in the rotational energy thereby bringing the object to a less excited state. This was first addressed by Burns and Safronov (1973) who gave an estimate for the damping timescale, $\tau_{damp}$, as

$$\tau_{damp} \sim \frac{\mu_{rig} Q_{damp}}{\rho_N K^2 r_N^2 \Omega_N^3} \qquad (2.54)$$

where $\mu_{rig}$ is the rigidity, $Q_{damp}$ is a quality factor of the material, and $K$ is a shape factor. The knowledge of the values of $\mu_{rig}$ and $Q_{damp}$ are poor while $K$ is related to the oblateness, H, by being approximately 0.1 $H^2$. De Pater and Lissauer (2015) gave a more easily evaluated approximation for asteroids using nominal asteroid parameters as

$$\tau_{damp} \sim \frac{0.7 \, 2\pi}{r_N^2 \Omega_N^3} \qquad (2.55)$$

where $r_N$ should be entered in units of [km] and the orbital period, computed through $2\pi/\Omega_N$, is in units of [day]. The timescale is then given in units of [Gy]. Using approximate numbers for 67P, we obtain 550,000 years which is long compared to the comet's lifetime in the inner Solar System and therefore unlikely to be of relevance.

The total rotational angular momentum vector of 67P has an obliquity of 52.3° (e.g. Brugger et al. 2016) which is highly significant for the surface energy balance. The sub-solar latitude as a function of time is shown in Fig. 2.12. This indicates that the southern hemisphere experiences more intense insolation but over a shorter period. Figure 2.13 shows the importance of this in the energy balance at the surface. The figure shows the total energy per square metre irradiating the surface of the

**Fig. 2.12** The sub-solar latitude on 67P through perihelion. Note the rapid transition of the Sun from northern latitudes to southern latitudes beginning around 1 year before perihelion

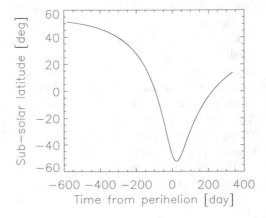

**Fig. 2.13** Total input solar energy onto a horizontal surface of a spherical nucleus with the orbit and obliquity of 67P

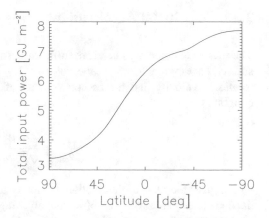

nucleus as a function of latitude. For simplicity, a smooth spherical nucleus has been assumed and only the time from about 1.5 years pre-perihelion to 1 year post-perihelion has been included.

Through perihelion, the south pole is constantly illuminated and receives the largest energy input. The total energy input is around a factor of 2.5 more than the north pole which is constantly illuminated when the comet is further from the Sun. The magnitude of the energy input can be compared to the latent heat of sublimation of water ice ($L_{H2O} \sim 2.84$ MJ kg$^{-1}$).

The loss of material from a cometary surface can be estimated by using the heat input from, for example, Fig. 2.13 for 67P, and assuming that all of that energy goes into sublimation. We can then divide by $L_{H2O}$ and obtain a mass of sublimed $H_2O$ and, using a density, derive an estimated depth of erosion through sublimation. For 67P such a calculation would result in eroded depths of 2 m at the north pole but 5 m at the south pole. This example illustrates the huge importance of the rotational characteristics and obliquity in the local surface energy budgets of comets.

The eroded depths from this calculation are factors of between 4 and 9 greater than the globally averaged eroded depth computed from the mass loss per apparition using Eq. (2.27). This is an illustration of the issue of nuclei needing to be only 10–20% active (with respect to free sublimation) in order to match the measured production rates (e.g. Keller et al. 1987).

As a definition of the integrated heat input to the surface, Fig. 2.13 does have limitations. The surfaces of comets are not smooth and surface roughness effects are of major importance. For an atmosphereless body, the orientations of surface facets can result in significant differences in local temperatures. A vertical cliff is an obvious example where the cliff face might be illuminated orthogonally while its base sees no illumination at all leading to temperature differences of several hundred Kelvin at perihelion over relatively short baselines. This effect will operate over all length scales (see also Sect. 2.9.3.6).

## 2.5   Centripetal Accelerations

The masses of cometary nuclei are sufficiently small that the rotational velocity at the surface can approach the escape velocity. By manipulating the equations for the rotational velocity and the escape velocity at the equator, one can arrive at the equation

$$\frac{v_r}{v_{esc}} = \frac{\Omega_N \, r_N^{3/2}}{\sqrt{2GM_N}} \tag{2.56}$$

which gives the ratio of the rotational velocity to the escape velocity and shows the dependencies on three variables, namely, angular velocity, nucleus radius and nucleus mass, for a spherical object.

In the case of rotational instability, the centripetal acceleration must approach the surface gravitational acceleration. For a spherical body without tensile strength, the force balance shows that the critical rotation period is

$$P_{crit} = 2\pi \sqrt{\frac{r_N^3}{GM_N}} = \sqrt{\frac{3\pi}{G\rho_N}} \tag{2.57}$$

For nucleus parameters appropriate for 67P, the ratio of $v_r/v_{esc}$ would be around 0.3 indicating a significant opposing force to gravity at the equator. Thus, while the nucleus of 67P is not yet close to rotational instability, the centripetal acceleration does lower, locally, the effective gravitational potential and the escape velocity.

## 2.6   Surface Reflectance

At this point, we have addressed, for unresolved nuclei, the geometric albedo, $p$, and have introduced $\varphi$ as the value of the (unresolved) phase function at the phase angle, $\alpha$ (Eq. 2.1). The directional-hemispherical albedo, $A_H$, has also been introduced as an important component in the determination of the heat balance at a surface element (Eq. 2.4). In resolved surface photometry, we must look at the variation in the surface reflectance as a function of the three photometric angles, i.e. the angle of incidence ($i$), the angle of emission ($e$) and the phase angle. These angles are shown in Fig. 2.14.

There is a vast body of literature about the photometry of surfaces including outside planetary research. Much of this is beyond the scope of this text (see for example, Li et al. 2015) but is interesting in itself as it has been driven by requirements for gaming software that seeks to be as realistic as possible. The nomenclature is often different from that used in planetary physics but, even in this field, a large

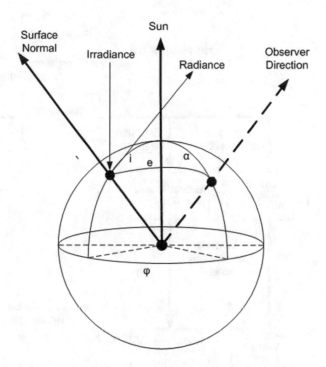

**Fig. 2.14** Definition of photometric angles for surface photometry

number of different names for concepts connected to intensity and flux are used in the literature. The nomenclature we will use is summarized in Fig. 2.15.

The radiance (often known as intensity in this field) is the fundamental quantity in the description of a radiation field and defines (Fig. 2.16) the energy, $dE$, at point O, passing through a surface element of area, $d\sigma$, within a frequency interval $(\vartheta, \vartheta+d\vartheta)$, a time interval, $dt$, and a solid angle interval, $d\Omega_s$ (Hovenier et al. 2004). Here, we call it a radiance in keeping with one of the most common usages in planetary surface photometry. Given that the energy flow may not be normal to the surface element, the angle of the flow direction with respect to the normal, $e$, must also be accounted for leading to the equation

$$dE = I \, \cos e \, d\vartheta \, d\sigma \, d\Omega_s \, dt \tag{2.58}$$

Note that we shall use the term flux for radiant emittance from a surface. The flux is the total energy (or total energy per wavelength or frequency interval) flowing in all directions per unit surface area, per unit time. Hence the flux from a surface is related to the radiance from that surface by an integration over solid angle so that

$$F = \int_{2\pi} I \, \cos e \, d\Omega_s \tag{2.59}$$

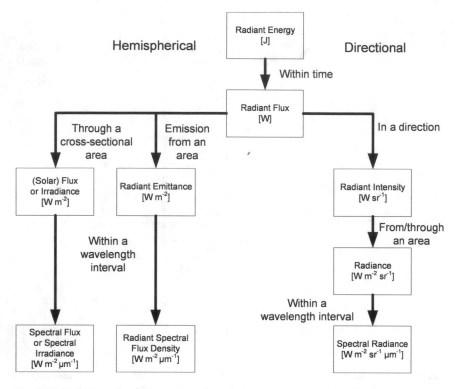

**Fig. 2.15** Definitions for photometric studies of planetary surfaces with typical units

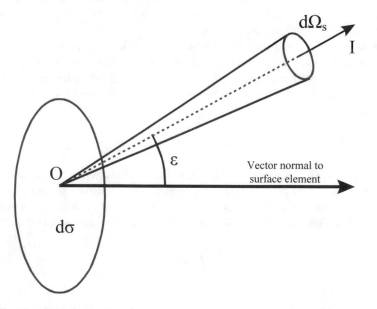

**Fig. 2.16** The definition of radiance, $I$, at a point O in a direction at an angle of emission, $e$, with respect to a surface element of area, $d\sigma$

where the integration occurs over a hemisphere ($2\pi$ steradians) when referring to surface emission.

There are several interpretational issues that are relevant and require explanation. We begin with some definitions.

The reflectance coefficient, $\rho_C$, is defined through the equation

$$\rho_C(\alpha, e, i) = \frac{\pi I(\alpha, e, i)}{\mu_0 \, S_{\odot}/r_h^2} \tag{2.60}$$

where $I$ is the measured radiance from the surface, $\mu_0 = \cos i$, and $S_{\odot}$ is the solar flux at 1 AU from the Sun (Eq. 2.5). When defined over a wavelength interval, $I$ becomes the spectral radiance and $\rho_C$ is a function of $\alpha$, $e$, $i$, and $\lambda$

$$\rho_C(\alpha, e, i, \lambda) = \frac{\pi I(\alpha, e, i, \lambda)}{\mu_0 \, F_{\odot}(\lambda)/r_h^2}. \tag{2.61}$$

$S_{\odot}\mu_0$ gives the integrated irradiance at the surface at 1 AU.

The fact that comets are irregular bodies frequently means that $\mu_0$ is difficult to determine. This is particularly true where no 3D shape model of the object can be determined. Hence, the reflectance factor is often used. This is given by

$$\rho_F(\alpha, e, i, \lambda) = \frac{\pi I(\alpha, e, i, \lambda)}{F_{\odot}(\lambda)/r_h^2} \tag{2.62}$$

This quantity is also referred to in the literature as REFF. In both cases the reflectance is a unitless property of the surface and one can see from this that the $\pi$ appears to have a unit, namely [sr]. In photometry papers, it is a common practise to define the solar flux as being $\pi F$ (where $F$ is obviously the solar flux divided by $\pi$) in order to remove the $\pi$ from the equations. However, this practise is not universal. It is this quantity that is usually referred to as "I over F" in planetary photometry (which would be algebraically correct if the solar flux is defined as $\pi F$). Finally, the bidirectional reflectance, as given in many texts, is I/F according to our definitions (see e.g. Hapke 1981). I have avoided using this definition because, strictly speaking, it is not unitless (having units of [sr$^{-1}$]) and can be confusing.

In the case of a true Lambertian surface, $\rho_C = 1$ independent of the photometric angles. When $I(\alpha, e, i) = $ constant, this is equivalent to a Lambertian surface but here the constraint of reflecting the entire incident light may not be applicable in which case $\rho_C$ will be $<1$. The synthetic fluoropolymer, Spectralon®, has near-Lambertian behaviour in the visible range (up to a wavelength of 1.5 µm) and is frequently used as a laboratory calibration standard for this purpose.

It should be noted that the reflectance factor can be greater than 1 for mirrors and specular reflections from surfaces. This might appear counter-intuitive but the radiance in the equation for $\rho_F$ contains the reciprocal of the solid angle which approaches zero for mirror reflections of a directional parallel beam.

$\rho_C$ is related to the geometric albedo, $p$, in the general case through the equation

$$p = \frac{1}{\pi} \int_0^{2\pi} \int_0^{\pi/2} \rho_c \ \sin i \ \cos^2 i \ di \ d\phi \qquad (2.63)$$

In the case of a Lambertian surface

$$p = \frac{2}{3}\rho_c \qquad (2.64)$$

and we can see here a relationship to $A_H$ for this specific assumption through Eq. (2.4).

Surface photometric models have been studied for many years and several simple analytical and semi-empirical models have been used extensively in the past, some of which remain useful today. The Lommel-Seeliger law is one example the full derivation of which is given in Hapke (1981) and Fairbairn (2005). This model assumes exponential attenuation of the light as it penetrates the surface and each volume element encountered scatters part of the beam isotropically. No multiple scattering is included. The resulting reflectance law is given by

$$\rho_F = \frac{\omega}{4} \frac{\mu_0}{\mu_0 + \mu} \qquad (2.65)$$

where $\omega$ is the single scattering albedo of the individual particles contributing to the surface and $\mu$ is cos $e$.

The weakness of the Lommel-Seeliger law is that it makes no account of multiple scattering within the medium nor does it account for the opposition effect. The opposition effect is a relative brightening of a rough surface when the illumination is moved to being directly behind an observer. It is particularly evident when considering the apparent brightness of the Moon. (Think about the Moon when it is a half-Moon. Is the brightness of two half-Moons put together equal to the brightness of the full Moon? It is not by a factor of several.) The most significant attempt to construct a semi-analytical scheme to remove these deficiencies and at the same time relate the reflectance to light-scattering parameters is that given in series of papers over more than 30 years by Bruce Hapke (see Hapke 1981, 1984, 1986, etc.).

Hapke first addressed the multiple scattering and derived the equation

$$\rho_F(i, e, \alpha) = \frac{\omega}{4} \frac{\mu_0}{\mu_0 + \mu} [4\pi \Phi_S(\alpha) + H(\mu_0)H(\mu) - 1] \qquad (2.66)$$

where $\Phi_S(\alpha)$ is the single particle angular scattering function. (This quantity and the single scattering albedo are discussed in more detail in the section on cometary dust. The factor $4\pi$ results from ensuring that the normalization of $\Phi_S(\alpha)$ is consistent between this section of the text and the section addressing dust particle scattering.) The H functions in this case are

$$H(\mu) = \frac{1 + 2\mu}{1 + 2\gamma_H \mu} \tag{2.67}$$

where

$$\gamma_H = \sqrt{1 - \omega} \tag{2.68}$$

The 2002 model provided improvements in the approximations of the H-functions and introduced the opposition effect. The resulting reflectance equation was

$$\rho_F(i, e, \alpha) = \frac{\omega}{4} \frac{\mu_0}{\mu_0 + \mu} [4\pi\Phi_s(\alpha)B_{SH}(\alpha) + M(\mu_0, \mu)]B_{CB}(\alpha) \tag{2.69}$$

where $M(\mu_0, \mu)$ was the new model for multiple scattering.

The two terms $B_{SH}$ and $B_{CB}$ were introduced to model two different phenomena associated with the opposition effect. The shadow-hiding opposition effect (SHOE) is the contribution to the opposition effect arising from the fact that as one moves towards zero phase angle, shadows produced by rough surfaces disappear. The coherent backscatter opposition effect (CBOE) results from the fact that waves travelling in opposite directions along the same multiply scattered path within a scattering medium interfere constructively with each other as they exit the medium near zero phase and consequently produce a relative peak in brightness (Hapke 2002). The CBOE has, in particular, been the subject of detailed laboratory investigation (Shkuratov et al. 2002; Nelson et al. 2002).

As will be shown below, the single particle angular scattering function is analytically difficult to determine for particles except simple spheres and the resulting function is frequently not trivial. Hapke attempted to circumvent this problem by using Henyey-Greenstein (H-G) functions which are non-isotropic phase functions specifically for simulating the scattering of materials. The single H-G function is given by Haltrin (2002)

$$\Phi_s(\alpha) = \frac{1 - \xi^2}{\left(1 + 2\xi \cos\alpha + \xi^2\right)^{3/2}} \tag{2.70}$$

where $\xi$ is sometimes called the asymmetry parameter and is the mean scattering angle of photons interacting with the individual particles of the medium. It is thus related to the mean phase angle by

$$\xi = \langle \cos\alpha \rangle \tag{2.71}$$

A single H-G function is shown in Fig. 2.17 and note that it has been expressed here as a function of scattering angle $(\pi-\alpha)$. The function should be normalised over solid angle so that

**Fig. 2.17** Examples of
Henyey-Greenstein
functions to describe single
particle scattering. Solid: A
single parameter (two-term)
H-G function ($\xi = 0.6$).
Dashed: A double H-G
function ($\xi = 0.6$, $c = 0.5$).
Note the backscattering
peak in the double H-G
function

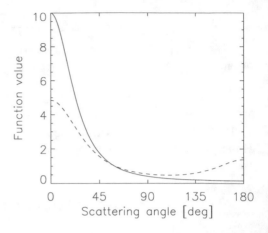

$$\int_0^{4\pi} \Phi_s(\alpha)\, d\Omega_s = 1. \tag{2.72}$$

Many particles show two peaks in the angular scattering function. The forward
scattering peak is well modelled by the single H-G function but the backscattering
peak (at low phase angles = high scattering angles) is ignored in this function. The
double H-G function is intended to account for this and takes the form

$$\Phi_s(\alpha) = \frac{1+c}{2}\ \frac{1-\xi_1{}^2}{\left(1+2\xi_1\cos\alpha+\xi_1{}^2\right)^{3/2}} + \frac{(1-c)}{2}$$

$$\times\ \frac{1-\xi_2{}^2}{\left(1+2\xi_2\cos\alpha+\xi_2{}^2\right)^{3/2}} \tag{2.73}$$

where $c$ is a variable. Hapke suggests reduction of the number of parameters from
three to two is adequate in many cases by setting $\xi = \xi_1 = \xi_2$. An example of how the
double H-G function generates the backscattering peak is also shown in Fig. 2.17.

The correction for macroscopic roughness, introduced into his scheme by Hapke
in 1984, is based on defining a mean slope angle for the surface under study. The
slope distribution function, $a_s(\theta)$, is normalized so that

$$\int_0^{\pi/2} a_s(\theta)\, d\theta = 1 \tag{2.74}$$

and the mean slope angle is then

$$\tan\bar{\theta} = \frac{2}{\pi}\int_0^{\pi/2} a_s(\theta)\,\tan\theta\, d\theta \tag{2.75}$$

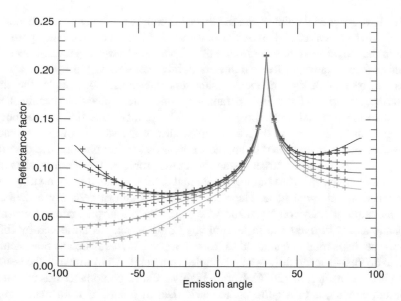

**Fig. 2.18** A Monte Carlo calculation of the reflectance factor of rough surfaces of various degrees (crosses). The calculation was performed for a fixed incidence angle of 30°. The calculation used the measurement reflectance of a flat, powdered piece of the Allende meteorite. This is compared to the reflectance expected using Hapke's equations for rough surfaces (lines). Black: smooth surface. Blue: 10° mean slope angle. Green: 20° mean slope angle, Red: 30° mean slope angle Pink: 40° mean slope angle, Orange: 50° mean slope angle (Credit: A. Gracia Berná and A. Pommerol)

if a Gaussian distribution can be assumed. Multiple scattering between facets of the rough surface was assumed to be negligible. The resulting equations to determine the effect on the reflectance are mathematically straightforward although rather long and they will not be repeated here. In recent years, however, it has become possible to check the validity of this approach by using Monte Carlo ray-tracing approaches.

In Fig. 2.18, we see the result of a calculation constructed to test Hapke's approach to determining the changes in reflectance caused by surface roughness. Here a piece of the Allende meteorite was powdered and a flat surface was constructed. The Hapke parameters were computed. Hapke's macroscopic roughness equations were used then to predict the reflectance factor for increasingly rough surfaces (indicated by lines of different colour). In parallel, a Monte Carlo calculation with ray tracing was used to produce the reflectance factor (indicated by the crosses) under the assumption of single scattering from the surface (applicable for a darkish surface such as that provided by the Allende powder). The results from the two approaches agree very well. For a typical dark surface on a cometary nucleus, this approach should therefore be sufficient. However, icy surfaces on planetary satellites and exposed icy surfaces on comets may require a multiple scattering approach (rays bouncing from one surface facet to another before finally arriving at the observer). At the moment only the Monte Carlo/ray-tracing algorithm would be able to perform this type of calculation.

We have reached a point where the theory accounts for many of the phenomena expected to be observed and, in one case, we have been able to verify the approach at least in the single scattering approximation. There are however several issues. The resulting theory has eight free parameters (which can be reduced if some assumptions can be made). Four of these parameters (the mean slope angle, the single scattering albedo, and the H-G parameters) have clear physical meanings with respect to the material under investigation. The primary question is whether the problem can be inverted to derive the properties of the surface from photometric measurements. It is important to appreciate here that fitting photometric data from multiple observations of a target surface is not the problem. The Finnish astronomer and photometry expert, Kari Lumme, is reputed to have said in this context, "give me 7 parameters and I can fit an elephant". (This was almost certainly a deliberate mis-quotation of John von Neumann who said, "With four parameters I can fit an elephant, and with five I can make him wiggle his trunk." as attributed by Enrico Fermi—students might be amused to learn that some mathematicians have actually tried to do this—successfully—with complex numbers). An example of the issues is given in Gunderson et al. (2006). Here, a highly accurate goniometer was used under optimum conditions to acquire photometric data of a surface from many angles. When fitting the data, correlation coefficients were determined and it was shown that there are correlations between several of the parameters (e.g. between $\omega$, $\xi$, and c) basically confirming the trends first reported by Helfenstein and Veverka (1987). Subsequently, Shepard and Helfenstein (2007) (who followed this further in 2011— Shepard and Helfenstein (2011), Helfenstein and Shepard (2011)) have looked closely at the relationship between the parameters and physical quantities and have assessed the mutual coupling between the parameters to provide better error determination.

At present, the inversion of Hapke parameters derived from remote-sensing data to produce meaningful material properties still seems challenging. One can also ask why. The fundamental assumption is that scatterers are sufficiently dispersed such that each particle interacts with the incoming beam in isolation. This might be valid if the particles were even only a few wavelengths apart. However, the particles are in contact and this affects the interaction of the wave.

One of the results illustrated later in Fig. 4.7 shows the single scattering albedo computed from Mie theory. Two points are relevant here. Firstly, the single scattering albedo is particle size dependent. The sizes of particles on cometary surfaces are highly unlikely to be a single size and the size distribution is poorly constrained. The size dependency is not treated in the Hapke formulation. (The asymmetry parameter is also size dependent.) Secondly, if we assume that the particles are typically larger than a few microns, then the single scattering albedo in Fig. 4.7 is an order of magnitude larger than the results for cometary surfaces derived so far and shown in Table 2.1. The large particle value of the single scattering albedo is influenced by the material properties (through the complex refractive index) but it is generally larger than 0.5. The reason is that diffraction plays a very strong role in determining the single scattering albedo and this is not influenced by the material. Hence, what is being derived is not the single particle single scattering albedo—despite the fact that

**Fig. 2.19** Reflectance curve of a surface element of 67P/C-G computed from the Hapke parameters derived by Fornasier et al. (2016). The emission angle is set to zero and i = $\alpha$

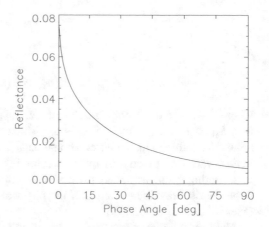

this is actually the starting point for Hapke's theory. What this value is physically, is unclear and its description as a single scattering albedo is highly misleading and should be avoided.

On the other hand, Hapke parameters are a very useful way of describing a data set as they can be used to fit the bidirectional reflectance distribution function (BRDF) to high accuracy (e.g. Pommerol et al. 2019). This is being used now in databases such as DACE (Data and Analysis Center for Exoplanets)[1] to provide the BRDF of materials measured in the laboratory to a wider community. The geometric albedo and hemispherical albedos can also be determined to high accuracy (e.g. Table 2.1).

Fornasier et al. (2016) determined the surface phase function of the nucleus of 67P by fitting OSIRIS observations of the nucleus. The long duration of the rendezvous allowed observations of individual areas in many observing geometries leading to well-constrained results. A reflectance factor curve for a surface element is shown in Fig. 2.19. Here, the emission angle has been set to zero and the reflectance plotted against the phase angle. Note that there is a factor of 10 decrease in the observed reflectance from opposition geometry ($\alpha = e = i = 0°$) to the terminator geometry ($\alpha = i = 90°$, $e = 0°$).

There are several other photometric functions that have been used in the literature over the years. These are usually much simpler in form and have their uses where data are sparse.

The Minnaert function is now probably only significant for historical reasons. It takes the form

$$\rho_F(\alpha, e, i) = \pi A_M \, \mu_0^{k_M} \, \mu^{k_M - 1} \tag{2.76}$$

where

---

[1]https://dace.unige.ch/dashboard/

$$\mu = \cos \varphi_M \cos \lambda_M \tag{2.77}$$

and

$$\mu_0 = \cos \varphi_M \cos (\alpha - \lambda_M) \tag{2.78}$$

where $\lambda_M$ is the photometric longitude, $\varphi_M$ is the photometric latitude and $0 < k_M < 1$ is a parameter. $A_M$ is the Minnaert albedo and typically one would adjust this value to match observational data. The Minnaert function has little basis in theory, is not derived from first principles, is restricted in validity to certain classes of scattering geometries, and, perhaps most importantly, cannot be used to interpret photometric behaviour in terms of the physical properties of the reflecting surface (Meador and Weaver 1975).

The Lunar-Lambert law used by McEwen (1991) takes the form

$$\rho_F(\alpha, e, i) = A_N \left[ 2k_E \frac{\mu_0}{\mu_0 + \mu} + (1 - k_E)\mu_0 \right] \tag{2.79}$$

where $A_N$ is the normal albedo and $k_E$ is a parameter. It has some uses as a simple empirical fit to photometric data especially where there are limited numbers of observations.

As pointed out by Shepard (2017), the definition of the normal albedo has some ambiguity. In some texts the normal albedo is defined as being the ratio of the radiance from a surface seen at zero phase angle and an emission angle of zero (I (0,0,0)) to that of a perfectly diffusing Lambertian surface in the same geometry ($I_L(0,0,0)$), i.e.

$$A_N = \frac{I(0,0,0)}{I_L(0,0,0)} \tag{2.80}$$

whereas other texts define the quantity at zero phase but with variable emission angle, i.e.,

$$A_N = \frac{I(e,e,0)}{I_L(0,0,0)} \tag{2.81}$$

There are subtle differences and it should be noted that techniques such as laser altimetry lead to use of the latter (more imprecise) definition because surface slopes may not be known a priori. However, in the case of McEwen's formula, the difference in definition would usually be of little significance.

Fitting of the bidirectional reflectance over the hemisphere allows the derivation of the directional-hemispherical albedo, $A_H$, introduced in Eq. (2.4), by using its defining equation

$$A_H = \frac{1}{\mu_0 \frac{S_\odot}{r_h^2}} \int_{2\pi} I(\alpha, e, i) \cos e \, d\Omega_s \qquad (2.82)$$

where $I(\alpha,e,i)$ is again the radiance from the surface element typically in units [W m$^{-2}$ sr$^{-1}$]. The integral is over all solid angles, $\Omega_s$. Note that the integral on the right-side is the flux from the surface as given by Eq. (2.59). A more general form would also include the wavelength dependence of reflectance

$$A_H = \frac{r_h^2}{\mu_0} \int_0^\infty \frac{1}{F_\odot} \int_{2\pi} I(\alpha, e, i, \lambda) \cos e \, d\Omega_s \, d\lambda \qquad (2.83)$$

with $I$ expressed in [W m$^{-2}$ sr nm$^{-1}$] and the solar flux is now $F_\odot$ in units [W m$^{-2}$ nm$^{-1}$] as shown in Eq. (2.5). If the surface is Lambertian, then $I(\alpha,e,i)$ is independent of the viewing geometry (i.e. it is constant) and Eq. (2.82) reduces to

$$A_H = \frac{\pi \, I(\alpha, e, i)}{\mu_0 \left(\frac{S_\odot}{r_h^2}\right)} \qquad (2.84)$$

which is itself equal to the radiance coefficient, $\rho_C$, as noted above. This illustrates that $A_H$ expresses the ratio of scattered to incident light at a surface element with the scattered light being integrated over all directions.

We note in passing that this quantity is occasionally referred to casually as the "Bond albedo". This is actually erroneous. The Bond albedo is the fraction of power incident on an entire body, integrated over all wavelengths, that is scattered back out into space and is used to determine, for example, a predicted value for the equilibrium temperatures of the giant planets. The concept here is similar because we are seeking the total energy reflected over all wavelengths. However, here, it is being applied locally on a surface element at an incidence angle specific to that element. Furthermore, $A_H$ can vary with the angle of incidence because of the asymmetry of the phase curve. Hence, the use of the term Bond albedo should be avoided.

Figure 2.20 shows the reflectance factor for observations of the nucleus of 1P/Halley at the time of the Giotto encounter. Unlike the observations of 67P, the Giotto imaging system (HMC) only observed the nucleus from one geometry (a phase angle of 107°) and hence it is difficult to compare the results. However, the geometric albedo of 1P/Halley was estimated to be about 0.04 and the reflectance near the terminator in Fig. 2.20 is around a factor of 10–12 less than this and so it agrees roughly with the Rosetta observations. There seems little reason to doubt that the photometric behaviour of the two surfaces are broadly similar.

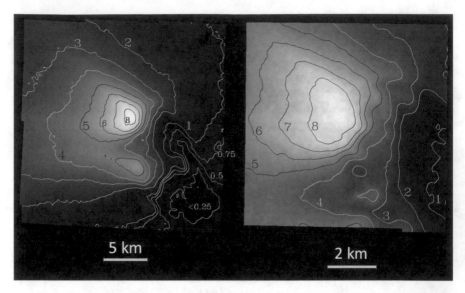

**Fig. 2.20** The reflectance factor ("I/F") of 1P/Halley as derived from Giotto/HMC data in the HMC clear filter ($\lambda = 652.9$ nm, $\Delta\lambda = 372.6$ nm; Thomas and Keller 1990a). The values of reflectance have been multiplied by 1000 to produce the contour labels. The observations were acquired at a phase angle of 107°. Left: HMC image 3457. Right: HMC image 3491

## 2.7   Visible Colour

The reflectance factor of most (solid) planetary surfaces is wavelength dependent and absorptions characteristic of minerals can be found in spectra. Indeed, reflectance spectroscopy in the wavelength range 0.2–5 µm has proven to be the most powerful technique for determining surface mineralogical composition by remote sensing although thermal emission spectroscopy in the range 5–50 µm is also now an established technique. For comets, however, spectra in what is loosely referred to as the visible wavelength range (from 0.2 to 1 µm) appear to be fairly featureless. We shall defer discussing composition and compositional changes using the full wavelength range to a later sub-section. But for the optical reflectance here it is necessary to appreciate that the reflectances of cometary nuclei at optical wavelengths also show wavelength dependence and that this dependence seems to vary between comets.

In Fig. 2.21, we can see an image of the surface of 67P acquired by the Rosetta/OSIRIS instrument. Three areas have been isolated corresponding to morphologically different types of material. The image sequence contained images over the full wavelength range of OSIRIS. The reflectances for the three areas are shown in Fig. 2.22 and a linear fit has been made to the data for the dusty material.

**Fig. 2.21** Three areas marked by boxes (corresponding to different types of material, rocky, dusty, and broken boulders) on the nucleus of 67P have been investigated using an OSIRIS imaging sequence acquired at 2014-12-16T05.54.04 (Image number: N20141216T055404431ID30F22)

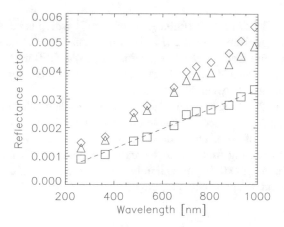

**Fig. 2.22** Reflectance factors for the three areas in Fig. 2.21 (square: dusty; triangle: rocky; diamond: broken material). A linear fit to the dusty material is also shown

It can be seen that the trends are basically linear with a gradient indicating that the nucleus is red with respect to a solar spectrum. Interpretation of the departures from linearity seen would require detailed analysis of the relative errors in the calibration.

Many unresolved observations of asteroids show similar fairly featureless trends. The asteroid, (21) Lutetia, visited by Rosetta on its way to 67P is an example (see e.g. Busarev 2016). There are, however, exceptions (e.g. (4) Vesta). The linear trend has led to use of a spectral gradient to define the spectra at optical wavelengths. The most common form is

$$S'\left(\lambda_1, \lambda_2, \lambda_{ref}\right) = \frac{1}{\rho_{ref}} \frac{\rho_{\lambda_2} - \rho_{\lambda_1}}{\lambda_2 - \lambda_1} \tag{2.85}$$

where $\lambda_{1,2}$ are the wavelengths of the observations in units of [100 nm] and $\rho_{\lambda1,2}$ are the reflectance factors. $\rho_{ref}$ is the reflectance factor at a reference wavelength which is used to normalize the spectrum so that the resulting gradient can be expressed as a percentage (i.e. in units of [% (100 nm)$^{-1}$].

There is no universal consensus on the choice of the wavelength at which $\rho_{ref}$ should be taken and hence care should be taken in comparing values. For example, ground-based observers have tended to use 0.55 μm (e.g. Licandro et al. 2018) whereas Fornasier et al. (2016) used the central wavelength of the green filter at 0.535 μm for spectrophotometric analyses of Rosetta/OSIRIS data. In these cases, this is probably of little significance but older papers (e.g. Thomas and Keller 1989) used very different reference wavelengths and hence comparisons should take this into account. The conversion from one reference wavelength ($\lambda_{refA}$) to another ($\lambda_{refB}$) is simply

$$S'\left(\lambda_1, \lambda_2, \lambda_{refA}\right) = S'\left(\lambda_1, \lambda_2, \lambda_{refB}\right) \frac{\lambda_{refA}}{\lambda_{refB}} \tag{2.86}$$

The fit in Fig. 2.22 gives 20.1($\pm$0.8)% (100 nm)$^{-1}$ which is consistent with the pre-perihelion maps from August 2014 given in Fornasier et al. (2016). For comparison, Thomas and Keller (1989) determined 7($\pm$3)% (100 nm)$^{-1}$ (converted to the same reference wavelength as Fornasier et al.) for 1P/Halley indicating that substantial differences between comets exist.

Licandro et al. (2018) have acquired spectra of asteroids in what are considered to be comet-like orbits and hence one can argue that the spectra provide good signal to noise observations of bare, albeit dead or dormant, nuclei. All of their spectra showed linear increasing trends with increasing wavelength in the range 3–15% (100 nm)$^{-1}$ referenced to 0.55 μm with only weak evidence for absorptions. Hence, while cometary nuclei are usually red, they do have demonstrably different degrees of reddening.

## 2.8   Interior Structure

### 2.8.1   Large-Scale Structure

It might seem curious to address interior structure at this point because relatively little is known of cometary interiors. However, conceptual ideas about the interior have had a tendency to influence our interpretation of other data and hence it is necessary to introduce some of these concepts (Fig. 2.23) and discuss some of the rather limited constraints.

The first issue is whether cometary nuclei are single solid structures or whether they are composed of sub-nuclei. It was Whipple in 1950 that imagined a solid nucleus as the source of cometary activity—a concept that was totally vindicated by the observations of Halley's nucleus. On the other hand, this paper was also responsible for the "dirty snowball" concept implying a single nucleus as envisaged in Fig. 2.23 with an albedo strongly influenced by that of water ice. Whipple himself was well aware of the fact that, in the early Solar System, comets had to grow through some form of accretion mechanism and looked closely at Giotto data to try to establish the typical size of cometesimals that came together to form Halley's nucleus. The evidence was also pointing towards lower albedo values for the

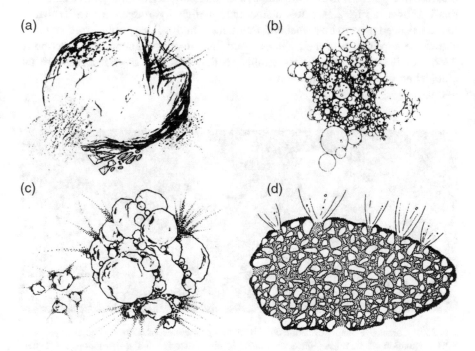

**Fig. 2.23** This (now famous) diagram shows artist's conceptions of four conceptual models of cometary nuclei. (**a**) Whipple's icy conglomerate model, (**b**) the fractal aggregate model of Donn et al. (1985), (**c**) the rubble pile model of Weissman (1986) and (**d**) the icy-glue model of Gombosi and Houpis (1986). (Reprinted/adapted by permission from Springer out of the work by Donn 1991)

cometary nucleus before the Giotto encounter that subsequently led Keller to suggest that nuclei were more like icy dirtballs. Nonetheless, the question remained whether the tensile strength between these cometesimals after accretion (including any modification as a result of the accretion process and subsequent evolution) was sufficiently high that the nuclei could be considered structurally as one object.

The "icy-glue" conceptual model of Gombosi and Houpis (1986), for example, had been imagined as a single solid structure in which separate chunks of non-volatile material are cemented together by icy material forming an interior that is homogeneous on a large scale but inhomogeneous on smaller scales.

Two observations place concepts of single solid nuclei in doubt. Firstly, the lobate shapes seen at 103P/Hartley 2 and, in the even more extreme case of 67P, strongly suggest that large sub-nuclei have come together at some stage to produce the presently observed shapes. The recent observations of the KBO, (486958) Arrokoth (formerly 2014 $MU_{69}$) by NASA's New Horizons spacecraft is a further demonstration of the need to consider bi-lobate structures as typical for primitive bodies (Stern et al. 2019).

These observations have prompted considerable interest in means of producing bilobate structures. Recent modelling work has indicated that the formation of bi-lobed structures is a natural outcome of low energy, sub-catastrophic collisions that have the potential to alter the shape of a small body significantly. An example of this is shown in Fig. 2.24 where a smoothed particle hydrodynamics (SPH) calculation has been used to track material following a high velocity collision. The figure shows the appearance of the fragments and their re-accumulation over a period of 27 h. A bilobate structure very similar to that observed at 67P is seen at the completion of the re-accumulation.

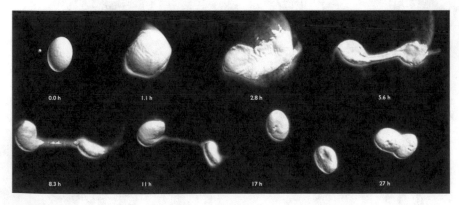

**Fig. 2.24** Comet 67P shape formation by sub-catastrophic collisions. Shown is an example of an SPH calculation of an impact on a rotating ellipsoid. After the initial disruption, subsequent re-accumulation leads to the formation of two lobes. This processes may include the possible formation of layers. The two lobes are gravitationally bound and collide with each other within ~1 day forming a bi-lobed structure (Credit: Jutzi and Benz 2017, reproduced with permission © ESO)

This work also suggests that the "head and body" appearance of 67P is unstable over the lifetime of the solar system (Jutzi and Benz 2017) and is the result of multiple sub-catastrophic impacts experienced by the body after the formation of a precursor object.

An alternative idea with a collision between two parent bodies of the sizes seen in the head and the body of 67P is less attractive because the collision velocity must be fairly low for the bodies to remain gravitationally interacting and not catastrophically disrupted after the initial impact.

The second observation of importance was the remarkable break-up of comet D/Shoemaker-Levy 9 in 1992 (Fig. 1.8) prior to the fragments impacting Jupiter itself in 1994. Asphaug and Benz (1996) gave the equation

$$\sigma_T \approx \frac{GM_p}{R^3} \rho_N r_N^2 \tag{2.87}$$

for the tidally induced stress on the nucleus when making a close encounter with a planet (where $M_p$ is the mass of the planet and $R$ is the distance of the object from the centre of the planet) and estimated that breaking-up in the Jovian gravitational field required D/Shoemaker Levy 9 to have a tensile strength of <6.5 Pa and were able to model the relative motions of the fragments with a strengthless single precursor. This has been taken as strong support for the "rubble pile" theory of Weissman (1986) as shown in Fig. 2.23c. Here, the sub-nuclei are gravitationally bound or weakly bonded requiring very little force to disrupt or re-structure the nucleus. The relatively frequent observation of the splitting of cometary nuclei such as 73P/Schwassmann-Wachmann 3 (Boehnhardt et al. 1996; Dello Russo et al. 2007) has been taken as further evidence of low tensile strength although the exact mechanism provoking splitting in interplanetary space is far from understood.

In the rubble pile concept, the size distribution of the sub-nuclei is not specified. Furthermore, the sub-nuclei do not necessarily have the same initial composition or bulk density. Consequently, the concept is rather flexible. The fractal aggregate concept as shown in Fig. 2.23 appears artificial. This is unfortunate because the concept suggests a quantitative method of describing the components of a nucleus through the fractal dimension which might provide a useful supporting concept for the rubble pile model. However, it is not apparent how one can specify the fractal dimension, $D_f$. For 54 fragments of 73P/Schwassmann-Wachmann 3, a cumulative size distribution has been computed showing $N(>\gamma_{sn}) \sim \gamma_{sn}^{-1.1}$ (Fuse et al. 2007; Fernandez 2009).

The evidence from the Rosetta observations of 67P is tenuous. From the propagation time and form of the signals acquired by Rosetta's bistatic radar experiment, CONSERT, the upper part of the "head" of 67P is homogeneous on spatial scales greater than a few metres (Hérique et al. 2019). The Philae lander did not operate for long enough to provide data on large scales. On the other hand, the measurements of the gravity field provided no evidence for internal voids or large scale heterogeneity (Pätzold et al. 2016).

The individual components of split comets can provide additional information on homogeneity. In the case of 73P/Schwassmann-Wachmann 3, for example, Schleicher and Bair (2011) using visible emissions concluded that, to within the uncertainties, the largest components had the same composition and that this composition was consistent with that measured in the pre-fragmented nucleus. A study by Harker et al. (2011) in the mid-infrared noted the similarity in mineralogy and grain properties between the two major fragments again implying homogeneity in composition and structure between sub-nuclei. Some differences in grain size in the comae between fragments have been noted but varying grain sizes in a cometary coma can be the result of many processes and is certainly not a clear indicator of inhomogeneity.

## 2.8.2   The Surface Layer and the Strength of Cometary Material

The disruption of D/Shoemaker-Levy 9 (Fig. 1.8) provided a clear indication that comets are weakly bound objects. Asphaug and Benz (1994, 1996), for example, showed that the break-up of D/Shoemaker-Levy 9 could be modelled as the separation of a strengthless rubble pile of smaller cometesimals. The resulting behaviour was dependent upon the density of the material but the preferred value of about 0.6 g cm$^{-3}$ was remarkably close to that found for 67P. The impact of metre-sized cometary material into the Earth's atmosphere producing "fireballs" also led Ceplecha (1994) to conclude that material of cometary origin is weak. It is nonetheless important to distinguish between large-scale tensile strength, small scale tensile strength, and compressive strength. This is illustrated by the fact that consolidated material generally shows decreasing strength at larger scales, following a power law proportional to $d^{-q}$, where $d$ is a length scale and the exponent, $q$, $\sim 0.6$ for water ice (Petrovic 2003; Attree et al. 2018). Here, Rosetta has played a major role in increasing our knowledge.

Attree et al. (2018) have studied overhangs to estimate the tensile strength of the material comprising 67P. The rough irregular surface provides several examples of overhangs. One is shown in Fig. 2.25. The tensile strength needed to support these overhangs is remarkably small because of the low gravitational acceleration and values for the stress of around 20 Pa were derived (where the stress is computed from the maximum force per unit area). On the other hand, the compressive strength may be significantly higher.

There is no straightforward relationship between tensile and compressive strength. An everyday example is reinforced concrete. Concrete has a high compressive strength making it good to stand on. But its tensile strength is low. Consequently, reinforcing the concrete with steel bars that have low compressive strength but high tensile strength leads to very strong structures.

**Fig. 2.25** An example of an overhang in the Geb region of the nucleus. (Image number: N20160709T032854309ID10F22). The scarp is actually in shadow. However, the reflected sunlight from other parts of the nucleus provides sufficient illumination for structures to become visible if a non-linear stretch is applied to the data. (Modified following Attree et al. (2018), A&A, reproduced with permission, © ESO)

The Philae lander provided information on the compressive strength of the surface layer but this is open to some interpretation. Philae initially impacted and bounced rather than anchoring itself to the surface as originally planned before reaching its final resting place. The first rebound off the comet's surface, at a site named Agilkia, was at a speed of ~0.38 m s$^{-1}$. After this impact, the lander bounced but it left behind three depressions in the surface which are inferred to be from the feet of the lander. These depressions were imaged by the orbiter imaging system and are estimated to be around 20 cm deep using the SPC technique. This was almost identical to a prediction by Biele et al. (2009). The compressional strength of the surface material was estimated to be between 1.5 and 2.0 kPa (Roll et al. 2016) and 10 kPa, while its Young's modulus is of the order of a few MPa (Möhlmann et al. 2018). A precursor signal in the accelerometer data suggested a very soft surface layer of ~20 cm thick. This uppermost layer is probably desiccated because obser- vations of infrared radiation consistently show surface temperatures on comets when inside 1.5 AU that are well above the free sublimation temperature of water ice (Emerich et al. 1987; Groussin et al. 2013).

At the final landing site, called Abydos, the MUPUS experiment on Philae was able to deploy an arm with a penetration device. The device encountered very hard material just below the surface. The resulting stress was measured by the device and resulted in compressive strengths >2 MPa (Spohn et al. 2015; Boehnhardt et al. 2017) and therefore 2–3 orders of magnitude higher than that found from the rebound. However, it needs to be noted that the orientation of the lander at this time was far from nominal (Spohn et al. 2015).

These results might be reconciled if the sub-surface ice content and its spatial variability are important and may be evidence of recondensation of volatiles emitted from the surface and forced back into the nucleus surface layer by the gas pressure over the surface. This effect was identified in the "KOSI" ("KOmeten-SImulation") experiments in the late 1980s (e.g. Thiel et al. 1989) and was modelled by, for example, Prialnik (1991). In the experiments, the crust thickness reached 7 cm with a mechanical strength exceeding 5 MPa while some degree of porosity was maintained. Hence, this seems to be a way to produce a more dense and therefore potentially harder ice layer immediately under a desiccated surface layer of relative low thermal conductivity. Evidence from the SESAME-PP and CONSERT instruments also indicates that the porosity within the first metre of the surface layer is lower than at depth (Brouet et al. 2016a). Hérique et al. (2019), analysing CONSERT data, concluded that porosity variations could not exceed 10% and that larger porosity variations needed to be restricted to at most metre-scales within the interior. This is probably still consistent with the analysis of Brouet et al. (2016a).

The Deep Impact experiment on 9P/Tempel 1 also showed through infrared spectroscopy that the uppermost layer of the surface was lacking in water ice but, after the impact, water emissions (including evidence for long-lived ice particles) were detected in the ejecta. Analysis of the crater produced by the impactor suggested it was 50 m in diameter surrounded by a low rim about 180 m in diameter. The impacting mass was 372 kg of which nearly 50% was copper in a spherical cap at the front of the impactor. At an impact velocity of 10.3 km s$^{-1}$, the energy delivered was just under 20 GJ or 5 tons of TNT (A'Hearn 2008). Schultz et al. (2007) compared laboratory simulations to observations of the Deep Impact event and concluded that some of the features of ejecta pattern landing uprange of the crater produced by the oblique impact were consistent with experiments using layered targets. The observations, including subsequent observations of the final impact crater from the Stardust-NExT mission to 9P/Tempel 1, were interpreted as possibly arising from impact into a layered target with a loose particulate surface about 1–2 m deep over a slightly more competent substrate (Schultz et al. 2013).

These diverse observations suggest that locally the sub-nuclei have significant compressive strength just below the surface and that this material contains water ice (Fig. 2.26). This is covered with a variably thick layer of a dusty component. Moving further in, this hard layer is over a more porous and homogeneous interior. Our knowledge of the interior beyond these aspects is limited to modelling work based on numerous assumptions and specifically how energy is input to or lost from the interior via the surface.

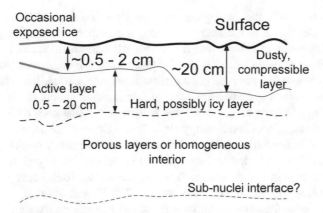

**Fig. 2.26** Schematic diagram attempting to reconcile the information we currently have of the interiors of cometary nuclei

## 2.9   Surface Processes

### 2.9.1   Introduction

Our understanding of cometary surfaces has been improved enormously by spacecraft missions. However, the details of how cometary surfaces are shaped and evolve are still quite poorly understood. The level of complexity can be appreciated by inspection of Fig. 2.27 which summarizes the most important processes actively influencing surface evolution when a comet enters the inner solar system.

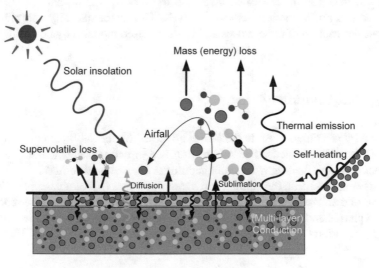

**Fig. 2.27** Schematic diagram of processes influencing cometary surfaces

Solar insolation is the driving energy source for activity and heats the surface. Some of this energy is re-radiated back into space as thermal emission but part of it initiates sublimation of volatile material. Water is the dominant volatile by mass. But more volatile species such as $CO_2$ are also present and the interaction between these species to produce the observed gas emission is not understood. It has been inferred that, in a well-mixed system, some form of fractionation should occur as modelled by, for example, Herny et al. (2020), but it is not clear that such a type of process actually does take place. Some energy is conducted through the uppermost layer into the interior. The thermal conductivity is low but almost certainly non-negligible. The relative absence of water ice on the surfaces of nuclei (at least in pure form) suggests that sublimation is from below the surface with the surface itself being hot (above the free sublimation temperature of water ice of ~200 K) and desiccated. Even where water ice has been found on the surface, measured temperatures were well above the free sublimation temperature (Groussin et al. 2013). Conduction, probably through layers of differing conductivity, may therefore be a necessary part of the picture. Diffusion of subliming gas from below, through the surface layer, is now assumed to be important and this gas flow through a porous layer produces numerous physical phenomena that complicate interpretation of the surface-coma interaction. Describing the surface of a nucleus as flat and homogeneous is no longer adequate. The surface topography influences the heat balance not only through shadowing but also through thermal re-radiation (sometimes referred to as self-heating), an effect that is particularly important where steep slopes are prevalent.

In addition, dust is emitted from the surface. This is also an energy loss but the importance of dust transport across the nucleus (sometimes referred to as "airfall" or "dust hail"), resulting from particles not being sufficiently accelerated to escape the nucleus into the coma, has now been established as a significant process affecting cometary evolution (Sect. 2.10.8). The contradiction between sub-surface sublimation through a porous dessicated layer and the loss of that layer to produce dust emission has not been satisfactorily resolved. The mathematically and physically accurate description of these processes in a generalised model is by no means trivial.

## 2.9.2   Sublimation of Ices

The activity of comets close to the Sun is governed by the sublimation of ices from the surface of the nucleus. The sublimation is a direct consequence of heating by sunlight although there may be small additional internal heat sources.

In its simplest form, the equilibrium vapour pressure ($p_s$) of a gas in equilibrium with its solid and/or liquid phase can be derived by equating the Gibbs free energy of the two phases and substituting the change in entropy ($\Delta S$) between the two states with the equation

$$\Delta S = \frac{L_S}{T} \tag{2.88}$$

where $L_S$ is the enthalpy associated with the sublimation from the solid phase into vacuum. This results in the Clausius-Clapeyron equation,

$$\frac{dp_s}{dT} = \frac{L_S}{T(V_2 - V_1)} \tag{2.89}$$

which defines the slope of the vapour pressure curve. Using the ideal gas law and assuming the expansion from solid to gas phase is large gives

$$\frac{dp_s}{dT} = \frac{p_s L_S}{N_A k T^2} \tag{2.90}$$

where $N_A$ is the Avogadro number and, following integration,

$$p_s = p_0 e^{-L_S/N_A k T} \tag{2.91}$$

which shows that $p_s$ is an exponential function temperature. It is perhaps more intuitive to write this equation as a ratio of the saturation vapour pressure at two different temperatures, $T_1$ and $T_2$ so that

$$\ln \frac{p_{s2}}{p_{s1}} = -\frac{L_S}{N_A k} \frac{1}{(T_2 - T_1)} \tag{2.92}$$

This equation is, however, an idealised one and empirical fits to experimental data are used more frequently. The most common form is

$$\ln p_s = A - \frac{B}{T} \tag{2.93}$$

where $A$ and $B$ are constants. For $H_2O$, $A = 28.9$ and $B = 6141$ K are commonly seen values. However, fits with more terms are available and are listed for the most common cometary species using

$$log_{10} p_s = A - \frac{B}{T} + C \ln T + DT \tag{2.94}$$

as a standard form (Huebner et al. 2006) in Table 2.4. As can be seen, caution needs to be exercised because not all authors use natural logarithms.

The change in enthalpy for sublimation into vacuum is also temperature dependent (as is shown by Huebner et al. (2006) from which the most precise values can be calculated). However, we give in Table 2.4 indicative values that can be used for an

**Table 2.4** Constants for fits to the equilibrium vapour pressure of the major molecules in comets (from Huebner et al. 2006)

| Molecule | $L_S$ [J kg$^{-1}$] | A | B | C | D |
|---|---|---|---|---|---|
| $H_2O$ | $2.84 \times 10^6$ | 4.07023 | $-2848.986$ | 3.56654 | $-0.00320981$ |
| CO | $2.16 \times 10^5$ | 53.2167 | $-795.104$ | $-22.3452$ | 0.0529476 |
| $CO_2$ | $4.5 \times 10^5$ | 49.2101 | $-2008.01$ | $-16.4542$ | 0.0194151 |

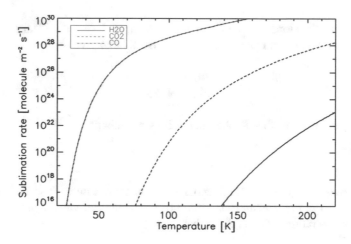

**Fig. 2.28** The temperature dependence of the sublimation rates of $H_2O$, $CO_2$, and CO derived using the Hertz-Knudsen equation and the equilibrium vapour pressures given in Table 2.4

initial assessment. This also illustrates that the heat required for the phase change is much larger for water than for $CO_2$ or CO.

In equilibrium, the number of gas molecules striking the solid surface equals the number being emitted from the surface and hence the flux of molecules crossing a unit cross-section of the surface can be used to derive the emitted flux. The 3D Maxwell-Boltzmann distribution can be used for this purpose resulting in the Hertz-Knudsen equation which provides the maximum gas emission rate, $Z(T)$, from an ice surface into vacuum,

$$Z(T) = \frac{p_s}{\sqrt{2\pi mkT}} \qquad (2.95)$$

in [molecule m$^{-2}$ s$^{-1}$] (e.g. Steiner et al. 1991). Clearly, substitution for $p_s$ and $m$ can result in this equation having slightly different forms depending upon preference. Figure 2.28 shows $Z(T)$ for the three main volatile species in comets. It indicates that emission rates differ by several orders of magnitude between the species and that emission rates are strongly temperature dependent. This can also be seen in Fig. 2.29 which shows the change in sublimation rate per degree change in temperature for

**Fig. 2.29** The rate of
change with temperature of
the sublimation rate of water
from a pure ice surface into
vacuum

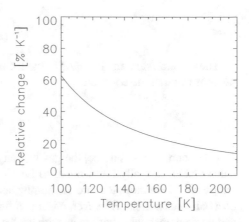

water ice. This illustrates that subtle changes in surface temperature can produce
very large changes in the gas emission rate.

This rapid change in sublimation rate with temperature has led several authors to
write about the "free sublimation temperature" of ices with values of roughly 200 K
for $H_2O$, 125 K for $CO_2$ and 26 K for CO with the implication that surfaces need to
be roughly at these temperatures for the respective ice to sublime. Strictly speaking,
this is inaccurate because sublimation occurs at all temperatures but these provide a
"rule-of-thumb" for the temperatures at which emitting ices sublime strongly under
vacuum conditions found in the much of the Solar System.

Laboratory work on water ice has shown that it may be necessary to introduce an
additional factor, a sublimation coefficient, $k_s$, in these equations

$$Z(T) = k_s \frac{p_s}{\sqrt{2\pi mkT}}.$$  (2.96)

Observations suggest that this coefficient varies between 0.2 and 0.8.

In Eq. (2.96), $Z$ is an effusion rate from the surface. However, if there is
significant pressure above the surface then the effusion rate will be partially balanced
by returning molecules. Although comets as a whole are in vacuum, there are at least
two cases where this effect might need to be accounted for. Firstly, if the sublimation
is intense, molecules will collide with each other after release from the surface
providing a return flux (sometimes called back pressure in the literature) at the
surface. The net effusion rate from the surface is therefore reduced as is the net
energy lost. Secondly, although we have discussed surfaces up to this point, the solid
material may be (and in fact almost certainly is) porous. Sublimation from points
below the actual surface are therefore possible with the emitted molecules finding
their way through a porous structure to space. The porous structure here provides a
resistance to flow and hence pressure build up above the subliming surface can
occur. The Hertz-Knudsen equation can be modified by accounting for this pressure
using

$$Z(T) = \frac{p_s - p}{\sqrt{2\pi mkT}}. \tag{2.97}$$

The gas emission rate is linked to the local gas density, $n_g$, and velocity, $v_g$, at the nucleus through the equation

$$Z(T) = \frac{1}{4} n_g v_g. \tag{2.98}$$

For a non-porous surface, the gas velocity distribution function (VDF) of molecules would normally be assumed to be an equilibrium distribution at the surface temperature. For emission into the hemisphere above a cometary surface this would be a half-Maxwellian. However, it is not entirely clear that this is necessarily the case and some numerical experiments with $\cos^n$ distributions have been performed (Liao et al. 2016). The initial distribution is, however, unstable to collisions and so this subtlety is probably only of importance for very low gas emission rates, close to the surface, if at all.

### 2.9.3   Energy Balance

#### 2.9.3.1   Simple Surface Energy Balance with Sublimation and Conduction

When a surface is in thermal equilibrium, the energy input is equal to the energy lost such that

$$E_{in} - E_{out} = 0 \tag{2.99}$$

which we have used in discussing the integrated thermal emission in Sect. 2.1.

The surface energy input on comets is completely dominated by solar insolation. An inactive surface which is perfectly thermally insulating (i.e. the thermal conductivity, $\kappa$, is zero) loses this energy through black-body radiation so that one can write for an inactive surface in our Solar System

$$\frac{S_\odot (1 - A_H)}{r_h^2} \cos i = \varepsilon \sigma T^4 \tag{2.100}$$

where we have incorporated the thermal emissivity, $\varepsilon$. It should be noted that this equation assumes the Sun to be a point source—an assumption that will not hold for objects such as sun-grazers with perihelia inside a few solar radii.

Sublimation provides an additional energy loss for the surface so that we can increase the complexity of Eq. (2.100) thus

$$\frac{S_{\odot}(1 - A_H)}{r_h^2} \cos i = \varepsilon\sigma T^4 + Z(T)L_S \tag{2.101}$$

which shows the coupling between the energy input and the gas emission. For $H_2O$, $L_S$ is $8.56 \cdot 10^{-20}$ J/molecule (cf Table 2.4). Equation (2.101) in combination with Fig. 2.29 illustrates why we have not placed a great deal of emphasis on the sublimation coefficient, $k_s$. If $k_s < 1$, $Z$ is reduced. To maintain equilibrium with the energy input, the temperature must rise but this rise does not have to be large because $Z$ is such a strong function of $T$ as we have seen in the description of the Clausius-Clapeyron equation. Thus, error in the knowledge of the sublimation coefficient results in a very small change in the ratio between the radiative and sublimation energy loss terms which, for most observational conditions, would be very hard to identify.

Conduction increases the complexity a step further. The recognition of its importance was a direct consequence of the Giotto observations of 1P/Halley when it was seen that the surface was dark and inferred to be mostly free of pure ice subliming directly to space. Conduction was a means of transporting heat from the dark, possibly inert, surface to sub-surface volatiles.

At the surface, conduction can be entered into the heat balance equation in steady-state with an additional term describing the heat flux into the interior, viz,

$$\frac{S_{\odot}(1 - A_H)}{r_h^2} \cos i = \varepsilon\sigma T^4 + Z(T)L_S + \kappa\frac{dT}{dz} \tag{2.102}$$

where $dT/dz$ is the temperature gradient with depth, $z$, at the surface. The term can be positive or negative depending upon how you define $z$.

The conductivity introduces time-dependency because comets rotate and hence $\cos i$ varies with time. There is considerable evidence to suggest that the thermal conductivity of cometary surfaces is extremely low but this should not be taken to imply that the term can be neglected in all cases and its significance for activity should not be underestimated.

The general equation for thermal conduction is given by

$$\nabla^2 T - \frac{1}{d_t}\frac{\partial T}{\partial t} = 0 \tag{2.103}$$

where $d_t$ is the thermal diffusivity given by

$$d_t = \frac{\kappa}{\rho_N c_{SHC}} \tag{2.104}$$

where $\rho_N$ is the mass density and $c_{SHC}$ is the specific heat capacity of the material. Analytical solutions for the temperature dependence of a surface with a time variable source in the presence of thermal inertia exist for simplified conductive systems (e.g. Turcotte and Schubert 2002). However, for real applications, we rapidly reach situations where only numerical solutions are possible. It is, however, normally sufficient to work in 1D if a plane-parallel approximation can be assumed (which is often the case) although full 3D solutions for conduction have been applied to cometary nuclei in the past (Guilbert-Lepoutre and Jewitt 2011) to study effects of local inhomogeneity.

The amount of heat conducted into the interior during the day and returned to the surface at night is an important quantity for calculations of sublimation rates of volatile (and especially highly volatile) species. Hence, these calculations are of considerable interest.

Following Spencer et al. (1989), an important quantity is the thermal skin depth, $x_1$, which is the scale length over which the amplitude of a thermal wave produced at the surface reduces by a factor of $1/e$. It is given by

$$x_1 = \sqrt{\frac{2d_t}{\Omega_N}} \tag{2.105}$$

Another interesting quantity is the surface heat content. This gives a characteristic number describing how much heat we have in a surface layer.

$$H_D = x_1 T \rho_N c_{SHC} = \sqrt{\kappa \rho_N c_{SHC}} \frac{T}{\sqrt{\Omega_N}} \tag{2.106}$$

The thermal inertia of a surface, $\Gamma$, and defined as

$$\Gamma = \sqrt{\kappa \rho_N c_{SHC}} \tag{2.107}$$

represents the ability of the surface to respond to the temperature forcing provided by the solar illumination. It is typically given in units of [J m$^{-2}$ K$^{-1}$ s$^{-\frac{1}{2}}$] although because this is rather tedious to write out in full, we will use [TIU] ("thermal inertia units") as an abbreviation (e.g. Williams et al. 2015). A surface with a low thermal inertia will heat rapidly when illuminated and cool off quickly during the night. A porous dust surface would normally have a low thermal inertia. Conversely, a surface with a high thermal inertia will heat slowly during the day but takes much longer to cool at night. Silicate-dominated rocks have high thermal inertia. This property is used to identify dust-free surfaces on Mars (e.g. Fergason et al. 2006).

We can also define a characteristic time scale for radiating the amount of heat described by the heat content, $H_D$, by dividing the heat content by the radiation equation to obtain

$$t_R = \frac{\Gamma}{\sqrt{\Omega_N} \varepsilon \sigma T^3} \tag{2.108}$$

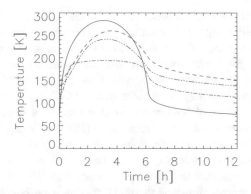

**Fig. 2.30** Surface temperatures for a set of simple cases combining thermal inertia and sublimation. An equatorial insolation distribution has been chosen with a period of 12.406 h at a heliocentric distance of 2 AU. Solid line: No sublimation with a thermal inertia of 4 TIU. Dashed line: No sublimation with a thermal inertia of 126 TIU. Dash-dot line: Surface sublimation of $H_2O$ with a thermal inertia of 40 TIU. Dash-triple dot line: Surface temperature for a thermal inertia of 40 TIU with sublimation of $H_2O$ from a layer 3.33 cm below the surface

A dimensionless quantity which relates this time to the rotation period (through the angular velocity) and called the thermal parameter is then given by

$$\Theta = \frac{\Gamma\sqrt{\Omega_N}}{\varepsilon\sigma T^3} \tag{2.109}$$

In the absence of sublimation, objects with the same value of $\Theta$ should exhibit similar thermal properties (Spencer et al. 1989). As $\Theta$ approaches 100, the temperature at constant latitude on the object becomes isothermal.

We can now combine these elements to show how the temperature varies with local time in an idealised system. Figure 2.30 shows surface temperatures for a set of simple cases combining thermal inertia with water sublimation. The solid line shows a very low thermal inertia case. The thermal inertia here is 4 TIU which is close to a balance between insolation and re-radiation in the absence of conduction and sublimation. Totally eliminating conduction would, of course, result in a temperature of 0 K on the nightside. $A_H$ was set to 0.04 and $\varepsilon = 0.9$. The rotation period of 67P was used and the profiles display the temperatures at the equator with the Sun at zenith. As the thermal inertia increases (dashed line), the nightside temperature rises appreciably, the local time of the maximum temperature shifts towards later (afternoon) times, and the maximum temperature is reduced.

Introduction of water sublimation at the surface has a major influence on the temperature of the dayside. The temperature drops below 200 K to the free sublimation temperature of water ice. Most of the insolation energy, in this case, drives sublimation rather than re-radiation. The temperature change with time on the dayside is also reduced.

The thermal inertia of cometary material is not particularly well constrained. Groussin et al. (2013) concluded from analysis of infrared observations from the

Deep Impact spacecraft that the thermal inertia was <250 TIU for 103P/Hartley 2 and <45 TIU for 9P/Tempel 1. For both comets, the temperature of the regions with exposed water ice was more than 100 K above the sublimation temperature of water ice indicating that the thermal emission was dominated by dust. Furthermore, the water ice could not be intimately mixed with dust at the scale of observations.

Similar results can be obtained for the nucleus of 67P. We can illustrate this using the MIRO measurements of the brightness temperature of the continuum thermal emission. This was measured at two frequencies (188.2 GHz and 562.8 GHz). These measurements do not give the temperature at the actual surface. It is often assumed that the two frequencies sample two discrete depths below the surface but although this is good approximation it is not completely accurate. The effective temperature, $T_{eff}$, is computed by integrating the thermal emission contribution of each sublayer below the surface, weighted by a radiative transfer function, which expresses the extinction of the thermal emission as it propagates towards the surface. $T_{eff}$, assuming a scatter-free homogeneous layer under conditions of local thermodynamic equilibrium, is determined by

$$T_{eff} = \frac{1}{d_{el}\,\cos e_t} \int_0^{z_{max}} T(z) e^{-z/(d_{el}\,\cos e_t)}\ dz \qquad (2.110)$$

where $\cos e_t$ is the transmission angle in the sublayers of the surface (in essence, the angle of emission) and $d_{el}$ is the electrical skin depth of the layer at a given wavelength. Schloerb et al. (2015) suggested that the data from September 2014 acquired from the Imhotep region were quantitatively consistent with very low thermal inertia values of between 10 and 30 TIU with the 0.5 mm emission arising from 1 cm beneath the surface and the 1.6 mm emission from a depth of 4 cm.

Schloerb et al. used Fourier series to analyse the data but we can look at the results with the rather more physical model described above. In Fig. 2.31, brightness temperatures from the sub-mm and mm receivers of MIRO are plotted as individual points. A sub-set of data that looks at the Imhotep region (Table 2.5) close to equinox post-perihelion (sub-solar latitude = $-5 \pm 2°$) and covers the period 12 February 2016 to 3 March 2016, has been extracted. Imhotep was chosen as a large flat region on the nucleus and the period selected to get a large thermal wave—the Sun being close to zenith seen from Imhotep during this period. There was sufficient data in this period to bin the points in bins of 4 min in local time to provide mean values which are shown as the histograms superposed on the individual points in Fig. 2.31.

We can now model the observations under the assumption of no sublimation. The calculation of $T_{eff}$ was found to have relatively little effect here compared to other issues and has been ignored. The results of a model using a thermal inertia of 32.5 TIU and depths of 1 cm and 3.5 cm for the sub-mm and mm receivers respectively (similar to that of Schloerb et al.) are shown as the dashed lines in Fig. 2.31.

The key point to note here is that while the shapes of the model curves fit the data well, the absolute values do not. It is informative to discuss why it is clear that there

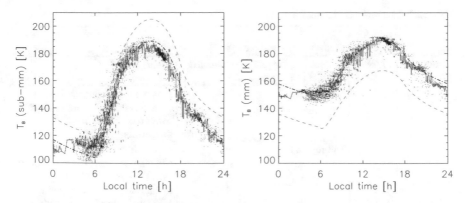

**Fig. 2.31** Sub-mm (left) and mm (right) brightness temperatures of Imhotep close to equinox acquired by the MIRO experiment. The actual data are shown as small dots and 4 min averages are shown by the histograms. The dot-dashed lines show fits to the data based on model calculations. The fits indicate that the sub-mm receiver was sensing depths of 1 cm and the mm receiver was sensing 3.5 cm. The fit here used a thermal inertia of 32.5 TIU

is a problem. The interior of the nucleus is expected to be very cold (e.g. 30 K). So there is a negative temperature gradient from the surface to the interior. The temperature is an expression of the heat content at each depth. The mean temperature of the mm receiver observations is around 25 K higher than the mean temperature in the sub-mm receiver. This cannot be physical unless there is an internal heat source. If we assume that there are valid reasons for the observations to be simply offset from the actual temperatures and we allow those offsets to be free parameters then we get the dash-dotted lines which fit the data set very well. Clearly, this is not exhaustive and other solutions may be possible. A combination of the MIRO and VIRTIS data on Rosetta has suggested thermal inertia in the range of 20–80 TIU with local variability (Marshall et al. 2018; Groussin et al. 2019) and fully consistent with observations of other comets. This leads to diurnal skin depths of the order of 2–3 cm.

These results can be compared to the thermal inertia of compact water ice. The thermal conductivity of water ice is temperature dependent as is known from the simple empirical formula used by Klinger (1981)

$$\kappa = \frac{567}{T} \tag{2.111}$$

where $\kappa$ is in units of [W m$^{-1}$ K$^{-1}$], while the specific heat capacity of ice is around 2.1 kJ kg$^{-1}$ K$^{-1}$. Combined with the well-known density, this gives a thermal inertia of nearly 2000 TIU at 200 K and thus up to 2 orders of magnitude above observation. Attempts to reconcile this problem have a long history.

In the early 1980s, it was recognized that the total water production rate from a comet could be combined with the free sublimation rate of water to derive a lower limit for the emitting surface area and thus the radius of the nucleus. Keller (1990)

**Table 2.5**  The regional classification of 67P

| Region | Rough. index | Characteristics and area [km²] | |
|---|---|---|---|
| Atum (Am) | 17.09 | Complex region with consolidated material and very rough. | 1.9497 |
| | | Sub-region a | A very rough topographic high (with respect to its surroundings) with boulders and some lineaments. |
| | | Sub-region b | A smooth fractured surface adjoining Khonsu. On the Khonsu side, there is a cliff leading to rough fractured terrains possibly indicating loss of this smooth layer. It is topographically at slightly higher elevation than the adjoining Anubis region with a distinct step evident at the boundary. |
| | | Sub-region c | An undulating terrain with inter-mediate roughness. It is bounded by Anubis and Anhur on the north and south side respectively and by a steep cliff to the east that forms the Geb region. |
| Khonsu (Kn) | 18.83 | Complex region with a mixture of smooth and rough terrains. | 2.16872 |
| | | Sub-region a | This sub-region is at an angle with respect to the rest of the region. It also contains small scale roughness and a lot of boulders. |
| | | Sub-region b | Very rough terrain on many scales adjoining the Apis "face" and showing the side of the rougher part of Atum. |
| | | Sub-region c | Very rough and in places pitted terrain with fractures. Topographi-cally low compared to adjacent Atum sub-regions. |
| | | Sub-region d | Adjoining Atum, this region is very complex. There are flatter areas (dust deposits) but with rough outcrops. |
| | | Sub-region e | A small sub-region which is dom-inated by flat, apparently dusty material |
| Apis (Ap) | 12.77 | Consolidated and fractured but topographically smooth. Topo-graphically stands out above Ash | 0.39798 |

(continued)

**Table 2.5** (continued)

| Region | Rough. index | Characteristics and area [km$^2$] | |
|---|---|---|---|
| Imhotep (Im) | 15.14 | Smooth "dusty" depression surrounded by more consolidated material. Circular features at the edges of the smooth terrain | 4.90446 |
| | | Sub-region a | Smooth material at the centre of the region. Observed to change dramatically over the mission. Bounded by Ash to the north. On two sides there are steps upwards to rougher terrain (sub-region b) while on the remaining side there are layers downwards to sub-region c with a more gradual transition than elsewhere. |
| | | Sub-region b | Rim of sub-region a. Contains layered terrain incorporating a large circular structure. |
| | | Sub-region c | Rougher terrain inside the rim of Imhotep. Includes all the small quasi-circular structures. Adjoins the smooth terrain. At the boundary there are indications of layering. |
| | | Sub-region d | Clearly rocky at its edge but covered with smooth material in depressions. Evidence of surface changes in places similar to those observed in sub-region a. Boundary to smooth surface (sub-region a) often associated with a clear scarp. Similar to Khepry although topographically lower. |
| Anubis (Ab) | 11.63 | Smooth surface probably not consolidated and has undergone surface modification possibly similar to that observed in Imhotep. | 0.92241 |
| Bes (Be) | 17.15 | Multiply-layered terrain bordering the scarp into the southern part of the neck. | 2.42084 |
| | | Sub-region a | Topographically lowest level. Covered in boulders in some places. |
| | | Sub-region b | Separated from a by a cliff. Contains a diamond-shaped structure surrounding a surface with large boulders |

(continued)

**Table 2.5**   (continued)

| Region | Rough. index | Characteristics and area [km$^2$] | |
|---|---|---|---|
| | | Sub-region c | Adjoins Imhotep and appears to be at a level intermediate between sub-regions a and b although it has no contact with a. Generally smooth with no major topographic features. |
| | | Sub-region d | A steep cliff separates this level from sub-region c. It is at a higher topographic level—similar to b or possibly slightly higher. |
| | | Sub-region e | The uppermost level. Separated from d by a significant change in slope. The steep cliff down to Anhur sub-region c is strongly apparent in the shape model. |
| Seth (Se) | 16.16 | Consolidated, possibly more brittle in nature when compared to other more strongly consolidated regions. Dominated by circular and semi-circular structures and talus. | 4.66022 |
| Ash (As) | 15.74 | Covered with a presumed sedimentary deposit producing smooth surface. Occasional exposures of more consolidated but brittle material below. | 6.25734 |
| | | Sub-region a | Adjoining Babi at an edge and the Aten depression via a sharp change in slope, this sub-region adjoins an adjacent sub-region at a rough hummocky interface. The sub-region is mostly smooth with some smaller depressions and small cliffs covered in dust. |
| | | Sub-region b | Adjoining Seth, this sub-region is smooth. Its boundary to Seth is characterized by a transition to rougher terrain and a substantial change in slope. |
| | | Sub-region c | Adjoining Aten, this is rougher terrain. It is topographically higher than sub-region b and where it meets sub-region b there are arc-shaped cliffs. |
| | | Sub-region d | Dust coated. Smoother region. |
| | | Sub-region e | Sub-region containing the large circular structure which may be the result of impact. Possibly related material outside the putative rim is included. |

(continued)

**Table 2.5**  (continued)

| Region | Rough. index | Characteristics and area [km²] | |
|---|---|---|---|
| | | Sub-region f | Smooth sub-region with a small pit and some scarps. Intermediate in character. |
| | | Sub-region g | Seth like. Adjoining Atum. |
| | | Sub-region h | Adjoining Apis. Rock-like surface with a slight depression. Topographically separated from the rest. |
| | | Sub-region i | Large-scale rough terrain. Dust covered but with exposed layering in many places. Transitions to the Imhotep region at a boundary between very rough terrain and that of intermediate character. |
| | | Sub-region j | Borders Aten and is also a depression but not as deep as Aten. There is a ridge dividing two sections of the sub-region. The bases of the depression on both sides of the ridge are smooth. |
| Aten (An) | 16.75 | Depression with little or no sedimentary deposits. Interior mainly dominated by talus resulting from progressive rim failure. | 1.12758 |
| Babi (Bb) | 12.60 | Covered with a deposit producing a smooth surface. Occasional exposures of more consolidated but brittle material below. Topographically separated from Ash. | 1.45666 |
| | | Sub-region a | Topographic high with cliffs on three sides. Uppermost surface is dust covered. |
| | | Sub-region b | Topographically low and strongly sloping. Bounded by Khepry, Seth and Ash. Some spur-like structures possibly originating from sub-region a are evident. |
| Geb (Gb) | 12.27 | Consolidated material | 1.02767 |
| | | Sub-region a | Large numbers of depressions on a steep slope. |
| | | Sub-region b | The neck side of Geb. Covered in boulders. |
| | | Sub-region c | Smoother fractured surface similar to that seen in Anhur and Bes. |
| Khepry (Kp) | 15.71 | Consolidated and fractured material but rather smooth with ponded deposits. | 1.63087 |

(continued)

**Table 2.5** (continued)

| Region | Rough. index | Characteristics and area [km²] | |
|---|---|---|---|
| | | Sub-region a | Flat but rocky-like sub-region with ponded deposits |
| | | Sub-region b | A small sub-region with a prominent cliff. Adjoins Bes with similar characteristics. |
| | | Sub-region c | Topographically almost at right-angles to sub-region a. Highly complex sub-region with rough, rocky terrain, smoother coatings it places and boulders. Talus from collapse of material from Ash is also evident. |
| Anhur (Ah) | 24.47 | Consolidated material with significant intermediate scale roughness | 1.87013 |
| | | Sub-region a | Plateau with extreme intermediate roughness including isolated ridges. Includes some pits. |
| | | Sub-region b | Cliffs descending from sub-region a to the neck. Surface texture similar to that in sub-region a. |
| | | Sub-region c | With respect to the roughly ellipsoidal shape of the body, topographically on same level as Bes sub-region a which it adjoins but with the face being at a large angle to Bes sub-region a. |
| Aker (Ar) | 14.10 | Strongly consolidated material similar to the adjacent region, Khepry. Contains a large complex fracture system near a steep topographic slope that descends towards Hapi. It has four distinct faces. | 0.87022 |
| | | Sub-region a | Contains a large set of tectonic fractures and a smooth bottomed shallow depression. |
| | | Sub-region b | Topographically distinct from sub-region a but has some similarities. It adjoins Anhur where there is a change is slope and surface roughness. |
| | | Sub-region c | Comprises a cliff that drops sharply to the boundary with Babi at its base. Significant evidence of collapse is evident along the face. |
| | | Sub-region d | Interfaces primarily with Hapi and is a steep fractured cliff. |

(continued)

**Table 2.5** (continued)

| Region | Rough. index | Characteristics and area [km²] | |
|---|---|---|---|
| **Total Body** | | | **31.66** |
| Hapi (Hp) | 8.36 | Smooth, probably non-consolidated surface | 1.98356 |
| Sobek (So) | 21.40 | Consolidated material, texturally very rough | 0.83735 |
| | | Sub-region a | Set of quasi-parallel steps/small scarps |
| | | Sub-region b | Boulder-covered terrain |
| **Total Neck** | | | **2.82** |
| Anuket (Ak) | 17.01 | Consolidated, "rocky" appearance. Smooth on large scale but with some large knobs and significant small scale roughness. | 2.0523 |
| Neith (Ne) | 18.63 | Mainly comprising the cliff separating Wosret and Sobek. Significant intermediate scale roughness covering the whole region. | 1.60746 |
| Maftet (Mf) | 16.92 | Weakly consolidated material dominated by arcuate-shaped depressions and associated talus. | 0.67813 |
| Serqet (Sq) | 14.13 | Mix of strongly consolidated material with substantial vertical relief and a smoother dusty deposit area at the base of a cliff. | 1.03333 |
| | | Sub-region a | Vertical fractured cliff adjoining Anuket |
| | | Sub-region b | Flat dust covered surface with ripples possibly of gas driven origin adjoining Nut. |
| | | Sub-region c | Transitional sub-region with rocky material becoming increasingly similar to Maftet-like morphology at the Maftet boundary. |
| Nut (Nu) | 14.33 | Depression possibly similar to Aten but significantly shallower. | 0.47264 |
| Wosret (Wr) | 18.26 | Consolidated material that appears highly fractured with occasional pits | 2.35911 |
| | | Sub-region a | An apparently flat "face" with ponded materials and knobby textured terrain. |
| | | Sub-region b | Topographically lower than sub-region a and displaying long fracture systems. |

(continued)

**Table 2.5**  (continued)

| Region | Rough. index | Characteristics and area [km²] | |
|---|---|---|---|
| | | Sub-region c | Rougher terrain with numerous quasi-circular structures and non-aligned ridges and pits. |
| Ma'at (Ma) | 12.99 | Covered with a deposit producing a smooth surface on small scales. Occasional exposures of more consolidated but brittle material below. Similar to Ash but with some pits. | 3.81651 |
| | | Sub-region a | Smooth dust-covered shallow depression with knobs |
| | | Sub-region b | Smooth dust-covered shallow depression with knobs and an irregular-shaped ridge-like structure at its centre. |
| | | Sub-region c | Topographically lower with significant numbers of depressions and quasi-circular/arcuate depressions. |
| | | Sub-region d | A plateau at a lower elevation that Ma'at sub-regions around it. Bounds Bastet at a cliff. |
| | | Sub-region e | Large-scale roughness dominated substrate with dust-covering. |
| Bastet (Bs) | 14.76 | Consolidated material with texturally rough surface and limited amounts of dust coating. | 1.98781 |
| | | Sub-region a | Smoother terrain adjoining Hatmehit and Wosret. |
| | | Sub-region b | Undulating terrain on a face at an angle with respect to a. Pock-marked in places. |
| | | Sub-region c | Fractured consolidation terrain. Parts of this sub-region show similarity to Hathor which adjoins it. |
| Hathor (Hh) | 19.6 | Consolidated, but fractured material on a gravitationally steep slope. Comprises most of the cliff separating Ma'at and Hapi. | 2.16217 |
| Hatmehit (Hm) | 13.79 | Large circular depression with a smooth interior (some rocks) surrounded by more consolidated material at the rim. | 1.08561 |
| | | Sub-region a | The floor of the circular depression. This is generally smooth and flat with a small ridge running roughly through the centre. Some talus from fracturing is evident at the margins. |

(continued)

**Table 2.5** (continued)

| Region | Rough. index | Characteristics and area [km$^2$] | |
|---|---|---|---|
| | | Sub-region b | The south and west sides of the rim of the depression adjoining Maftet and Wosret. Contains quasi-circular depressions. The rim of Hatmehit is less pronounced. |
| | | Sub-region c | The north and east sides of the rim of Hatmehit adjoining Bastet and Maat. The steepest parts of the rim are included in this sub-region. The interior of the rim is fractured in many places. |
| **Total Head** | | | **17.26** |

The regions are indicated by the left column with their surface area. Each region is sub-divided (where feasible) into sub-regions. The total surface areas for the head, neck and body regions are also shown. The characteristics of the region and of the sub-regions are given in each case. We also include unique abbreviations for each region to simplify display. The roughness index gives an indication of the surface roughness of each region relative to the other regions. Hapi (roughness index = 8.36) is the smoothest region, Anhur (24.47) is the roughest (Thomas et al. 2018)

gives an equation under the assumption that re-radiation is negligible (appropriate for an active comet) as

$$R_N = r_h \sqrt{\frac{Q_g L_S}{\pi S_\odot (1 - A_H)}} \qquad (2.112)$$

where $Q_g$ is the gas production rate in [kg s$^{-1}$] and $L_S$ is the latent heat of sublimation in [J kg$^{-1}$]. It was immediately apparent from the Halley fly-bys that this equation substantially underestimates the radii of cometary nuclei and its use, in combination with observations of the brightness of nuclei at high heliocentric distance, also led to huge overestimates in albedos prior to the Halley encounters (Keller 1990). Keller et al. (1987) concluded that 10% of the surface of 1P/Halley was active (cf discussion in Sect. 2.4) and this has led to the concept of an effective active fraction (eaf) which describes, empirically, the outgassing rate of a surface as a fraction of the free sublimation rate. By using the eaf, it is not necessary to understand the actual physics of the emission. While this is unsatisfactory, it nonetheless avoids actually having to specify the physics of the gas emission process.

One can envisage two simple cases as possible extremes. One possibility is that the surface may be composed of a mixture of ice and non-volatile material in the ratio of the eaf so that the ratio of the emitting surface to the non-volatile surface really does correspond to the eaf. Alternatively, the volatile material may be below a

**Fig. 2.32** The sublimation rate at the equator of a rotating nucleus with a thermal inertia of 80 TIU. The subliming water ice surface has been varied. Solid line: Water at the surface. Dashed line: The subliming surface is 4.2 cm below the actual surface. Dot-dashed line: The sublimation is from 8.2 cm below the surface

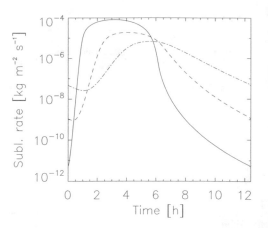

thin, inert, insulating layer so that the gas emission is throttled by the thermal inertia of the inert surface layer and has to pass through this surface layer to escape. In this case, the emitting (sub-) surface area can be (much) larger than the eaf. This has the attractive implication that ices would be more homogeneously distributed in the sub-surface.

In Fig. 2.32, the influence on the local production rate of setting the subliming surface at depth is illustrated for a system with a thermal inertia of 80 TIU. The figure compares the sublimation rates over one rotation if the subliming surface is at the actual surface with the rates if the subliming surface is at two different depths (4.2 and 8.4 cm) below the actual surface. No internal pressure build-up has been included here (cf. Eq. 2.97). If the layer above the subliming surface insulates then the sublimation profile is shifted in time and the peak is reduced. The integral over the rotation period is also reduced. For surface sublimation, the integral is 1.94 kg m$^{-2}$ rotation$^{-1}$ whereas sublimation at depth produces only 0.17 kg m$^{-2}$ rotation$^{-1}$ or less than 10%. Hence, this is a viable mechanism for reducing the production rate without using an eaf but there is also a clear shift in the phase of the maximum of emission relative to the maximum in illumination which should, in principle, be measureable. On the other hand, there are also physical difficulties with such a model. In particular, the depth of the inert layer above the subliming surface would grow with time as the ice sublimes and retreats into the nucleus. This would rapidly reduce the sublimation rate and choke further emission unless the inert surface layer can be disrupted. In effect, there are positive feedback loops in this process that must be stabilised for the mechanism to be viable. It is this problem that leads gas dynamics modellers to continue using eafs as this avoids introducing, as yet, unsubstantiated models of the source process.

A final point is that the depth at which sublimation needs to occur to equate to the eaf is directly proportional to the thermal conductivity. Hence, a lower thermal conductivity leads to sublimation needing to be closer to the surface but an identical thermal inertia can still be achieved by making a compensating increase in the heat capacity or the material density (or both). This is important because it implies that simply knowing the thermal inertia does not give you the depth of any sub-surface sublimation front. You must also establish either the thermal conductivity of the inert layer or its total heat capacity (cf. Shi et al. 2016).

The thicker and denser the inert layer, the more resistance to the gas flow. This builds an internal gas pressure within the layer that slows the sublimation from the sub-surface sublimation front and Eq. (2.97) cannot, in general, be ignored. This process, however, needs to be calculated with a gas dynamics calculation. But here again, although this limits the depth at which sublimation can occur and still match the eaf, it does not avoid that fact that, without separate knowledge of the thermal conductivity and the total heat capacity, the modelled depth of the sublimation front will be non-unique.

### 2.9.3.2 Volume Absorption, Solid-State Greenhouse Effect, and Porosity

The mathematical description of the thermal balance can be extended to include the influence of a large number of additional phenomena. We shall look at some of these effects which might be significant in real cases.

High porosity at the surface can be interpreted as implying that there are significant voids in the surface layer. Solar photons are therefore able to penetrate this surface to a depth that may be significant compared to the thermal skin depth. The conduction equation must then be modified. Here we reduce to 1D and describe the sub-surface absorption through

$$\frac{d^2T}{dz^2} - \frac{1}{d_t}\frac{\partial T}{\partial t} = \kappa \frac{dF_\odot(z)}{dz} \tag{2.113}$$

where $F_\odot(z)$ is the insolation at depth (Urquhart and Jakosky 1996) which is also time dependent and can be expressed as

$$F_\odot(z,t) = \frac{S_\odot(1 - A_H)}{r_h^2} \cos i \sin \Omega_N t \, e^{\frac{-z}{\delta_P(i)}} \tag{2.114}$$

for

$$\frac{\pi}{2} \leq \Omega_N t \leq \frac{3\pi}{2}$$

and

$$F_\odot(z,t) = 0$$

for the remainder of the rotation (the nightside). Here, $\xi_p$ is a penetration scale length. In the Urquhart and Jakosky formulation, a dependence on the angle of incidence is included and, in the simplest case, there would be a relationship to the cosine of $i$. (They also substitute $\Omega_N$ with $2\pi/P$ where $P$ is the period.) Given the low thermal inertia, it is conceivable that this type of volume absorption could play a role if $\xi_p$ is significantly greater than $x_1/10$. This type of process was first implemented in comet models by Davidsson and Skorov (2002a, b).

Solid-state greenhouse effect (Matson and Brown 1989) occurs as a consequence of ice in its purest form being translucent. This results in the impinging solar energy being deposited below the physical surface. It is now well established that this phenomenon can lead to unusual effects. The most spectacular is the formation of geyser-like activity on Mars arising from basal sublimation of $CO_2$ in response to sunlight penetrating a 1 m thick layer of $CO_2$ ice (Kieffer et al. 2006). In this example, thermal re-radiation from below the ice layer is effectively trapped by the $CO_2$ ice above and generates a sort of pressure cooker until the pressure is released via a crack in the ice with subsequent violent outgassing. There is no evidence that metre-thick translucent ice layers are present on comets but the principle may operate on smaller length scales. One can imagine ices being semi-translucent over millimetre scales so that basal heating of an ice layer can occur. This might be a mechanism for more explosive loss of larger amounts of material. This would probably have had to be on sub-decimetre scales to avoid detection by Rosetta's imaging system.

The high porosity of cometary nuclei has effectively been confirmed by Rosetta observations of the bulk density (Sect. 2.3). On the other hand, the high porosity refers to the nucleus as a whole and it is not completely clear that this holds for the uppermost surface layer. However, if it does, there are implications for the thermal behaviour of the surface layer.

The thermal conductivity of a structure is reduced as voids within the structure become more and more significant as the porosity, $\Psi$, increases. Porosity is defined through

$$\Psi = \frac{V_v}{V} \tag{2.115}$$

where $V_v/V$ is the ratio of the volume of the voids to the volume of the whole. A useful example is the porosity of a skeleton of spherical particles of radius $a$, which is simply

$$\Psi = 1 - N_p \frac{4}{3}\pi a^3 \tag{2.116}$$

where $N_p$ is the number of particles per unit volume. However, the effect on thermal conductivity is mitigated by thermal re-radiation within the porous structure. Hence, the pore size distribution within the structure also plays a role. This results in the effective conductivity becoming a complex function of the radiation field. The conductivity can be described (see Russell 1935; Espinasse et al. 1993; Marboeuf et al. 2012) as

$$\kappa_m = \frac{\kappa_s\left[\Psi^{2/3}\kappa_p + \kappa_s\left(1 - \Psi^{2/3}\right)\right]}{\kappa_s\left[\Psi - \Psi^{2/3} + 1\right] - \kappa_p\Psi^{2/3}\left[\Psi^{1/3} - 1\right]} \tag{2.117}$$

where $\kappa_p$ is the radiative conductivity across the pores as described by Squyres et al. (1985) through the equation

$$\kappa_p = 4\varepsilon\sigma r_p T^3 \tag{2.118}$$

and $\kappa_s$ is the conductivity of the solid phase of the components which can be the result of a mixture of materials with individual conductivities (Marboeuf et al. 2012).

Laboratory measurements of the thermal conductivity of porous but consolidated materials are now fairly commonplace using hot plates or methods such as the transient hot wire technique (e.g. Smith et al. 2013). However, measurements of fragile, porous, dust aggregates in the laboratory is not trivial. Techniques have been developed by Krause et al. (2011) using the thermal IR emission observed following controlled illumination. They obtained values for three types of porous dust samples, consisting of spherical, 1.5 μm-sized $SiO_2$ particles, with volume filling factors in the range of 15–54% (Fig. 2.33). This work shows order of magnitude changes in the thermal conductivity over a restricted range in porosity. Modelling of these data by Arakawa et al. (2019) including the radiative transfer suggests that extension to other materials and structures may be possible.

It should be noted that this type of work is of considerable importance in other fields (including the oil and gas industry e.g. Huang (1971)) so that many studies (see Wang and Li (2017) and Askari et al. (2017) for typical recent examples) have been performed. One convenient equation given by Kou et al. (2009) is

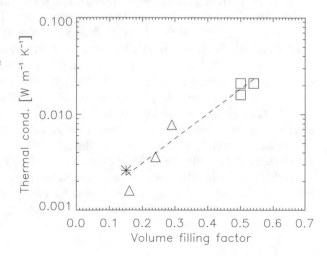

**Fig. 2.33** Thermal conductivity measurements of $SiO_2$ spheres with different filling factors. Note the strong dependence on the volume filling factor (Re-plotted from Krause et al. 2011)

$$\frac{\kappa_e}{\kappa_g} = \frac{(2 - D_p)\Psi \left(1 - \Psi^{\frac{D - D_p + 1}{2 - D_p}}\right)}{(D_t - D_p + 1)(1 - \Psi)}\left(\frac{\lambda_{max}}{L_0}\right)^{D_t - 1} + (1 - \Psi)\frac{\kappa_s}{\kappa_g} \qquad (2.119)$$

which describes the dimensionless thermal conductivity for a porosity, $\Psi$, a pore area fractal dimension, $D_p$, a tortuosity fractal dimension, $D_t$, a maximum pore size, $\lambda_{max}$, and a characteristic length scale, $L_0$. Here, $\kappa_s$, $\kappa_g$ and $\kappa_e$ are the thermal conductivities of the solid material and the gas and the effective thermal conductivity respectively. The tortuosity fractal dimension takes a value of 1 for a straight capillary and 2 for a highly tortuous path.

In addition to radiative heat transport, the gas flow through a porous surface layer can lead to heat transport (Fanale and Salvail 1984). The sublimed gas from the sub-surface can be heated by interaction with a warmer surface layer (a phenomenon that will be looked at in more detail later) while the structure can be cooled by this flow. Should the surface layer be colder than subliming gas, as might be the case for higher thermal inertia regions near the evening terminator, then the gas can provide a heat source for the layer. These cases can be treated numerically but the number of free parameters rises rapidly. A relatively simple expression was presented by Mendis and Brin (1977) using the emissivity of the material, the intergrain spacing, and the Knudsen number of the flow as free parameters but other treatments using, for example, the tortuosity of a more mathematical treatment of the pore structure, have also been studied in the context of comets (e.g. Skorov et al. 2011).

Internal porosity can also allow more volatile ices to sublime at depth as the heat wave penetrates. This could result in a type of fractionation (Fig. 2.34) where the sublimation front of water ice is close to the surface but $CO_2$ sublimation occurs from 10 to 20 cm below the surface and CO sublimation occurs from even deeper in the structure. The porosity is sufficient to allow the sublimed gas to escape.

This is further complicated, however, by the possibility of condensation in the porous layer. There are two cases of possible significance. First, when the surface temperature drops (e.g. at sunset or through afternoon shadowing) below the free sublimation temperature, this leads to the uppermost layer becoming a cold trap for the gas and warming of the structure through release of latent heat as the gas condenses. This further modifies the thermal conductivity through both the temperature change and the reduction in porosity caused by the condensed gas and can be extremely dynamic at least at some scale. Second, subliming gas at the surface not merely generates a pressure gradient outwards away from the nucleus but also inwards through the porous layer into the interior. Condensation in the interior adds to the enthalpy sub-surface and also modifies the conductivity of the sub-surface through modification of both the porosity and the conducting material, as was recognized in the late 1980s at the time of the KOSI experiments (e.g. Spohn et al. 1989; Seiferlin 1991). As we have seen above (e.g. Fig. 2.33), if water ice is introduced, the potential changes in thermal conductivity can be huge allowing heat transport deeper into the interior. The MUPUS observation of hard consolidated

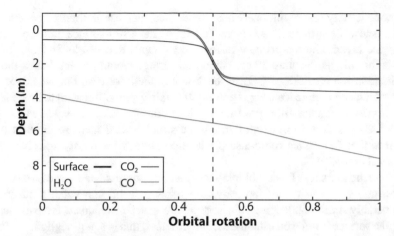

**Fig. 2.34** The erosion of the surface caused by sublimation over one orbital period of the nucleus of 67P using the model of Marboeuf et al. (2012). If the surface layer is porous, fractionation of the ices occurs because highly volatile species can sublime at depth and the gas can escape. In this example, CO is subliming around 4 m below the surface. Because of the high sublimation rate of water ice near perihelion, the subliming $CO_2$ front is closest to the surface around perihelion (Credit: C. Herny, pers. comm.)

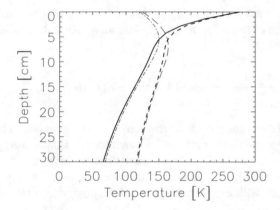

**Fig. 2.35** Temperature with depth for two models using the 1D heat equation. The dashed lines show the dayside (heavy, dashed) and nightside (light, long dashed) temperature distribution for a model with an increase in thermal conductivity of a factor of 20, 10 cm below the surface. The solid and dot-dashed lines show the same model but with a constant low conductivity with depth. The jump in thermal conductivity is intended to simulate a sub-surface ice layer. The internal temperature in both cases was set to 50 K

material just below the surface at the Abydos resting place of the Philae lander (Spohn et al. 2015) might indicate that this process is important.

We can illustrate this further by using a two-layer model for the thermal conductivity in the 1-D heat transfer equation. An example result is shown in Fig. 2.35. Here, we compare the temperature profile with depth of a case where there is a single low conductivity layer with one where there is an increase in thermal conductivity 10 cm below the surface. This is intended to simulate a sub-surface more conductive water ice layer. The temperature difference at depth that results is large. In this simulation, the temperature 30 cm below the surface would be well below the free sublimation temperature of $CO_2$ in the low conductivity case but would be well above it if a conductive ice layer is present. It is fairly straightforward to imagine that sub-surface re-condensation producing a less porous water ice layer would both block $CO_2$ gas flow from depth through the structure and increase the heat transferred to that $CO_2$. This could also produce conditions for quasi-explosive events (Agarwal et al. 2017).

While the mass loss from subliming ices can be computed and turned into a depth by assuming a density and porosity, the mass loss of non-volatile material is considerably more challenging to assess as the exact mechanism for particle loss from the surface is not well established. Nonetheless, this is a non-negligible element of the surface energy balance because the non-volatile material will be hot and has a heat capacity and thus carries away energy if it is ejected making this a more complex version of Stefan's problem. Haruyama et al. (1993) gave an equation for the specific heat capacity of an ice-silicate mixture as

$$c_{mix} = 8.9 \, (0.28 + 0.72 \, f_{ice}) \, T \qquad (2.120)$$

in [J kg$^{-1}$ K$^{-1}$] where $f_{ice}$ is the mass fraction of ice (see also Sirono 2017). One can see the magnitude of this effect by crudely including this additional energy loss term in Eq. (2.101) using

$$\frac{S_\odot \, (1 - A_H)}{r_h^2} \, \cos i = \varepsilon \sigma T^4 + Z(T)(L + \chi \, c_{mix} \, T) \qquad (2.121)$$

where $\chi$ is the dust to gas mass loss ratio which can be assumed to be $(1/f_{ice}) - 1$ for simplicity. The gas production rate per unit area for this simple energy balance at the sub-solar point is shown in Fig. 2.36 for three different heliocentric distances. It can be seen that non-volatile mass loss has a measureable, but relatively small, effect on the energy balance with the gas production rate being reduced by about 25% as $\chi$ is increased from 0 to 10.

Brin and Mendis (1979) used the balance between the acceleration by gas drag and the gravitational acceleration (this is related to the problem of the largest liftable mass and will be addressed later) to determine mass loss of a dust particle distribution. Fanale and Salvail (1984) extended this approach by including a capillary model of the solid porous medium. In other words, they assumed the porosity would lead to funnelling of the gas thereby increasing the gas drag. They included a factor of $3t_m/\Psi$ in the acceleration term with $t_m$ being the tortuosity to account for this. They also incorporated additional physics to provide partial blow-off of a dust mantle formed by sublimation and retreat of a sublimation front.

**Fig. 2.36** The gas production rate per unit area for a simple energy balance without conduction but including heat loss arising from the heat capacity of non-volatile material. All calculations are for the sub-solar point. Solid line: 1.3 AU, dashed line: 1.6 AU, dot-dashed line: 2.0 AU

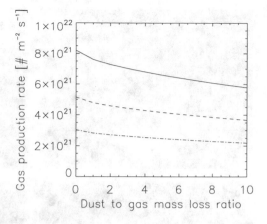

### 2.9.3.3  Multi-volatile Models and Ice Fractionation

Multi-species models of the surface layers can be constructed numerically but require numerous assumptions. The work of Marboeuf et al. (2012) is a recent example. Here, dust and water ice are coupled so that sublimation of water ice releases dust grains at the surface. If the dust grains are large, the gas drag is insufficient to remove the dust and a dust mantle starts to form as in the Fanale and Salvail approach. If drag is sufficient, the dust escapes. The super-volatiles such as CO and $CO_2$ are not driving dust emission but pass through the porous layers of the nucleus when heated by conduction. Hence, these species have sub-surface sublimation fronts. An example of the evolution of the sublimation fronts over one orbital period using parameters for 67P is shown in Fig. 2.34. The x axis gives the fraction of the orbital period with perihelion defined at 0.5. It can be seen that the CO sublimation front is typically 2–4 m below the surface whereas the $CO_2$ front is significantly below the surface for much of the orbit but is closer to the surface near perihelion when water sublimation is at its most intense.

The Marboeuf et al. model treats porosity, changes in pore size with condensation and gas transport, variable conductivity arising from the pore property changes, and the amorphous to crystalline water ice transition (which is discussed in the following sub-section). For the 2012 version of the model, clathrates were also investigated. The model is detailed and numerically sophisticated. Some alternative descriptions of the surface layer were studied showing that the evolution and outgassing rates are sensitive to the structure chosen. Other initial structures might also be envisaged (e.g. coupling of volatile species) that could influence the results. The model is also 1D and hence both latitudinal and local spatial dependencies can arise in this model. Nonetheless, the model illustrates that the sublimation fronts of super-volatiles may be at considerable depth (Fig. 2.34).

Seiferlin (1991) pointed out the importance of sintering in water ice for thermal conductivity. The rate of heat transfer in a porous medium depends partially on the sizes of contacting area in the structure. Even here, there is the potential for dynamical effects. Ice particles that come into contact produce ice bridges that

**Fig. 2.37** Scanning-electron-microscope micrographs of water ice particles from Jost (2016). Images (**a**), (**c**), and (**d**) were produced with the same method but are shown at different stages of their evolution. (**a**) was a fresh sample, (**c**) was left at 30 °C in a chest freezer for 1 h while (**d**) was imaged after storage for 17 h. Panel (**b**) show an SEM micrograph of water ice particles produced by a different method

show a temperature-dependent evolution often referred to as sintering. This can be seen in cryo-scanning electron microscope images (Fig. 2.37) and can also influence the photometric properties of the material as was observed by Jost et al. (2013) using photo-goniometer measurements of a highly controlled sample of micron-sized ice particles (Fig. 2.38). The evolution of the ice bridges will influence the thermal conductivity given sufficient time.

The diurnal thermal skin depth is an estimate of the depth of penetration of the heat wave from insolation over one rotation period and can be computed from Eq. (2.105). However, as we have seen, $\kappa$ is not easily established. Most estimates suggest $x_1$ for diurnal timescales is of the order of 2–3 cm but it can take values from 20 cm (for non-porous ice) and may be spatially inhomogeneous. The annual skin depth describes the penetration depth of the heat wave over the orbital period with values likely to be in the range 1–3 m. Material a few skin depths deep experiences only a weak heat wave, with the heat flow becoming constant over an orbital period at depths of $\gtrsim 10\ x_1$. Consequently, the deep internal temperature of the nucleus is barely affected by the diurnal or orbital insolation. Using the usual estimate for the thermal diffusive timescale, $\tau_{\text{diff}}$,

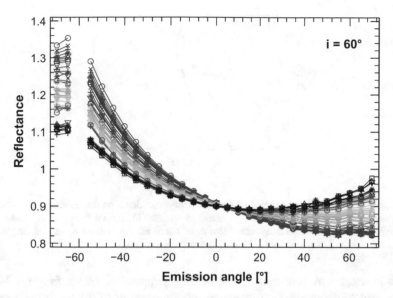

**Fig. 2.38** Temporal evolution of the photometric behaviour of a water ice sample at 60° illumination angle (from Jost et al. 2013). The first measurement appears in red. The highest reflectance is located in the back-scattering region (low phase angle) and the lowest reflectance is located in the forward-scattering region (high phase angle). The black curve was measured 41 h later. Between these two measurements, the temporal evolution is characterized by a continuous decrease of reflectance at low phase angle and a continuous increase of reflectance at high phase angle

$$\tau_{diff} = \frac{L^2}{d_t} \left( = \frac{r_N^2}{d_t} \right)$$

(2.122)

where $L$ is a length scale that we can assume to be the radius of the nucleus, $r_N$, we can estimate that the nucleus may need to be in a stable state for ~5 million years to equilibrate. This implies that, even for objects that have completed many revolutions in the inner Solar System, the internal temperature may be a residual from when the comet was in an earlier, and perhaps its original, reservoir and may be different from comet to comet. This produces uncertainty in the boundary conditions for thermal calculations because the internal temperature should be specified for short-term calculations. The value assumed influences the temperature gradients within a few skin depths of the surface and this, in turn, can influence the energy input to potentially important sub-surface super-volatile reservoirs.

This is illustrated in Fig. 2.39 which compares identical simple thermal models in which two different internal temperatures, 40 K and 100 K, have been chosen. The resulting temperature structure at midday and midnight are shown for the two cases. As might be expected, the temperature structure, 15 cm and below, is almost independent of local time. However, the different internal temperatures influence the temperature right up to the surface, especially on the nightside. This can influence the loss of $CO_2$ on the nightside, for example, as the diurnal thermal

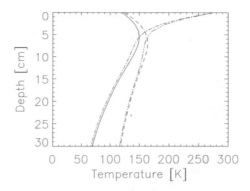

**Fig. 2.39** The effect of the assumed internal temperature at depth on the temperature structure in the near-surface layers. Solid line: Local midnight with 40 K internal temperature. Dash: Local midnight with 100 K internal temperature. Dot-dash: Local midday with 40 K internal temperature. Dot-dot-dot-dash: Local midday with 100 K internal temperature

wave penetrates. A constraint on the internal temperature comes from the MIRO instrument on Rosetta which measured the temperature of the south polar region when it was experiencing constant darkness (the constantly unilluminated pole resulting from the obliquity). The temperature measured was between 25 and 50 K although the exact depth at which the sub-millimetre and millimetre wave measurements were being made was somewhat uncertain (Choukroun et al. 2015). Nonetheless, this suggests that the lower temperature used in Fig. 2.39 would be more appropriate. In Sect. 3.9, we will see that observations of the ortho-to-para ratio of certain gas species also suggest low internal temperatures.

### 2.9.3.4    The Amorphous-Crystalline Transition of Water Ice

Ice exists in three distinct solid configurations at low pressures (Grundy and Schmitt 1998). Amorphous ice is a non-crystalline solid form of water ice and can be produced by rapid cooling of water to temperatures below about 136 K. If the cooling is fast enough, nucleation of crystals fails to occur and the amorphous phase is reached. On heating amorphous ice from lower temperatures to temperatures between 110 and 150 K, it transforms irreversibly to a cubic crystal structure. This transition temperature is considerably higher than is inferred for the interior of 67P from the work of Choukroun et al. (2015).

Another irreversible transformation takes place above 190 K, to the hexagonal crystal structure found at temperatures of 273 K (Jenniskens et al. 1998). Ice solidified from liquid water always crystallizes with the hexagonal structure, and it maintains that structure, even when cooled to cryogenic temperatures. There are many other stable forms of crystalline ice with varying densities (see Mastrapa et al. 2013).

Experiments have indicated that there are two main forms of amorphous ice but the formation and structure is highly dependent on deposition conditions. The high density phase (called $I_ah$) has a density of about 1.1 g cm$^{-3}$ and exists at temperatures <70 K. The low density phase (~0.94 g cm$^{-3}$) is called $I_al$ and exists between 70 and 120 K. An amorphous phase ($I_ar$) can exist above 120 K coexisting with a cubic crystalline phase.

In a series of experiments, Jenniskens and co-workers (Jenniskens and Blake 1994; Jenniskens et al. 1995) have demonstrated amorphous ice formation by gas phase deposition onto cold substrates with the amorphous phase dependent upon the deposition rate. The importance for comets is that cold dust particles in the proto-solar nebula should form a substrate for low temperature deposition of water molecules and these particles may then be incorporated into the cometary nucleus in this form. The low temperatures of nuclei in the Oort cloud and the Kuiper Belt would then be sufficient to maintain the ice in this state to the present day. The presence of amorphous ice on interstellar grains has been established since Jenniskens et al. (1995) and it is now assumed that amorphous ice is the dominant form of water in the universe, even though it does not occur naturally on Earth (Loerting et al. 2015).

The possible presence of amorphous ice in the nucleus has provoked considerable discussion because the transition from the amorphous phase to the crystalline phase is exothermic, producing $9 \times 10^4$ J kg$^{-1}$, and is therefore an internal heat source. The amorphous-crystalline transition has been invoked as an explanation for observed activity of comets at high heliocentric distance (see Tancredi et al. 1994; Gronkowski 2007; Hosek et al. 2013).

Theoretical work (e.g. Kouchi et al. 1994) shows that the crystallization rate (from the amorphous to crystalline phase) is strongly dependent upon temperature. At temperatures below about 70 K, the low density form of amorphous ice ($I_al$) can be preserved over timescales comparable to the age of the Solar System. However, the timescale for transition drops by roughly an order of magnitude every 5 K (see Mastrapa et al. (2013) for a detailed plot). Consequently, if amorphous ice, interior to the nucleus, is warmed to about 90–100 K, theory suggests the transition must occur producing additional heat.

The presence of amorphous ice can have significant implications for heat transport within the nucleus. Although there is uncertainty in this value at very low temperatures, Haruyama et al. (1993), following Kouchi et al. (1992), used

$$\kappa = 7.1 \ 10^{-8} \ T \tag{2.123}$$

for the thermal conductivity of amorphous water ice (in units of [W m$^{-1}$ K$^{-1}$]) which, when compared with Eq. (2.111), can be seen to be appreciably lower than that of crystalline ice.

In principle, the different forms of water ice can be determined through infrared spectroscopy (Kokaly et al. 2017). The absorption coefficient for water ice at 190 K

**Fig. 2.40** Near-infrared absorption coefficients of $H_2O$ ice (dashed line) and $CO_2$ ice (solid)

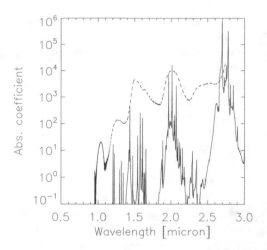

from the GhoSST[2] database is shown in Fig. 2.40 and compared to the $CO_2$ ice absorption coefficient which has considerably more fine structure in the near-IR.

The exact positions of the features are dependent upon the exact conditions in which the ice finds itself. Temperature dependence was shown by Grundy and Schmitt (1998) after laboratory studies had shown in the early 1980s that the infrared features display temperature effects in band shape and position. Ehrenfreund et al. (1996) showed that the infrared features of $H_2O$ ice when contained within matrices of other molecules were also modified and presented a "fingerprint" of the matrix. Mastrapa et al. (2008) also analysed the differences between amorphous and crystalline $H_2O$ ice spectra and identified weakening of bands and shifting of those bands to shorter wavelength in amorphous $H_2O$ ice spectra.

Following Grundy and Schmitt (1998), the strength of a 1.65 μm absorption feature in crystalline $H_2O$ ice is now used as a diagnostic for the presence of crystalline ice over the amorphous form. The presence of crystalline $H_2O$ on objects remote from the Sun (surface equilibrium temperatures $\lesssim 80$ K) is then assumed to indicate either past closer proximity to the Sun or evidence for internal heating followed by cryo-volcanism. For example, Jewitt and Luu (2004) examined spectra of the KBO (50000) Quaoar to conclude the latter. Terai et al. (2016) also detected crystalline ice on (136108) Haumea and (90482) Orcus but also on (42355) Typhon and 2008 $AP_{129}$, both of which are smaller than the minimum size for inducing cryo-volcanism (and therefore amorphous to crystalline phase transition) through thermal evolution resulting from the decay of long-lived isotopes.

Furthermore, activity was detected in the long period comet C/2017 K2 (PANSTARRS) at $>23$ AU by Jewitt et al. (2017), who argued that this could not be explained by the amorphous-crystalline transition. Nucleus temperatures at 23 AU are too low (60–70 K) either for water ice to sublimate or for amorphous ice to crystallize, requiring another source for the observed activity (Jewitt et al. 2017).

---

[2]https://www.sshade.eu/db/ghosst

Some previous work on irradiation of amorphous water ice has indicated that bombardment may provoke amorphous to crystalline transitions at low temperature (Hudson and Moore 1992; Moore and Hudson 1992). Conversely radiation may also have the opposite effect (Dartois et al. 2015). Consequently, it is hard to generalise the implications of low temperature ice phase transitions for comets on the basis of what is currently known.

### 2.9.3.5  Thermal Emission from Resolved Surfaces

The observations of the Halley Armada showed that the gas production rate was inconsistent with a uniform, freely subliming, pure water ice, surface (Keller et al. 1987). The surface area of the nucleus was around a factor of 10 greater than that required to match the observed gas production. Furthermore, the observations of thermal emission indicated that the surface temperature at the time of the Vega 2 encounter exceeded 350 K (Emerich et al. 1987)—almost double that expected for an illuminated freely subliming water ice surface (~200 K; Sect. 2.9.2). There are three physically different means of reconciling the observations.

– The nucleus may have active (mostly ice) and inactive (mostly dust) regions that are large in scale relative to the size of the nucleus (Fig. 2.41 left). In this case, contamination of the icy region with dark material would be necessary to reduce the albedo to observed values and increase the heat input to drive the sublimation but the contamination would only need to be a few percent.
– The nucleus may have active and inactive patches at scales that are small compared to the nucleus and possibly below the resolution of observing instruments (Fig. 2.41 centre).

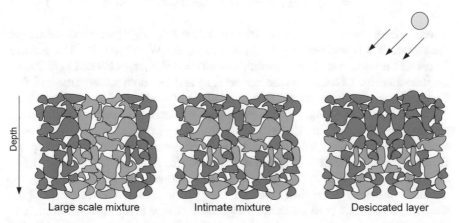

Large scale mixture          Intimate mixture          Desiccated layer

**Fig. 2.41**  Schematic diagram of three possible distributions of water ice (blue) and non-volatile dust (red) material to explain the reduction gas production relative to that from a pure water ice surface layer

– The surface layer may be uniform over the nucleus but the uppermost part might be desiccated with gas emission from below the surface at a depth where the heat available for sublimation is substantially below that available at the surface (Fig. 2.41 right).

A combination of these three end-members is also conceivable.

Because the reflectance of dust-laden ice can appear to be similar to ice-free surfaces at optical wavelengths, these three models may look similar in imaging observations. One might think that at thermal wavelengths, the temperature contrast between dusty and icy surfaces could help distinguish between these models. However, there are significant ambiguities.

We have seen the black-body emission from a surface at a temperature, $T$, in Eq. (2.9). When the integrated flux from a surface is normalised to 1, i.e.

$$\int_0^\infty F_N(\lambda)\, d\lambda = 1 \tag{2.124}$$

Then

$$F_N(\lambda) = \frac{h^4 c^4}{\pi^4 k^4 T^4} \frac{15}{\lambda^5 e^{hc/\lambda kT} - 1} \tag{2.125}$$

The surfaces in Fig. 2.41 have two components that will have different temperatures. At 1 AU the dark, dusty, non-volatile component when at the surface and illuminated may have a temperature exceeding 350 K, whereas the water ice surfaces will be at 200 K because of vigorous sublimation. One can then compute the emitted (thermal) flux from a two-component surface

$$F_N(\lambda) = F_{N1}(\lambda)\sigma T_1^4 A_1 + F_{N2}(\lambda)\sigma T_2^4 (1 - A_1) \tag{2.126}$$

where $A_1$ is the fractional area of the surface at $T_1$ and $(1-A_1)$ is the fractional area of the surface at a temperature $T_2$. $F_N(\lambda)$ is in units of [W m$^{-2}$ nm$^{-1}$]. The radiated fluxes for the two terms of this equation using $T_1 = 350$ K, $A_1 = 0.9$ and $T_2 = 200$ K, are shown in Fig. 2.42. One can see immediately that the contribution to the total flux from the cold surface is small. This is similar to the effect of surface roughness discussed in relation to Fig. 2.1. The brightness temperature of the total surface area can then be calculated. In the case shown, the integral over all wavelengths corresponds to 341.81 K—which should be a detectable difference compared to $T_1$. However, it is important to note that the thermal emissivity, $\varepsilon$, has been set to 1 in Eq. (2.126). This quantity is not well enough known and simply changing the assumed emissivity of the dusty material by a factor equal to the reciprocal of the fractional area would produce an almost identical change in the observed brightness temperature. It is this combination of the low fractional area of the subliming material and the lack of knowledge of $\varepsilon$ (even ignoring the further complication arising from the beaming parameter, $\eta_{th}$, discussed below) that limits the usefulness

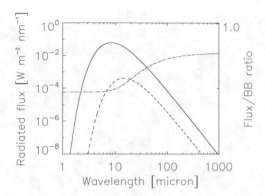

**Fig. 2.42** Thermal fluxes from surfaces at 350 K (solid line) and 200 K (dashed line) with 90% contribution to the flux from the warm surface. The dot-dashed line shows the ratio of the total flux to that from a black-body at the warm temperature (axis to the right). It takes a value of 0.9 at 1 micron wavelength. At short wavelengths, the cold surface hardly contributes but as the wavelength increases the colder temperature component contributes more and more to the total flux from the surface

of low resolution brightness temperature measurements in constraining the actual thermodynamic temperature of the surface when more than one temperature component is present. It is also evident that small effective active fractions (eafs) of subliming surface water ice have little, unambiguous, effect on the radiated flux from a surface in the 1–10 micron wavelength range. The absence of a thermal infrared mapping spectrometer on Rosetta to measure surface temperatures at high spatial resolution precisely has compromised more detailed study in the case of 67P.

### 2.9.3.6  Surface Roughness, Infrared Beaming, Self-Shadowing and Self-Heating

The bilobate structure of 67P and the evidence that other primitive objects have a bilobate appearance has led to an increased emphasis on defining the exact heat budget on surface facets of these objects. The effects of self-shadowing on larger scales was rarely considered in cometary models prior to the imaging of 67P despite both 19P/Borrelly and 103P/Hartley 2 showing evidence for concave surfaces between two larger lobes. The major influence of shadowing on the Hapi region of 67P (Fig. 2.43) has led to this now being an important element in any surface heat budget calculation although the implementation for irregularly-shaped bodies remains computer intensive if the number of facets in any 3D shape model is large.

Small scale roughness adds further complexity to the problem. The absence of an atmosphere and the inferred low thermal conductivity implies that rough surfaces will show large temperature differences depending upon the orientation of surface facets to the Sun including shadowing. This problem was addressed rigorously by Lagerros (1998) although there were several studies prior to this.

**Fig. 2.43** 3D shape model illustrating the self-shadowing of the nucleus of 67P. The image to the right shows the nucleus with the Sun to the left. The large lobe casts a shadow over the neck between the two lobes (the Hapi region) and influences the local surface energy budget (Liao 2017)

Thermal emission is "beamed" into the sunward direction so that, at low phase angles, higher flux levels are observed at an elevated apparent colour temperature. This effect is referred to as thermal-infrared beaming and is accounted for by an additional parameter, $\eta_{th}$, the infrared beaming parameter (Lebofsky et al. 1986), entering in the radiative loss term in the energy balance equation leading to an equation

$$F_{th} = \frac{2}{3} \, \eta_{th} \varepsilon \, S_{\odot} (1 - A_H) \frac{r_N^2}{\Delta^2 r_h^2} \qquad (2.127)$$

for the thermal flux at zero phase angle geometry. This constitutes the refined standard thermal model (STM) for asteroids. Lagerros (1998) attempted to provide a physical explanation of this quantity by studying regular-shaped and stochastic surface roughness. The implementation of this type of approach on local scales for resolved observations would be challenging as there would be a need to define the small scale roughness within the measurement footprint.

This is linked to the need to account for re-absorption of thermal emission (self-heating). This has been shown to be of importance for objects which are highly irregularly shaped, such as 67P, and can lead to shadowed areas receiving sufficient heat to modify sublimation profiles (Keller et al. 2015). Because of the low effective active fraction and low thermal conductivity, the thermal re-radiation from a cometary surface is close to the solar insolation. Hence, a shadowed cliff orthogonal to an infinitely large, fully illuminated, surface will receive close to half the solar flux in radiated power. To put this in perspective, in this idealised case, if the illuminated surface is at 300 K, the unilluminated cliff will be at 252 K and thus well above the free sublimation temperature of water ice. On the other hand, realistic cases have shown this effect to be small on global scales (Marschall 2017; Marschall et al. 2017). Nonetheless, locally there may still be need to take such effects into account for specific geometries (Höfner et al. 2017).

## 2.10 Surface Appearance and Cometary "Geology"

Arguably, it is in the field of cometary geomorphology that the Rosetta mission has made the biggest contribution to cometary research. The rendezvous and near-nucleus mapping allowed ultra-high ($<1$ m px$^{-1}$) imaging of the surface. An example showing a range of surface textures is shown in Fig. 2.44. The diversity

**Fig. 2.44** Rosetta/OSIRIS image of 67P showing the Khonsu region in the foreground, the smooth terrains of Anubis at centre right and the rougher surface of Anuket on the small lobe in the background. The scalebar is 200 m (Image number: N20160210T122332750ID30F22)

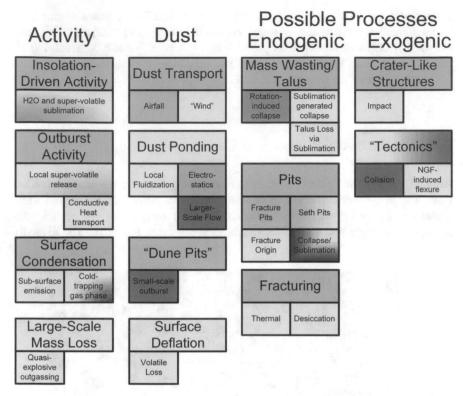

**Fig. 2.45** Processes that are probably or possibly occurring on the nucleus of 67P. The upper box indicates the process while the small boxes below suggest mechanisms. The colours (from green to red) indicate whether the suggested mechanism has been modelled plausibly (green), reasonably understood (yellow) or still to be properly quantified (shades of red). There are some grounds for argument on these assignments. They have also been loosely grouped

of processes that are probably or possibly occurring on the nuclei of 67P to produce these types of surface appearance is summarized in Fig. 2.45 and we will discuss these processes at length. However, the contributions from previous fly-by missions should not be underestimated.

The observations of 1P/Halley were of sufficient quality to allow identification and crude interpretation of surface features on the nucleus. Topographic irregularity down to 500 m scales could be seen over roughly ¼ of the nucleus (Fig. 2.46). This effectively marked the start of the science of "cometary geomorphology" and some of the observed structures are probably related to structures seen subsequently on other nuclei at higher resolution.

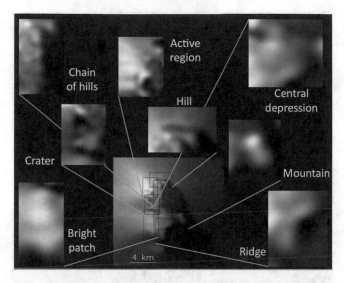

**Fig. 2.46** Features on the nucleus of 1P/Halley derived from Halley Multicolour Camera data (following Keller et al. 1994)

## 2.10.1   Regional Classification

The range of topography on 1P/Halley was intriguing but said relatively little about the local surface texture. As resolution on cometary nuclei has improved, it has become apparent that surface texture varies such that globally surfaces are inhomogeneous in appearance. The observation of 19P/Borrelly (Fig. 2.47) hinted at this and the nucleus was divided into regions with different surface textures. The descriptions were rather crude and the interpretation of some of the features as "mesas" can be challenged (cf. Britt et al. 2004; Thomas et al. 2013a) but these data were important as they indicated that areas of different texture or structure were present on individual nuclei that could, in turn, be interpreted in morphological terms. The Deep Impact observations at 9P/Tempel 1 (Fig. 2.48) and subsequently at 103P/Hartley 2 (Fig. 2.49) provided considerably more evidence for surface textural inhomogeneity (Thomas et al. 2013a, b). In the case of the latter, a smooth neck-region between two rougher lobes could be observed. The appearance of smooth terrain at topographic/gravitational lows seems to be a recurring theme on nuclei and has been seen on 103P/Hartley 2, 67P, and, arguably, 1P/Halley.

The only resolved nucleus that appears to be more uniform in texture is that of 81P/Wild 2 (Fig. 2.2). Here the surface is characterized by quasi-circular depressions up to nearly 2 km across. These depressions covered the visible surface although two different morphologies (*pit halo* and *flat floor*) could be identified (Brownlee et al. 2004).

In the case of 67P, the textural inhomogeneity is extreme although one needs to be careful in that the more than 10-fold improvement in resolution has, almost

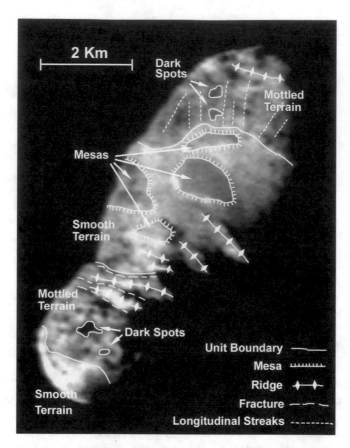

**Fig. 2.47** Interpretation of the MICAS observations of 19P/Borrelly acquired from onboard NASA's Deep Space 1. (From Keller et al. (2004) in Comets II by Michel C. Festou, H. Uwe Keller, and Harold A. Weaver Jr. © 2004 The Arizona Board of Regents. Reprinted by permission of the University of Arizona Press)

inevitably, led to a more detailed view of local surface textures. This high resolution has allowed an attempt to classify surface textures and the relationships between them on the nucleus.

In a regional classification the surface is divided into distinctive coherent regions based on clear and major morphological and/or topographical boundaries. This is essential in the case of small irregularly-shaped bodies as these objects generally lack the well-established geographical reference system needed both to describe the orientation of observations and to map various features (El-Maarry et al. 2015a). For example, the neck region of 67P (Fig. 2.2) is a particularly challenging cartographic problem because of the ambiguity arising if simple longitude/latitude reference systems are used.

A regional classification should be kept separate from geological mapping where units that display similar structural and stratigraphic position as well as a common

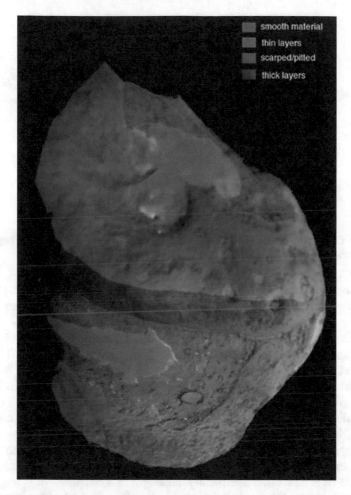

**Fig. 2.48** This classification shows the colour-coded distribution of different terrain types on 9P/Tempel 1. The green regions are very rough and pitted. The brown sections appear to be thin layers parallel to the surface. The olive green areas are smooth, with the upper one including the ice patches. Finally, the blue ribbon marks a ridge of exposed layers (Credit: NASA/UM/Cornell/Peter Thomas)

possible origin are grouped into one unit irrespective of their spatial distribution. In a geological mapping, structural elements such as major faults and fractures are required to be present, surface deposits should be distinguished and morphological features such as scarps, dunes, and layers need to be identified (e.g. Fig. 2.47).

Figure 2.48 shows an example of how different types of surfaces on 9P/Tempel 1 have been identified to form a regional classification. The type of nomenclature used is similar to that used for 19P/Borrelly. At 103P/Hartley 2, the observations were also sufficient to show smooth terrain at the neck but with rougher material on both lobes (Fig. 2.49). Although the shape is visually rather similar to 19P/Borrelly, the

**Fig. 2.49** Comet 103P/Hartley 2 observed during the Deep Impact Extended Investigation (EPOXI) mission (Credit: NASA/JPL-Caltech/UMD/Brown University)

distribution of textures on the nucleus surface appears to be very different. However, the inferior resolution of the 19P/Borrelly data may be misleading (Thomas et al. 2013b).

By grouping similar terrain and looking at distinct topographic discontinuities, a total of 25 regions have been classified on 67P (El-Maarry et al. 2015a, 2016, 2017a; Thomas et al. 2018). The regions are shown in Fig. 2.50 where they have been mapped back onto a 3D shape model of the nucleus. The number of distinct regions on the surface is large and, as a result, the individual regions were given names to allow easier reference to a particular area. The names and their characteristics are given in Fig. 2.50 and Table 2.5.

By mapping the regional boundaries back onto the 3D shape of the nucleus, the surface area of each region can be estimated. This is not an entirely straightforward calculation (it is rather like determining the length of a coastline—it is longer if the resolution is higher) but with a nucleus model comprising 12 million facets (Preusker et al. 2017), the values are shown in Table 2.5. These numbers are useful for comparative purposes. A roughness index is also included in the table to indicate the comparative surface roughness of the regions (at 2–3 m resolution). A low index indicates the dominance of smoother terrains (e.g. Hapi) while a high index (e.g. Anhur) indicates significant metre-scale roughness (Thomas et al. 2018).

The most recent versions of the 3D reconstruction of the nucleus of 67P has allowed further sub-division in 71 distinct (either morphological or topographical) sub-regions. An example is shown in Fig. 2.51 where the Aker region (Ar) has been split into four sub-regions (region c is not evident from this viewing geometry). On the left is an image from the Rosetta camera, OSIRIS, and on the right is the sub-region definition. Aker has three distinct faces that are topographically at different orientations. At the bottom, the surface texture changes abruptly from the smoother terrain of Aker b to the much rougher terrain of Anhur. The topographical and morphological boundary between Bastet (Bs c) and the Hapi region (which

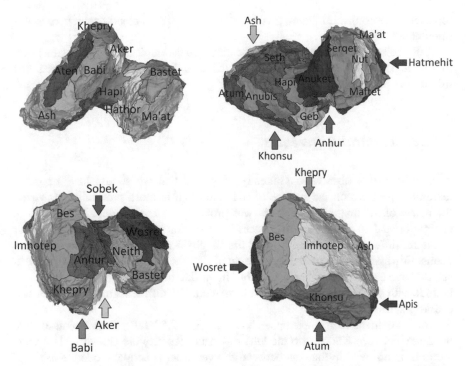

**Fig. 2.50** Montage of four orientations of the nucleus of 67P showing the region definitions (Reprinted from Thomas et al. (2018), with permission from Elsevier)

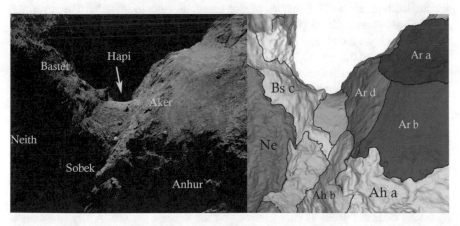

**Fig. 2.51** Left: OSIRIS image (N20141122T105253805ID10F22). Right: The sub-region definition showing in particular the face of Aker leading down to the Hapi region in the neck (Reprinted from Thomas et al. (2018), with permission from Elsevier)

represents the northward facing part of the neck of 67P) is clearly seen here in this orientation.

With the help of the regional classification, we can look at the different surface morphologies, the relationships between them, and the individual processes that might have produced them.

## 2.10.2 Textural Differences

The 9P/Tempel 1 observations already suggested that we should have expected textural differences on the surface of 67P and often in close proximity. However, the extent of the textural differences was probably underestimated. It is of course straightforward to argue that differences in insolation will produce textural and/or structural differences even on a body that is physically and chemically homogeneous. However, the regional classification of 67P shows that adjacent surfaces, that should have experienced identical energy input (i.e. almost identical longitude and latitude with similar orientation), often have markedly different surface textures and characteristics.

This is evident in the example shown in Fig. 2.52 left. Here, points on the boundary between Khonsu (to the left) and Atum (centre) are shown with arrows. Atum is a narrow strip that sits between the very rocky, active, eroded, and rough Khonsu region to the left and the smooth surface of Anubis to the top right. The surface of Atum also appears "rocky" but is relatively smooth. It is important to be careful in the use of terms such as rocky. The low density of the nucleus cannot be reconciled with rock-like densities of, for example, silicate material. However, the surface cannot be solid or porous ice alone, even with a high degree of structural strength, because the albedo is low and significant erosion (sublimation) of this area was not observed during the Rosetta mission. To avoid this issue, it has usually been preferred to refer to the surface as appearing "consolidated" which is a more neutral word with respect to composition.

Although there is change in elevation between the three areas (the surface of the section of Khonsu shown in Fig. 2.52 left is topographically lower), the surfaces have roughly the same orientation and are close to the same latitude. Yet their surface textures could not be more different. Similarly, in Fig. 2.52 right, we see the surface of Apis which forms an almost planar "face" in the northern hemisphere next to the large Ash region. Despite the whole surface being almost identically illuminated over both the comet's rotation and its orbital motion, the surface has distinctly different areas as characterized by the surface texture. A key question is therefore whether this is a consequence of origin and formation or whether local effects in the past have strongly influenced surface evolution.

Homogeneity is an attractive concept allowing us to make simplifying assumptions about cometary formation and evolution. However, it remains to be explained how these many surface textures can all be produced from one initial physical and chemical composition. Consequently, the development of surface evolution models

**Fig. 2.52** Left: Atum sub-region c is in the centre of this image. To the top right are the smooth plains of Anubis. To the left is the rocky surface of Khonsu large numbers of lineaments. (Image number: N20140919T045716392ID30F22). The two arrows mark points on the boundary between Khonsu and Atum. Right: The Apis "face" is seen from the centre of the image towards the lower right. (Image number: N20160130114149685ID10F22). Note the change in texture from fractures and lineaments below (indicated by the green arrow) through a more consolidated material in the centre to a coating type material (green arrow top) that appears to be being eroded. A point on the Ash-Apis boundary is marked by the yellow arrow

has become rather important with particular emphasis on studying changes arising from subtle differences in the initial conditions. This research direction is however in its infancy.

### 2.10.3   Impact Cratering

Atmosphere-less bodies in the Solar System generally show evidence of impact cratering and the basic mechanics is described in, for example, Greeley (2013). Even small bodies such as asteroids have surfaces that are strongly influenced by impact processes. The Near Earth Asteroid Rendezvous (NEAR) spacecraft was renamed NEAR Shoemaker after its 1996 launch. The spacecraft completed a rendezvous with the asteroid (433) Eros and orbited it several times completing a full mapping before touching down on the surface of the asteroid on 12 February 2001 (Veverka et al. 2001). Close-up observations of (433) Eros before landing revealed a high surface density of impact craters (Fig. 2.53).

Surface modification to erase or bury ancient impact craters has been seen on inactive small bodies. An example is the asteroid (21) Lutetia (Fig. 2.54) which was

**Fig. 2.53** This mosaic of two NEAR Shoemaker images, taken 20 August 2000, from an altitude of 49 km shows an oblique view of Eros' surface. The shadows inside the craters to the left result from the low solar elongation angle. The dense population of impact craters indicates this part of Eros has undergone few other types of surface modification (Mosaic of images 0142203174 and 0142203236) (Credit: NASA)

**Fig. 2.54** Rosetta/OSIRIS observation of the asteroid (21) Lutetia obtained on 10 July 2010. Note the heavily cratered surfaces in the Noricum and Achaia regions and the comparative absence of craters in the North Polar Crater Cluster (NPCC) area within the Baetica region

the subject of a fly-by by the Rosetta spacecraft 4 years before the spacecraft encountered 67P (Thomas et al. 2012). Even at 60 m px$^{-1}$, the surface of the asteroid is covered with the evidence of impacts in most areas. The only exception is the lower surface density of impact structures in a region referred to as Baetica (and specifically the North Polar Crater Cluster or NPCC within this region) which is to the left in Fig. 2.54. The region is assumed to have been impacted relatively recently (within the past 1 billion years) and hence its surface is much younger.

As discussed in Sect. 2.4, crude energy balance calculations show that the surface of a Jupiter family comet can lose between 1 and 10 m of material locally per apparition although the average erosion over the surface is required to be a factor of 10 less than this to match the observed mass loss. The more detailed calculation in Fig. 2.34 also indicates values of this order for 67P while Fig. 2.13 shows that latitude-correlated variations in the erosion rates across the nucleus are to be expected and this should be the case for all nuclei.

Larger impactors should result in the complete destruction of the nucleus although gravitational re-accumulation of disrupted material has been shown to be potentially important (Fig. 2.24) if the impact energy is not too high. On the terrestrial planets (e.g. Mercury) craters smaller than about 10 km in diameter are bowl-shaped and referred to as "simple" (see Pike 1988). The depth/diameter ratios are, in these cases, between 0.15 and 0.2. Hence, if there was initially a 500 m diameter crater on a comet, one could expect it to be eroded and lost on timescales of perhaps 10–200 apparitions. For such a large crater in relation to the nucleus dimensions and for the comet's lifetime inside the orbit of Jupiter, this is an extremely short time. The almost complete absence of unambiguously identifiable impact craters on the surfaces of cometary nuclei is therefore unsurprising. Nonetheless circular structures that resemble conventional impact craters have been found on 9P/Tempel 1 (Thomas et al. 2007) and 67P (Thomas et al. 2015a). Flat-bottomed depressions were seen in abundance on 81P/Wild 2 but these are probably not impact structures (at least not conventional ones). Potentially similar structures were also seen on 67P (Fig. 2.94).

The simulation of impacts in solid surfaces is already at a somewhat advanced level. Codes such as iSALE (impact-Simplified Arbitrary Lagrangian Eulerian) are now available (Amsden et al. 1980). iSALE is a multi-material, multi-rheology shock physics code (sometimes called a hydrocode). It is now well-established and has been used in studies of the formation of large impact craters on the Earth. This has been extended to planetary physics and used to investigate the influence of target property variations on crater formation, the influence of a water layer on crater formation, as well as investigating the mobility of large rock avalanches. It has been expanded over the past 30 years and a 3D version of the code is freely available[3] for non-commercial use (Elbeshausen et al. 2009).

There are however some issues with applying hydrocodes to cometary nuclei. The principal problem is that the material properties of the target are very poorly

---

[3]https://isale-code.github.io

understood. As discussed earlier, the tensile strength of whole nuclei may be very low but the compressive strength of the surface layers is potentially much higher and the compressive strength of the deeper interior is essentially unknown. The importance of material strength can be illustrated by looking at crater scaling laws.

The volume of an impact crater, $V_{imp}$, will depend upon impactor radius, velocity, and density as well as the mass density of the target, the surface gravity and the strength of the impacted material. Scaling laws have been developed to establish the relationship between these quantities. Two regimes, the strength regime and the gravity regime, can be distinguished. In the strength regime, gravity can be ignored and target strength is dominant. The proportionality is then given by

$$V_{imp} \propto \frac{m_{imp}}{\rho_N} \left( \frac{\rho_N U^2}{Y} \right)^{3\mu_{imp}/2} \left( \frac{\rho_N}{\delta_{imp}} \right)^{1-3\nu_{imp}} \tag{2.128}$$

where $m_{imp}$ and $U$ are the impactor mass and velocity, $Y$ is a measure of the strength of the impacted material, $\delta_{imp}$ and $\rho_N$ are the mass densities of the impactor and (nucleus) target respectively, and $\nu_{imp}$ and $\mu_{imp}$ are coefficients to be determined.

In the gravity regime, the event is large and gravity's influence dominates. The strength of the material is then eliminated and the scaling law becomes

$$V_{imp} \propto \frac{m_{imp}}{\rho_N} \left( \frac{g a_{imp}}{U^2} \right)^{-3\mu_{imp}/(2+\mu_{imp})} \left( \frac{\rho_N}{\delta_{imp}} \right)^{(2+\mu_{imp}-6\nu_{imp})/(2+\mu_{imp})} \tag{2.129}$$

where $a_{imp}$ is the impactor radius and $g$ is the gravitational acceleration (Holsapple 1993, 1994). With gravity being extremely low on small bodies, the strength regime is the most likely to be applicable and $V_{imp}$ increases proportionally to $Y^{-3\mu_{imp}/2}$. Use of the equations requires knowledge of several parameters. Examples for materials of possible interest are shown in Table 2.6 where one can see the material strength parameter, $Y$, varies by two orders of magnitude for different materials. With similar values of $\mu_{imp}$ for the two given materials, the crater volume varies as $Y^{-33/40}$ and thus almost inversely to the material strength.

The high porosity inferred for comets is often (incorrectly) associated with low material strength. Man-made, light, porous structures with significant strength are now well known but strong, porous structures can also occur naturally. Examples come from the gas industry where porosities greater than 50% can be found for

Table 2.6 Crater scaling law constants for materials of possible interest on comets (from https://www.lpi.usra.edu/lunar/tools/lunarcratercalc/theory.pdf, retrieved 8 May 2019)

| Constant | Soft rock | Cold ice |
|---|---|---|
| $\mu_{imp}$ | 0.55 | 0.55 |
| $\nu_{imp}$ | 0.33 | 0.33 |
| $Y$ [MPa] | 1 | 0.014 |
| $\rho_N$ [kg m$^{-3}$] | 2100 | 930 |

materials. These can have non-negligible compressive strength with 35% porosity sandstones have a compressive strength in excess of 14 MPa (Palchik 1999). Nonetheless, investigations of impacts into low compressive strength material have also been performed with particular emphasis on ejectaless craters (Housen and Holsapple 2012).

The term, ejecta blanket, describes the material from the crater launched by the impact in such a way that it is deposited outside the periphery of the crater depression. For cometary-sized bodies, the weak gravity field is insufficient to prevent the ejecta escaping the body so that the mass retained by the object in any ejecta blanket is small. However, Housen and Holsapple (2012) showed that impacts into highly porous materials primarily drive material downward and outward into the floor of the expanding transient crater causing permanent compaction of pore spaces. As a percentage of crater volume, much less material is ejected upward than for impacts into materials of lower porosity. Penetration into the interior is therefore more significant. If the material is highly compressible, impactors will penetrate, boring an impactor diameter-sized hole into the target. This effect has also been illustrated by de Niem et al. (2018).

In the event that the target has a harder layer (e.g. a "crust") above a lower strength material (a possible model for comets), the harder layer reduces the amount of ejected material (e.g. Jutzi et al. 2013) and the surface manifestation of the impact is reduced. Such a scenario would be consistent with Deep Impact observations as discussed in Sect. 2.8.2.

Impacts also generate heat and the presence of volatiles in comets including what are sometimes referred to as super-volatiles (e.g. CO) might lead to interesting consequences. However, the temperature rise resulting from an impact, while non-negligible, is not very large. Even if all the kinetic energy of the impactor (excluding any other losses) is made available to sublime water ice, a 2 km s$^{-1}$ impactor only sublimes roughly a mass of water equivalent to the mass of the original impactor. The more significant influence on the comet of the impact is the resulting compaction and local damage. Figure 2.55 shows the density change and temperature rise seen in an impact simulation (2 km s$^{-1}$ impact velocity at 45° with respect to the surface normal). The images show the values 4s after first contact.

The density increase within the crater is around a factor of 4 while the temperature rise is only of the order of 10 K locally. In the case of water ice, restructuring of the material might lead to sintering with time and a harder, icy, layer (or layers) within the crater.

The best studied examples of possible impact craters on 67P can be seen in Fig. 2.56. The structure on the left is in the Ash region and is a circular depression around 350 m in diameter with no obvious ejecta blanket. The rim to the lower left is layered with local slopes approaching 90°, indicative of some structural strength. The structure shown in Fig. 2.56 (right) is in the Imhotep region. The central depression is roughly circular and rimmed with the southern rim being more pronounced. Layering is evident and gives the impression that the surroundings have been eroded leaving a structure similar to a pedestal crater.

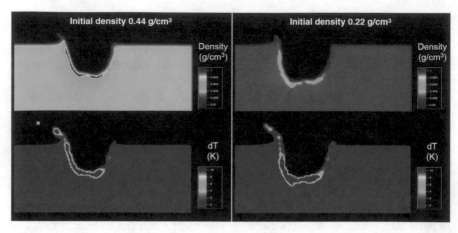

**Fig. 2.55** Density and temperature changes seen in smooth particle hydrodynamics (SPH) simulations of impacts into two targets of different density. The temperature rise in the base of the crater is only locally greater than 10 K (Courtesy of Martin Jutzi)

**Fig. 2.56** Three examples of possible impact structures on 67P. Left: Large structure in the Ash region. Arrow A points to what appears to be a degraded rim with vertical fractures. Arrow B points at talus possible resulting from collapse of the internal rim (Image number: W20160903T005801763ID10F12) Centre: Small but circular structure in Ma'at probably covered with dust deposits. (Image number: N20160902T232834745ID10F22). Right: An irregular circular feature in Imhotep resembling a pedestal crater. Arrow C points at the rim. Layers are also evident (D and E) (Image number: N20151212T205138158ID10F22)

It is tempting to interpret these structures as indicating impact and compression of porous ice-bearing material that has subsequently sintered or re-frozen to produce more dense and stronger near-surface layers. The compaction also increases the thermal conductivity. Hence, the sublimed mass from a compacted crater is lower than for the surroundings and production of pedestal structures can arise. However, the evidence supporting this concept is solely morphological.

The third crater in Fig. 2.56 (centre) is a small circular feature in Ma'at just 35 m in diameter and appears to be covered by dust deposits. The observed depth/diameter ratio has been used to place an upper limit on the depth of the dust deposit blanketing the crater and will be discussed later. Some transient craters resulting from airfall were also seen at 67P (Fig. 2.82).

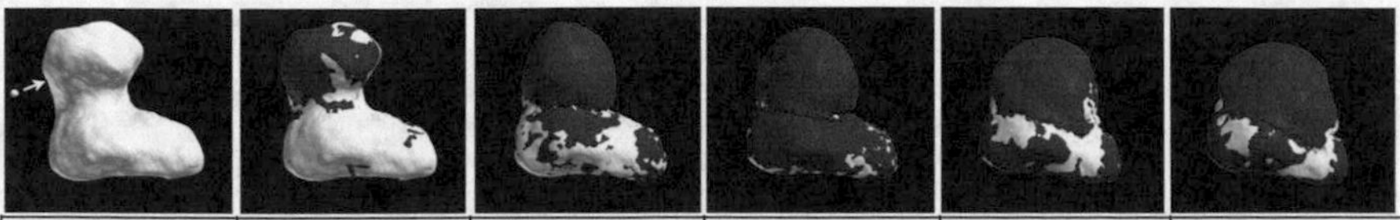

| v (m/s) | 10 | 30 | 50 | 80 | 100 |
|---------|-----|------|------|------|------|
| Q (J/Kg) | 0.01 | 0.08 | 0.23 | 0.59 | 0.92 |

**Fig. 2.57** The resulting shape of the nucleus of 67P after a collision with a 100 m sized impactor travelling at different velocities with respect to the nucleus (Credit: Jutzi et al. 2017, A&A, reproduced with permission © ESO)

This shows that although impact cratering dominates the surfaces of inactive objects such as asteroids, its effect on cometary surfaces is more subtle. Impact cratering might be important locally in modifying the density and structure at or close to the impact site but their surface manifestations are erased by sublimation on short timescales.

As noted above, impacts may also result in complete disruption and reaccumulation (Fig. 2.24). This was investigated further by Jutzi et al. (2017) who used their SPH code to study the collision of a 100 m impactor at varying velocities with a 67P-shaped nucleus. The result of this calculation is shown in Fig. 2.57. The final shape of the nucleus is shown as a function of impact velocity with the energy imparted given in [J kg$^{-1}$] below. The initial condition is to the left. It can be seen that disruption of the shape occurs for values of the input energy above about 0.2 J kg$^{-1}$ where this value depends upon the material strength ($Y$ in Eq. 2.128) which was assumed to be of the order of 10 Pa.

This suggests that while the bulk shape of the nucleus is influenced by formation and subsequent collisions, other, more rapid, processes dominate the intermediate scale morphologies on much shorter timescales. It should also be noted that impact may be a source of "activity" on less volatile objects (Table 6.1).

The change in brightness of an unresolved body (e.g. seen from the ground) resulting from an impact was estimated by Jewitt (2012) based upon scaling laws for the ejected mass and its speed (Housen and Holsapple 2011) and the size distribution of the resulting ejecta. The uncertainties are however substantial because of the absence of knowledge of the impactor properties (principally the impact velocity) and the size distribution of ejecta particles in hyper-velocity impacts.

## 2.10.4   Depressions, Pits, and Other Quasi-Circular Structures

The imaging observations of 81P/Wild 2 from the Stardust spacecraft (Fig. 2.2) revealed a number of flat-floored quasi-circular depressions (Brownlee et al. 2004).

**Fig. 2.58** Circular steep walled depressions in the Seth region of 67P (Image number: N20150305T003807453ID10F22). Examples of steep walls are indicated by the arrows. The flat area to the lower left is part of the Hapi region. Steep cliffs form the boundary between Seth and Hapi in this area

Similar morphologies were seen on 67P specifically in the Seth region (Fig. 2.58) and hence this is probably not unique to 81P/Wild 2. The formation mechanism of these depressions is unknown but sublimation remains a possibility and there is some morphological similarity to the sublimation depressions known as "Swiss cheese" at southern latitudes on Mars although the reflectance of the comet surface is far lower. The generation of a depression through a sublimation process is of course a trivial proposal but the quasi-circular appearance of the depressions is less so. Furthermore, some of the depressions are remarkably deep. The most active depression on 67P was pit-like with a diameter of 220 m and depth of 185 m (Vincent et al. 2015). Three effects may be of importance.

On an icy comet surface, sublimation maintains surface temperatures close to the free sublimation temperature. But once the ice is depleted, the equilibrium temperature of the surface can be much higher when the comet is in the inner Solar System. This increases the energy input to the surroundings via lateral conduction thereby speeding up the sublimation around the initially depleted point leading to a circularly expanding heat wave. This type of behaviour has been observed in the laboratory. But one must caution that the length scales here are appreciably different and any significant lateral inhomogeneity will break the symmetry.

Secondly, small topographic depressions can expand and deepen by insolation effects. Once a small depression is initiated, the sides sublime less quickly than the floor until self-shadowing of the structure becomes significant. This has been illustrated by Thomas et al. (2008) for a comet-like case. This illustration was, however, in an ideal mathematical model but again it is not obvious that this would scale in a real world case unless the material was remarkably homogeneous. Furthermore, the high depth-diameter ratio observed at 67P limits the illumination of the pit floor and is therefore a natural throttling mechanism and challenging to the concept (although Mousis et al. (2015) have argued otherwise).

Third, the compressive strength of cometary material might be very low—at least locally. If the material really does have a strength in the range of 10s of Pascal, then the overburden pressure on sub-surface material might be sufficient to induce collapse or promote collapse following a trigger. Using the density and gravitational acceleration of 67P as an example, 20 Pa pressure would be generated at a depth of ~230 m and comparable to the tensile strength (Attree et al. 2018). The deepest pits are of this order. However, the difference between compressive strength and tensile strength is of particular importance here and, as we noted above, the compressive strength is extremely ill-defined. It is conceivable that the Seth region is highly porous with low compressive strength and could not support the overburden. Elsewhere it is either stronger or it has already collapsed to a level at which the internal strength can support it.

Mechanisms for initiating collapse were explored by Vincent et al. (2015) including the possibility of having large scale voids in the interior. Three concepts for producing voids were proposed namely

- Voids might be primordial inherited from formation,
- Voids may result from evolutionary processes through direct sublimation of super volatiles,
- Voids may come from sublimation at depth triggered by a secondary source of energy such as the amorphous to crystalline transition.

However, the sounding experiment on Rosetta found no evidence for voids larger than metre-sized near the Philae landing site (Hérique et al. 2019) and the Radio Science Investigation (RSI) has found no evidence for large-scale inhomogeneities in the interior (Pätzold et al. 2016). This is in direct contradiction to the morphological evidence that shows these pits are concentrated in specific regions.

Vincent et al. also speculated that there may be an age-depth relationship with deep pits being younger. The unpitted surface gradually erodes with time because of normal sublimation processes to reach the pit-bottom level when the structure is fully evolved. This would suggest pits in Ma'at and parts of Seth as being young, the circular pit-like structures in Seth and Ash as being older and the circular structures in regions such as Maftet and parts of Hatmehit (Fig. 2.59) being very old. The pits seen on other comets would fit in this chronology.

The material under the floor of the depressions has been shown to be ice-rich (Pajola et al. 2017) and characterized by an albedo >0.4. An outburst event on 67P observed by the Rosetta spacecraft navigation camera shortly before perihelion was subsequently shown to be related to the loss of material from the edge of a cliff in the Seth region (Fig. 2.60). The cliff seems to have arisen from the removal of part of one side of a circular depression roughly 650 m in diameter known as "Aswan". The floor was a candidate landing site for the Philae lander in preliminary discussions. The structure gave the appearance of being cut in half creating a steep cliff around 150 m high from the base of the pit down to an intermediate level before descending

**Fig. 2.59** Pit-like structures in the Maftet and Hatmehit regions are shallower but the rims are also close to vertical and therefore similar in basic form to those in Seth and Ash. It is conceivable that these are older, more eroded, structures (Image number: N20160110T153451390ID10F22)

further to the Hapi region in the neck. Prior to the outburst event, a fracture was clearly visible in close-up images (Fig. 2.60 upper left). After the event, the fracture was no longer present and what appeared to be newly exposed material on the cliff face was markedly brighter and inferred to be more ice-rich. The brighter surface gradually faded in brightness with time over the next months (Fig. 2.60 bottom). The depth of the dust deposit on the pit floor was thus seen to be thin but structures in deposited dust close to the cliff edge show that it was not negligible.

This observation is important in two ways. Firstly, it shows that the material below the depression's floor is probably ice-rich. Secondly, it is apparent that the timescale for the surface to return to its original appearance is actually very long.

The eroded depth through sublimation can be calculated from

$$s_d = \frac{1}{\rho} \int_{t=t_1}^{t=t_2} \frac{dm(t)}{dt}\, dt \qquad (2.130)$$

where $dm(t)/dt$ is the mass loss per unit area per unit time and varies with time as the comet rotates and the heliocentric distance changes. For water ice, values approaching of ~1 mm/h are conceivable. Hence, the low rate of change in the

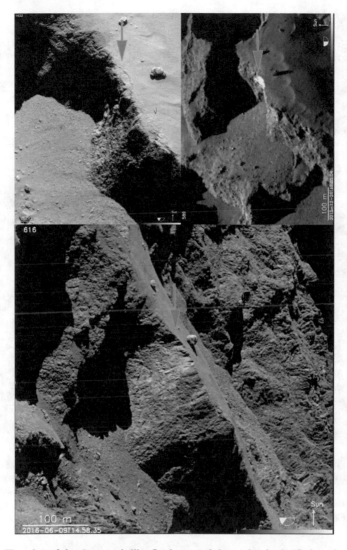

**Fig. 2.60** The edge of the Aswan pit-like flat-bottomed depression in the Seth region. Top left: Image from 21 September 2014 showing the crack close to the edge of the cliff that drops down to Hapi. Top right: Image from 26 December 2015 showing the cliff shortly after the collapse of the cleaving material. Bottom: Side on image from 9 September 2016 showing that the fresh material that originally appeared bright (top right) had faded in brightness with time. Note that the Sun direction for the top two images is very different (Image numbers: N20140921T023400377 ID10F22, N20151226T17050437ID10F22, N20160609T145835754ID10F22)

brightness either indicates that mass loss was much slower in this region or that the build-up of a lag on the surface was very slow. The latter explanation is probably preferable. The Aswan fracture exposed material on a steep cliff so that non-volatile loose material would fall to the base of the cliff rather than forming a lag deposit

**Fig. 2.61** Pit in the Seth region adjacent to the Aswan feature. Top left: Context image identifying the pit. Top right: Stretched version of the image showing activity in the form of dust jets emanating from the pit (e.g. position A) (W20141020T081550752ID10F18). Bottom: A high resolution image of the structure of the pit wall (N20140921T010901354ID10F22). Fractures can be seen and there is evidence of polygonal structure (position B) and what appear to be nodular-like structures (position C)

directly on the cliff. Consequently, the build-up of an ice-poor layer took longer than had a flat surface been exposed.

The pit adjacent to Aswan was well observed on several occasions and jet activity was seen emanating from its interior—probably from its sides (Fig. 2.61 top right). The internal surfaces of the pits could be seen with oblique imaging and have a mottled appearance (referred to originally as "goose-bumps") with a typical scale of the order of 3 m (Fig. 2.61 bottom). The goose-bumps were at first discussed in terms of the characteristic scale of pebbles within a pebble accretion mechanism (Sierks et al. 2015). However, their appearance may be related to fracturing and polygon formation (e.g. Hérique et al. 2019) again suggesting ice-rich material.

## 2.10.5  Fracturing

The resolution on the nucleus obtained by the OSIRIS instrument on Rosetta has been unique in that it has allowed the identification of fractures in consolidated material. Following a detailed analysis, El-Maarry et al. (2015b) suggested six different morphologies were present on the nucleus. Here we group these into two main types.

### 2.10.5.1  Torque-Induced and Other Possible Tectonic Fractures

Previously, we have seen how the angular velocity of the nucleus of 67P increased during the 2015 perihelion passage. This increased the centripetal acceleration and increased the stress at the neck. The asymmetric and time-varying torque on the two lobes of the nucleus as a result of the reaction forces arising from sublimation and mass loss must also place a stress on the neck region. Figure 2.62 presents evidence of the significance of these stresses for cometary evolution. The image shows a 500 m long fracture of crack in the surface in an area close to the boundary between the small lobe and the neck (the Anuket region). This crack may be evidence that 67P is starting a splitting process.

**Fig. 2.62** ~500 m long crack (marked with yellow arrows) in the Anuket region close to the interface between the small lobe and the neck of the nucleus of 67P (Image number: N20160606T141233677ID10F22)

**Fig. 2.63** The Aker fracture system ends at a cliff that drops down to the Babi region. The image (**a**) shown here indicates that a collapse has occurred and that the surface has been disrupted. The fractures seen (**b**) are part of Aker sub-region a. The steep cliff (Aker sub-region c) down to Babi is to the left in this view (Image numbers: N20160213T190740736ID10F22, N20160213T065857712ID10F22)

Hviid (pers. comm.) has performed a more detailed calculation specifically for 67P and estimated that the centripetal acceleration will overcome gravitational acceleration when the spin period becomes less than ~3.7 h, in close agreement with the more general solution given by Eq. (2.36). Under the assumption of near-zero tensile strength (Groussin et al. 2015a) splitting at this time would appear to be inevitable.

The stresses in the neck region may also be enhanced by the asymmetric torques from activity. The mass loss from the head is phase shifted with respect to the body with the resultant reaction forces acting along vectors that are time variable. In addition to modifying the rotational properties, these forces act to stress the neck region and may also contribute to the observed fractures.

The Aker fracture system (Fig. 2.63 right) seems to be unique on 67P. The main fracture is around 200 m in length and is associated with what appears to an uplift. The fracture is part of the Aker a sub-region and adjoins a steep cliff that forms most of the Aker sub-region c. The surface of Aker sub-region a, near the cliff, is highly fractured with a set of almost parallel fractures evident. The cliff itself shows evidence of collapse in places where the uppermost layer of Aker sub-region a appears to have been disrupted (Fig. 2.63 left). There is presently no adequate explanation for the formation of this fracture system. The quasi-parallel nature of the fractures indicates some level of coherence in the structure of the material. This is rather surprising given that the formation mechanism for 67P is widely assumed to be a stoichastic processing with little subsequent processing of the accreted material. There is considerable evidence for aligned lineaments elsewhere on the nucleus. This type of morphology is usually associated with stress perpendicular to the lineament (e.g. Anderson faulting). But the source of this stress on 67P is not known.

### 2.10.5.2 Thermal Fractures

Close observation of 67P showed that the more consolidated material was often heavily fractured (Thomas et al. 2015a; El-Maarry et al. 2015b) even down to the smallest scales. An example is shown in Fig. 2.64 but there are many images from all consolidated regions on 67P showing similar types of surface morphology. Individual rocks down to 5 m in size and below could also be seen to be fractured (Fig. 2.65). Given the huge temperature ranges and temporal gradients likely to be experienced by the surface materials over diurnal and orbital timescales, the most

**Fig. 2.64** Consolidated material in the Wosret region of 67P. Note the fractured appearance (e.g. position A) with cross-cutting fractures (position B). A pond-like deposit (Sect. 2.10.10) is also seen (position C) (Image number: N20160130T062841932ID10F22)

**Fig. 2.65** Two examples of fracturing in consolidated material on 67P at 5 m scales (Image numbers: N20160902T183736416ID10F22, N20160902T203747847ID10F22)

plausible mechanism for crack production based on terrestrial studies is thermal insolation weathering (Hall and Thorn 2014). Four of the fracture types noted by El-Maarry et al. (2015b) were thought to be thermal in origin. The term insolation weathering covers two processes—thermal fatigue and thermal shock. Thermal fatigue is the process whereby the continuous heating and cooling cycle leads to cracking whereas thermal shock describes fracturing from a large temporal temperature gradient. Production of fresh regolith by thermal fatigue fragmentation is now thought to an important process for the rejuvenation of the surfaces of near-Earth asteroids with decimetre-size rocks being broken on timescales of $10^3$ years at 1 AU (Delbo et al. 2014). In the case of comets, spatial temperature gradients can also be high because of the low thermal inertia of the surface layer.

It is widely assumed that a decrease in temperature leads to contraction. However, Britt and Opeil (2017) have performed experiments looking at the thermal conductivity, heat capacity and thermal expansion of five CM carbonaceous chondrites (Murchison, Murray, Cold Bokkeveld, NWA 7309, Jbilet Winselwan) at low temperatures (5–300 K). The mineralogy of these meteorites is dominated by abundant hydrous phyllosilicates, but they also contain anhydrous minerals such as olivine and pyroxene found in chondrules. Although this material is probably of asteroidal origin, it is interesting to note that the thermal expansion measurements for all these CMs indicate a substantial increase in meteorite volume as the temperature decreases from 230 to 210 K followed by linear contraction below 210 K. Such transitions are unexpected and are not typical for anhydrous carbonaceous chondrites or ordinary chondrites. Hence, using linear thermal contraction to initiate fractures is an assumption that may not be universally applicable.

An example of the magnitude of the temperature extremes on a surface layer is shown in Fig. 2.66. The plot shows how the temperature changes with time for an equatorial surface on an object rotating with a 12.4 h period. The surface is assumed to move into shadow exactly at midday to illustrate the magnitude of the temperature change with time that can be produced by such geometries. A heliocentric distance

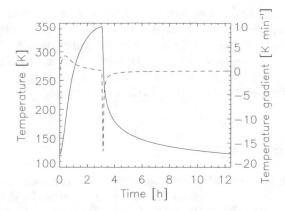

**Fig. 2.66** Temperature excursions and gradients in a simple model with insolation, thermal radiation and thermal conductivity only for a heliocentric distance of 1.3 AU using a thermal inertia of 50 J $K^{-1}$ $m^{-2}$ $s^{-1/2}$ and a rotational period of 12.4 h. The model assumes that the surface passes into shadow precisely at midday with the surface illuminated at zero incidence. This produces a large temporal temperature gradient. Solid line: The temperature of the surface (left axis). Dashed line: The temporal temperature gradient in [K $min^{-1}$] (right axis)

of 1.3 AU and a thermal inertia of 50 TIU has also been assumed. We can see here temperature changes of more than 200 K within 3 h and gradients, when crossing into shadow, of more than 15 K $min^{-1}$. Studies of rocky material on Earth suggest that gradients exceeding 2 K $min^{-1}$ are sufficient to initiate fractures while Delbo et al. (2014) have demonstrated that thermal weathering may be a key factor in debris production on asteroids as a result of fracturing caused by temperature gradients that are almost a factor of 5 less than shown here.

The spatial gradients in temperature with depth close to the surface are also substantial in these models typically reaching 5 K $mm^{-1}$. The sign of this gradient can change rapidly if the surface passes into shadow. Hence, cracks have a potential to initiate along sub-surface planes in quasi-homogeneous layers.

Once fractures are initiated they might develop significantly because of the brittle nature of the surface materials. They are even expected to propagate to depths of up to 25 times that of the diurnal thermal skin depth (estimated to be 1–2 cm by Gulkis et al. (2015) using MIRO data, cf. Sect. 2.9.3.1) assuming an ice-rich silicate substrate (Maloof et al. 2002).

Despite this apparently straightforward explanation, there do remain some issues. For example, the almost linear propagation of fractures (well seen in Fig. 2.64) suggests quasi-aligned lines of weakness in the material or preferred propagation directions once fractures have initiated. In standard tectonics, extension is indicative of stress perpendicular to the fracture but in the case of the Wosret fractures in Fig. 2.64, the source of any such stresses is unclear.

It is also evident in Fig. 2.64 that the surface is not merely fractured but that some depressions are visible that may be the result of local mass loss. Fracturing can, of course, expose fresh material and can therefore be a source of new activity. Höfner et al. (2017) have looked at how thermal re-radiation from hot mouths of cracks can promote the propagation of heat into a fracture even if the deeper sections of the crack are in shadow. If $CO_2$ (or CO) ice is available at depth then this becomes a very simple means of initiating new activity and might lead to ejection of larger blocks of material thereby explaining structures similar to those seen in Wosret. Under optimum conditions, water ice may serve the same purpose.

### 2.10.5.3   Polygonal Networks

The observation of polygonal networks on 67P was not expected but it is a common on Mars, for example, where sub-surface water ice is present. Figure 2.67 shows an example of periglacial features in the northern lowlands of Mars. The large scallop-shaped structures (typically 500 m to 1 km across) have been interpreted as the result of progressive sublimation of sub-surface water ice from equator-facing slopes (Lefort et al. 2009) although a thermokarst mechanism has also been presented in the literature (Séjourné et al. 2011). At smaller scales, the HiRISE image show <10 m sized polygonal networks that are thought to be the result of thermal contraction and ice wedging (Baker 2001). Desiccation polygons, arising from evaporation of liquid water, are also seen on Mars in some areas (El-Maarry et al. 2015c) but have no relevance here.

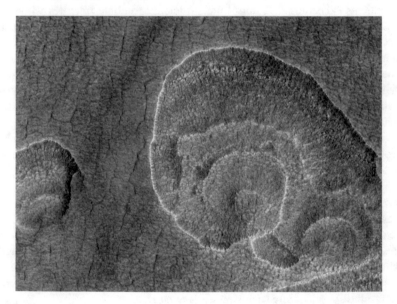

**Fig. 2.67** Periglacial features in Utopia Planitia on Mars. The scalloped terrain is a characteristic of periglacial terrain on Mars and is thought to be related to sublimation of near-surface ice. Close inspection of the image shows polygonal networks at <10 m scale (Credit: NASA/JPL/University of Arizona)

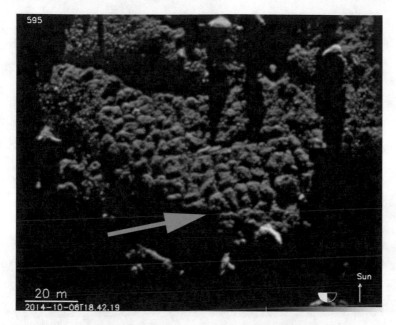

**Fig. 2.68** Putative thermal contraction polygons on the nucleus of 67P (after Auger et al. (2018) with permission from Elsevier) (Image number: N20141006T184219554ID10F22)

Looking closely at putative polygonal structures on 67P (Fig. 2.68), it has been noted that polygons have a homogeneous size across the nucleus, with 90% of them in the size range 1–5 m and a mean size of $3.0 \pm 1.4$ m. Their presence is consistent with diurnal or seasonal temperature variations in a hard (MPa) and consolidated sintered layer of water ice, located a few centimetres below the surface (Auger et al. 2018).

## 2.10.6 Heat Trapping

A consequence of the low albedo and low thermal conductivity of nucleus surface layers is that low incidence angle, direct sunlight leads to high surface temperatures. These surfaces are strong radiators. The idea that re-emission from the surface can contribution to the energy balance of other surfaces (self-heating) has already been introduced. If surfaces are enclosed on several sides, this can produce a reasonably effective heat trap as has been demonstrated for fractures (Höfner et al. 2017) but can also act on larger scales. The most extreme examples on 67P are the deep pits. Here, if sunlight strikes the bottom of a 160 m deep pit, then much of the heat input will be close to orthogonal to the base, maximising the heat input, while much of the radiation will fail to escape directly. This provides the possibility of enhancing local mass loss although the depth and form of the pits prevents this from being a

**Fig. 2.69** An alcove on the Hathor face of 67P. The alcove is on the Hathor side of the interface to Anuket. The smooth area to the bottom right forms part of Hapi. The base of the cliff has an accumulation of talus. Some bright material in the talus is marked by arrows (Image number: N20140820T014254569ID30F22)

continuous process. This illustrates, however, that specific geometries could lead to enhanced sublimation producing locally higher surface erosion rates. One area that may be evidence of this effect is an alcove at the Hathor-Hapi interface (Fig. 2.69). A section of the Hathor cliff seems to be undergoing mass loss. Notice in particular that there are bright spots both on the surface of the cliff in the alcove and within the talus at its base. Talus deposits (sometimes called scree) are collections of broken rock fragments, particles, and dust that have accumulated as the cliff face erodes. They typically have a concave form when accumulation via rockfall over many events occurs. It has been suggested that the bright spots seen here are ice-rich by analogy with other areas on 67P (Thomas et al. 2015a). This is not seen further to the left of the interface between the cliff and the base. The smooth, dust covered, area has been defined as part of the Hapi region. The surface of the dusty area curves upwards towards the cliff. The alcove forms a trap for thermal emission from the surface of Hapi and this may lead to locally enhanced surface erosion.

## 2.10.7  Activity Induced Mass Wasting

Fractures and cracks were often seen on cliff faces as well as cliff tops on 67P (Fig. 2.70) and this has led to the accumulation of talus at the bases of cliffs in many

**Fig. 2.70** Looking down on a fractured cliff top (B) in the Bes region of 67P. Lower ground is towards the right and a smooth surface can be seen at the base of the cliff (A) (Image number: N20160130T173323717ID10F22)

places on the nucleus. There are many examples on 67P and we show several here illustrating various aspects of the process.

Figure 2.71 shows the interface from the smooth terrain of Hapi to the consolidated material of Seth. The material at the base of the cliffs of Seth has the characteristics of a talus deposit. We have seen previously (e.g. Fig. 2.60) that cliffs in the Seth region have cleaved and collapsed, exposing fresher, brighter material. This shows that this is a currently active process influencing the surface properties.

The brighter material in the talus (e.g. Fig. 2.69) has been shown to be bluer in colour at several positions on the nucleus (Pommerol et al. 2015). A further example is seen in Fig. 2.72 which shows a cliff at the interface between Khepry and Imhotep in an enhanced colour image. The images from OSIRIS through different filters show that much of the blocky material is spectrophotometrically bluer and likely to be water ice-rich. The presence of water ice in this material was confirmed with the VIRTIS experiment (Barucci et al. 2016). One edge of the cliff also appears to be relatively blue that may indicate the origin of one of the most recent collapses.

The remarkable brightness of the newly exposed material can be seen even more clearly through plotting the ratio of observed radiance to that of a control area. In Fig. 2.73, we can see boxes marking three areas (A, B, and C) where low resolution spectra have been extracted from Rosetta/OSIRIS data. A fourth area, D, is very small but includes some of the brightest material in the scene.

Figure 2.74 shows the wavelength dependence of the ratios of the reflectance in areas A, B, and D to the control area, C. The relative brightness of the bright material is increasingly greater than the control area as we move towards the blue and is a factor of 7 brighter than the control area at near-UV wavelengths (around 270 nm).

**Fig. 2.71** The Seth-Hapi interface showing that erosion of the rock-like material from cliffs (A) comprising Seth falls into the Hapi depression producing talus (B). Note also the ripple-like structures marked C that will be discussed in Sect. 2.10.8 (Image number: N20151202T123631647ID10F22)

**Fig. 2.72** The collapse of cliffs at the Khepry-Imhotep boundary has exposed bright, ice-rich boulders in the talus seen here in colour as more blue material (First image in sequence: N20140916T150900365ID30F22)

**Fig. 2.73** Image of the Imhotep region showing four areas where spectra have been extracted from OSIRIS data. A: Rocky area. B: Dark dusty area. C: Lighter-toned dusty area. D: A bright block (Image number: N20140916T150900365ID30F22)

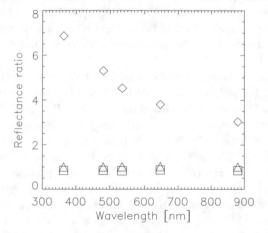

**Fig. 2.74** The reflectance ratio of areas, A (triangles), B (squares), and D (diamonds) to that of C shown in Fig. 2.73. Note that the bright talus material increasingly stands out at blue wavelengths and is a factor of 7 brighter at 270 nm than the control area

**Fig. 2.75** One of the highest resolution images of the nucleus obtained by OSIRIS at 67P. The image was acquired shortly before the Rosetta spacecraft impacted the surface. Note the very bright spots adjacent to some of the rocks (some marked by the green arrows). This bright material can be identified in several images in this final imaging sequence before the spacecraft impacted (Image number: N20160930T100706777ID30F32)

The disrupted material produces a large surface area for sublimation and can lead locally to a large increase in gas production post-collapse. The presence of ice-rich material in the talus may therefore be important in the total mass loss budget although we have little information on the frequency of these collapses.

Bright individual particles were also detected scattered across the surface shortly before Rosetta was crash-landed on the nucleus at the end of September 2016 (Fig. 2.75). These images provided the highest resolution images of the surface and show small bright dots scattered across the surface but between the rocky material.

One of the most interesting areas to study these processes is in the Bes region. An example image is shown in Fig. 2.70. The top of a cliff is covered with an extensive fracture network. The base of the cliff is dust covered. By stretching images where the background surface is in shadow or the background is towards deep space, one can see weak or modest activity even if the optical depth is very low. This technique has been used in Fig. 2.76 to show that the cliff of Bes or its immediate vicinity appears to be actively emitting dust. Individual dust particles (which must be relatively large to reflect sufficient light) can be seen. This implies that activity from cliffs is an erosional mechanism and it has been argued that active cliffs may be the dominant mechanism for activity (Vincent et al. 2016a). This is, however, difficult to prove because the optical depth at 67P was, in almost all but the most unusual localised activity phenomena, very much less than 1. Consequently, dust emission from many other surfaces cannot be seen against the bright background of

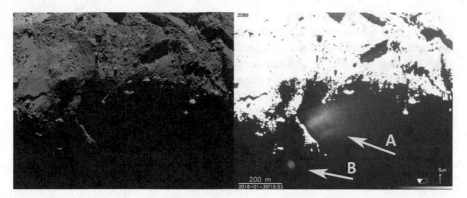

**Fig. 2.76** By stretching an image strongly where the background is in shadow or is towards dark space, one can see modest levels of activity. Here the cliff in Bes seems to be active and individual large particles can be seen. On the left is the original image showing the context. To the right is the stretched image revealing the activity (arrow A). Note how the activity appears to start away from the cliff. The dust particles may be emerging into sunlight after having left the cliff face. Note also the dust particle marked B which is close to the spacecraft and therefore out of focus (Image number: N20160130T190323815ID10F22)

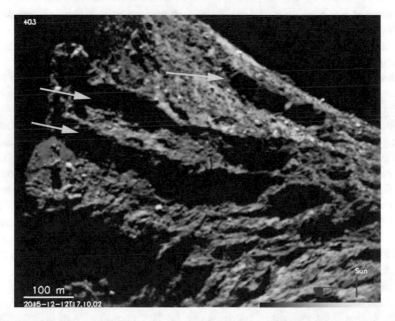

**Fig. 2.77** Quasi-aligned cliffs (marked with arrows) in the Sobek region in the southern hemisphere part of the neck of the nucleus (Image number: N20151212T171002165ID10F22)

the nucleus itself. A particular striking set of quasi-parallel cliffs can be seen in the southern hemisphere section of the neck (Fig. 2.77) and these have been observed to be active when the geometry has allowed.

The retreat of cliffs through progressive erosion on 67P seems to have resulted in multiple landslide deposits suggesting that cliff collapse is an important process in reshaping cometary surfaces (Lucchetti et al. 2020, in press). It has also been proposed that rotational instabilities can induce avalanche-like behaviour. Steckloff et al. (2016) have argued that such avalanches on 103P/Hartley 2's were sufficient to excavate down to $CO_2$-rich material and activate the small lobe of its nucleus producing the observed $CO_2$ outgassing.

## 2.10.8    Sedimentary Processes

One of the most important findings of the Rosetta mission was that re-accumulation of non-escaping particles is a major process influencing the properties of the surface layer (Thomas et al. 2015a). The process had been predicted in the 1990s by Moehlmann (1994) where it was described as a "dust hail" phenomenon. This was somewhat of a misnomer as hail arises from condensation of volatiles (water) in an atmosphere. Here, dust particles are emitted from the surface, returning on quasi-ballistic trajectories to impact the surface at positions often remote from their origin. The process has been described as "airfall" in analogy with volcanic processes on Earth but the mechanism itself was accurately described by Moehlmann.

The process arises from there being a wide range of dust emission velocities with some of these being lower than the escape velocity. Prior to Rosetta, it was accepted that return of non-escaping particles to the surface would occur and may even act to choke activity in the immediate proximity of an active region. However, the magnitude of the effect was totally underestimated.

There were several relevant observations made by Rosetta at 67P. First, the north-facing surfaces were usually smooth with an almost conformal coating of material with particle sizes close to and below the resolution limits of the cameras. These surfaces (most notably in the Ash and Ma'at regions) were bounded by rougher material and the change in texture at the interface was abrupt. A good example of this is shown in Fig. 2.78 where the smooth surface to the top left of the image almost certainly arises from airfall deposits. The smooth appearance of the putative impact crater in Ma'at (Fig. 2.56 centre) is a further indication of dust deposition and was used to estimate the depth of the deposit to be around 5 m (although there are many caveats associated with this number).

There have been high resolution observations of these surfaces (e.g. Fig. 2.79) that show that the apparently smooth surface is made up of smaller particles close to the resolution limit. Individual particles can be seen and a range of brightnesses is evident. This implies that many of the particles are in the decimetre size range—a size that has interesting consequences for dust ejection models as will be discussed below.

Observations post-equinox showed that surfaces with faces directed towards the south were mostly devoid of these smooth deposits. Keller et al. (2017) suggested that the stronger activity from south-facing surfaces around perihelion, produced by

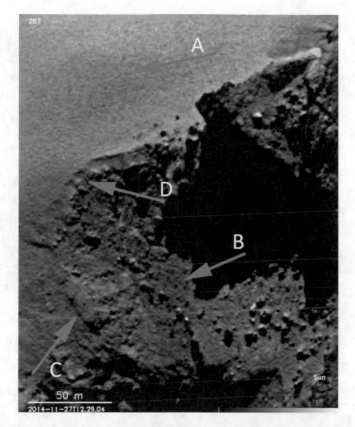

**Fig. 2.78** The smooth terrain of Ash (to the top left of the image; A) is almost certainly the result of "airfall". The surface of Seth (centre and right) is rougher with boulders (B) and consolidated but fractured material (C). The boundary between Ash and Seth (D) is very sharp with transition is texture occurring over ~10 m (Image number: N20141127T122904841ID10F22)

the obliquity of the rotation axis, would lead to south-facing surfaces being devoid of deposits and produce transport of material from the south to the no longer illuminated north-facing surface facets. Hence, much of the material seen on the surfaces of Ash and Ma'at may actually have been ejected from the southern hemisphere of the nucleus. This would add to (and possibly even dominate) deposits arising from dust emission from the highly active north-facing Hapi region pre-equinox.

The second evidence for airfall comes from the trajectories of particles seen during rapid imaging sequences by OSIRIS. These fall into three categories. The trajectories of some near-nucleus particles within a specific imaging sequence were described by Agarwal et al. (2016). Adding together multiple images from such a sequence reveals particle motion very clearly. An example is shown in Fig. 2.80.

The projected velocities of some of the lower velocity particles in Fig. 2.80 are below 0.35 m s$^{-1}$ and hence nearly a factor of 2 below escape velocity. Consequently, it is highly likely that particles are re-impacting the surface even if the

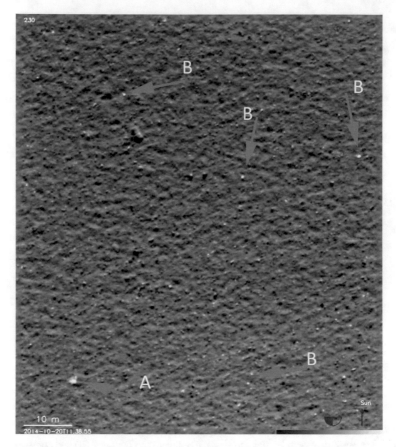

**Fig. 2.79** The surface texture of airfall deposit areas indicates some bright individual particles (some are marked B). There is quite a wide range of brightness evident. The areas are smooth at the 10 m scales but probably particulate at sub-decimetre scales. A larger bright object is marked A (Image number: N20141020T113855625ID10F22)

velocity components of these particles into or out of the image plane are relatively large.

In Fig. 2.81, one can see further evidence of low velocity particles. The jet-like structure to the lower right of the nucleus appears strongly curved (Lin et al. 2016). The curvature in projection here is towards the Sun direction which is the opposite of what one would expect if gas drag (another potential process to produce non-radial flow) or solar radiation pressure were responsible. The combination of low velocities and nucleus rotation is therefore the most likely explanation. The sub-spacecraft latitude at the time of this observation was 76.6° south implying that we are looking almost pole on so that projection effects should be small. Hence, the outflow velocity can be estimated directly from the curvature giving $\gtrsim 1.1$ m s$^{-1}$. Marschall et al. (2019) showed for one example on 5 May 2015 that, while the dust velocity was low, it was above escape velocity and accelerating presumably because of gas drag.

**Fig. 2.80** Ten pairs of images (each pair taken 6 s apart and each pair separated by 40 s) have been added together to reveal particle trajectories directly above the nucleus. The colour table has been inverted (as in Agarwal et al. 2016) for easier recognition. Some individual tracks are marked but many more can be identified in this field alone with detailed inspection (First image in sequence: N20160106T070140589ID30F22)

Consequently, detailed modelling would usually be required for such estimates. The radial nature of other jet-like features indicates far higher velocities as had been predicted through gas drag calculations for small (<100 μm) particles.

Deposition leading to burying of an impact crater and the creation of a new impact crater can be seen in the montage of the Serqet region in Fig. 2.82. A 5 m diameter crater would be typically 1–2 m deep suggesting that re-surfacing to that depth is possible (although dust mobilization to smooth out the depression cannot be ruled out as a mechanism). The recent impact shows that larger chunks of material are probably impacting the surface after being launched by activity.

We have given several examples because the importance of this process cannot be underestimated. There are numerous implications.

The deposition of material, in a process that can be referred to as sedimentary, can act as an insulating layer throttling activity from below. This layer may be quite porous and of low conductivity so that the sedimentary layer need only be quite thin to choke activity entirely. When first deposited, however, it is not obvious that the material is inert. When ejected from the surface, the particles will heat up but larger particles may have sufficient size that the outgassing of volatiles is not completed before the particles re-impact the surface. Hence, they could be initially a source of outgassing while sitting on a relatively inert layer.

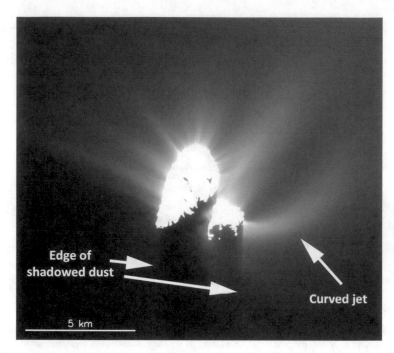

**Fig. 2.81** Evidence of a curved jet structure seen on 31 May 2015 at 03:16:49 UT (Image number: W20150531T031649472ID30F18)

**Fig. 2.82** The same region on 67P viewed at three different epochs by Rosetta at roughly the same scale. Left to right: September 2014, March 2015, June 2016. The arrows indicate observed changes (Image numbers: N20140914T054255346ID30F22, N20150328T052640179ID30F22, N20160616T001920568ID30F22)

The whole process will affect the dust size distribution in the surface material. Smaller particles are more likely to reach escape velocity when ejected. Hence, particles falling back will have a size distribution strongly skewed towards larger particles thereby changing the size distribution in the smooth surfaces. It is important here to note that the word "smooth" has been used without referring to a dimension. The surfaces showing evidence of sedimentary deposition may appear smooth at decametre scales but at decimetre and centimetre scales (below the resolution of imaging systems) the surface may be rough as a consequence of the particle size distribution of the deposit.

Compared to active regions, the layer itself will become enriched in the non-volatile component. One could envisage sampling these areas at a later date but it should be apparent that the properties of the layer will bear little resemblance to the original pristine properties of the material. Furthermore, the layer will contain material originating from several different areas on the nucleus. It is not clear that the source regions would have identical properties. Indeed a key question in cometary science is how homogeneous are the cometesimals making up the nucleus. But the deposits will have become homogenized by the sedimentation process and thus the interpretation of the properties of the layer is increased in complexity.

Many of the exposed cliffs on the nucleus show evidence of discrete layers. An example of this layering is shown in Fig. 2.83 which is an image of the Ash region on the side towards Imhotep. Layers can be seen frequently in the Seth region on the

**Fig. 2.83** Layering in the Ash region of the nucleus of 67P. A wide angle shot of the nucleus on the right shows a box indicating the position of the expanded image on the left. The image was taken at quite a low phase angle where contrast was quite low (Image number: N20140816T145914556ID30F22)

sides of the quasi-circular depressions that are so prevalent there. There is a suggestion that the layers have different spectral properties (Ferrari et al. 2018). It has also been suggested that the layering is indicative of a, possibly primordial, internal layering that passes through the body and that the two lobes exhibit different orientations of this layering (Massironi et al. 2015). This would be quite remarkable in that it would contend that the object formed in such a way that planes could arise (e.g. through large scale sedimentary processes). It is probably fair to say that this is not universally accepted as it would require a force field (e.g. a gravitational field) to produce defined layers over 2 km distances. This would suggest that the (sub-)nuclei were formed within a far larger body that disrupted. On the other hand, there are alternatives. Thermal processing on diurnal and annual scales may provide a means of processing the surface layer to a certain depth (see e.g. Sunshine et al. 2016). Changes in the orbit modify those depths and thus you have a mechanism for modifying the surface down to a fixed depth that changes occasionally after a close encounter with Jupiter. The sedimentation/airfall mechanism may also play a role by generating layers that are given years to consolidate by a mechanism such as sintering. This would support layering being a local phenomenon rather than a global one.

### 2.10.9   Activity in Dust-Covered Areas

One might expect that the airfall deposit should contribute to the slowing and eventual choking of activity on the northern hemisphere of 67P. Modelling of the outgassing from the nucleus, however, suggests that, while there were inhomogeneities, the outgassing from the northern hemisphere was broadly speaking insolation-driven (i.e. proportional to the solar insolation; Bieler et al. 2015a, b). However, evidence of surface changes in the smooth terrain was clearly evident as seen in Fig. 2.84. This figure shows an area of Ma'at close to some of the pits and compares observations made more than 7 months apart. In the later image (acquired in March 2015), the surface has taken on a mottled texture in several (marked) places when compared to the data acquired in August 2014 (panel a). Panels c and d show the bottom of a pit from the March 2015 observation in detail. Here, too, surface changes are visible (panel c) and, when saturating the image to exposure low intensity levels, it can be seen that the surface was locally active (panel d).

Dust coverage has not extinguished or prevented activity and if this material is part of the original surface (it is hard to imagine otherwise) then the depth of the airfall deposit must be very small. Indeed, stretching Fig. 2.84 panel a, the surface changes seen later seem to be at a place that we observe to be slightly brighter in August 2014. This argues that the airfall deposit is really very thin which would seem to agree with what we have seen in Fig. 2.60.

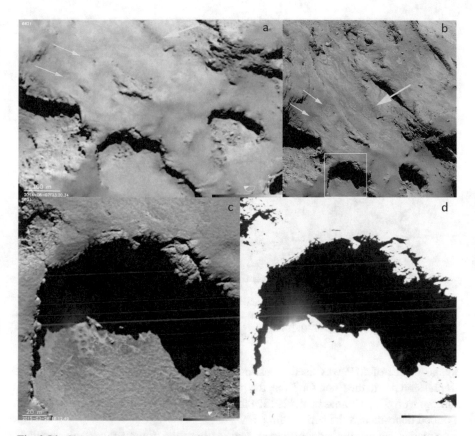

**Fig. 2.84** Changes in the appearance of the surface of Ma'at through the perihelion passage. (**a**) Image acquired on 7 August 2014. (**b**) The same area acquired on 28 March 2015. Arrows mark the areas where significant changes have occurred in the smooth terrain. (**c**) A sub-image of (**b**) (marked by the box) shows an exposure of unusual material in a region that previously (image **a**) was much smoother in texture. (**d**) The same sub-image but stretched reveals activity from this area against a shadowed background (Image numbers: N20140807T232034548ID30F22, N20150328T161249393ID10F82)

## 2.10.10   Dust Ponding

The close-up observations of (433) Eros by the NEAR spacecraft revealed the presence of "ponded deposits". These are flat-floored, smooth surfaces within an irregularly-rimmed depression. Morphologically, the deposits sharply embay the bounding depression in which they sit (Dombard et al. 2010) although Roberts et al. (2014) state that less than half the pond candidates on (433) Eros have clearly flat floors.

There are several surfaces in the Khepry and Aker regions on 67P that appear to be similar to ponded deposits (Fig. 2.85). The features on 67P are up to 160 m diameter and therefore similar in size to those seen on Eros (Roberts et al. 2014). The

**Fig. 2.85** Two ponded deposits (marked A and B) in the Khepry region of 67P. One of three very large (20 m scale) boulders that are prominent features in the Khepry region is marked C (Image number: N20160210T174957774ID10F22)

shape model of 67P was used to estimate a maximum depth of 35 m from the depression rim to the floor. On Eros, the ponded terrain was relatively blue in colour, although on 67P, Thomas et al. (2015b) found no significant differences between the ponded deposits and the surroundings.

Smaller flat, smooth deposits are seen in-between rougher, possibly eroded, materials elsewhere on the nucleus. An example in Imhotep is shown in Fig. 2.86 and there are other examples in the Wosret region. In Fig. 2.86, note that adjacent smoother terrain (top left) is not really smooth but dotted with boulders and with a rougher texture. This seems to suggest that there is indeed a specific process at work producing these flatter surfaces.

Four mechanisms for ponded deposit production have been proposed. These are seismic shaking, erosion of a central peak or boulder, electrostatic levitation and re-impact, and fluidization of the surface material. This is of some interest because these types of surfaces may form potential landing sites because of their inherent smoothness. Hence, we examine each possible mechanism briefly. In judging these mechanisms, it should be noted that the particle size distribution of the material in the ponds is unknown and each of the mechanisms discussed here will act on different particle sizes with different effectiveness.

Cheng et al. (2002) proposed that ponded deposits are the result of seismic shaking from impacts. Settlement of the surface material can arise from consolidation or failure of the material under the surface, densification of dust or sand layers caused by the ground shaking and liquefaction of the surface material. On 67P, there are undoubtedly some seismic effects associated with the stresses arising from the

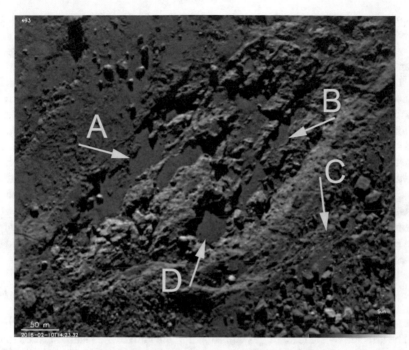

**Fig. 2.86** Ponded deposits are observed within rocky terrain areas. This is in Imhotep. Arrows A and B point to flat, apparently dusty material. C indicates some layered terrain. One can see boulders on the uppermost layer. Arrow D points at ponded deposit but here there are some small boulders apparently lying on the surface (Image number: N20160210T142332710ID10F22)

non-gravitational forces acting on the nucleus. The low porosity would imply that settling is quite possible. However, on 67P the ponded deposits are fairly localised and limited in areal extent which seems contrary to the more global effects of NGFs.

Dombard et al. (2010) have suggested that the ponds form as a consequence of thermal disaggregation of boulder material within the depression in a type of insolation weathering driven by the repeated day/night cycling (see also Sect. 2.10.5). The flattening is produced by seismic shaking of ponds in response to impact. Roberts et al. (2014) have criticized this by showing that the pond material follows the underlying topography which is inconsistent with the material originating by erosion of central boulders.

Electrostatic levitation of dust and transport has been proposed and investigated by several authors. Poppe et al. (2012) have pointed out that there is now significant evidence for electrostatically-induced dust grain transport above the lunar surface and they extended previous modelling work to include the ponded deposits of Eros and the trapping efficiency of dust grains by craters. They showed that grains will tend to accumulate within crater boundaries as a consequence of the presence of complex fields at crater rims with larger grains being trapped more efficiently. The main problem, however, is the absence of a well-defined launch mechanism.

Micrometeoroid impact has been proposed but found to be insufficient in the case of
Eros (Colwell et al. 2005). For electrostatic lofting, cohesive forces need to be
account for and this leads to preferential lifting of intermediate-sized (15 μm) grains
(Hartzell et al. 2013). In the case of 67P, this problem may not exist because grains
are being levitated by the sublimation process and sedimenting through airfall.
Hence, only the preferential transport of these grains into depressions needs to be
clarified. The Poppe et al. (2012) paper appears to demonstrate that this is feasible
although we note the relatively small scale of the modelled crater (7 m diameter) in
their work compared to the observed deposits on 67P. The application to 67P is
nonetheless an extremely complicated physical problem. There are numerous effects
at work. For the dust particles themselves, impacts of electrons and ions can transfer
charge to directly the grain, ultraviolet radiation from the Sun can lead to photo-
emission of electrons, and recombination with free electrons from the dust grain
environment can occur. In addition, the surface itself can become positively charged
as a result of the release of photo-electrons and a dayside-nightside electric field can
arise. The resultant force on the dust particle will be

$$F_E = q_c E \qquad (2.131)$$

where $E$ is the electric field vector and $q_c$ is the particle charge. All of these processes
will be affected by the level of gas emission and the ratio of the resultant force to that
of other forces (such as the drag force) may be (and probably is) highly time variable.
Hence, while an equation of motion can be determined under specific assumptions,
the uncertainties with respect to real cases are likely to be enormous. Nordheim et al.
(2015) constructed a model addressing many of these issues and concluded that
particles of $<50$ nm in size can be levitated electrostatically. Very high negative
potentials can also be reached in shadowed areas of the nucleus and on the nightside.
The local electric field strength can also reach values of $\sim100$–$1000$ V cm$^{-1}$ over
centimetre scales. Piquette and Horanyi (2017) have shown that asymmetric surface
topography producing such field strengths can have a significant effect on dust
dynamics when particles are charged. Topographic highs can for example have a
sunlit and an unilluminated side leading to high local field strengths and localized
transport effects.

Finally, Sears et al. (2015) have recently suggested that fluidization associated
with degassing should also be considered as a possible explanation. As this effect
may be of importance elsewhere on the nucleus, we will devote the next sub-section
to this particular effect.

### 2.10.10.1  The Surface Fluidization Mechanism

Fluidization is the process of stationary solid particles being brought into a dynamic
"fluid-like" state by an upward stream of fluid (gas or liquid) (Fan and Zhu 2005).
This process has been studied extensively because of its industrial applications
(e.g. Wang et al. 2015). When a gas flow is introduced into the bottom of a bed

packed with solid particles, the gas will move upwards through the gaps between the particles. When the gas velocity is low, the drag force on each individual particle is also low, and the bed remains in a fixed state. By increasing the velocity or density of the flow, the aerodynamic drag forces begin to counteract the gravitational forces, causing the bed to expand in volume as the particles move away from each other. Further increasing the velocity, it will reach a critical value at which the upward drag forces will exactly equal the downward gravitational forces, causing the particles to become suspended within the fluid. At this critical value, the bed is said to be fluidized and will exhibit fluid behaviour.

In industrial cases, the pressure gradient, $\Delta p$, across a bed of depth, $z$, can be controlled so that the weight of the particles can be equated with the buoyancy provided by the gas flow. This leads to the equation

$$\Delta p = z\,(1 - \Psi)\,(\rho_s - \rho_g)g \qquad (2.132)$$

where $\rho_s$ is the density of the solid, $\rho_g$ is the gas density, $\Psi$ is the porosity and $g$ is the gravitational acceleration. It is straightforward to calculate that for 67P, a pressure gradient of around $10^{-2}$ Pa across a 10 cm bed would be sufficient to fluidize the material in the absence of cohesive forces. This very low value arises from the low gravitational acceleration, of course. Note that the gas flux through a porous layer as a consequence of a pressure gradient is discussed in Sect. 3.4.7.

There are numerous regimes that can arise from pressure gradients across a bed. An example is bubbling fluidization. If the gas velocity is high, bubbles form near the gas emitting surface. They rise up and coalesce. The local mean bubble size increases rapidly with increasing height above the emitting surface. As they reach the surface of the bed, the bubbles burst, ejecting particles.

This effect can be seen in laboratory simulants of cometary material. Figure 2.87 shows two stills from a video made of water ice-bearing charcoal in a vacuum

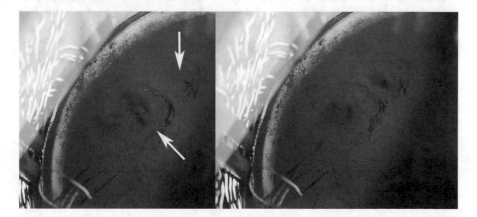

**Fig. 2.87** Water ice-charcoal mixture in the University of Bern's simulation chamber. Left: Earlier. Right: A few seconds later. The top arrow indicates the position of a small plume that starts between the two frames. The lower arrow indicates a site of more continuous particle emission

chamber at the University of Bern. The simulant has been illuminated by a bright light representing the Sun. It is apparent that sub-surface sublimation of the ice occurs at depth and a type of bubbling fluidization of the charcoal is seen as the gas passes through the more desiccated surface layer. The bubbles eject charcoal particles that fall back on to the surface. This experiment is, in principle, more severe than expected at a comet. The gravitational acceleration is 5 orders of magnitude higher than on the comet and charcoal should be extremely cohesive. And yet this process is strong enough to counter these higher restraining forces and would be sufficient to eject particles from the surface of any comet.

## 2.10.11   Surface Changes in Smooth Terrains

Groussin et al. (2015b) first noted the significant changes evident in the smooth terrain of Imhotep. Figure 2.88 shows example images of before, during and after the changes had occurred. The data show the expansion of quasi-circular structures across Imhotep beginning shortly before perihelion. Looking closely at the centre image of Fig. 2.88, brighter material can be seen at the edges of two of the areas exhibiting changes. Figure 2.89 illustrates this using higher resolution images. Groussin et al. (2015b) determined that observed brighter spots were also bluer and concluded that this must be exposed ice.

What is remarkable is that post-perihelion, many of the changes that occurred appear to have been smoothed out. The surface has mostly returned to its original appearance. There are differences but these differences are much less profound than the changes that occurred during July 2015. Groussin et al. (2015b) pointed out that the erosion rate from sublimation was too slow to match the expansion rate of the quasi-circular features during July 2015. Hence, the whole process lacks a credible explanation at this point. The process must be related to the heat input (and presumably sublimation of ices) as the main changes occur when the solar flux

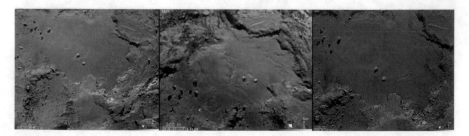

**Fig. 2.88** The smooth region of Imhotep seen at three different times during the Rosetta mission (left to right: 7 May 2015, 11 July 2015, 10 February 2016). Note the remarkable changes seen in the 2 month period between May and July 2015 and the apparent return to a smoother appearance by February 2016. Note also that the central view is rotated by about 40° anti-clockwise with respect to the other two images (Image numbers: N20150507T2014163781D10F22, N20150711T12442803910F22, N20160210T1423327101D10F22)

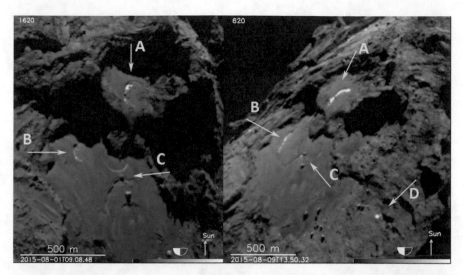

**Fig. 2.89** Bright features and their evolution in the Imhotep region of 67P in August 2015. Left: 1 August 2018 09:08 (N20150801T090848829ID10F22). Right: 9 August 2018 13:50 (N20150809T135032829ID10F22). The brightest areas are saturated in the OSIRIS images. The bright area marked A is not on the floor of Imhotep but a smooth elliptically-shaped plateau. The bright area appears to have evolved between the two images. The quasi-circular depression B is seen to be bright but possibly with less evolution between the images. Using the small boulder as a reference point, it can be seen at point C how the quasi-circular depression has expanded in the time between the two image acquisitions. Point D shows a transient exposure of bright, presumably icy, material

reaches a maximum. However, it is notable that Imhotep is one of the regions on the nucleus inferred to have a low total production rate (e.g. Marschall et al. 2016). It is possible to imagine a mechanism whereby fluidization of the surface layer is followed by settling. The return to a fairly flat surface as fluidization slows could be the result of localized redistribution of dust as seen in the laboratory example above (Fig. 2.87).

An alternative explanation for the return to a flat appearance is that it is a consequence of the equipotential surface. Non-escaping particles fall back to the nucleus preferentially into local potential minima. The Imhotep smooth surface forms an equipotential surface as can be seen in Fig. 2.90. Effects of "splash" (impact particles sputtering other particles from the surface) would simply increase the transport and local smoothing. There are two difficulties with this model. First, there are subtle albedo differences across the surface that would be difficult to explain with a homogenized "rain" of particles from above. Second, there are local points that are topographically lower than the smooth surface as can be seen by oblique views of the edge of the smooth surface on Imhotep (e.g. Fig. 2.91). The lower gravitational potential of this area is also evident in Fig. 2.90 (bottom left). To maintain this as a viable explanation, a method to keep lower surfaces "clean" of deposits is needed with the rather obvious idea of unseen gas emission being a prime

**Fig. 2.90** Images of 67P from different orientations distorted such that equipotential surfaces are at the same radial distance from the centre of the nucleus. The images show that smooth surface, such as Hapi and the smooth surface in the centre of Imhotep, form equipotential surfaces (Credit: R.M. Marschall)

candidate. This idea has been discussed as the reason why the southern part of the neck (region Sobek) is not dust covered despite being deep in the gravitational well—the area is highly active during southern summer through perihelion. However, the concept of preventing dust accumulation locally by gas emission in Imhotep remains unproven.

The edge of the smooth surface in Imhotep drops via a set of layers (terraces) down to rougher terrain (Imhotep sub-region c). This appears to be similar to the smooth material seen on 9P/Tempel 1 (i.e. the olive green terrains in Fig. 2.48). Hence this type of surface may be generic but there are subtle differences. For example, the edges of this layered terrain on 9P/Tempel 1 were seen to be a source of jet activity (Farnham et al. 2007, 2013). This has not been observed at Imhotep.

While Imhotep provides the most remarkable evidence of quasi-circular structure development and evolution in a dusty surface, it is by no means the only example. Structures of similar appearance were seen in the Hapi and Anubis regions (Fig. 2.92). The structures in Hapi began to appear 8 months before perihelion. At this time the northern hemisphere was well illuminated although the total heat input was still quite low because of the heliocentric distance. Nonetheless, the appearances of the structures are quite similar to those seen in Imhotep. There does appear to be evidence of wall collapse in both the Hapi and Anubis structures with talus seen as the bases of the steepest sides.

**Fig. 2.91** Oblique view of the edge of the smooth surface and the quasi-circular crater-like feature in Imhotep. Note the cliff from the smooth surface down to the region with small circular structures (Image number: N20160130T141149683ID10F22)

Structures similar in shape and form to those seen in Fig. 2.92 were also evident in more consolidated regions associated with activity. An example is shown in an image of the Bes region (Fig. 2.93).

Circular, expanding depressions often seem to be related to sublimation processes. Malin et al. (2001), for example, have shown that pits on the south residual cap of Mars are expanding at rates that are measurable with repeated high resolution imaging. These pits are almost perfectly circular but in this case the subliming material is almost pure $CO_2$ while on 67P we appear to have a complex mixture of $H_2O$, $CO_2$ and refractory materials.

## 2.10.12   Other Circular Structures

In the Imhotep region, we can see a series of circular structures in positive relief with rims (Fig. 2.94). Deep Impact observations of 9P/Tempel 1 show pits with rims (Thomas et al. 2013a) that look very similar to these structures. Thomas et al. (2013a) concluded that volatile loss and transport from depth with on-going

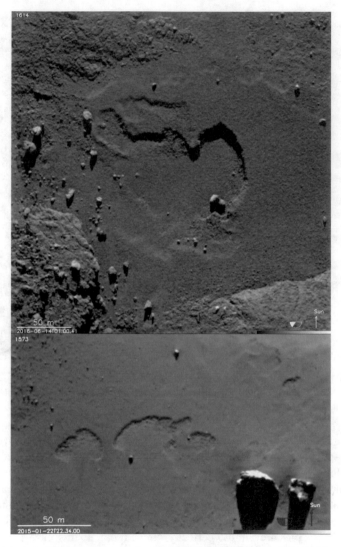

**Fig. 2.92** Quasi-circular depressions in Anubis (top; N20160614T010041718ID10F22) and Hapi (bottom; N20150122T223400384ID10F22)

sublimation at the surface is a plausible explanation. They also raised the question about whether these features were being exhumed or whether they were being created by current activity. El-Maarry et al. (2017b) showed that some of these features appeared on the nucleus of 67P at some time during the perihelion passage. The surface changes that occurred are seen in the pre- and post-perihelion images in Fig. 2.94. Boulders near the features marked with the yellow arrows appear to have been exhumed during the perihelion passage suggesting that material surrounding

**Fig. 2.93** Image of the Bes region of 67P showing an irregular depression (marked E) between two cliffs (marked A and B). This suggests that the irregular shape of depressions produced by activity is common. Smooth material at the base of cliff B is marked (C) and talus at the base of cliff A is also marked (D). (Image number: N20160130T180323711ID10F22)

the features has been removed. However, this is possibly not conclusive as creation of the structures may have been part of the surface change process. Furthermore, because of surface dust transport, we have no way of knowing if the material removed had been only recently emplaced. Consequently, the formation process and its timing have not been established.

## 2.10.13  Surface Dust Transport

One of the most remarkable observations on 67P was the evidence of what appear to be dune ripples (Fig. 2.95). This is remarkable because, prior to these observations, the gas velocity vector was widely assumed to be primarily radial whereas a strong

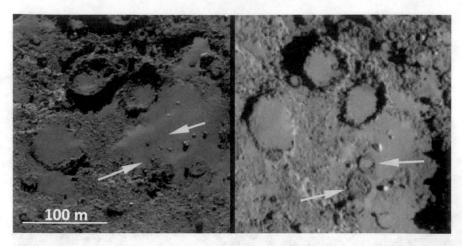

**Fig. 2.94** A part of the Imhotep region of 67P showing almost circular structures that are in positive relief and occasionally show rims. Left: Image from 2014-11-22T06:52:53. (N20141122 T065253908ID30F22). Right: Image from 2016-02-10T14:23:32 (N20160210T142332710 ID30F22). Changes that occurred during the perihelion passage are evident and particularly at the positions of the arrows (Modified following El-Maarry et al. 2017b)

component parallel to the surface would normally be required to generate gas-driven ripple structures. (We avoid the use of the word "aeolian" here because the driving gas flow is clearly not the result of a pressure gradient in a stable atmosphere.) Furthermore, when the ripples were first detected, it was assumed that the gas densities were insufficient to overcome dust particle cohesive forces and hence generation of saltation or particle motion across the surface via reptation (Fig. 2.96) was thought to be challenging. This was illustrated by Thomas et al. (2015b) who used a simple expression from Shao and Lu (2000) that fit experimental data to determine the fluid threshold under assumed gas pressures at the surface. The fluid threshold, $u_{th}$, is the critical wind speed above which the drag and lift forces exerted by the fluid are sufficient to lift some particles from the surface. These particles are accelerated by the fluid but brought back to the surface by gravity.

Bagnold (1941) gave an approximation for $u_{th}$ as

$$u_{th} = A_{fric}\sqrt{2\sigma_{ps}ga} \tag{2.133}$$

where $A_{fric}$ is a dimensionless threshold friction velocity, $g$ is the gravitational acceleration, $a$ is the particle radius and

$$\sigma_{ps} = \frac{\rho_s - \rho_g}{\rho_g} \tag{2.134}$$

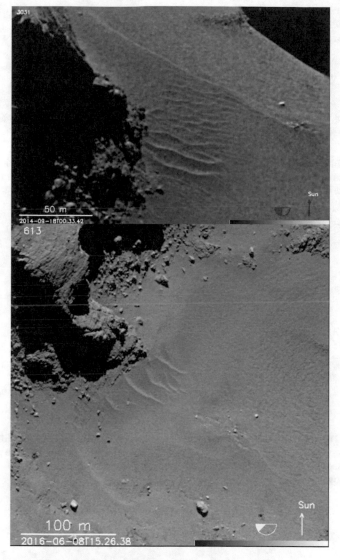

**Fig. 2.95** Observations of ripples in the Hapi region of the nucleus of 67P. Top: Image on 18 September 2014 pre-perihelion (N20140918T003342230ID10F24). Bottom: Image from 8 June 2016 post-perihelion (N20160608T152638721ID10F22). Although the observation geometry is different, it is quite apparent that significant changes have occurred

with $\rho_s$ being the particle density and $\rho_g$ being the gas density. However, this equation for $u_{th}$ is invalid for grains smaller than about 100 μm because of cohesive forces. Shao and Lu's formulation is

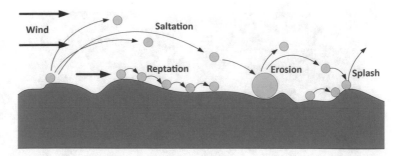

**Fig. 2.96** Illustration of saltation and reptation across a surface driven by a surface wind. Saltating particles can also produce erosion of large static particles by impact. Impact on to small particles may levitate them via a "splash" mechanism

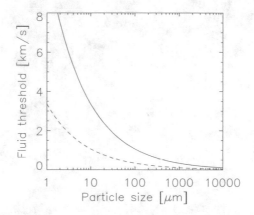

**Fig. 2.97** The fluid threshold velocity calculated using the Shao-Lu formulation with values for gravity appropriate for 67P ($g = 1.55 \times 10^{-4}$ m s$^{-2}$). The velocity rises rapidly as the particle size decreases because of cohesive forces. Gas pressures of 0.003 Pa (solid line) and 0.03 Pa (dashed line) are shown

$$u_{th} = \sqrt{A_V \left( 2\sigma_{ps}ga + \frac{\gamma_c}{2\rho_g a} \right)} \tag{2.135}$$

where $A_V$ is a modified dimensionless velocity threshold, and $\gamma_c$ is an empirical constant describing the cohesive forces. This revised equation leads to the values for the fluid threshold at 67P, shown for two different surface pressures, in Fig. 2.97. With local gas velocities unlikely to exceed a few hundred m/s, motion via reptation and/or saltation can only occur for larger particles in fairly dense flows. The pressures here are difficult to interpret because of the non-equilibrium nature of the gas flow but local densities could reach these values close to a source. The flow parallel to the surface might arise at positions close to the terminator where dayside to nightside near-surface flows are to be expected. Pähtz and Durán (2016), in an abstract guided by a simulation of sediment transport in a Newtonian fluid using the numerical model by Durán et al. (2012), predicted threshold wind shear velocities that were also fairly extreme (e.g., $u_{th} = 45$ m s$^{-1}$ for $a = 0.5$ cm) but such values are feasible. Using a large number of assumptions, Jia et al. (2017) constructed a self-consistent model that also illustrated the feasibility of ripple production under 67P-like conditions.

Figure 2.95 shows that the ripples are dynamic on orbital timescales. Furthermore, they are not small features. A digital terrain model reconstruction of the site pre-perihelion is shown in Fig. 2.98.

The stoss (windward slope) slope of a ripple is the less steep side and is transverse to the wind direction. Here it is to the south (left) of the crest with the slipface (the leeward side) to the right of the crest indicating that the primary direction producing the dune was from south to north.

The heights of the ripples and the magnitude of the changes are illustrated in Fig. 2.99 which shows height relative to an arbitrary zero position pre- and post-

**Fig. 2.98** Digital terrain model reconstruction of the Hapi ripples pre-perihelion (courtesy of L. Jorda) using the combined SPG and SPC technique called MSPCD (Capanna et al. 2013)

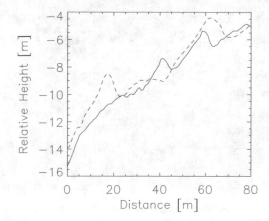

**Fig. 2.99** Cut through the digital terrain model of the Hapi ripples pre- (solid line) and post-(dashed) perihelion showing metre-scale changes in the positions and heights of the ripples

perihelion. Changes of over 1 m in height can be seen indicating local motion of tonnes of material.

The remaining point of discussion concerning this phenomenon is the actual mechanism that initiates movement. Wind-driven reptation of large particles appears feasible but does require fairly extreme lateral gas flow that is, in some ways, counter-intuitive in a vacuum environment. The cohesive forces can be much higher than the gravitational force and hence, once the cohesive force is overcome, the lifted particle should be lost. The conclusion is either cohesive forces are weak implying large particles, or the lifting mechanism is not purely wind-driven. Flow of sub-surface sublimed gas through the bed to levitate particles may be effective in a fluidization-type mechanism (Jia et al. favoured a mechanism similar to this). Alternatively, non-escaping particles from active regions may provide a source by impact (a 'splash' mechanism) although it is difficult to envisage this mechanism being sufficient to levitate tonnes of material locally as observed. There are numerous subtle effects that may also play a role. Saffman lift force, for example, results from the shear flow of gas over a surface although the usual cases where this is significant normally involve sub-micron particles.

There is substantial evidence for other ventifacts elsewhere on the nucleus. Figure 2.100 shows what appears to be a dune exhibiting a bifurcation. The stoss side here is north-facing. In the adjacent Ma'at and Maftet regions there are many other elongated, smooth mounds that appear to be aligned in the north-south

**Fig. 2.100** A bifurcated dune-like structure in the Nut region of 67P. Note the bright spots on the stoss side of the structure (position A) and also the boulders on the putative slip-face (B) (Image number: N20141004T204134574ID10F22)

direction. These have been categorized as wind-tails by Keller et al. (2017). However, this interpretation should be treated with some caution because there are rarely
strong relationships of these features to an obstacle retarding the flow. It should also
be noted that to generate wind-tails, the obstacle and the tail itself should have
dimensions considerably larger than the mean free path and it is not clear that this
condition is met even at times of high activity in the southern hemisphere. The
presence of large particles (boulders) is also of note.

Smooth surfaces seen preferentially on one side of boulders (e.g. Fig. 2.101) may
also indicate an area of stagnant flow in the wind-shadow of an obstacle. However,
Fig. 2.96 illustrates that erosional processes provide an alternative explanation where
impact by wind-carried particles erodes the surface of the boulder over time to
produce a deposit at its base. Although it is unclear which explanation is correct in

**Fig. 2.101** Smooth surfaces preferentially to one side of boulders (marked with arrows) in the Hapi
region of 67P (Image number: N20141210T062855791ID10F22)

the case of Fig. 2.101, the important conclusion is that dust transport across the surface is an important process in defining local surface morphology.

## 2.10.14   Dune Pits

Another feature seen at 67P as a result of the high resolution was the "dune pit" (Fig. 2.102). These features are characterized by smooth dunes/ripples being disrupted, often apparently progressively from one side, by pit-like depressions. The production mechanism of these features is unclear but this might again point to transmission of heat to a sub-surface ice pocket that escapes explosively as a result of pressure build-up.

It is sometimes argued that, because comets are highly porous, there is almost no resistance to gas flow and pressure cannot build-up to generate catastrophic disruption of the medium through which the gas flows. This is, at best, naïve and almost certainly does not reflect reality. The sub-surface sublimation rate will be strongly

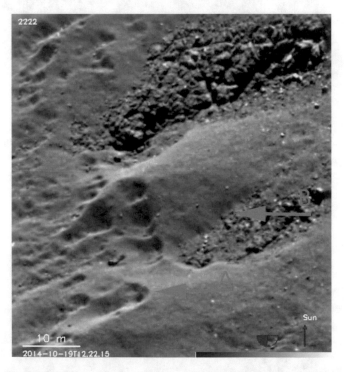

**Fig. 2.102** Pits (position A) seen in the smooth material may be indicative of explosive release of sub-surface volatiles. This example is from the Maftet region and nicely compares the rockier material with the smooth dune-like material that has been modified by the pit production mechanism to the left. Note the interface between the substrate and the dustier material at position B (Image number: N20141019T122215525ID10F22)

related to the energy input while the gas loss rate is related to the gas diffusion coefficient, $D_g$. In the simplest case of fluid diffusion in one dimension, this can be represented by Fick's law

$$j_x = -D_g \frac{dn_g}{dx} \qquad (2.136)$$

where $j_x$ is the gas flux. If the flux away from the source is slower than the source rate, then pressure must increase at the source. The source rate is related to the thermal conductivity and hence the source and loss rates are not entirely independent of each other. (A lower diffusion rate will, in general, result from good thermal contact.)

Assuming steady-state and a constant temperature across the porous layer, the 1D diffusion equation can be inverted to give the gradient in number density across a porous medium of depth. This allows us to calculate a pressure difference using the constant temperature assumption if the diffusion coefficient can be estimated. Typical values would be of the order of $10^{-5}$ m$^2$ s$^{-1}$ and, in simplified cases, is directly proportional to the porosity (Huebner et al. 2006). For a gas emission rate of $3 \times 10^{19}$ molecule m$^{-2}$ s$^{-1}$ at a depth of 5 cm (cf. Fig. 2.32), the pressure difference is over 400 Pa and therefore exceeds most estimates of the structural (tensile) strength of cometary material.

While this is a crude estimate with numerous simplifying assumptions, it illustrates that pressure can be built up under the right conditions to allow explosive events and could be a viable explanation for dust pit production and other surface disruption phenomena.

### 2.10.15 Ice Exposures

The search for exposed ice on the nuclei of comets has been somewhat frustrating. It was recognized at the time of the 1P/Halley fly-by that local variations in brightness on the nucleus were small and probably less than 50% at resolutions of ~100 m. Given the low overall albedo, it was clear at this time that pure, highly reflecting, large area, ice surfaces could not be present on the nucleus. Variations in brightness alone would, in any case, not have been sufficient to demonstrate the presence of surficial water ice. The first clear detection of exposed water ice (Fig. 2.103) was made by the Deep Impact spacecraft at 9P/Tempel 1 (Sunshine et al. 2006) using the strong infrared absorptions at 1.5 and 2.0 μm (Fig. 2.40). The total area of exposed water ice was far below that required to produce the observed water production rate and the visible brightness ratio of icy areas to non-icy areas was small. Hence other sources of subliming water ice were "hidden" from the remote sensing investigations.

Water ice was equally difficult to detect with the infrared spectrometer (VIRTIS) onboard Rosetta (Capaccioni et al. 2015) with first results noting the complete

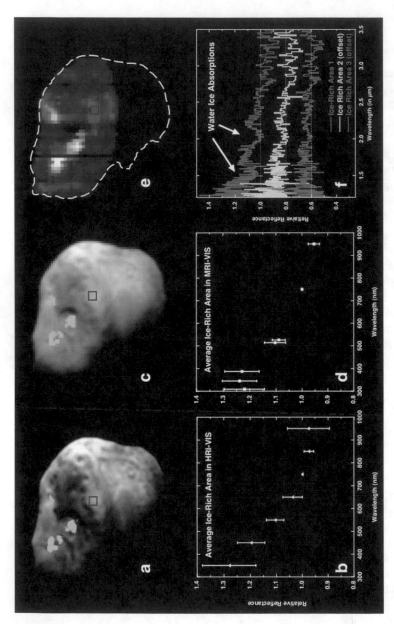

**Fig. 2.103** Maps and spectra of ice-rich areas relative to non-ice regions of the nucleus. (**a** and **b**) Deep Impact/MRI visible data (16 m pixel$^{-1}$). (**c** and **d**) Deep Impact/MRI visible data (82 m pixel$^{-1}$). (**e** and **f**) Deep Impact/IR data (120 m pixel$^{-1}$). Note that the IR scan at the highest resolution only covers IR data (120 m/pixel). Note that the IR scan at the highest resolution only cover the upper half of the nucleus, as shown. The ice-rich areas are mapped in the visible images as combinations of high 450 nm/750 nm relative reflectance (387 nm/750 nm for MRI) and low 950 nm/750 nm relative reflectance and in the IR images by the strength of absorptions at 2.0 μm. (CREDIT: NASA/UM/SAIC J. M. Sunshine et al., Science 311, 1453 (2006); published online 2 February 2006 (10.1126/science.1123632). Reprinted with permission from AAAS)

**Fig. 2.104** An eroding mesa in the Khepry region. The arrow lower left (A) shows small bright boulders that may have resulted from mass wasting of the mesa. This should be compared to the talus indicated by the arrow at the top of the image (B) (Image number: N20140903T064422578ID10F22)

absence of water ice absorption bands at 1.5, 2.0 and 3.0 μm. Unambiguous detections were finally made by combining imaging observations of bright clusters of material with infrared spectra (Pommerol et al. 2015; Barucci et al. 2016; Filacchione et al. 2016a).

Figure 2.72 shows an example of a bright cluster at optical wavelengths and a further example is shown in Fig. 2.104 in which rougher material to the left of the mesa shows similar properties. No other site in the image shows this effect. Pommerol et al. (2015) catalogued several sites where brightness differences with respect to the surroundings were sometimes as large as a factor of 5 (e.g. Fig. 2.72) prompting a search with VIRTIS that confirmed an abundance of water ice at specific bright cluster sites (Barucci et al. 2016; see also Filacchione et al. 2016a).

Areas of brighter material have rarely been seen in the Rosetta data set. In Fig. 2.105 left, we can see quasi-circular bright areas that appeared in the Hapi region in September 2014. It is conceivable that these areas have been exposed by more vigorous sublimation as the comet was approaching the Sun. The two patches seen in Bes close to the interface to Anhur (Fig. 2.105; right) were well observed and rapid changes were evident. Figure 2.106 shows patch C at two epochs less than 5 days apart. The representation is a colour composite. What is evident here is that the position of the bright patch is clearly different. Referring back to Fig. 2.105, one can see a depression. In the earlier observation, the left side of the depression was bright. Five days later, it was the right side. We were probably observing gradual exhumation of brighter material with a gradient in exhumation time from left to right.

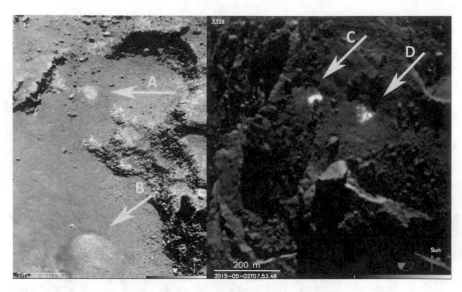

**Fig. 2.105** Bright areas on the surface of the nucleus that appeared during the approach to perihelion. Left: Bright areas in the Hapi region (A and B) in September 2014 (N20140921T133707408ID10F22). Right: Two transient bright patches (C and D) in the Bes region first seen in April 2015 (N20150502T075348404ID10F23)

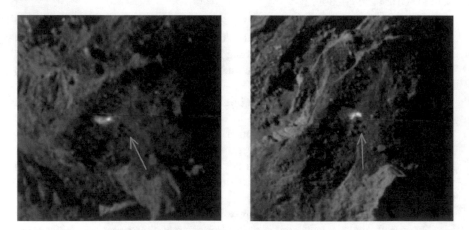

**Fig. 2.106** Transient bright material on the surface of 67P in the Bes region. This pseudo-RGB representation indicates that the material is somewhat bluer and therefore probably ice. The orientations of the two images (taken 5 days apart) are different but the same boulder is identified by the arrow. The bright material is closer to the cluster of boulders in the image to the right (Image sequences: N20150427T181810028ID30F41, N20150502T104252535ID30F41)

This did not occur at patch D. The whole surface of the brighter material in patch D seems mottled and non-uniform. For both C and D, the depressions left behind after the bright material had (presumably) sublimed were clearly visible.

The transient appearance of surface ices was also evident in infrared spectra. Filacchione et al. (2016b) identified $CO_2$ ice on the surface of a cometary nucleus for the first time but, as can be seen in Fig. 2.107, the $CO_2$ was transient and had disappeared less than 4 weeks later. 67P had not yet reached perihelion at this time and hence the heat input to the surface element was still increasing.

While "pore" ice (ice present within pores of a non-volatile matrices or condensed on the matrices) has frequently been offered as an explanation for the absence of pure ice surfaces, the bright patches are indicative that "massive" ice (volumes dominated by ice) do exist within cometary nuclei. This observation seems to be incompatible with a constantly eroding, compositionally homogeneous, surface layer.

## *2.10.16 Other Surface Changes*

For 67P, losing 0.1% of the mass (Sect. 2.3) implies a 0.1% loss in volume corresponding to around 0.6 m of loss in radius if the mass loss is equally distributed over the nucleus. The heat input as a function of latitude (Fig. 2.13) and the irregular shape (Sect. 2.2) imply that a homogeneous distribution of loss is not to be expected. We have seen evidence of surface changes during the Rosetta mission in, for example, Figs. 2.88, 2.89, 2.94, 2.95, and 2.105 (cf El-Maarry et al. 2019). However, there has not been any clear quantitative evidence of erosion presented to this point.

Figure 2.108 (left) shows part of the Anhur region which was studied by Fornasier et al. (2017, 2019). The colour enhancement of this image is a cruder version of that used by them. Fornasier et al. (2019) identified the formation of new scarps in the vicinity and showed evidence of surface erosion locally of $14 \pm 2$ m during the perihelion passage. The formation of a 14 m deep cavity has also been reported in the Khonsu region by Hasselmann et al. (2019).

Fornasier et al.'s observations also indicated significant colour differences. This is evident in the image (Fig. 2.108 left) and in low spectral resolution filter data from Rosetta/OSIRIS (Fig. 2.108 right). For the latter, the data from the sub-areas B to D have been ratioed to sub-area A and normalized at 986 nm (Fornasier et al. (2017) normalized at 535 nm). The strong blue colour comes from a 50% change in relative reflectance in sub-area B. The control areas, A and C shows only 2–3% variations with respect to each other which probably illustrates the accuracy with which the relative reflectance can be determined. Sub-area D shows around 10% relative absorption in the visible.

It is noteworthy that the blue material is in depressions although a cluster of bluer, more "rocky" material (marked by the arrow in Fig. 2.108 left) can also be seen. However, this blue material is highly localised and covers a relatively small total area of the surface. It is very tempting to interpret the bluer material as ice-rich but caution is warranted. Fine icy material should sublime fairly quickly (e.g. Fig. 2.106) so why is the material on the floor blue whereas the surrounding cliffs that are assumed to be sources of this material are not? Although higher ice content remains the most plausible explanation, other possibilities such as particle size effects in

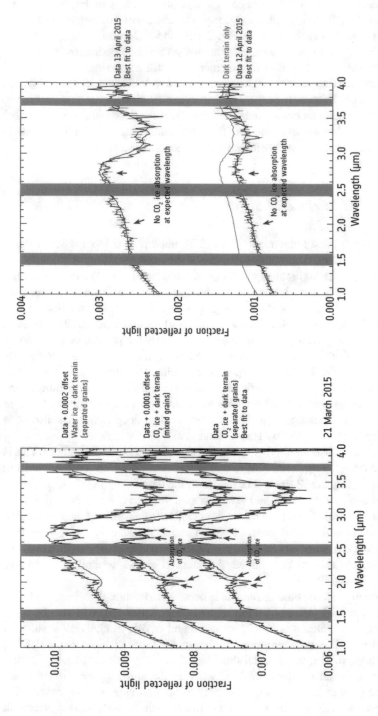

**Fig. 2.107** Spectroscopic identification of transient $CO_2$ ice in the Anhur region. Left panel: March 21, 2015 spectra showing spectral modeling results. Note that the modeling with water ice (top plot) fails to match the observed data. Right panel: April 12–13, 2015 spectra showing that the $CO_2$ ice features have disappeared. The greyed-out spectral ranges are affected by instrumental order sorting filters. From Filacchione et al. (2016b), Science, 354, 1563. Reprinted with permission from AAAS

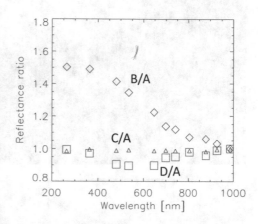

**Fig. 2.108**  Right: Enhanced colour composite of a part of the Anhur region identified by Fornasier et al. (2019) showing the relatively blue colour of several small depressions. Right: Colour ratios of the positions indicated in the image to the left showing the strong blue colouring of area B, the similarity of areas A and C, and subtle differences at position D (First image in the sequence, N20160210T081406243ID30F22)

smooth terrains should not be ignored. Fornasier et al. assumed a linear mixing model to obtain an ice/refractory mixing ratio of 20% but this should be treated with care (Yoldi et al. 2018) as will be shown in Sect. 4.14.

Quasi-circular surface changes on the smooth terrains of Imhotep and Anubis were discussed earlier (Figs. 2.89 and 2.92) but are these smooth areas accumulating or losing material in general? For another part of Anubis region, we appear to have an answer. Figure 2.109 shows part of Anubis pre- (upper left) and post- (upper right) perihelion. To the lower right, there is a DTM of the surface post-perihelion which has been published through the MiARD project.[4] Three profiles (A, B, and C) are marked. The profiles are compared in the plot to the lower left. Profile B passes through a rock which appears to have a depression adjacent to it post-perihelion marked by the arrow. This depression is not evident in the pre-perihelion data and appears in the comparison of the pre- and post-perihelion DTMs as a subtle depression post-perihelion. This is close to the limit of the capability of the MSPCD technique which is around 50 cm for this particular DTM. However, along profile A, we see significant changes with up to 2 m of material having been removed. This agrees with the visual impression one gets from the comparison of the two images in Fig. 2.109. Hence, the DTM is showing that the smooth terrain of Anubis has net mass loss up to a factor of 3 higher than the mean erosion expected. It is interesting to note here that there is no evidence for bright, icy, material in this area and there does not appear to be any obvious surface manifestation of a driving volatile.

---

[4]http://www.miard.eu/wordpress/wp-content/uploads/2018/10/D1.2_MiARD_localDTMs_v5.pdf

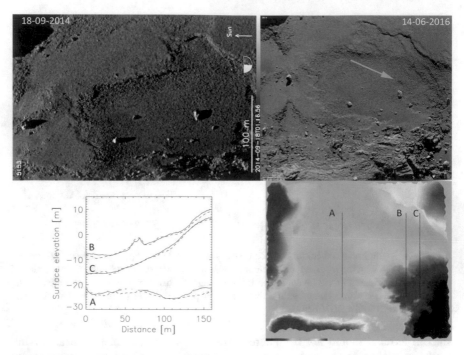

**Fig. 2.109** Part of the Anubis region of 67P. Top: Two images of the region taken before (left) and after (right) perihelion (N20140918T011656347ID10F22, N20160614T010041718ID10F22). Below right: A digital terrain model of the area derived using the MSPCD technique (Capanna et al. 2013; Credit: L. Jorda). Below left: Elevation profiles through the DTM at the positions marked on the right. Solid lines: Pre-perihelion. Dashed lines: Post-perihelion. The profiles have been offset with respect to each for clarity. Profile B passes through the rock and depression marked by the green arrow. No depression can be seen pre-perihelion. More significant mass loss is seen along profile A

## 2.10.17   Evidence for Large-Scale Mass Loss

In Fig. 1.3, we noted that 1P/Halley exhibited a large, rapid, increase in brightness several years after perihelion. Outbursts are seen relatively frequently in ground-based observations. Recent examples include C/2011 W3 (Lovejoy) (Sekanina and Chodas 2012), C/2015 $ER_{61}$ (PAN-STARRS) (Sekanina 2017), P/2010 V1 (Ikeya-Murakami), 217P/LINEAR (Sarugaku et al. 2010), C/2010 G2 (Hill) (Kawakita et al. 2014), 17P/Holmes (Watanabe et al. 2009), and 2P/Encke (Lamy et al. 2003).

These observations suggest that comets can lose mass in rapid, relatively large-scale, events. For the 17P/Holmes outburst in 2007, an isolated dust cloud was observed moving away from the nucleus suggesting an abrupt end to the mass loss rather than a slow decay as might be expected for a slow shutdown of a subliming area. The problem here lies in a lack of substantive knowledge of the mechanism(s) causing these outbursts although there are several hypotheses that have been reviewed by Gronkoski and Wesolowski (2016). Several of the early

**Fig. 2.110**  OSIRIS image showing most of the northern hemisphere of 67P. The arrow marks the Aten depression that may be a site of large scale mass loss (Image number: N20140805T214314596ID30F22)

theories were somewhat fanciful. The concepts worthy of more detailed investigation seem to be those where sub-surface volatiles are heated to produce internal pressure that destabilises the surface layer until the pressure is, perhaps violently, released.

The transition and subsequent energy release from amorphous to crystalline ice has been discussed as a means of producing outburst activity at high heliocentric distances. Given low local thermal conductivity, a runaway might occur because of the exothermic nature of the transition possibly leading to internal disruption provoking either collapse or larger scales ejections of mass. Other mechanisms might involve slow warming of CO or $CO_2$ pockets through conduction. But is there any evidence of large scale mass loss on the surface or indeed would we recognize it if we saw it?

One region that might bear further investigation is the Aten region of 67P (Fig. 2.110). This region is unusual in that it is a depression enclosed by steep-sided walls. It is neither pit-like (quasi-circular) nor is it a shallow depression structure. It is also not obviously at a junction of possible parent bodies. It gives the impression of being a "scar" in the surface. The formation of this structure has attracted relatively little attention but it is not easily explained without invoking locally inhomogeneous processes generating substantial loss despite the fact that 67P's outgassing is currently fairly close to being dominated by insolation-driven sublimation. One possibility is that it has arisen from local, larger-scale, activity.

# Chapter 3
# Gas Emissions Near the Nucleus

## 3.1 Fundamentals

We begin this chapter by looking at highly simplified models of the spatial distribution of gas in the inner coma. The simplest approach is to assume that

(a) gas emission from the nucleus is isotropic,
(b) the nucleus is a point source,
(c) the gas velocity is constant (which is another way of saying that collisions are negligible), and
(d) the gas species undergo no reactions.

We will see that close to the nucleus the first three of these assumptions are not tenable and at large distances from the nucleus, the fourth assumption isn't either! But nonetheless these assumptions are useful because their simplicity allows us to derive helpful relationships for density and column density that are straightforward but fundamental.

Under the above assumptions the local density is given by

$$n_g(r) = \frac{Q_g}{4\pi r^2 v_g} \qquad (3.1)$$

where $r$ is the cometocentric distance, $v$ is the outflow velocity and $Q$ is the total production rate. We will see below that similar equations can be used for the dust outflow and hence we will use a subscript, g, for the gas in this sub-section and subscript, d, for the dust in Chap. 4.1.

The column density along a line-of-sight can be determined by integration. Here, we define the impact parameter, $b$, as the minimum distance to the nucleus along the line-of-sight and $s$ as the distance along the line-of-sight. Hence,

© Springer Nature Switzerland AG 2020
N. Thomas, *An Introduction to Comets*, Astronomy and Astrophysics Library,
https://doi.org/10.1007/978-3-030-50574-5_3

$$r^2 = b^2 + s^2 \tag{3.2}$$

and the column density is then

$$N_g(b) = 2 \int_{s=0}^{s=\infty} n_g(r) \, ds = 2 \int_{s=0}^{s=\infty} \frac{Q_g}{4\pi \left(b^2 + s^2\right) v_g} \, ds \tag{3.3}$$

which is a standard integral leading to

$$N_g(b) = \frac{Q_g}{4bv_g} \tag{3.4}$$

By integrating the column density over a circle surrounding the nucleus at constant impact parameter, and multiplying by the impact parameter, we obtain

$$G_g = \int_0^{2\pi} N_g(b) \, b \, d\theta = \int_0^{2\pi} \frac{Q_g}{4v_g} \, d\theta = \frac{\pi Q_g}{2v_g} \tag{3.5}$$

which shows that, for free-radial outflow, the product of the impact parameter and the integral of the column density on a circle is a constant and independent of $b$. This equation can be integrated again from the nucleus outwards to a distance $b_{max}$ to give the total mass of gas within a cylindrical volume centred on the nucleus leading to

$$G_c = \frac{\pi}{2} \frac{Q_g}{v_g} b_{max} \quad . \tag{3.6}$$

This can be compared to the total mass of gas within a sphere of radius, $r_{max}$, centred on the nucleus, $G_s$, which is

$$G_s = \frac{Q_g}{v_g} r_{max} \quad . \tag{3.7}$$

The equations show a linear dependence on distance and are different by a simple constant.

The assumption of force-free radial outflow implies no interaction and consequently an angular distribution in the production rate from the point source can be incorporated into the equations in the form

$$n_g(r, \theta) = \frac{Q_g(\theta)}{r^2 v_g} \tag{3.8}$$

where $Q_g(\theta)$ describes the production rate at an angle, $\theta$, with respect to a reference direction in units of [molecule s$^{-1}$ sr$^{-1}$]. The total production rate from the source is then the integral over the solid angle, $\Omega_s$,

$$Q_g = \frac{1}{4\pi} \int_{4\pi} Q_g(\theta) d\Omega_s \qquad (3.9)$$

A prudent choice of $Q_g(\theta)$ can result in analytical solutions for the total production rate and column density. We look at when these assumptions become invalid in Sect. 3.4 but first we need to establish the species in the gas coma and the means by which they are observed.

## 3.2 Major Species and Their Emissions

It is now known that the compositions of the gas comae of comets are extremely rich and exhibit diversity between comets (Eberhardt 1999; Biver et al. 2002a; Rubin et al. 2019). There are many, sometimes rather complex, species present. However, the comae of comets are dominated by three major species; $H_2O$, $CO_2$ and CO (Table 3.1). There is variability in the relative abundances of these three molecules between comets. There is also variability with heliocentric distance and there is possibly variability arising from differences in the source locally on the nucleus. However, a good starting point is to assume that the relative abundances can be described as $H_2O:CO_2:CO = 0.88:0.04:0.03$ with around 5% attributable to other (minor) species. We shall look at variations with respect to gas composition later but it is important to note that there is increasing evidence for inhomogeneity in the compositions of individual comets including variations in the organics as indicated in C/2013 V5 (Oukaimeden) (DiSanti et al. 2018).

It should be noted here that the table from Le Roy et al. (2015) does not include $O_2$ which was reported by Bieler et al. (Bieler et al. 2015b). It was found at a mixing ratio of $3.80 \pm 0.85$ relative to water ($H_2O = 100$) and thus comparable to the mixing ratios of $CO_2$ and CO, making it a 4th "major species". The need for high resolution mass spectrometers in the investigation of coma chemistry is well illustrated by this observation (Fig. 3.1). Rubin et al. (2015) determined that the $O_2$ abundance was consistent with Giotto-based observations of 1P/Halley and the recent estimate of the bulk abundance by Rubin et al. (2019) essentially confirms this. This observation has provoked considerable discussion as to whether the $O_2$ is primordial (Mousis et al. 2016; Taquet et al. 2016; Eistrup and Walsh 2019) or produced in situ (Dulieu et al. 2017; Yao and Giapis 2017) because of the implications associated with a primordial origin. It is notable that Galli et al. (2018) irradiated pure water ice with energetic electrons (0.2 to 10 keV) and produced $O_2$ which suggests that at least surficial water ice could be an in situ modern day source.

**Table 3.1** Composition of the coma of 67P relative to water vapour obtained in October 2014, 10 km from the centre of mass (Le Roy et al. 2015)

| Species | Northern summer hemisphere [$H_2O = 100$] | Southern winter hemisphere [$H_2O = 100$] | Selected derived bulk abundance [$H_2O = 100$] |
|---|---|---|---|
| $H_2O$ | 100[a] | 100[a] | 100[a] |
| $O_2$ | | | $3.1 \pm 1.1$ |
| CO | 2.7 | 20 | $3.1 \pm 0.9$ |
| $CO_2$ | 2.5 | 80 | $4.7 \pm 1.4$ |
| $CH_4$ | 0.13 | 0.56 | $0.34 \pm 0.07$ |
| $C_2H_2$ | 0.045 | 0.55 | |
| $C_2H_6$ | 0.32 | 3.3 | $0.29 \pm 0.06$ |
| $CH_3OH$ | 0.31 | 0.55 | $0.21 \pm 0.06$ |
| $C_2H_5OH$ | | | $0.039 \pm 0.023$ |
| $C_3H_8$ | | | $0.018 \pm 0.004$ |
| $C_6H_6$ | | | $0.00069 \pm 0.00014$ |
| $C_7H_8$ | | | $0.0062 \pm 0.0012$ |
| $H_2CO$ | 0.33 | 0.53 | $0.32 \pm 0.10$ |
| HCOOH | 0.008 | 0.03 | $0.013 \pm 0.008$ |
| $CH_2OHCH_2OH$ | 0.0008 | 2.5e-3 | $0.011 \pm 0.007$ |
| $HCOOCH_3$ | 0.004 | 0.023 | $0.0034 \pm 0.0020$ |
| $CH_3CHO$ | 0.01 | 0.024 | $0.047 \pm 0.017$ |
| $N_2$ | | | $0.089 \pm 0.024$ |
| $NH_2CHO$ | <1e-4 | <1e-3 | |
| $NH_3$ | 0.06 | 0.15 | $0.67 \pm 0.20$ |
| HCN | 0.09 | 0.62 | $0.14 \pm 0.04$ |
| HNCO | 0.016 | 0.031 | $0.027 \pm 0.016$ |
| HNC | b | b | |
| $CH_3CN$ | 0.006 | 0.016 | $0.0059 \pm 0.0034$ |
| $HC_3N$ | <2e-5 | <5e-4 | $0.00040 \pm 0.00023$ |
| S | | | $0.46 \pm 0.36$ |
| $H_2S$ | 0.67 | 1.75 | $1.10 \pm 0.46$ |
| OCS | 0.017 | 0.098 | $0.041^{+0.082}_{-0.020}$ |
| SO | 0.004 | 0.0014 | $0.071^{+0.142}_{-0.037}$ |
| $SO_2$ | 0.011 | 0.041 | $0.127^{+0.254}_{-0.064}$ |
| CS | c | c | |
| $CS_2$ | 0.003 | 0.024 | $0.0057^{+0.0114}_{-0.0028}$ |
| $S_2$ | 0.0004 | 0.0013 | $0.0020^{+0.0040}_{-0.0010}$ |

The list contains all species found at comets prior to Rosetta and the corresponding variation in abundance measured by the ROSINA mass spectrometer above the northern and southern hemispheres of the comet. The right column provides the bulk abundance determined by Rubin et al. (2019) and references therein
[a]Definition
[b]Cannot be distinguished from HCN
[c]Cannot be resolved from $CO_2$

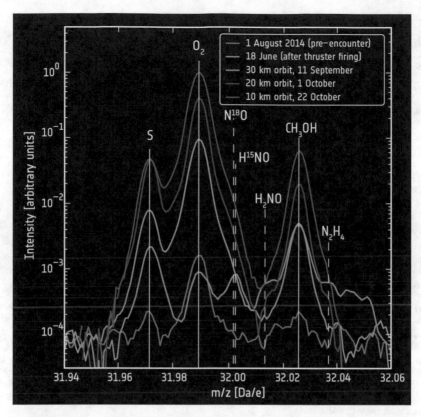

**Fig. 3.1** ROSINA observations of the inner coma of 67P focussing on the region of the spectrum near 32 atomic mass units per charge. Note the nice separation of S, $O_2$ and $CH_3OH$ at these resolutions which allows an accurate determination of the abundance of $O_2$ (http://blogs.esa.int/rosetta/2015/10/28/first-detection-of-molecular-oxygen-at-a-comet/) (Data from Bieler et al. 2015b)

The immediate source of a particular species in the coma is sometimes unclear. The nucleus is, of course, the source of all the mass in the coma. However, a particular gas species may be released from, for example, dust that has already left the nucleus. In such a case, the species is described as coming from an extended source. $H_2O$ and $CO_2$ are widely assumed to be emitted from the nucleus itself. In the case of water vapour, this assumption is not entirely straightforward. For example, there might be a minor contribution to the water vapour in the coma resulting from the release of water of hydration. The major concern, however, is in the release of $H_2O$ from icy particles or "chunks" emitted from the nucleus through activity. Observations of 103P/Hartley 2 demonstrated that emission of $CO_2$, which has a lower sublimation temperature, may drag out chunks of water ice (Fig. 3.2) which then sublime in the coma rather than at the nucleus itself. This emission from an extended source can have profound effects by both increasing the total gas production relative to simple surface sublimation (the subliming area is larger) and slowing

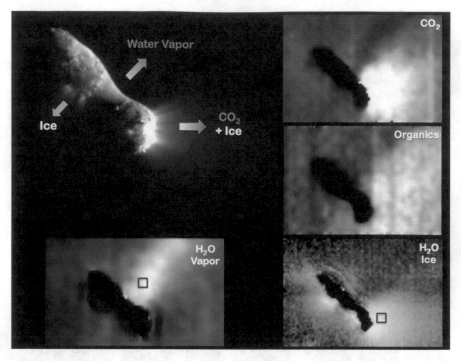

**Fig. 3.2** The spatial distributions of major species emitted from the nucleus of 103P/Hartley 2. Note the diverse distributions of the major species and the ejection of $H_2O$ ice in the direction where most $CO_2$ is being emitted. (From A'Hearn et al. 2011, Science, 332, 1396. Reprinted with permission from AAAS)

the radial outflow of the gas coma by dumping, fresh, slow moving molecules into a rapidly expanding flow field.

While $CO_2$ may be less influenced by these effects, evidence that a major fraction of CO in the coma was coming from an extended source was already apparent at the time of Giotto (Eberhardt et al. 1987) with other supporting observations seen at, for example, 29P/Schwassmann-Wachmann 1 (Gunnarsson et al. 2002) where the extended source was a factor of >3 higher than the nucleus source at distances out to 20,000 km from the nucleus. The splitting of polymerised organics on dust grains in the coma is widely assumed to be responsible although release of CO from dissociation of more complex, less volatile molecules subliming from the particles remains a plausible alternative (Brucato et al. 1997).

Probing of these major species can be achieved by spectroscopic measurements. The electronic states of atoms and molecules are, of course, quantized. Molecules with two or more atoms can also store energy through the additional degrees of freedom arising from vibration and rotation. The vibrational and rotational levels are also quantized. Relaxation from an excited state to a lower energy state can occur with release of a photon.

Transitions from one electronic state to a lower level will result in photo-emission at visible or UV wavelengths ($\lambda \lesssim 1$ μm). Vibrational transitions occur in the near to mid-infrared ($\lambda \sim 1$–100 μm) while rotational transitions result in emission in the far-infrared and microwave ($\lambda \gtrsim 100$ μm). In a first approximation, the Born-Oppenheimer approximation, these types of transitions can be treated separately (Lopez-Puertas and Taylor 2001).

The interpretation of the resulting emissions requires knowledge of the processes involved and the density regime that the emitting species find themselves in. These processes can be separated into collisional excitation and radiative processes (Bockelée-Morvan et al. 2004a). Coma gases have temperatures that are very much less than ~1000 K so that electronic and vibrational excitations occur mainly as a result of radiative processes (Combi 1996). Irradiance by the Sun is the main source of the radiation in this case although other processes, such as electron impact excitation, prompt emissions arising from dissociation reactions, and thermal collisional losses, are of importance.

In the innermost coma of an active comet, local thermal equilibrium (LTE) may be reached when there are sufficient collisions to thermalize the rotational populations of the ground vibrational states of the major species at the kinetic temperature of the gas (Zakharov et al. 2007). The collisions lead to a Maxwellian distribution for the velocity distribution function (VDF). We shall see in Sect. 3.4 that although the gas is tenuous, the VDF does indeed tend towards a Maxwellian distribution within a few kilometres of the nucleus surface for an active comet. However, with the gas density dropping rapidly with distance (Eq. 3.1) and the mean free path increasing (Fig. 3.19), collisions become infrequent and the VDF can no longer be described as a Maxwellian. Hence, with such low temperatures, all of the vibrational bands are in non-LTE everywhere in the coma and the concept of LTE also becomes meaningless for the rotational populations beyond a few tens of kilometres even for an active nucleus.

For active comets, optical thickness effects need to be accounted for. In particular, the rotational lines of $H_2O$ are subject to self-absorption effects in active comets and are important in Rosetta observations of the innermost coma of 67P. Consequently, it is necessary to make a short diversion and look at the radiative transfer equation for an optically thick medium before going into more detail.

In the general case, light can be removed from a beam through both absorption and scattering with extinction being the sum of the absorption and scattering cross-sections. Let us begin by assuming that scattering can be ignored. The decrease in radiance at a specific frequency when travelling along a path of interval, $ds$, through an absorbing medium is given by

$$dI_\vartheta = -\kappa_\vartheta I_\vartheta ds \tag{3.10}$$

where $\kappa_\vartheta$ is the absorption coefficient in units of $[\text{m}^{-1}]$. The absorption coefficient can be expressed in terms of the absorption cross-section, $\sigma_{abs}$, which is frequency dependent, through

$$\kappa_\vartheta = n_g \sigma_{abs} \tag{3.11}$$

One can immediately see here that if we were to include scattering then

$$\kappa_\vartheta = n_g(\sigma_{abs} + \sigma_{sca}) = n_g \sigma_{ext} \tag{3.12}$$

where $\sigma_{sca}$ and $\sigma_{ext}$ are the scattering and extinction cross-sections respectively and we would then call $\kappa_\vartheta$ the extinction coefficient. Returning to our simpler case, integration along a beam leads to the equation

$$I_\vartheta(s) = I_\vartheta^0 e^{-\int_0^s \kappa_\vartheta \, ds} \tag{3.13}$$

where the 0 indicates the initial radiance and $s$ is the point of interest along the path. The optical depth can be written here as

$$\tau_\vartheta = \int_0^s \kappa_\vartheta \, ds \tag{3.14}$$

Light can also be added to the beam by emission from the gas and that can be described as

$$dI_\vartheta = j_\vartheta ds \tag{3.15}$$

where $j_\vartheta$ is the emission coefficient. In thermodynamic equilibrium, Kirchhoff's law holds and this relates the two coefficients to a function that is solely dependent upon temperature, i.e.,

$$f_\vartheta(T) = \frac{j_\vartheta}{\kappa_\vartheta} \tag{3.16}$$

In a more general case, other than thermodynamic equilibrium, a source function is required such that

$$J_\vartheta = \frac{j_\vartheta}{\kappa_\vartheta} \tag{3.17}$$

and hence the emission term, adding light to the beam, becomes

$$dI_\vartheta = \kappa_\vartheta J_\vartheta ds \tag{3.18}$$

Combining the absorption and emission terms, we get

$$\frac{dI_\vartheta}{ds} = -n_g \sigma_{abs}(I_\vartheta - J_\vartheta) \tag{3.19}$$

which is the radiative transfer equation. The emissions complicate the formal solution for the radiance at point $s$ along the path because the attenuation of the additional radiance produced along the path must also be accounted for. In the case of $H_2O$ in a cometary coma this is a significant issue. The basic equation is then given by

$$I_\vartheta(s) = I_\vartheta^0 e^{-\int_0^s n_g(s')\sigma_{abs}(s')\, ds'} + \int_0^s n_g(s')\sigma_{abs}(s')J_v(s')\, e^{-\int_{s'}^s n_g(s'')\sigma_{abs}(s'')\, ds''} ds' \tag{3.20}$$

where we use $'$ to indicate integration along the complete path from 0 to point s, and $''$ to indicate integration from a point somewhere between 0 and $s$ to $s$. Both Chamberlain and Hunten (1987) and Lopez-Puertas and Taylor (2001) give clear descriptions of how to reach this equation. In most physically realistic cases, it is necessary to solve for the radiance numerically.

In the case of LTE, the source function is given by Planck's radiation law

$$J_\vartheta = B_\vartheta(T) \tag{3.21}$$

where $B_\vartheta(T)$ is defined through Eq. (2.8). The rotational lines are evident at sub-mm wavelengths and, consequently, simplification by using the Rayleigh-Jeans law,

$$B_\vartheta(T) = \frac{2\vartheta^2\, kT}{c^2} \tag{3.22}$$

is also possible when $h\vartheta/kT << 1$. Note also that the brightness temperature can be computed here as shown in Eq. (2.21).

Individual lines are broadened by three processes

- Natural broadening,
- Doppler broadening and,
- Pressure broadening.

Natural line broadening is usually of little consequence and can be ignored for this application. For Doppler broadening, the frequency dependence of the absorption cross-section is given by (Lopez-Puertas and Taylor 2001)

$$k_\vartheta(\vartheta) = \frac{S_s}{\alpha_D \sqrt{\pi}} e^{-\frac{(\vartheta-\vartheta_0)^2}{\alpha_D^2}} \tag{3.23}$$

where $S_s$ is the spectral line intensity and the Doppler width, $\alpha_D$, is given by

$$\alpha_D = \frac{\vartheta_0}{c} \sqrt{\frac{2R_g T}{M_M}} = \frac{\vartheta_0}{c} \sqrt{\frac{2kT}{m_g}} \tag{3.24}$$

where $R_g$ is the ideal gas constant and $M_M$ is the molar mass in [kg mol$^{-1}$]. The Doppler width is often expressed slightly differently as the full width (or half width) at half maximum (FWHM) which is

$$\Delta\vartheta_D = 2\frac{\vartheta_0}{c} \sqrt{\frac{2kT \log_e 2}{m_g}} \tag{3.25}$$

These equations are sometimes seen in an alternative but analogous form with the absorption cross-section written as

$$\sigma_{abs} = S_s\, \varphi(\vartheta, \vartheta_0) \tag{3.26}$$

where $\varphi(\vartheta, \vartheta_0)$ is the line-profile function that takes into account the Doppler broadening. Here, $\vartheta_0$ is the rest frequency of the emission line. The function can be written as

$$\varphi(\vartheta, \vartheta_0) = \frac{c}{v_{th}\vartheta_0\sqrt{\pi}} e^{\frac{-c^2(\vartheta-\vartheta_0)^2}{\vartheta_0^2 v_{th}^2}} \tag{3.27}$$

where $c$ is the speed of light and $v_{th}$ is the most probable velocity given by

$$v_{th} = \sqrt{\frac{2kT}{m_g}} \tag{3.28}$$

A variable similar to $\alpha_D$ can be defined for pressure broadening so that (Lopez-Puertas and Taylor 2001)

$$\alpha_L(STP) = \frac{1}{2\pi c t_{STP}} \tag{3.29}$$

where $t_{STP}$ is the mean time between collisions at standard temperature and pressure (STP). Then

$$k_\vartheta(\vartheta) = \frac{S_s}{\pi} \frac{\alpha_L}{(\vartheta - \vartheta_0)^2 + \alpha_L^2} \tag{3.30}$$

where

$$\alpha_L(p, T) = \alpha_L(STP)\frac{p}{p_{STP}}\sqrt{\frac{T_{STP}}{T}} \tag{3.31}$$

Here $p_{STP}$ and $T_{STP}$ define STP conditions ($p_{STP} = 10^5$ Pa and $T_{STP} = 273.15$ K). Combining the Doppler and pressure broadening gives the Voigt profile

$$k_\vartheta(\vartheta - \vartheta_0) = \frac{S_s}{\alpha_D\sqrt{\pi}}\frac{y}{\pi}\int_{-\infty}^{+\infty}\frac{e^{-t^2}\,dt}{y^2 + (x - t)^2} \tag{3.32}$$

where we have defined two constants, $y$ and $x$, as $y = \alpha_L/\alpha_D$ and $x = (\vartheta - \vartheta_0) / \alpha_D$. Caution should be exercised over normalization (Huang and Yung 2004).

Returning to our specific application, if we consider the major species and their daughter products, many of the most interesting cases arise from rotational and vibrational spectral lines at sub-mm and infrared wavelengths, respectively. Each of the major species are different in that CO is a diatomic molecule, $CO_2$ is triatomic but linear and $H_2O$ is also triatomic but non-linear. Thus from a rotational point of view, CO is the simplest to study.

Assuming the molecules to be rigid rotators, the energy involved in a transition from one rotational level to another is given by

$$E_{J+1} - E_J = \frac{h^2}{4\pi^2}\frac{(J+1)}{I_m} \tag{3.33}$$

where $I_m$ is the moment of inertia and given by

$$I_m = \frac{m_1 m_2}{m_1 + m_2}r_i^2 \tag{3.34}$$

where $m_1$ and $m_2$ are the masses of the two atoms (in our case, C and O) and $r_i$ is the interatomic separation. $J$ is the total angular momentum quantum number and the selection rule states that $\Delta J = \pm 1$.

The rigid rotator model does not hold for highly excited states. The molecule can be thought of as "stretching" in response to centrifugal forces as J increases and more sophisticated treatments become necessary. In comets, however, the less excited CO $J(2-1)$ line has been observed, for example (Gunnarsson et al. 2002) and the CO $J$ $(5-4)$ line was included in one of the bands observed by the MIRO instrument on Rosetta (Gulkis et al. 2007). The main CO emissions in the millimetre wavelength range are shown in Table 3.2 where one can see the almost monotonic relationship in frequency between the transitions in these less excited states.[1]

Asymmetric molecules tend to have rotational lines at millimetre and sub-millimetre wavelengths and the study of $H_2O$ through these lines is of major importance. The lines are particularly interesting because of the existence of

---

[1]See also https://physics.nist.gov/cgi-bin/micro/table5/start.pl

**Table 3.2** Major CO and CS rotational emission line frequencies (Schöier et al. 2005; Gottlieb et al. 2003)

| CO | | CS | |
|---|---|---|---|
| Transition | Frequency [GHz] | Frequency [GHz] | Comment |
| $J = 1 \rightarrow 0$ | 115.271 | 48.991 | |
| $J = 2 \rightarrow 1$ | 230.538 | 97.981 | |
| $J = 3 \rightarrow 2$ | 345.796 | 146.969 | CS line discovered in interstellar medium by Penzias et al. (1971) |
| $J = 4 \rightarrow 3$ | 461.041 | 195.954 | CS line observed in 19P/Borrelly by Bockelée-Morvan et al. (2004a) |
| $J = 5 \rightarrow 4$ | 576.268 | 244.935 | CO line observed by MIRO at 67P. CS line observed from the ground by several observers |
| $J = 6 \rightarrow 5$ | 691.473 | 293.912 | |
| $J = 7 \rightarrow 6$ | 806.652 | 342.883 | |
| $J = 8 \rightarrow 7$ | 921.800 | 391.847 | |
| $J = 9 \rightarrow 8$ | 1036.912 | 440.803 | |
| $J = 10 \rightarrow 9$ | 1151.985 | 489.751 | |

techniques to provide very high resolution spectra (e.g. Hartogh 1997). This can then be used to determine the exact width of isolated lines. Water emission at 557 GHz ($18.58$ cm$^{-1}$) is one of the principal emissions of interest (Table 3.3) and, for this reason, a microwave radiometer experiment (MIRO) was selected for flight on the Rosetta spacecraft. This emission line is particularly strong and optical thickness effects must be taken into account for lines of sight passing close to the nucleus even at relatively high heliocentric distances. The use of Eq. (3.20) with the optical thickness terms is therefore required. However, observing the nearby spectral lines of the isotopologues of $H_2O$ can be used to remove (or reduce) optical thickness effects under the assumption that relative abundances of the isotopologues to the main isotopologue are known and invariant. The frequencies of the isotopologues of $H_2O$ for the 1(1,0)-1(0,1) transition are shown in Table 3.3 and were all observed by the Rosetta/MIRO experiment.

The FWHM of the thermally broadened line is given by Eq. (3.25) which shows that a resolving power of ~300 kHz is needed to determine the molecular temperature. The spectral resolution of the MIRO spectrometer on Rosetta was around 44 kHz.

When viewing water vapour in the coma above the limb of the nucleus, one sees the species in emission against the low temperature of deep space (nominally 2.7 K). On the other hand, the nucleus can provide an emission source with the water molecules in the line of sight from the nucleus to the instrument absorbing that emission because the water vapour is colder than the nucleus as a consequence of the initial expansion. Hence, in this case one sees the water vapour in absorption. This effect was first expressed in Kirchhoff's three laws of spectroscopy which are

1. A solid, liquid, or dense gas emits light at all wavelengths.

**Table 3.3** Frequencies of the 1(1,0)–1(0,1) transitions in water isotopologues

| Isotopologue | Frequency of 1(1,0)–1(0,1) transition [GHz] |
|---|---|
| $H_2^{16}O$ | 556.936002 |
| $H_2^{17}O$ | 552.020960 |
| $H_2^{18}O$ | 547.676440 |

2. A low density, hot gas seen against a cooler background emits an emission line spectrum.
3. A low density, cool gas in front of a hotter source of a continuous spectrum creates an absorption spectrum.

This is illustrated in Fig. 3.3 with some data acquired on 4 November 2014 by the MIRO experiment. The observation was relatively early in the mission with the comet at a fairly high heliocentric distance and a low water production rate (~2 kg s$^{-1}$). The sub-mm line of the main isotopologue of water vapour was therefore not optically thick. The lower line in the plot shows water vapour ($H_2^{16}O$) in emission and corresponds to the second of Kirchhoff's laws—a relatively warm gas viewed against a colder background. The upper line shows the same line but now in absorption corresponding to the third of Kirchhoff's laws—a relative cold gas viewed against the warmer background of the nucleus. The nucleus here was not very warm because it was being viewed from a phase angle of 114°, i.e. the sub-spacecraft point was above the nightside. However it was sufficiently warm to show the rapidly cooling gas in absorption.

Another point of note in Fig. 3.3 is the Doppler shift between the two lines. The observation towards to the nucleus shows a negative Doppler shift of the line—the gas being observed is mostly moving towards the instrument along the line of sight. Looking off the limb, however, we see that much of the gas is moving away from the observer. This is qualitatively consistent with most of the gas emission being towards the Sun with the observer being at a phase angle of 114°.

As the gas production rate increases, observations of $H_2^{16}O$ close to the nucleus become increasing susceptible to optical depth effects. We illustrate this with a model calculation using Eq. (3.20) and the output from a gas dynamics simulation. The latter provides the $H_2O$ density, rotational temperature and line-of-sight gas velocity as input (Fig. 3.4) for the radiative transfer model. A 110 K nucleus background has been assumed. The results (Fig. 3.5) show the effect of opacity on the $H_2^{16}O$ line and the appearance of the $H_2^{18}O$ line which is not saturated. The model here assumes LTE in which the source function is given by the Planck function.

Under the assumption of LTE, we can now attempt to fit MIRO data. An example of two fits to an observation from near the time of perihelion (10 July 2015) is given in Fig. 3.6. The self-absorption of the $H_2^{16}O$ line is evident in the centre panel. The $H_2^{18}O$ line is seen in the lower panel and is not saturated. The dashed lines were computed with a surface temperature model based on Eq. (2.101). The models suggest that the velocity of the gas is not high enough. For the dot-dashed lines, the surface temperature has been increased proportional to the cosine of the solar

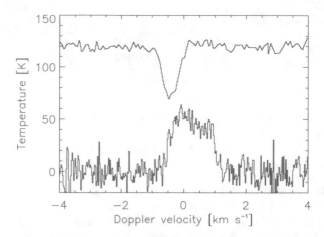

**Fig. 3.3** MIRO observations from 4 November 2014. The top plot shows an observation at 2014-11-04T03:20:00 when the instrument was pointing towards the nucleus. The lower plot shows an observation at 2014-11-04T12:48:37 (roughly 9.5 h later) when the line of sight was off the limb in approximately the sunward direction. The phase angle was similar for both observations and ~ 114°. The plot illustrates the difference between observing water vapour in absorption and emission

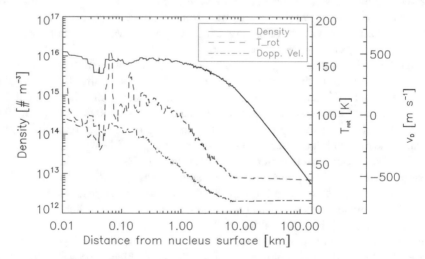

**Fig. 3.4** Output from a DSMC model giving the density, rotation temperature and Doppler velocity along the line of sight to the nucleus. These rather typical values have been used as input for a radiative transfer calculation of the shapes of the 557 GHz water lines. Note the increase in the gas velocity towards the observer and the rapid drop in temperature and density

zenith angle and provides a better result. This will be discussed later in the context of surface porosity and sub-surface sublimation. Note that the high temperature wing of the $H_2^{16}O$ line is seen in the higher temperature model (although the fit here is not ideal).

Marshall et al. (2017a, 2017b) used the isotopologues to produce estimated water production rates for 67P through the Rosetta mission by integrating over the

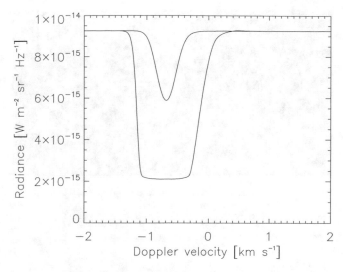

**Fig. 3.5** The modelled radiance from the nucleus using the data in Fig. 3.4 as input. The lower curve is for the $H_2^{16}O$ line. It is clearly saturated. The $H_2^{18}O$ line is the upper curve and is not saturated in this case because of the lower abundance of this isotopologue

observed line widths resulting in the production rates shown in Fig. 3.7. The subset of MIRO data used was spectra acquired in absorption with the instrument viewing the nucleus as the example in Fig. 3.3 shows.

In the resulting measurements by Marshall et al., there is significant variance in the resulting production rates which may be the result of the method. Much of the data was acquired close to the terminator where large gradients in the gas column density are to be expected. Nonetheless a strong latitudinal dependence of the gas emission was evident as can be seen in Fig. 3.8 where observations acquired near perihelion with the sub-solar point at high southern latitudes ($>32°$ S) have been extracted and plotted against sub-spacecraft latitudes. The north-south asymmetry is clearly evident.

The MIRO dataset is large and quite rich. It also includes measurements of CO ($J = 5 \rightarrow 4$) line as well as the ammonia (10–00) line (572.498 GHz) and three lines of methanol. A surprising result, however, was that the CO line is extremely weak (and mostly undetectable) in the data set.

Vibrational bands of the three major species (CO, $CO_2$, and $H_2O$) can be found in the 2.6–4.8 µm wavelength range. The vibration of the CO molecule is not as straightforward as the rotation because the stretching of the molecule is non-linear when compared to a mechanical spring and hence the harmonic approximations are inadequate. This can be overcome by setting up an expression for the potential energy as a function of the separation of the two atoms with the expression having sufficient terms to fit the anharmonicity. In the simplest case, the quantized energy is given by the equation

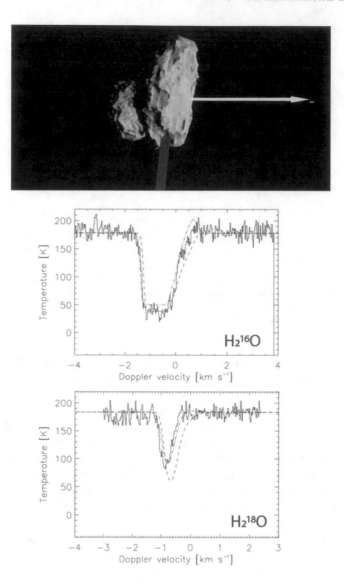

**Fig. 3.6** Fits to MIRO observations acquired on 10 July 2015 with the instrument looking at the nucleus as indicated by the diagram in the uppermost panel. (The Sun is to the right.) The centre and lower panels are for the $H_2^{16}O$ line and the $H_2^{18}O$ line respectively. Here the solid histogram style line gives the observations. The dashed lines are LTE models using the temperature as given by Eq. (2.101) for surface sublimation. The dot-dashed lines assume that the gas is emitted from the subsurface and is heated as it passes through a porous desiccated surface layer. This gives more energy to the gas and produces a better fit to the data

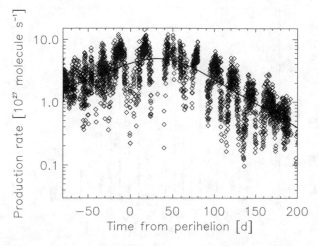

**Fig. 3.7** MIRO measurements of the water production rate at 67P through perihelion. A fit has been made to the data giving a peak production rate of $5 \ 10^{27}$ molecule s$^{-1}$ around 38 days after perihelion. (Data from Marshall et al. 2017b)

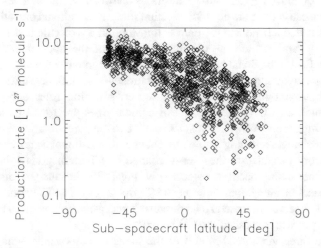

**Fig. 3.8** MIRO observations from the 60 day period when the sub-solar latitude was further south than $-32°$. The plot shows production rate measurements as a function of sub-spacecraft latitude and indicates the strong asymmetry in water production between the northern and southern hemispheres at this time. A crude polynomial fit is also shown indicating a factor of 4–5 difference between the south and the north at mid-latitudes. (Data from Marshall et al. 2017b)

$$E = \left(v_q + \frac{1}{2}\right)v_v - \left(v_q + \frac{1}{2}\right)^2 v_v X \qquad (3.35)$$

where $v_q$ is the vibrational quantum number, $v_v$ is the harmonic wavenumber and X is an anharmonicity constant (e.g. Lopez-Puertas and Taylor 2001). Transitions that are forbidden according to a harmonic oscillator model do occur because of the non-linearities and thus the vibrational spectrum becomes more complex as $\Delta v_q$ can

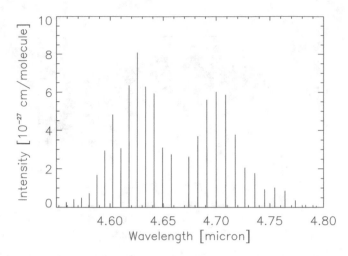

**Fig. 3.9** The band structure of the ro-vibrational transitions of CO computed for 100 K from HITRAN database (Rothman et al. 2005)

take values of $\pm 1$, $\pm 2$, $\pm 3$, etc. Transitions between vibrational levels are not constrained because there are no strict selection rules. A vibrational transition between two states is called a band and the precise transition in a molecule is denoted as ($v_q'$, $v_q''$) where $v_q'$ and $v_q''$ are the quantum numbers of the upper and lower states respectively. Each vibrational band can be further split into rotational levels.

Transitions involving changes in both vibrational and rotational states are referred to as ro-vibrational transitions. Here, however, selection rules apply. In linear molecules such as CO (and $CO_2$), the rotational lines are found as simple progressions at higher and lower frequencies relative to the pure vibrational frequencies. The energy change resulting from the change in rotational state can be either subtracted from or added to the energy change of vibration giving rise to the P- and R- branches of the spectrum, respectively (Fig. 3.9). The names, P and R branch, are of historical origin (Lopez-Puertas and Taylor 2001). The CO band at 4.67 μm was not well observed in Rosetta data because it was weak compared to hot bands of water at 4.7 μm which masked the CO signal.

There are three vibrational modes of the linear $CO_2$ molecule—the symmetric stretch ($v_1$), the bending motion ($v_2$), and the asymmetric stretch ($v_3$). There are animations of these modes available on numerous websites.[2] The $v_1$ vibration of $CO_2$ is inactive because there is no net change in the dipole moment associated with the motion. (It is, however, Raman active although this is of little relevance here.) The other stretching motion ($v_3$) is active but has a restriction in that any transition must be accompanied by a rotational transition. The band is centred around 2340 cm$^{-1}$ (4.27 μm). The $v_2$ bending motion does not have this restriction. This band is centred around 667 cm$^{-1}$ (15.0 μm) and is at a wavelength close to the peak of the thermal

---

[2]E.g. the bending motion is seen at https://www.youtube.com/watch?v=H-aUVqKrybw

**Table 3.4** Vibrational band origins of the main modes for water vapour and its isotopologues and $CO_2$ (Tennyson et al. 2009, 2010, 2013, 2014)

| Motion | Symmetric stretch | Bending | Asymmetric stretch | Relative abundance [%] |
|---|---|---|---|---|
| Isotopologue | $\nu_1$ [cm$^{-1}$] | $\nu_2$ [cm$^{-1}$] | $\nu_3$ [cm$^{-1}$] | |
| $H_2^{16}O$ | 3657.05 | 1594.75 | 3755.93 | 99.729 |
| $H_2^{17}O$ | 3653.14 | 1591.32 | 3748.32 | 0.0370 |
| $H_2^{18}O$ | 3649.69 | 1588.28 | 3741.57 | 0.20394 |
| $HD^{16}O$ | 2723.68 | 1403.48 | 3707.47 | 0.0299 |
| $D_2^{16}O$ | 2671.65 | 1178.38 | 2787.72 | 2.245 10$^{-6}$ |

**Table 3.5** The g-factors for infrared bands of the three major species (Debout et al. 2016)

| Molecule | Band [micron] | g-factor at 1 AU [photon s$^{-1}$ molecule$^{-1}$] |
|---|---|---|
| $H_2O$ | 2.69 | 3.16 10$^{-4}$ |
| $CO_2$ | 4.27 | 2.69 10$^{-3}$ |
| CO | 4.67 | 2.50 10$^{-4}$ |

emission from the Earth's surface making it an important transition for studies of the Earth's greenhouse effect.

In the case of water vapour, the symmetric stretch is active because the molecule is not linear. The main vibrational transitions of water vapour and its isotopes are shown in Table 3.4 (Tennyson et al. 2009, 2010, 2013, 2014) where the frequencies of the band origins are given. The O-H symmetric stretch (in the absence of a deuterium atom) occurs at a wavelength of 2.69 μm. It is typically the three wavelengths indicated in Table 3.5 that are targeted by infrared spectroscopy to study the coma emissions of the major species. Figure 3.2 was constructed by observing the respective bands of $CO_2$ and $H_2O$ in the vicinity of the nucleus of 103P/Hartley 2.

Polyatomic molecules exhibit a Q-branch which corresponds to transitions where $\Delta J = 0$. The factor determining whether a vibrational transition can occur without any rotational transition is whether the component of the dipole moment along the axis of symmetry changes when the vibrational state changes (Lopez-Puertas and Taylor 2001). $CO_2$ is an example where the $\nu_2$ bending motion has a Q-branch but the $\nu_3$ asymmetric stretch does not.

In true thermodynamic equilibrium, the radiative field is blackbody radiation and the source function is given by the Planck function. In LTE, the source function is still given by the Planck function but the radiative field can differ from a blackbody. In order for this to hold, in LTE, it is necessary to assume that a kinetic temperature can be defined and that this temperature is maintained locally by collisions. It remains possible that the rotational and vibrational temperatures of the molecules are different and LTE can then still be assumed. However, if there are insufficient collisions, the source function will differ from a Planck function in the presence of a non-zero radiative field.

Collisional processes dominate in the innermost comae of active comets and lead to LTE (e.g. Bensch and Bergin 2004; Sect. 3.4). We shall see that for 67P, this could be assumed within a few nucleus radii of the surface once production rates exceeded 20–40 kg s$^{-1}$ (Fig. 3.19). Beyond this and in the outer coma, LTE cannot be assumed and the source function must be found by solving the radiative transfer equation together with a statistical equilibrium equation that gives the populations of electrons in each energy level of the molecule and the associated emission rates. The calculation needs to take into account molecular collisions, electron-molecule collisions, fluorescence, and IR pumping.

Zakharov et al. (2007) compared two approaches to addressing non-LTE. The escape probability method (Sobolev's method) converts the global problem of solving the radiative transfer equation to a local one by assuming that each point within the gas field is coupled radiatively with the region surrounding it. A local region can be defined such that a photon has a finite probability of being absorbed along a line within the region and beyond which there is a vanishing probability of absorption. For a spherically symmetric medium expanding at constant velocity (which approximates cometary atmospheres as in Eq. 3.1) the escape probabilities are given by Litvak and Kuiper (1982) and these are required to compute the average intensity received from all angles within the local region (see also Rybicki 1984; Bockelée-Morvan 1987). Unfortunately, realistic comae present radial and azimuthal variations in gas density and velocity and this is especially the case when studying the innermost coma. In such cases, a Monte Carlo approach is necessary (Zakharov et al. 2007) at the cost of a major increase in computing time.

In the Monte Carlo approach, a domain is divided into a large number of cells with constant physical properties (density, velocity, temperature, etc.). It is assumed that the molecular excitation in the cell is uniform. For each cell, the average intensity received is approximated by the summation over a random set of rays from elsewhere which enter the cell and contribute to the radiation field within it. Knowing this average intensity and, assuming statistical equilibrium, the level populations can be determined from classical balance equations (Zakharov et al. 2007).

Yamada et al. (2018) have recently published a new open source code for the investigation of rotational line emission in non-LTE systems which implements a deterministic method for the solution of the multi-level non-LTE problem in a 1D spherically symmetric atmosphere and is able to treat optically thick transitions in both static and expanding atmospheres accurately.

Non-LTE is particularly important for infrared emissions as we shall now illustrate using the $H_2O$ molecule. Figure 3.10 shows the fine structure of the ro-vibrational band of $H_2^{16}O$ at 2.69 microns. The plot gives the spectral line intensity, $S_s(T,p)$. This calculation (produced using the HITRAN database) assumes that the ground-state populations are in LTE and follow a Boltzmann distribution at the specified temperature. The population is given by.

$$P_l = \frac{w_l}{Z_T} e^{-\frac{E_l}{kT}} \tag{3.36}$$

where $w_l$ is the statistical weight and $Z_T$ is the partition function given by

**Fig. 3.10** The fine structure of the $H_2{}^{16}O$ ro-vibrational band at 2.7 microns at temperatures of 70K and 200K (assuming LTE). The plot at 70K has been offset vertically to allow ease of comparison of the band changes significantly with temperature. This should be compared with Fig. 3.12

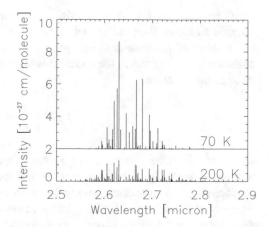

$$Z_T = \sum_l w_l e^{-\frac{E_l}{kT}} \tag{3.37}$$

The temperature dependence is significant (Villanueva et al. 2012). The rapid expansion of the gas in the inner coma leads to a cooling that reduces the temperature within a few radii of the nucleus (Fig. 3.4). Figure 3.10 illustrates the effect of this by comparing the spectral line intensities at 200 K (representative of the sublimation temperature) and 70 K (representative of temperatures remote from the surface but in regions where collisions are still sufficient to maintain a single kinetic temperature). Note in Fig. 3.10 that the unit of the spectral line intensity on the ordinate is [cm/molecule]. This is perhaps more straightforward to understand if it is described as [cm$^{-1}$/(molecule cm$^{-2}$)] which is a frequency (in wavenumbers) per column density. Unfortunately, for the strongest fundamental vibrational bands, the effects of collisions are insufficient (Bockelée-Morvan et al. 2004a) and this computation is inadequate.

In radiative excitation, a photon is absorbed from a source. If the photon is of a specific energy it raises the atom or molecule into a higher internal energy mode. The atom or molecule can then relax back to its original state by the release of a photon of the same energy. This process is usually referred to as resonant fluorescence. This has some similarity to a scattering process and has consequently been referred to as resonant scattering. However, in general, scattering implies non-isotropic emission whereas resonant fluorescence is normally isotropic. Resonant fluorescence can lead to an excited electronic, vibrational or rotational state depending upon the wavelength of the absorbed photon as outlined above. The Sun is weak in the UV and thus parent species are rarely identified or studied using their electronic bands (Bockelée-Morvan et al. 2004a). CO and $S_2$ are the most notable exceptions. On the other hand, radicals such as CN are monitored through these transitions and we shall use this when discussing daughter products.

If the main excitation mechanism is resonant fluorescence excited by solar infrared radiation, then the excitation rate of a molecular band is characterized by the number of photons absorbed per molecule per second. The intensity of the emission from the band in the optical thin case is then related to the column density through the equation

$$I = \frac{h\vartheta}{4\pi} \frac{g_f}{r_h^2} N_g \tag{3.38}$$

where $\vartheta$ is the central frequency of the band and $g_f$ is the band emission rate (often referred to as the "g factor" while some authors refer to it as a "photon scattering coefficient" in some applications). $g_f$ has units of [photon $s^{-1}$ molecule$^{-1}$] with the intensity in [W m$^{-2}$ sr$^{-1}$]. The g-factor incorporates the solar flux producing the fluorescence and hence it needs to be specified at a heliocentric distance (which is normally 1 AU). Note that literature often expresses Eq. (3.38) without the $r_h^2$ dependence and defines the g-factor at the heliocentric distance of the observation.

The integrated band strengths of the principal infrared emissions from the three major species in the optically thin case are given in Table 3.5. The vibrationally-excited levels have a short radiative lifetime (Crovisier 1984) and hence the levels will not remain populated. These values allow inversion of measured intensities into the column densities. The bands are however made of individual lines as shown in Fig. 3.10. More importantly, there are non-resonant fluorescence processes in which infrared photons are emitted to the ground vibrational state through branching into intermediate vibrational levels rather than directly to the ground state (Villanueva et al. 2012) as illustrated in Fig. 3.11 for the infrared lines of $H_2O$.

Other effects also need to be considered when looking in detail at the production of specific lines. For example, "hot bands" are transitions between two excited states. In the harmonic approximation, these transitions would not be distinguishable from fundamental transitions but the anharmonicity influences this and the transitions appear red shifted with respect to the fundamental transitions. For comets, this was first identified by Bockelée-Morvan and Crovisier (1989). These are distinct from combination bands which involve changes in the vibrational quantum numbers of more than one normal mode. Although these effects result in relatively weak emissions, they can be useful when, for example, Earth atmospheric water is a strong contaminant in spectral observations (Dello Russo et al. 2004).

To address the fine structure, detailed statistical equilibrium calculations are required. Computation of line-by-line fluorescence efficiencies (g-factors) entails construction of a full quantum mechanical model of the molecule. This requires a complete characterization of the rotational structure (energy levels) for all vibrational levels involved (both high-energy levels pumped by sunlight and lower levels involved in the subsequent cascade), along with statistical weights, selection rules, perturbations (e.g. Coriolis effects, splittings, vibration-vibration coupling, and tunneling) and band emission rates. Not only is this task extremely complex but it

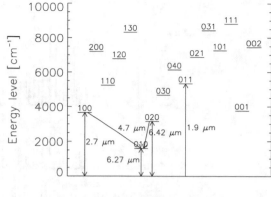

**Fig. 3.11** The vibrational levels of $H_2O$ (after Crovisier 1984 and Lopez-Puertas and Taylor 2001)

Ground state [000]

still has some limitations. Information for most hot bands for example, is not available in community spectral databases (Villanueva et al. 2012).

To illustrate the influence on the infrared fine structure, we can use an on-line resource called the Planetary Spectrum Generator (PSG)[3] which is now available to support analyses (Villanueva et al. 2018). This is a radiative transfer suite allowing synthesis of spectra for a broad range of planetary targets given appropriate input. The tool has a specific template for 67P, allowing the generation of synthetic spectra in the 0.1 μm–100 mm wavelength range. Figure 3.12 shows the result of a calculation for a weak comet so that the optical depth is zero. Two different temperatures for the gas are shown. Comparing this with Fig. 3.10, it can be seen that the structure of the lines is appreciably different if non-LTE has to be assumed.

Figure 3.13 uses PSG to model VIRTIS-H point spectrometer observations of 67P at a time when optical thickness effects could be neglected. The plot shows the sum of 21 observations were acquired between 24 Dec 2014 and 25 Jan 2015 (cf Bockelée-Morvan et al. 2015) with the spacecraft pointing off the limb. The effective resolution of the instrument at this wavelength was around 3.3 nm and the model calculation has been degraded to this resolution for comparison. Rosetta was between 2.5 and 2.7 AU from the Sun at this time (pre-perihelion) with the column density such that almost all the individual lines could be assumed to be optically thin (see Fig. 3.15 later). The online model agrees fairly well with the data giving a similar result to that seen in Bockelée-Morvan et al. (2015).

Collisional excitation is another primary means of excitation. The colliding species may be electrons, ions (although their densities in the innermost coma are far below that of neutrals) or other molecules (primarily $H_2O$ although CO may be of relevance at larger heliocentric distances or in CO-rich comae).

An example of electron collision excitation is given in Fig. 3.14 which shows the cross-section for excitation of the water molecule from the rotational ground-state to the J = 1 level (Itikawa and Mason 2005) and is from the Quantemol database

---

[3]https://psg.gsfc.nasa.gov

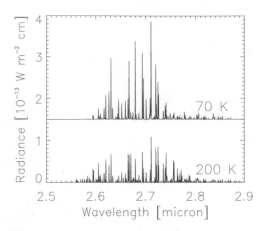

**Fig. 3.12** Calculations of the radiance of $H_2O$ at two different gas temperatures for weak cometary activity (optical depth = 0) using the Planetary Spectrum Generator (Villanueva et al. 2018). The plot at 70K has been offset vertically to allow ease of comparison

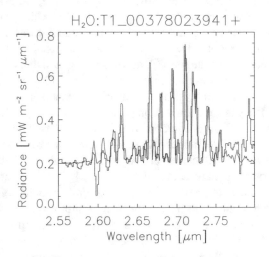

**Fig. 3.13** Solid line: VIRTIS-H observation of water vapour above the limb of 67P from the summation of 21 observations between 24 Dec 2014 and 25 Jan 2015. Dashed line: Model of the emission obtained using the Planetary Spectrum Generator (Villanueva et al. 2018) with a column density of $5 \ 10^{19}$ m$^{-2}$ at a temperature of 100 K. A heliocentric distance of 2.6 AU was assumed. The calculation was made at 1.1 nm resolution and subsequently degraded to 3.3 nm by smoothing

(Tennyson et al. 2017).[4] The cross-sections for excitation to the first vibrationally excited level are around two orders of magnitude smaller (see also Faure and Josselin 2008) than this as shown by the example for CO excitation from the $v = 0$ to $v = 1$

---

[4]https://www.quantemoldb.com Note that some services on this site are commercial.

**Fig. 3.14** Electron collision cross-sections for rotational excitation to the first excited state of H₂O (solid line) and the first vibrational excited state of CO (dashed line)

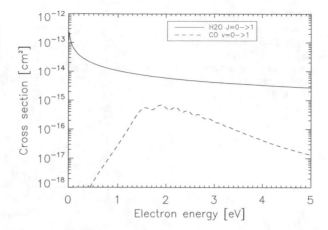

state (Laporta et al. 2012; see also Itikawa 2015). The collisional excitation rate in units of $[\text{m}^{-3}\,\text{s}^{-1}]$ is given in this case by the equation

$$C_{ex} = n_e n_N \sigma_{col} v_{rel} \tag{3.39}$$

where $n_e$ is the electron density, $n_N$ is the local density of the neutral, $\sigma_{col}$ is the collision cross-section at a given electron temperature and $v_{rel}$ is the relative velocity. The product $\sigma_{col} v_{rel}$ is the collision rate coefficient which is itself dependent upon the collision strength. For electron impact excitation of simple ions (e.g. $O^+$) at a constant electron temperature, collision strengths are frequently tabulated and equilibria can be calculated between excitation and collisional de-excitation (see e.g. Osterbrock 1989). However, this is far less straightforward in the case of water in the innermost coma of comets where outflow and gas drag are important and the properties of the electron distribution are strongly varying with the gas production rate with rapid electron cooling occurring with cometocentric distance (Cravens and Korosmezey 1986; Engelhardt et al. 2018). It continues to be assumed that excitation rates for H₂O due to this process are very low within a few thousand kilometres of the nucleus (Xie and Mumma 1992) although the sharp rise in electron temperature beyond the contact surface (see below) probably leads to changes in the populations of rotational energy states for several species through this mechanism at larger cometocentric distances (see e.g. Biver et al. 1999).

H₂O-H₂O collisional de-excitation cross-sections have been presented by Buffa et al. (2000) and an equation similar to that shown in Eq. (3.39) can be used (e.g. Lee et al. 2011). But, in general, there is a lack of fundamental atomic data to describe collisions. There are numerous topics that remain only partially addressed in the literature with several processes (e.g. molecular collisions with CO) taken as being insignificant for the sake of simplicity (e.g. Zakharov et al. 2007).

A further critical issue close to the nucleus is the influence of optical depth. As can be seen in Fig. 3.15, the absorption cross-section of water vapour within some lines can exceed $10^{-21}$ m² molecule⁻¹. Using the equation for column density (Eq. 3.4), water production rates greater than a few $10^{27}$ molecule s⁻¹ can produce

**Fig. 3.15** The absorption
cross-section of water
vapour at 200 K within the
$v_1$ band. This can be
compared to the column
density used to generate
Fig. 3.13

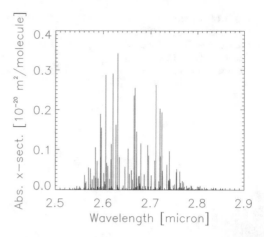

**Fig. 3.15** The absorption cross-section of water vapour at 200 K within the $v_1$ band. This can be compared to the column density used to generate Fig. 3.13

optical depths of 1 or more within 10–20 km of the nucleus for typical values of the outflow velocity. Hence, the observed infrared emission from the coma will be influenced by absorption in an active comet and, for non-isotropic outflow, will be dependent upon the observing geometry. This effect leads to a major increase in the complexity of the interpretation of the water vapour lines requiring determination of the spatial distribution, the gas velocity, and the gas temperature to determine the opacity from the Sun to the position of interest and from that point to the observer. The usual equations of radiative transfer can be used in constructing the solution if the coma properties are known. However, it is often precisely these quantities that we are interested in determining from the observed intensity, not the other around!

The intensity of a band is typically given in units of [W m$^{-2}$ sr$^{-1}$]. It remains somewhat unfashionable in cometary physics but the Rayleigh is a unit that is a useful way to describe the luminosity of a column of an emitting species but has the units of surface brightness. It was first proposed by Hunten et al. (1956) and defines 1 Rayleigh (R) as $10^6/4\pi$ photon s$^{-1}$ cm$^{-2}$ sr$^{-1}$. Hence 1 W m$^{-2}$ sr$^{-1}$ corresponds to $6.33 \times 10^9 \lambda$ Rayleigh where $\lambda$ is the wavelength in [μm].

We saw earlier that the total change of intensity within a beam requires inclusion of absorption and a source function, $j_\vartheta$, and we can re-write Eq. (3.19) as

$$\frac{dI_\vartheta}{ds} = -\kappa_\vartheta I_\vartheta + j_\vartheta \tag{3.40}$$

For the infrared bands, when optical depth is significant, the source function can be written as

$$j_\vartheta = \frac{h\vartheta_0}{4\pi} g_f \varphi(\vartheta, \vartheta_0) \tag{3.41}$$

where the value of the g-factor in Eq. (3.41) requires knowledge of the partition function as well as the optical depth to the source position (assuming that the

**Fig. 3.16** Rosetta VIRTIS-H observation of $H_2O$ at 2.7 µm from 67P acquired in an integration between 2015-07-08 21:10:50 and 2015-07-09 00:48:03 with the line of sight off the limb and 3.01 km from the centre of the nucleus (cube: T1_00395011055)

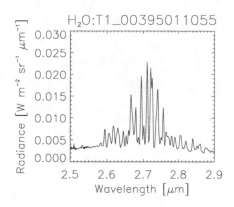

excitation mechanism is resonance fluorescence). As a result of the combination of emissions and absorptions, line profiles in a non-isothermal system can become extremely complex requiring detailed modelling. An example of $H_2O$ emission at 2.7 µm from 67P acquired with VIRTIS-H is shown in Fig. 3.16. At this time, with 67P just over a month from perihelion, this band was optically thick affecting the relative line strengths.

One way to reduce the complexity of the optical depth is again to look at isotopologues of the major species. The abundances of the heavier isotopologues are, of course, much reduced relative to the main isotopologues. Hence the corresponding lines in the heavier isotopologues of saturated lines in the main isotopologue may be far from saturation. Abundances of the isotopologues of water relative to total water are given in Table 3.4 although these relative abundances are approximate. Isotopologically heavier molecules have lower partial vapour pressures than the main isotopologue, leading to isotopologic fractionation between the phases during condensation and evaporation processes. It should be noted that on Earth, measurement of the isotopologic composition of atmospheric water vapour is used to derive information on the water cycle. Hence, one should not speak of a standardized value for the relative abundances. Schroeder et al. (2018) report an 11% enrichment of $^{18}O$ at 67P compared to a calculated terrestrial value (Baertschi 1976). The most common isotopologue of $^{16}O^{12}C^{16}O$ is $^{16}O^{13}C^{16}O$ at an abundance of roughly 1% of main species.

Figure 3.17 shows an observation of the $CO_2$ 4.27µm band again obtained with the VIRTIS-H. The data were acquired in an integration between 2015-07-08 21:10:50 and 2015-07-09 00:48:03 with the line of sight off the limb at a distance of 3.01 km from the centre of the nucleus with a position angle roughly 10° from the Sun direction (Bockelée-Morvan et al. 2016). The $CO_2$ emission band is superposed on a continuum arising from thermal emission of dust. The band structure determined using PSG is also shown and it can be seen that the individual lines were not resolved by the VIRTIS-H instrument. An extended wing longward of the band is also evident in processed data where the continuum has been subtracted. (This is also evident by close inspection of Fig. 3.16) Bockelée-Morvan et al. (2016), in a

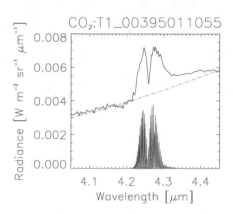

**Fig. 3.17** Solid line: Rosetta VIRTIS-H observation of $CO_2$ from 67P acquired in an integration between 2015-07-08 21:10:50 and 2015-07-09 00:48:03 with the line of sight off the limb and 3.01 km from the centre of the nucleus (cube: T1_00395011055). The dot-dashed line indicates a continuum level following a Planck function at 280 K. The $CO_2$ band structure is shown below. It should be noted that many lines in this $CO_2$ band were optically thick near perihelion

thorough analysis, attributed this to fluorescence emission of two hot bands of $CO_2$ and emission from the isotopologue, $^{13}CO_2$ ($^{16}O^{13}C^{16}O$).

## 3.3   Minor Species

Table 3.1 shows some of the species found by the ROSINA instrument at 67P (Le Roy et al. 2015; Rubin et al. 2019). Additional analysis of the mass spectrometer data is still on-going. In situ mass spectrometry provides the primary means of determining the parent species and the very high sensitivity allows identification of many molecules although some care is needed. For larger molecules, the ion source used to ionize the molecules coming into the experiment may also split the molecules leading to confusion with less complex molecules. However, testing on ground can be used to establish the branching ratios of the various fragmentation patterns (Gasc et al. 2017). Despite this, other spaceborne techniques are still useful. Infrared spectroscopy for example was used at 1P/Halley from onboard the Russian Vega spacecraft and Combes et al. (1988) reported not only detection of the major species ($H_2O$, $CO_2$, and CO) but also suggested that $H_2CO$ was present and possibly OCS. The advantage of such detections is that, in the best case, spatial distributions can be determined.

Detections from ground-based remote-sensing are also of importance and a summary of discovered molecules has been provided by Bockelée-Morvan and Biver (2017). In Fig. 3.18, a comparison of the ground-based measurements (from many comets) with those obtained by ROSINA at 67P is made. With some exceptions (notably ammonia), the agreement is rather good. The comet detections are in

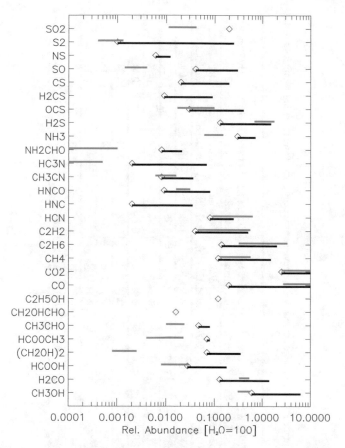

**Fig. 3.18** Remote sensing detection of molecules from many comets compared to the detections made by the ROSINA experiment at 67P. Ground-based values are shown by the black bars. The minimum values of all detected species are shown with the diamond. The grey bars show the ROSINA values from Table 3.1 where the bar indicates a range for the northern to southern hemisphere measurements. (Data from Le Roy et al. 2015, Rubin et al. 2019 and Bockelée-Morvan and Biver 2017)

essence a subset of detected interstellar molecules. About 200 species have been detected in the interstellar medium with lists being curated at several sites.[5] The abundances measured in comets are generally comparable to those measured in star-forming regions, suggesting that comets contain preserved material from the presolar nebula (Biver and Bockelée-Morvan 2019).

Molecular abundances relative to water measured in comae do show strong variations from comet to comet. (Variations with heliocentric distance will be discussed in Sect. 3.6). An example of this is given in Table 3.6 which gives the

---

[5]https://science.gsfc.nasa.gov/691/cosmicice/interstellar.html is an example.

**Table 3.6** Mixing ratios relative to water for species detected in the infrared for nine comets. Values have been averaged where no mean has been given in the reference

| Species | 45P/HMP | C/2017 E4 (Lovejoy) | C/2007 N3 (Lulin) | C/2012 K1 (PanSTARRS) | 21P/Giacobini-Zinner | C/2012 F6 (Lemmon) | C/2013 R1 (Lovejoy) | C/1999 H1 Lee | 103P/Hartley 2 |
|---|---|---|---|---|---|---|---|---|---|
| Reference | Dello Russo et al. (2020) | Faggi et al. (2018) | Gibb et al. (2012) | Roth et al. (2017) | DiSanti et al. (2013) | Paganini et al. (2014b) | Paganini et al. (2014a) | Dello Russo et al. (2006) | Dello Russo et al. (2011) |
| $H_2O$ | 100 | 100 | 100 | 100 | 100 | 100 | 100 | 100 | 100 |
| $CH_3OH$ | 4.50 | 1.62 | 3.6 | 2.69 | 1.22 | 1.48 | 2.75 | 1.6 | 1.25 |
| $C_2H_6$ | 0.83 | 0.39 | 0.67 | 0.87 | 0.14 | 0.29 | 0.82 | 0.71 | 0.74 |
| $NH_3$ | 0.74 | 1.86 | 0.28 | <1.8 | | 0.61 | 0.1 | | 0.61 |
| $HCN$ | 0.16 | 0.17 | 0.13 | 0.14 | | 0.19 | 0.28 | 0.20 | 0.24 |
| $C_2H_2$ | 0.072 | 0.14 | 0.067 | <0.11 | | <0.05 | | 0.24 | 0.14 |
| $H_2CO$ | 0.145 | 0.36 | 0.15 | <0.14 | | 0.54 | | | 0.11 |
| $CH_4$ | 0.97 | 0.34 | 1.15 | 0.46 | | 0.67 | 1.05 | 1.13 | |
| $CO$ | | 1.16 | 2.23 | 3.9 | | 4.03 | 9.89 | | |

abundances of several minor species measured for 9 comets including 103P/Hartley 2. All the observations were made by teams centred around M. Mumma and G. Villanueva. Variations in the relative abundances of factors of 3 or more are seen for some species. Although visually the coma observed during the EPOXI fly-by of P/Hartley 2 was unusual because of the large numbers of water ice particles seen, the chemical composition seen from the ground was not particularly unusual (Fig. 3.18).

It was already inferred from Giotto data that there are substantial amounts of hydrocarbons in the gaseous emissions from comets. The repetitive appearance of specific peaks in the mass spectra (alternating peaks 16 and 24 Daltons apart) were interpreted as being indicative of a specific polymer, polyoxymethylene (POM), which is polymerized formaldehyde (Huebner 1987) although Mitchell et al. (1989) later showed that this regular pattern is generally characteristic of any kind of CHO-bearing molecules that include POM-like structures. Wright et al. (2015) reported detection of CHO-bearing molecules at the surface of 67P using the PTOLEMY instrument onboard the Philae lander but it was considered unlikely that this was POM by Altwegg et al. (2017).

There are possibly issues with POM itself being a parent of the observed fragmentation pattern. It is known to decompose into formaldehyde just above room temperature and its lifetime in the gas coma of a comet is expected to be short (Le Roy et al. 2012). However, other laboratory studies (Butscher et al. 2019) continue to suggest that POM is a plausible organic component of the surface material and that specific forms can remain sufficiently stable to be detected in the inner coma. Garrod (2019) has noted that these complex organic molecules are likely to be a product of cosmic-ray processing of surface layers well before entry into the inner Solar System.

PTOLEMY measurements also indicated an apparent absence of aromatic compounds such as benzene, a lack of sulfur-bearing species, and very low concentrations of nitrogenous material although many of these species are evident in the higher sensitivity measurements made by ROSINA on the Rosetta orbiter (e.g. Schuhmann et al. 2019). There was however a positive detection of toluene from PTOLEMY data acquired at the surface and confirmed with ROSINA data (Altwegg et al. 2017). The low abundance of $NH_3$ in coma mass spectra may be attributable to the integration of ammonia into ammonium salts which now appear to have been unambiguously detected in data from two instruments (ROSINA and VIRTIS) on Rosetta (Altwegg et al. 2020; Poch et al. 2020).

Of the simple organics, methanol was detected from ground through its infrared emissions (Hoban 1993; Davies et al. 1993) and has been extensively monitored at sub-mm wavelengths (e.g. Biver et al. 2002b). Formic acid (HCOOH), another organic identified in both comets and the interstellar medium (ISM), has now been detected (Favre et al. 2018) along the line of sight towards the TW Hydrae protoplanetary disc with ALMA (cf Fig. 1.1) indicating that another link in the chain between the ISM and comet formation can now be verified by measurement. We will return to the subject of organics in Sect. 4.14 where we shall also discuss the detection of glycine in the coma by Altwegg et al. (2016).

## 3.4  Gas Expansion

### 3.4.1  The Initial Conditions

At the beginning of this section, we made four assumptions about the gas distribution. The first three were that there is a point source, that gas emission from the nucleus is isotropic, and that the gas velocity is constant. Close to the nucleus, none of these assumptions is valid. Within ten nucleus radii, the nucleus can no longer be considered a point source. The outgassing is driven by solar insolation and significant inhomogeneity in emission may be superimposed upon this. Finally, the gas must be accelerated to its final velocity from its initial state at the surface of the nucleus.

The investigation of the properties of the gas distribution in the immediate vicinity of the nucleus now forms an important tool in trying to understand the outgassing properties of the nucleus. The idea is that by understanding the gas distribution in the innermost gas coma, one can deduce the surface outgassing distribution and from that derive properties of the nucleus surface.

This appears logical and straightforward but the mathematical tools needed to address specific questions may change depending upon the flow regime. Figure 3.19 is a schematic diagram of the different flow regimes possible near the source. We begin by discussing an active comet.

The mean free path is defined as

$$\lambda_{MFP} = \frac{1}{\sqrt{2}\sigma_{col}n_g} \tag{3.42}$$

where $\sigma_{col}$ is the cross-sectional area of the species and $n_g$ is the local density. For water molecules, the collisional cross-section is around $2.5 \ 10^{-15}$ cm$^2$ (Crovisier 1984). Eq. (3.1) can be substituted for $n_g$ in Eq. (3.42) to obtain a mean free path with cometocentric distance. However, a more realistic value is obtained by assuming a uniform sphere emitting gas proportional to the cosine of the incidence angle. The mean free path at the sub-solar point is then given by

$$\lambda_{MFP} = \frac{\pi v_g r_N^2}{\sqrt{2}Q_g\sigma_{col}} \tag{3.43}$$

where $Q_g$ is the total gas production rate and $v_g$ is the outflow velocity. But this still overestimates $\lambda_{MFP}$ in the region where gas acceleration is important. A crude approximation to $v_g$ is given by

$$v_g = v_\infty\left(1 - \frac{r_N}{r}\right) \tag{3.44}$$

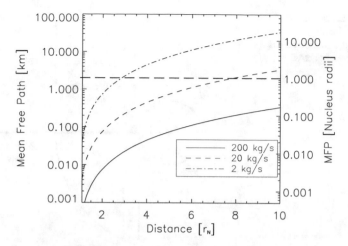

**Fig. 3.19** Mean free path changes with cometocentric distance for different production rates from a spherically symmetric nucleus taking into account the gas acceleration using a crude approximation. This provides an estimate of the size of the region within which collisions dominate

where $v_\infty$ is the terminal velocity, and can partially account for this effect. This substitution leads to the increases in $\lambda_{MFP}$ with cometocentric distance for different production rates seen in Fig. 3.19.

For an isotropically emitting 2 km radius sphere, the mean free path at the surface is comparable to the radius of the sphere for production rates of the order of $1.5 \ 10^{25}$ molecule $s^{-1}$ or around 0.4 kg $s^{-1}$. If the gas production is significantly higher than this or non-isotropic, then collisions are important close to the nucleus and the gas may equilibrate. In this case, the molecules can be described by a Maxwell-Boltzmann distribution and the flow is described as an equilibrium flow. The gas within this regime can be assumed to be in local thermal equilibrium (LTE). For such a gas in equilibrium, the velocity distribution function (VDF) is often described as being "Maxwellian" and the VDF in 3D is written as

$$f\left(v_g\right) = \left(\frac{\eta_d}{\pi}\right)^{\frac{3}{2}} e^{-\eta_d\left(v_g-u_g\right)^2} \tag{3.45}$$

where $u_g$ is the bulk or drift velocity of the flow and $\eta_d = m_g/2kT$. The proof that the equilibrium distribution is a Maxwellian is given in Cercignani (2000).

The gas is however expanding away from the source into vacuum. Hence, if we assume force-free radial outflow, the density drops as $1/r^2$ and the mean free path increases rapidly to the point where collisions become insignificant. The region where the mean free path becomes much larger than the nucleus itself is referred to as the region of free molecular flow (Fig. 3.20) and LTE can no longer be assumed on scales comparable to the nucleus size. Molecules in free molecular flow regimes are

**Fig. 3.20** Schematic
diagram of the different flow
regimes close to the nucleus
for a moderately active
comet

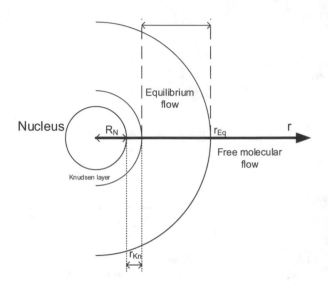

essentially independent of each other and respond to individually applied forces and reaction chemistry (e.g. photo-dissociation).

The origin of the gas is sublimation from the nucleus and molecules emitted from the surface have velocity vectors with components directed away from the nucleus. (The subliming gas cannot have a velocity component in a direction towards the nucleus otherwise it would not leave the surface). The VDF is therefore non-Maxwellian. The exact form of the velocity distribution is not known but a half-Maxwellian can be assumed as an approximation. This is obviously not in equilibrium and, even if the gas density is sufficient that it quickly equilibrates to produce an equilibrium flow, there is a non-LTE layer directly above the surface (Fig. 3.20). This layer is referred to as the Knudsen layer.

The VDF can be computed using Monte Carlo techniques as will be described below. However, for illustration purposes, Fig. 3.21 shows how the VDF changes with distance from the surface. This was computed using Bird's (1994) Direct Simulation Monte Carlo (DSMC) code. To the left we see a low production rate case roughly equivalent to a production rate of 40 g s$^{-1}$ from a spherical, isotropically emitting, 2 km-sized nucleus. The calculation is for water vapour. Just above the nucleus surface (top panel left), most of the molecules have a velocity away from the nucleus. Even at 1/1000th of a source radius some backscattered molecules (with negative velocity and therefore directed back to the nucleus) can be seen. However, the VDF is best described by a hemispherical distribution in velocity-space and we are clearly within the Knudsen layer.

In the top right, we see that for a factor of 200 higher production (roughly 8 kg s$^{-1}$), the VDF is almost circular and therefore closer to a Maxwellian near the source. This shows that the thickness of the Knudsen layer ($d_{Kn}$ in Fig. 3.20) can be small and, as one might expect, is strongly dependent on the production rate. In the denser

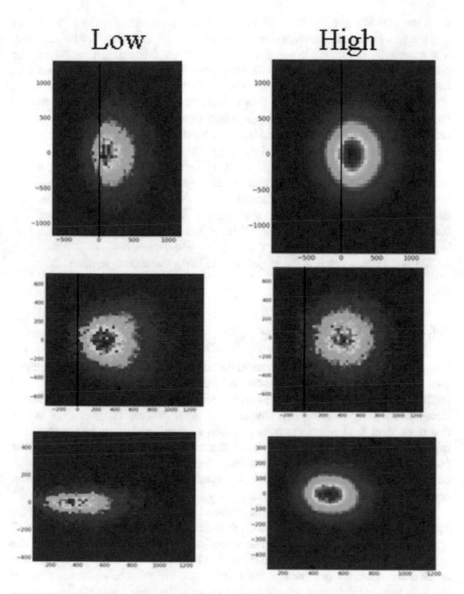

**Fig. 3.21** Velocity distribution functions in idealised cases. The rows indicate the VDF at different distances (1.001, 1.6, 6.7 source radii) from the source. The two columns are for different production rates leading to different initial surface densities (left: 1.81 $10^{14}$ molecule m$^{-3}$, right: 3.61 $10^{16}$ molecule m$^{-3}$). The black vertical lines indicate zero velocity relative to the source. The velocities are in [m/s]

flow, the VDF is a drifting Maxwellian 1.5 source radii from the centre of the source (middle panel right). However, the more tenuous flow (middle panel left) is distinctly non-spherically symmetric in velocity-space. Clearly, at this distance, the flow on the left has still not reached equilibrium. By the time the gas has reached 6.7 source radii from the source (lower panels), the low production rate source (left) shows an elongated, almost conical, VDF. This is indicating that we have already entered the free molecular flow regime. For the higher production rate source at 6.7 source radii, the VDF is close to spherical but with a little elongation indicating that the transition to free molecular flow is beginning. This is consistent with Fig. 3.19.

This description shows that the sizes of the Knudsen and equilibrium regions ($r_{Kn}$ and $r_{Eq}$ in Fig. 3.20) are strongly dependent upon the production rate. $r_{Kn}$ can become very small for high production rate cases while for low production rate cases, the size of the equilibrium flow regime can go to zero so that the gas emission can transit from a Knudsen regime to a free molecular flow without ever reaching LTE.

We now look in detail at the gas expansion using these three regimes.

## 3.4.2   The Knudsen Number

In gas dynamics, the macroscopic model describes gas as a continuous medium and uses terms such as velocity, pressure and temperature to describe the flow. The mathematical equations used to describe a continuum flow are the Navier-Stokes equations for viscous fluids. The simpler Euler equations can be used if the flow is that of an ideal fluid where viscosity and thermal conductivity can be ignored. The microscopic model, on the other hand, describes the gas in terms of discrete molecules and follows their individual behaviour as a result of interactions such as collisions. In the microscopic model, the equation to be solved is the Boltzmann equation.

The microscopic and macroscopic approaches have advantages and disadvantages. As we shall see, the microscopic model can, in principle, be used for all applications but numerical issues arise when local densities become too high. The macroscopic approach is, in general, faster but the continuum description begins to break down when the gradients in the macroscopic variables becomes too steep. If the scale lengths of changes are of the same order as the mean free path, $\lambda_{MFP}$, in the gas then the continuum equations become inaccurate and LTE can no longer be assumed. As we have seen above, the expansion of the gas implies that this accuracy limit will be reached at some distance from the nucleus source even if the production rate is high. A rough estimate for its distance from the nucleus, $r_{HD}$, can be obtained using the approximation

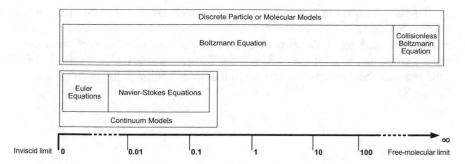

**Fig. 3.22** The relationship between the Knudsen number and the appropriate mathematical tool (after Bird 1994)

$$r_{HD} \sim \frac{\sqrt{2}\sigma_{col}Q_g}{\pi v_g} \tag{3.46}$$

where $\sigma_{col}$ is the collisional cross-section. (This is a slightly modified form of the expression used by Wallis 1974).

The quantity used to establish the validity or otherwise of the continuum description is the Knudsen number ($Kn$) which is the ratio of $\lambda_{MFP}$ to a characteristic dimension, $L_{Kn}$,

$$Kn = \frac{\lambda_{MFP}}{L_{Kn}} \ . \tag{3.47}$$

It is usually assumed that if $Kn > 0.2$ then the microscopic model must be used (Fig. 3.22). The characteristic dimension needs to be defined with respect to local macroscopic gradients in the flow such that

$$L_{Kn} = \frac{X_{Kn}}{dX_{Kn}/dx} \tag{3.48}$$

where $X_{Kn}$ is a macroscopic variable such as density (see Bird 1994).

In cometary gas emission studies, there are on-going debates about the distribution of sources on the surfaces of cometary nuclei. If all surfaces are active and driven by the input insolation, then the gradients in the gas flow will be substantially less than if activity is inhomogeneous with local active and inactive regions. The value of $L_{Kn}$ therefore depends on the outgassing model. A way to avoid any confusion is to define a source Knudsen number, $Kn_0$, as

$$Kn_0 = \frac{\lambda_{MFP}}{L_0} \tag{3.49}$$

where $L_0$ is the characteristic dimension of the source.

The mean free path in space can become extremely large so that $Kn$ can approach infinity. This is the free-molecular flow regime when individual molecules no longer interact with each other. The free molecular approximation has its uses in cometary coma studies and, while it is invalid close to the nucleus for typical comets $<3$ AU from the Sun, there are cases where its simplicity is helpful at larger cometocentric distances.

### 3.4.3   Fluid Expansion

#### 3.4.3.1   Fluid Equations for Equilibrium Flow

For low values of the Knudsen number ($Kn \lesssim 0.01$), flows in the inner gas coma can be described by the Euler equations for inviscid flow. The three Euler equations are conservation equations for mass, momentum, and energy. In the compressible form, the mass conservation equation is given by

$$\frac{\partial \rho_g}{\partial t} + \nabla \cdot \left( \rho_g v_g \right) = Q_s \qquad (3.50)$$

which is the continuity equation but allowing for a source term, $Q_s$. It describes the variation of gas density, $\rho_g$, as a result of expansion. In the absence of local gas production in the coma, $Q_s$ becomes zero. If required, the vector identity

$$\nabla \cdot ( fA) = f(\nabla \cdot A) + A \cdot (\nabla f) \qquad (3.51)$$

can be used to expand the second term.

It should not be forgotten that extended sources of gas may arise from the sublimation of ice particles so that $Q_s$ should not be neglected in the general case. Furthermore, if reaction chemistry is significant (e.g. through photo-dissociation) then sources and sinks of individual species may need to be accounted for.

In an incompressible flow,

$$\frac{\partial \rho_g}{\partial t} + v_g \cdot \nabla \rho_g = Q_s \qquad (3.52)$$

where the first term, the unsteady term, describes the local change in density and the second term, the advection term, describes density changes as the fluid moves. Incompressible flow does not imply that the fluid itself is incompressible. It merely implies that the density remains constant within a parcel of fluid that moves with the flow velocity. Assuming incompressibility does, however, allow significant simplification. There are two conditions for being able to make this assumption. Firstly, in a steady flow, the speed of the fluid must be small compared to the sound speed and specifically

$$Ma^2 \ll 1$$

where $Ma$ is the Mach number and defined through

$$Ma = \frac{|v_g|}{c_s} \tag{3.53}$$

where $v_g$ is the local flow velocity. $c_s$ is the speed of sound defined through

$$c_s = \sqrt{\frac{\gamma p}{\rho_g}} = \sqrt{\frac{\gamma kT}{m_g}} \tag{3.54}$$

assuming an ideal gas law (where $m_g$ is the mass of a molecule) and is typically 200–300 m s$^{-1}$.

Secondly, in an unsteady flow, the distance travelled by a sound wave in a characteristic time interval must be much larger than the distance over which changes occur. In effect, this is requiring that the propagation of pressure transients is rapid compared to changes in the flow.

The momentum conservation equation is given by

$$\frac{\partial (\rho_g v_g)}{\partial t} + (v_g \cdot \nabla)\rho_g v_g + \rho_g v_g (\nabla \cdot v_g) + \nabla p = \rho_g F \tag{3.55}$$

and describes how pressure gradients are balanced in the presence of external forces (e.g. gravity) (Schmidt et al. 1988). Note here the vector identity

$$(f \cdot \nabla) = \left[ a_x \frac{\partial}{\partial x} + a_y \frac{\partial}{\partial y} + a_z \frac{\partial}{\partial z} \right] \tag{3.56}$$

that produces a scalar field. Again, this can be simplified if the gas is incompressible and if there are no external forces so that

$$\frac{\partial v_g}{\partial t} + v_g (\nabla \cdot v_g) = -\frac{\nabla p}{\rho_g} \tag{3.57}$$

which is the incompressible Euler momentum equation.

The energy conservation equation (e.g. A'Hearn and Festou 1990) can take several forms depending upon the assumptions made. Following Schmidt et al. (1988) we have

$$\frac{\partial \left( \rho_g \frac{v_g^2}{2} + \rho_g \epsilon \right)}{\partial t} + \nabla \cdot \left( \rho_g v_g \left( \frac{v_g^2}{2} + h_e \right) \right) = \rho_g E \qquad (3.58)$$

where $E$ is a source-sink term that might arise from, for example, absorption or emission of photons. The first term on the left is the time derivative of the energy density where $\epsilon$ is the specific internal energy and is related to the specific enthalpy through the equation

$$h_e = \epsilon + \frac{p}{\rho_g} \qquad (3.59)$$

The enthalpy is proportional to the specific heat capacity at constant pressure ($C_p$) and, assuming an ideal gas, to the ratio of the specific heats, $\gamma$. Using the ideal gas law, this is simply

$$\epsilon = \frac{R_g T}{\gamma - 1} = \frac{p}{\rho_g} \frac{1}{(\gamma - 1)} \qquad (3.60)$$

where $R_g$ is the ideal gas constant. As $\gamma$ is a function of the number of degrees of freedom of the molecule, this implies differences in the energy distribution when comparing, for example, water vapour and $CO_2$ outflows. Given that the inter-molecular separation is large in a cometary coma and that the volumes are large compared to the size of the molecules themselves, the use of the ideal gas equation is normally quite sufficient.

The Euler equations are a special case of the Navier-Stokes equations in which the viscosity and the thermal conductivity are both zero. In the Navier-Stokes equations, the equation of continuity is unchanged but the momentum conservation equation must account for viscosity as follows

$$\frac{\partial \rho_g v_g}{\partial t} + (v_g \cdot \nabla) \rho_g v_g + \rho_g v_g (\nabla \cdot v_g) + \nabla p - \nabla \tau_v = \rho_g F \qquad (3.61)$$

where $\tau_v$ is the viscous stress tensor. If the viscosity is constant, the tensor is simplified and the momentum equation becomes

$$\frac{\partial \rho_g v_g}{\partial t} + (v_g \cdot \nabla) \rho_g v_g + \rho_g v_g (\nabla \cdot v_g) + \nabla p - v_k \rho_g \nabla^2 v_g = \rho_g F \qquad (3.62)$$

where $v_k$ is the kinematic viscosity given through

$$v_k = \frac{\mu_v}{\rho_g} \qquad (3.63)$$

and $\mu_v$ is the coefficient of dynamic viscosity. Further simplification can be made by assuming incompressibility and no external forces so that

$$\frac{\partial v_g}{\partial t} + v_g(\nabla \cdot v_g) - v_k\nabla^2 v_g = -\frac{\nabla p}{\rho_g} \tag{3.64}$$

The energy equation is also modified but here both viscosity and thermal conductivity appear

$$\frac{\partial\left(\rho_g\frac{v_g^2}{2} + \rho_g\epsilon\right)}{\partial t} + \nabla \cdot \left(\rho_g v_g\left(\frac{v_g^2}{2} + h\right)\right) - \varphi_\nu + q = \rho_g E \tag{3.65}$$

where $q$ is the heat flux and related to the thermal conductivity, $\kappa$, through the usual heat conduction equation

$$q = -\kappa\nabla T \tag{3.66}$$

$\varphi_\nu$ represents conversion of the bulk motion of the fluid into internal energy via viscous dissipation. It is the viscous analogue of heating by pdV work. By combining the kinetic energy with the internal energy, one can obtain a total (specific) energy, $e_s$, such that

$$e_s = \frac{v_g^2}{2} + h_e - \frac{p}{\rho_g} \tag{3.67}$$

which can be substituted back into Eq. (3.65) to obtain

$$\frac{\partial(\rho_g e_s)}{\partial t} + \nabla \cdot \left(\rho_g v_g\left(e_s + \frac{p}{\rho_g}\right)\right) - \varphi_\nu + q = \rho_g E \tag{3.68}$$

These equations form the basis for detailed calculation of the gas flow field. Their derivations are described in detail in Gombosi (1994).

The Navier-Stokes equations are computationally more time-consuming than the solution of the Euler equations and previous work has shown that the Euler equations are sufficient when $Kn < 0.01$ (see Kitamura 1986; Crifo and Rodionov 1999). However, discontinuities (arising from, for example, strong inhomogeneities in the gas source) can lead to inaccuracy in the Euler solution. Crifo and Rodionov (1999) noted that, when the VDF is Maxwellian, the Euler equations are sufficient but when the gas is non-Maxwellian then pressure is no longer isotropic and this generates dissipative effects such as viscosity thereby requiring Navier-Stokes solutions. It is important to recognize here that it is not the Knudsen number per se (despite it being convenient) that defines whether the Euler equations are adequate but the degree to which the VDF is non-Maxwellian (Crifo and Rodionov 1999) and the more subjective issue of how necessary it is to define the exact properties of discontinuities in the flow.

Simplification by assuming incompressibility is probably inadequate and the flow in the inner coma clearly breaches the required conditions. Authors such as Knollenberg (1994), Schmidt et al. (1988), and Crifo et al. (2005), who used the Euler equations, adopted the compressible formulation. However, there does not appear to have been a published, direct comparison between the compressible and incompressible formulations for cometary applications.

### 3.4.3.2   The Initial Evolution of the Velocity Distribution Function

In order to initiate the flow field, an initial condition is required. Physically, the source in the cometary case is the nucleus surface. The process provides sublimation at a given rate and an initial temperature of the gas. This can be converted to a source density, $n_0$, and a source temperature, $T_0$. However, even in the fluid case, the change from a half-Maxwellian to a full-Maxwellian cannot be ignored (the Knudsen layer) because the energy to go from the initial distribution to the full Maxwellian is, effectively, extracted from the gas itself and is no longer available to drive the flow. This leads to the concept of a macroscopic "jump" in density and temperature between the source and the equilibrium flow regime.

Anisimov (1968) first provided calculations of the magnitude of this jump by assuming that the Mach number becomes unity at large distances from a strongly subliming surface. He obtained values for the jump in the case of a monatomic gas. Cercignani (1981) extended this to polyatomic species resulting in the values in Table 3.7.

In Table 3.7, the macroscopic density and temperature jump is the difference between $n_0$ and $T_0$, respectively and the parameters of the equilibrium distribution at some distance from the source where the flow velocity is assumed to reach $Ma = 1$ (i.e. the sonic point). The density and temperature at this point are given by $T^*$ and $n^*$, respectively.

**Table 3.7** The magnitude of the macroscopic jump in temperature and density for a surface subliming into vacuum given for different gases characterized by their degrees of freedom. $F_b$ is the fraction of backscattered molecules

| Degree of freedom | | $T^*/T_0$ | $n^*/n_0$ | $F_b$ |
|---|---|---|---|---|
| 3 | Monatomic | 0.669 | 0.308 | 0.184 |
| 5 | Diatomic and linear polyatomic | 0.781 | 0.301 | 0.212 |
| 6 | Polyatomic non-linear | 0.814 | 0.299 | 0.219 |

### 3.4.3.3 Analytical Solutions for the Fluid Equations

In the general case, the solution of the fluid equations requires a numerical model. There are, however, analytical expressions in the fluid approach that are limited in applicability but useful for illustrative and verification purposes. For a Mach number $Ma < 1$, there is no analytical solution and so the applicability of the solutions requires assumptions. If one assumes that collisions are so frequent that the gas is always in thermal equilibrium then one can imagine that the mean free path is much smaller than any length scale of the gas flow which then allows a set of analytical solutions.

The principal assumption is that the gas flow is at $Ma = 1$ at the subliming surface—in essence, one is assuming that the Knudsen layer is infinitely small. The number density, the drift velocity, and the temperature at a distance, $r$, can then be determined using (Sone and Sugimoto 1993)

$$\frac{r}{R_N} = \frac{1}{\sqrt{Ma}} \left( \frac{1 + \frac{2}{\gamma-1}}{Ma^2 + \frac{2}{\gamma-1}} \right)^{-\frac{\gamma+1}{4(\gamma-1)}} \tag{3.69}$$

$$\frac{n_g}{n_g{}^*} = \left( \frac{1 + \frac{2}{\gamma-1}}{Ma^2 + \frac{2}{\gamma-1}} \right)^{\frac{1}{(\gamma-1)}} \tag{3.70}$$

$$\frac{T}{T^*} = \frac{1 + \frac{2}{\gamma-1}}{Ma^2 + \frac{2}{\gamma-1}} \tag{3.71}$$

$$\frac{v_g}{v_g{}^*} = Ma \sqrt{\frac{1 + \frac{2}{\gamma-1}}{Ma^2 + \frac{2}{\gamma-1}}} \tag{3.72}$$

where the asterisk indicates that the quantity should be taken at the sonic point using the macroscopic jump condition. The theoretical maximum gas velocity can be computed from

$$v_g^{max} = \sqrt{\gamma \left( 1 + \frac{2}{\gamma - 1} \right) \frac{kT^*}{m}} \tag{3.73}$$

and is approached asymptotically with increasing distance from the source. In comparing with the literature, it is useful to recall that

$$\frac{x + 1}{x - 1} = 1 + \frac{2}{x - 1} \tag{3.74}$$

as authors can express the basic equations in $\gamma$ using either of these forms. Consequently, in the case of three internal degrees of freedom (e.g. water), $\gamma = 4/3$, and,

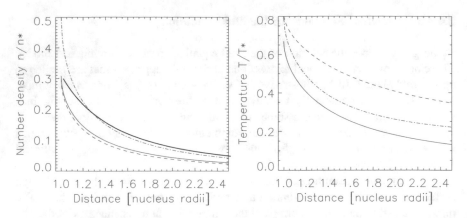

**Fig. 3.23** Left: The number density (relative to the source) as a function of distance from the centre of a spherical nucleus using the analytical solution for a fluid. Solid line: Monatomic molecules. Dashed line: Molecules with three internal degrees of freedom (e.g. $H_2O$). Dot-dashed line: The free expansion (collisionless) solution. The thick solid line shows a $1/r^2$ profile. Right: The temperature of the fluid. Solid line: Monatomic molecules. Dashed line: Molecules with three internal degrees of freedom (e.g. $H_2O$). Dot-dashed line: The free expansion (collisionless) solution. (Following Finklenburg 2014)

$$\frac{r}{R_N} = \left(\frac{6 + Ma^2}{7}\right)^{\frac{7}{4}} \frac{1}{\sqrt{Ma}} \tag{3.75}$$

$$\frac{n_g}{n_g{}^*} = \left(\frac{7}{6 + Ma^2}\right)^3 \tag{3.76}$$

$$\frac{T}{T^*} = \frac{7}{6 + Ma^2} \tag{3.77}$$

and using

$$Ma = v_g^* = \sqrt{\gamma \frac{k}{m} T^*} \tag{3.78}$$

$$\frac{v_g}{\sqrt{2\frac{k}{m}T^*}} = \sqrt{\frac{2}{3}} Ma \sqrt{\frac{7}{6 + Ma^2}} = Ma \sqrt{\frac{14}{3(6 + Ma^2)}} \tag{3.79}$$

In Fig. 3.23 (left) we can see solutions to these equations for the number density. The fluid solution is given for two different values of $\gamma$ (1.667 and 1.333). A similar plot for the temperature is given in Fig. 3.23 (right).

The two plots show a series of fairly trivial but nonetheless important results. Firstly, the gas cools extremely rapidly. Assuming water sublimation at 200 K, the gas cools to well below 100 K within one nucleus radius of the surface. We can recall here the technique used to determine the water vapour in a column between the nucleus and the MIRO instrument on Rosetta (e.g. Fig. 3.4). The water vapour could be a cold gas illuminated from behind by a warmer nucleus source if the nucleus is in the field of view. Secondly, the cooling rate is dependent upon the number of degrees of freedom of the gas. $CO_2$ may be initially colder when subliming from the surface because its free sublimation temperature is lower than that of water. However, it also cools more quickly as there is, relatively, less energy in the rotational degrees of freedom. Thirdly, the expansion and velocity increase leads to a density drop with distance that is far faster than the $1/r^2$ expected from force-free radial outflow. This density drop is close to being independent of the species in the fluid case.

The absence of collisions in the free molecular flow regime allows simplification. The equation for force-free radial outflow from a point source (Eq. 3.1) was shown at the beginning of this section. However, despite this being counter-intuitive, this equation does not give the same result as free expansion from a finite, homogeneously subliming, sphere as was demonstrated analytically by Sone and Sugimoto (1993). The solutions were calculated relative to an imaginary sub-surface reservoir of gas molecules with a temperature, $T_0$ and a number density, $n_0$. The number density at a position in the coma is then given by

$$n(r) = \frac{n_0}{2} \left[ 1 - \sqrt{1 - \left( \frac{R_N}{r} \right)^2} \right] \quad (3.80)$$

where $R_N$ is the radius of the spherical source (i.e. the nucleus) and $r$ is the distance from the centre of the spherical source. Note here that directly above the subliming surface, the density $n$ is not $n_0$ but $n_0/2$. This arises from the way that the source is defined and it is important to account for this when determining the production (or loss) rate from the subliming source in this calculation. The density profile is shown in Fig. 3.23 for comparison with the fluid solution and a $1/r^2$ profile as provided by Eq. (3.1).

The drift velocity of the molecules in this description is given by

$$v_g(r) = \sqrt{\frac{2kT_0}{\pi m_g}} \left[ 1 + \sqrt{1 - \left( \frac{R_N}{r} \right)^2} \right] \quad (3.81)$$

As noted above, the absence of collisions and the non-Maxwellian nature of the initial VDF means that the gas is not in equilibrium. We define the symbol $\parallel$ to indicate the direction parallel to the flow and therefore normal to the surface of the sphere. $\perp$ is the direction orthogonal to this. The temperatures in these two directions are not identical and can be written as

**Fig. 3.24** The ratio of $T_{\parallel}$ to $T_{\perp}$ for the free expansion model. Note the extremely rapid change in the ratio close to the nucleus

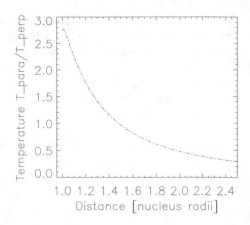

$$T_{\parallel}(r) = T_0\left(1 - \frac{2}{\pi}\right)\left[1 + \left(\frac{R_N}{r}\right)^2\right] - T_0\sqrt{1 - \left(\frac{R_N}{r}\right)^2} \tag{3.82}$$

and

$$T_{\perp}(r) = \frac{T_0}{2}\left[1 + \left(\frac{R_N}{r}\right)^2 - \sqrt{1 - \left(\frac{R_N}{r}\right)^2}\right] \tag{3.83}$$

It should be noted that there is no description of temperature in the force-free radial outflow case (Fig. 3.24).

As an aside, Knollenberg et al. (2016) noted that the stationary solution of the expansion of a free axisymmetric supersonic gas jet into a vacuum can well be approximated in the far field by a so-called virtual source flow, where all streamlines are directed radially away from a common source point. This description applies in the far field, at distances larger than ~10 times the source diameter (Koppenwallner et al. 1986).

### 3.4.4   The Knudsen Layer and Low Density Flow

#### 3.4.4.1   The Boltzmann Equation and the Direct Simulation Monte Carlo Method

At the interface between the subliming ice surface and the vacuum of space, the VDF cannot be Maxwellian because the surface itself prevents particle motion in one direction. As was noted above, it is often assumed that the VDF at this interface on a comet can be described by a half-Maxwellian (see Fig. 3.21). As the gas expands

from the surface, collisions act to change this initial distribution and it evolves into a
Maxwellian equilibrium distribution (Cercignani 2000).

In reality, the VDF at the surface of the nucleus is not known. It is possible to
imagine that the VDF is more collimated or jet-like than a half-Maxwellian if, for
example, gas is ejected through a porous medium at the surface. Distributions such
as $\cos^x$ for the VDF (where x can take a value between 1 and 9) have been proposed
for Earth applications. It seems, however, that different assumptions for the VDF
will not produce large differences on the final flow field at comets (Liao et al. 2016)
but it should nonetheless be borne in mind particularly when sources are highly
inhomogeneous.

The modern approach to addressing non-equilibrium flows is to solve the
Boltzmann equation. Here we follow the description of Mohamad (2011). A distribution function is defined as a function of seven parameters, $f(r,v,t)$, where $r$ is the
position vector of the molecule, $v$ is the velocity vector and $t$ is the time. $f$ at a specific
time, $t$, is sometimes referred to as the state vector in, for example, celestial mechanics.
$f(r,v,t)$ defines the number of molecules, at time $t$, positioned between $r$ and $r + dr$
and having a velocity between $v$ and $v + dv$ (Mohamad 2011). In the absence of
collisions and no external forces then

$$f(r + vdt, v, t + dt) \, dr \, dv - f(r, v, t) \, dr \, dv = 0 \qquad (3.84)$$

which simply indicates that molecules travel in straight lines unless there are external
influences. If an external force, $F$, (e.g. gravity) is present then

$$f(r + vdt, v + Fdt, t + dt) \, dr \, dv - f(r, v, t) \, dr \, dv = 0 \qquad (3.85)$$

However, if collisions occur there will be a net difference in the number of
molecules in the interval $drdv$. The rate of change between the final and initial
state of the distribution is referred to as the collision operator (or collision integral),
$\Omega(f)$, and the equation becomes

$$f(r + vdt, v + Fdt, t + dt) \, dr \, dv - f(r, v, t) \, dr \, dv = \Omega(f)dr \, dv \, dt \qquad (3.86)$$

If this equation is now divided by $dr \, dv \, dt$ and as $dt$ goes to zero, we find that

$$\frac{df}{dt} = \Omega(f) \qquad (3.87)$$

$f$ is function of $r$, $v$ and $t$. However, it is clearly desirable to obtain the time derivative.
This can be addressed by taking the partial derivative of $df$ and dividing by $dt$ so that

$$\frac{df}{dt} = \frac{\partial f}{\partial r}\frac{dr}{dt} + \frac{\partial f}{\partial v}\frac{dv}{dt} + \frac{\partial f}{\partial t}dt \qquad (3.88)$$

where one can see that substitutions for velocity ($dr/dt$) and acceleration are possible. Further substitution for acceleration (Newton's second law) leads to the equation

$$\frac{\partial f}{\partial t} + \frac{\partial f}{\partial r}v + \frac{F}{m}\frac{\partial f}{\partial v} = \Omega(f) \tag{3.89}$$

For a system without external forces, this equation can be written as

$$\frac{\partial f}{\partial t} + v \cdot \nabla f = \Omega(f) \tag{3.90}$$

which is the Boltzmann equation. The key problem in solving this equation is the determination of the collision operator. Physically, the equation implies that if $\Omega$ becomes zero, collisions have no effect on the distribution function. One can invert this by saying that if the system is collisionless then

$$\frac{\partial f}{\partial t} + \frac{\partial f}{\partial r}v + \frac{F}{m}\frac{\partial f}{\partial v} = 0 \tag{3.91}$$

which is a form of the Vlasov equation used as a starting point for many plasma physics problems. It is also called the Collisionless Boltzmann Equation (Cravens 1997). In the plasma physics case, $F$ is often substituted using the Lorentz force

$$F = q_c(E + v \times B). \tag{3.92}$$

and we shall see the usefulness of this later. On the other hand, if collisions occur but are of no importance to the solution, the gas can then be assumed to be in LTE (Davidsson 2008) and it can be shown that this only occurs if $f$ is a Maxwellian.

The Direct Simulation Monte Carlo (DSMC) method (Bird 1994) solves the Boltzmann equation by looking at individual particle collisions to produce a velocity distribution and an effective motion of the particles. The form of the Boltzmann equation used is

$$\frac{\partial n_g f}{\partial t} + \frac{\partial n_g f}{\partial r}v + \frac{F}{m}\frac{\partial n_g f}{\partial v} = \int_{-\infty}^{\infty}\int_0^{4\pi} n_g^2\left(f_1^* f_2^* - f_1 f_2\right)v_r\sigma_{col}\,d\Omega_s\,dv_2 \tag{3.93}$$

which, when compared to Eq. (3.89), shows that the right-hand side is the collision operator. $f_1$ and $f_2$ are two values of the VDF, $f$, at velocity values, $v_1$ and $v_2$. $\Omega_s$ is the solid angle over which we integrate and $v_r$ is the relative velocity between colliding particles. $n_g$ is the number density. The superscript * indicates post-collision properties. The algorithm is essentially that shown in Table 3.8.

The motion between collisions is ballistic and collision models, both between particles and with the boundaries, need to be specified. For example, collisions with solid boundaries may be thermal (giving the gas molecule a diffuse reflection at the temperature of the solid wall), specular, or a combination of the two.

| **Table 3.8** Basic algorithm for DSMC calculations (Alexander and Garcia 1997) | Initialize system with particles |
|---|---|
| | Loop over time steps |
| | Create particles at open boundaries |
| | Move all the particles |
| | Process any interactions of particles and boundaries |
| | Sort particles into cells |
| | Sample statistical values |
| | Select and execute random collisions. |
| | Output particle properties per cell |

The implementation that we will use here is called ultraSPARTS and is based on a published code called PDSC++ (Su 2013). This, in turn, was based on the PDSC (Parallelized Direct Simulation Monte Carlo Code) program developed by Wu and co-workers (Wu and Lian 2003; Wu et al. 2004; Wu and Tseng 2005). Other implementations have been used by, amongst others, Combi (1996), who was the first to use DSMC for cometary research, Skorov and Rickman (1999), Crifo et al. (2002a), Davidsson (2008), and Tenishev et al. (2011). Other publicly available codes that might also be used include SMILE,[6] OpenFOAM[7] (Scanlon et al. 2010; White et al. 2018) and Bird's codes.[8] A derivative of SMILE was used by Skorov et al. (2006) to study cometary comae.

The expansion of the gas produced at a cometary surface implies that, even if the gas is collisional at the surface, eventually it becomes collisionless with distance and the time to transfer energy between the rotational and translational degrees of freedom increases. This is, in part, one of the reasons why DSMC is a useful tool in analysing cometary comae. On the other hand, the number of collisions needed to equilibrate the gas (the collisional relaxation number) is not well established with values between 1 and 8 having been used by different authors. Liao et al. (2016) looked at the significance of this number and found that, for cometary expansion, the exact value is of limited importance in defining the final flow field although details close to the nucleus may be influenced.

In comparing DSMC and the fluid description, Crifo et al. (2002a) noted that complete consistency between the approaches is by no means straightforward to attain because the properties of the molecules must be represented consistently. In the case of complex molecules such as $H_2O$, approximations are unavoidable, and they are formulated differently in the two approaches. In the fluid case, a kinematic viscosity law is used whereas in DSMC a collision model is used. Consistency to within 10% in density for any complex flow should be considered perfectly reasonable. There are standard test cases to verify DSMC implementations and new codes should be checked against these.

---

[6]http://lnf.nsu.ru/en/smile.html

[7]https://www.openfoam.com

[8]http://www.gab.com.au

Finklenburg et al. (2011) compared the Euler approach with DSMC but showed that the subsonic region close to the nucleus often needs to be simulated with DSMC. As soon as smaller scale structures in a rarefied gas appear, the local Knudsen number increases and the Euler equations become increasingly unreliable.

Finally, it should be noted that, like the Euler equations, there are conservation equations associated with the Boltzmann equation. These relate the microscopic distribution function to the macroscopic variables of mass (density), momentum, and energy. The equations are

$$\rho(r, t) = \int m_g f(r, v, t) dv \tag{3.94}$$

$$\rho(r, t) \, u(r, t) = \int m_g \, v f(r, v, t) dv \tag{3.95}$$

$$\rho(r, t) \, \epsilon(r, t) = \frac{1}{2} \int m_g \, (v - u)^2 f(r, v, t) dv \tag{3.96}$$

where $u$ defines the fluid velocity vector, $\epsilon$ is (as before) the specific internal energy, and $m_g$ is the molecular mass (e.g. Bodenheimer et al. 2006). Note that we have ignored the subscript g for the velocity components in this description for clarity.

### 3.4.4.2 The Knudsen Penetration Number

The relative importance of jets from cometary surfaces compared to more homogeneous insolation-driven activity remains a subject of considerable debate. However, if jets are important, the interactions between them can be complex and also depend upon the density of the flows. It is useful here to discuss the Knudsen penetration number, $Kn_p$. If we assume that there are two sources, then the Knudsen penetration number is the ratio of the mean free path of particles from the one source penetrating the other to the distance from the interaction plane to the centre line of the second source, $r_{Knp}$. This (particularly opaque) definition is illustrated in Fig. 3.26.

The computation of $Kn_p$ in a cometary context requires a number of assumptions. However, a simple analytical result can be obtained by determining the number density at a point on the interaction plane. If we assume emission from two equal sources separated by a distance, $z$ (Fig. 3.25), and allow the expansion to be described by a $1/r^2$ dependence with a constant velocity, then we only need to establish an angular distribution for the outflow relative to the jet axis of symmetry. If the number density is proportional to the cosine of the angle to the jet axis of symmetry, $\varphi_A$, then the number density, $n_g$, is given by

$$n_g(r, \varphi) = \frac{Q_T \cos \varphi_A}{4\pi^2 r_{Knp}^2 v_g} \tag{3.97}$$

**Fig. 3.25** The basic
geometry for an analytical
determination of $Kn_p$

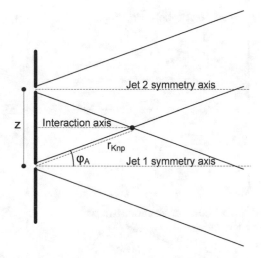

where $Q_T$ is the total production rate from the source (integrated over the full hemisphere) and $r_{Knp}$ is the distance to the interaction axis from the source at the angle, $\varphi_A$. The number density is related to the mean free path, $\lambda_{MFP}$, via

$$\lambda_{MFP} = \frac{1}{4\sqrt{2}\ r_m^2 \pi n_g} \tag{3.98}$$

where $r_m$ is the molecular radius introduced by expansion of the collision cross-section. The Knudsen penetration number is the ratio of the mean free path to the distance from the interaction axis to the centre line of the second source, which also happens to be $r_{Knp}$. Hence

$$Kn_p = \frac{\lambda_{MFP}}{r_{Knp}} \tag{3.99}$$

$r_{Knp}$ is related to the distance between the sources, z, through the angle $\varphi_A$. After some trivial manipulation, one obtains

$$Kn_p = \frac{1}{\sqrt{2}d_m^2}\ \frac{2\pi z v_g}{Q_T \sin\varphi_A \cos\varphi_A} \tag{3.100}$$

where $d_m$ is now the molecular diameter. For reasonable values of the parameters, a fairly wide range of values for $Kn_p$ can occur.

The consequences of source interactions are illustrated in Fig. 3.26. When the gas flux from two adjacent but separated sources is very low, we are in a free molecular flow regime. The molecules travel without colliding and so the gas from one jet penetrates the region occupied by the other jet without any noticeable disturbance. This is the free penetration regime and $Kn_p >> 1$. As the gas density in the plumes

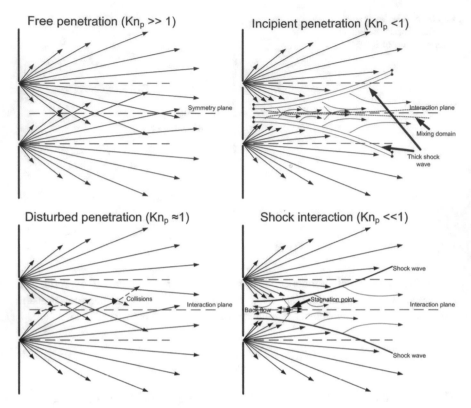

**Fig. 3.26** The strength of the interaction between two plumes of gas is described via the Knudsen penetration number. The different regimes are illustrated. From Marschall et al. (2020), modified from Dankert and Koppenwallner (1984)

rises, $Kn_p$ approaches 1. Collisions begin to occur and the flow becomes disturbed near the axis of symmetry of the system. A further increase leads to the formation of shocks on either side of the plane of symmetry. Incipient penetration and mixing occurs. As the density increases further such that $Kn_p \ll 1$, there is essentially no mixing between the gas from the two plumes. They are separated from each other. Furthermore, significant back flow to the inert surface between the jets occurs.

Gas dynamics simulations using Direct Simulation Monte Carlo show these different penetration regimes extremely well. In Fig. 3.27, we can see examples of the change in the interaction. Here the distance between the sources is constant but the distance is referred to the mean free path as given in the caption. At the top, the flow is rarefied whereas the bottom example has a much higher source strength with a small mean free path. On the left, the temperature is shown colour-coded while the Mach number within the flow is given by the contours. The degree of penetration can be seen on the right. Here the number of molecules from the left hand source at each position in the domain is colour-coded from 0 to 100%. One can see how the interaction develops as the Knudsen penetration number decreases from top to

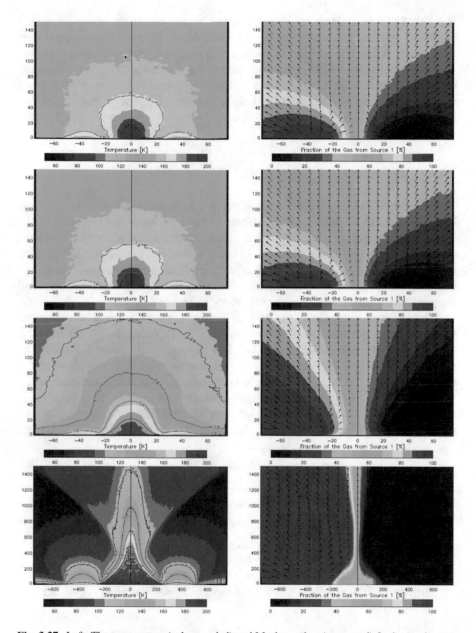

**Fig. 3.27** Left: The temperature (colour-coded) and Mach number (contoured) for interacting gas jets with different separations. Top to bottom: Separation, $d = 6\,10^{-5}\,\lambda_{MFP}$, $6\,10^{-2}\,\lambda_{MFP}$, $60\,\lambda_{MFP}$, and $3\,10^{3}\,\lambda_{MFP}$. Right: The fraction of molecules from source 1 at each position in the field corresponding to the four cases on the left

bottom. When the sources are strong, a shock region builds along the interaction plane and cross-contamination of the flow is limited. This has interesting consequences for in situ composition analysis. When $Kn_p \gtrsim 1$, then molecules from one source can be detected within another implying that sources with different chemical compositions are much harder to separate chemically than if the $Kn_p \ll 1$. For a mission such as Rosetta in which 67P passed through a large range of $Kn_p$, this can be significant for interpretation.

### 3.4.4.3  The Influence of Composition

The description in the previous sections has said little about composition although the gas species is evident in the equations in both the mass and the ratio of the specific heats, $\gamma$ (e.g. Eq. 3.73). Relatively little numerical modelling work has been carried out investigating the influence of composition and compositional variation on the flow field. On the other hand, there is irrefutable evidence that the parent molecule composition in the innermost coma varies strongly both spatially and temporally (e.g. Bockelée-Morvan et al. 2015; Hässig et al. 2015).

Figure 3.28 shows the local gas number density of the three major species in the innermost coma as a function of time measured by the ROSINA dual-focussing mass spectrometer (DFMS) onboard Rosetta. The measurements cover the month of November 2014. Although the effects of spacecraft –comet distance, phase angle, and sub-spacecraft latitude are still included in this raw result, the variations in the ratios of $CO_2$ and CO to $H_2O$ are clearly evident with the $CO_2/H_2O$ ratio being approximately 0.03 on occasions but ~0.3 on others. Hence, the assumption of a single-species is at times questionable. The Deep Impact observations at 9P/Tempel

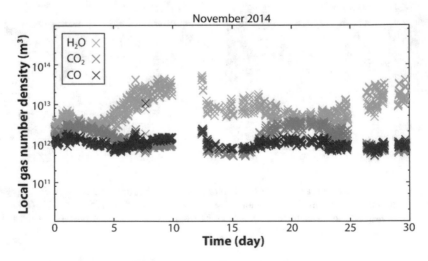

**Fig. 3.28** Local gas number density evolution of $H_2O$, CO and $CO_2$ measured by the instrument ROSINA/DFMS onboard Rosetta in November 2014 (from Herny et al. 2020, submitted)

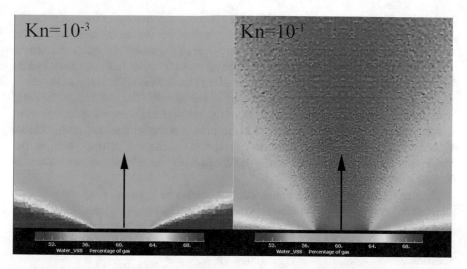

**Fig. 3.29** Jets arising from a 50/50 $CO_2$-$H_2O$ mixture at the source. Left: A low Knudsen number case. Right: An intermediate Knudsen number case. The results were calculated with a 2D-DSMC code. The percentage of water in the flow is colour-coded. The plot shows that fractionation of the gas species can occur. (Adapted from Finklenburg 2014)

1 (Feaga et al. 2007) and at 103P/Hartley 2 (see Fig. 3.2) provide further proof of this.

This compositional heterogeneity is perhaps not surprising given the difference in the free sublimation temperatures of the three species and one would intuitively expect $CO_2$ to dominate $H_2O$ near the terminator and CO to be the dominant molecule emitted from the nightside when the residual heat arising from thermal inertia is only sufficient to sublime highly volatile ices (Sect. 2.9.3). Nonetheless, the effect on the dynamics of the flow remains a relatively unexplored issue. We show here illustrations of two potential effects that might influence the interpretation of data.

Figure 3.29 shows the water mixing ratios for jets produced by a subliming 50:50 $CO_2$-$H_2O$ ice mixture with two different production rates resulting in Knudsen numbers of around $10^{-3}$ (an equilibrium flow case) and $10^{-1}$ (a case where non-LTE effects might be present). The jets are directed vertically upwards (indicated by the arrow), are centred, and have widths of ¼ of the shown width of the domain. The mixing ratio of water is shown as a percentage using the colour code. The deviation from 50% in the centre of the jet results from the dynamics of the expansion. In the hydrodynamic limit (applicable for $Kn = 10^{-3}$), the flux from a reservoir into vacuum is given by

$$j_{flu} = \left(\frac{2}{\gamma + 1}\right)^{\frac{\gamma+1}{2(\gamma-1)}} n_0 \sqrt{\frac{\gamma k T_0}{m_g}} \qquad (3.101)$$

and as $\gamma$ is different for water and $CO_2$, the resulting output ratio of the water flux to the total flux is around 61% (as seen in the centre of the jet). In the kinetic case, the outflow flux ratio is controlled by the square-root of the molecular mass and should nominally show a similar ratio at source but the water molecules can expand more quickly through the more diffuse $CO_2$ resulting in a lower mixing ratio at source.

However, what is more important is that, in the wings of the jets (and particularly close to the surface in the fluid case), the water mixing ratio is appreciably higher. In other words, the gas is fractionating. This implies that for jet-like outgassing from a mixed source, the mixing ratios of gas species measured at one point in the innermost coma may be substantially different from the mixing ratios of the subliming sources.

It should also be pointed out (although it should be apparent from the previous discussion) that in low production rate cases, inhomogeneous composition can lead to differing outflow velocities between species and therefore local densities may not reflect the actual surface composition but be biased towards heavier, slower moving, molecules.

### 3.4.5   Examples of Near-Nucleus Gas Flow

#### 3.4.5.1   Cases with Spherical Sources

The observations of cometary nuclei show them to be highly irregular but calculations with spherical symmetry remain extremely useful for illustration of specific phenomena. In Fig. 3.30, we see a DSMC calculation for a spherical, uniformly emitting, nucleus of 2 km in radius. Water is the sole species and the production rate is 2 kg/s. Profiles of the number density (top right), gas temperature (bottom left) and velocity (bottom right) are presented out to 10 km from the centre of the sphere. At the top left of Fig. 3.30, is a 2D slice through the centre of the emitting sphere and shows the log of the number density colour-coded. Notice that at the surface, the temperature is marginally over 160 K and a factor of about 0.82 below the temperature of the subliming reservoir (cf Table 3.7). The temperature decrease is rapid, reaching 70 K within five nucleus radii. At the same time, there is a rapid acceleration (see bottom right) taking the bulk velocity to over 600 m/s very quickly. The radial velocity approaches the value for the bulk flow measured by Lammerzahl et al. (1987) between 1000 and 4000 km from the nucleus of 1P/Halley of 800 ($\pm$50) m $s^{-1}$. The final speed is not a strong function of production rate in this spherically symmetric case. This can be seen in Fig. 3.31 which shows the gas speed for total water production rates covering two orders of magnitude. The speed at five nucleus radii changes by <20% over this range of $Q_g$.

Another feature of Fig. 3.31 is that gas speeds are approaching their asymptotic values fairly close to the nucleus. The change in velocity between 8 and 9 km from the centre of the sphere (3–3.5 radii from the surface) is already <5% $km^{-1}$.

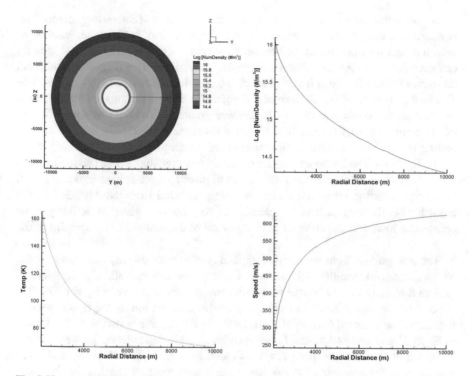

**Fig. 3.30** A DSMC calculation for a spherical isotropically emitting nucleus with a water production rate of 2 kg/s and an initial gas temperature of 200 K. Top left: The number density in a 2D slice through the centre of the 2 km radius nucleus. Top right: The number density along a radial profile from the surface. Bottom right: The bulk speed of the molecules along the same profile. Bottom left: The gas temperature along the profile

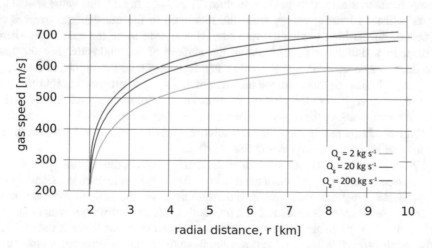

**Fig. 3.31** The gas speed from a 2 km isotropically emitting sphere with different water production rates. The final gas speed is a function of the production rate but not a strong one. (Courtesy of Raphael Marschall)

Breaking the symmetry of emission generates a number of interesting effects. The simplest and most realistic method to break the symmetry is to use an insolation-driven model for the initial condition on the dayside, i.e. using Eq. (2.101) with no emission from the nightside. For the purposes of this illustration (Fig. 3.32), the local production rates have been scaled to reach total production rates of 2, 20, and 200 kg $s^{-1}$. In the absence of emission from the nightside, flow is no longer radial close to the nucleus as the gas tries to balance pressure gradients. The top row shows the log of the number density for the three different production rates. Note that the colour-coding of the flow field has been scaled using the factors that are indicated in the plots. The three results appear rather similar. However, the lower production rate case has a lower relative density above the nightside of the sphere indicating that, relatively speaking, fewer molecules are being scattered from the dayside into the nightside hemisphere. A better impression of the effects is given by looking at the other three rows and in particular the magnitude of the non-radial component of the velocity.

The gas emitted from the dayside expands not merely radially but also laterally with a component parallel to the surface. Close to the nightside surface, the speed of the gas flow may be low but the non-radial component of the velocity is high. This concept of non-radial flow was used as a potential explanation for the ripples seen in Fig. 2.95. The speed of the (radial) outflow above the dayside is similar to that shown in Fig. 3.31 and shows the same (small) production rate dependence. However, close to the terminator, the gas speed drops off much more quickly in the low production rate case as the number of collisions experienced reduces quickly in the more rarefied flow. This results in a relatively lower number density above the nightside hemisphere in the low production rate case. Further evidence for the absence of collisions on the nightside in the lower production rate case can be seen in the temperature anisotropy plots (bottom row) which show the ratio of the rotational temperature to the translational temperature ($T_{rot}/T_{trans}$). In LTE this value should be 1 as collisions transfer energy from the rotational to the translational degrees of freedom. In the high production rate case, this is seen to be the case everywhere except in a thin layer directly above the surface of the nightside. For the low production rate case, almost the entire hemisphere above the nightside has $T_{rot}/T_{trans} \geq 3$ indicating that there are insufficient collisions to produce equilibrium and to transfer energy from the rotational degrees of freedom into speed. Lower radial outflow speeds above the nightside are a direct consequence of this. It is also evident that the condition required for the fluid solution (namely that $Ma = 1$ at the surface) is not met at or just above the nightside surface.

What we see here is that, even in a relatively simple case, the outflow speed of the gas is dependent upon position in the inner coma and that while there is a small speed dependence on production rate in the isotropic case, the insolation-driven case results in large speed dependencies on production rate away from the main direction of outflow. Of particular importance for the interpretation of Rosetta data is that near-nucleus dayside to nightside flows significantly affect the inner coma properties near the terminator. (Rosetta was frequently in terminator orbits about the nucleus and hence its in situ experiments were often sampling this region of the inner coma.)

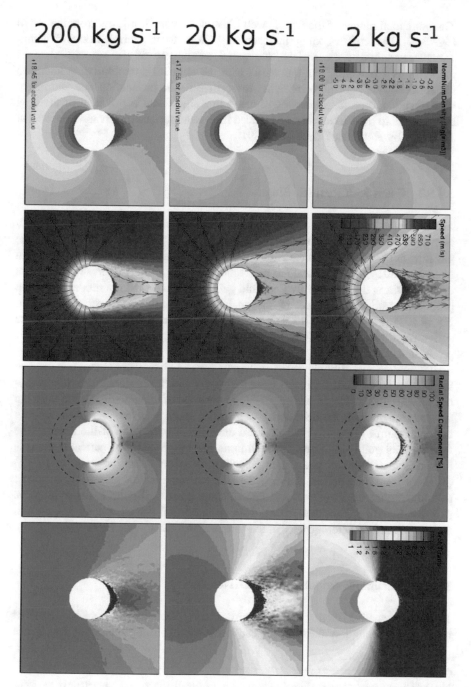

**Fig. 3.32** The gas flow field around a 2 km spherical source emitting water according to an insolation-driven model. The columns represent three different production rates. The rows show,

The non-radial nature of the flow is well illustrated by the third row from the top in Fig. 3.32. The colour-coding indicates that close to the terminator and within 1 nucleus radius of the surface, the radial component of the velocity changes markedly as a percentage of the whole with the remaining motion being lateral to the surface. This percentage is rather insensitive to the production rate and hence strong production rate gradients will result in a significant component of the flow being lateral to the surface as was deduced from near-surface observations of dust at 1P/Halley by Keller and Thomas (1989).

### 3.4.5.2  Cases with Realistic Shapes

The deviations of cometary nuclei from spherical symmetry influence the flow field. However, the effects can be subtle. In Fig. 3.33, we see a DSMC calculation of the gas flow field around the nucleus of 67P at one instant representative of a point in November 2014 when the total gas production rate was around 2 kg/s. The two plots to the left show the gas density (top) and gas speed (bottom) for a model where the production rate at each facet of the nucleus shape model was set according to Eq. (2.101) (and therefore insolation-driven) and multiplied by a single effective active fraction (Sect. 2.9.3.1) to provide consistency with the observed water production rate.

It is evident that although there are two regions of higher production rate (one on the head of the nucleus and one on the body), the gas flow smooths this inhomogeneity out so that by the time the gas reaches the edge of the domain, 10 km from the centre of the nucleus, the distribution of gas with azimuth is smooth and shows only one maximum in (roughly) the sunward direction. At the edge of the domain, the flow appears quite similar to that seen in an insolation-driven spherical case. This illustrates an important point. For modelling of observations at large ($>10$ $R_N$) distances from the nucleus, the irregular shape of the nucleus plays only a modest role in defining the flow field and hence, if the gas distribution is observed to be strongly deviating from that arising from an insolation-driven sphere, this implies that major surface heterogeneity in the source distribution must be suspected.

### 3.4.5.3  Deviations from Insolation-Driven Activity

The effective active fraction (eaf) was introduced in Sect. 2.9.3.1 as an hoc means of reducing the strength of the source because free sublimation from a water ice surface

---

**Fig. 3.32** (continued) from the top, the log number density, the speed, the magnitude of the radial component of the speed, and the temperature anisotropy. The colour-coding is the same for all panels in a row. Note that this has been achieved for the number density by using the scaling factor indicated in the plot. (Credit: Marschall, PhD. thesis, 2017)

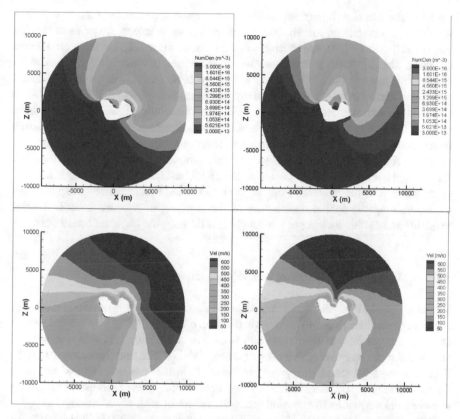

**Fig. 3.33** The number density and gas velocity around the nucleus of comet 67P using a realistic shape and a total production rate of around 2 kg/s. The left column shows an insolation-driven case. The right column shows a case when the activity from the neck region of the nucleus is enhanced by a factor 6

would provide 10–50 times more gas per unit surface area for insolation-driven activity than is typically observed when comparing total water production rates to nuclear radii. However, there is no particular reason to assume that the reduction in source strength implied by an eaf is constant over the whole nucleus. Indeed, given the morphological diversity of the surface appearance of 67P seen earlier, it would be rather surprising if the eaf were constant and evidence from gas density measurements in the coma of 67P suggests that the eaf cannot be constant. Modifying the eaf with position provides additional free parameters allowing fits to deviations from insolation-driven activity.

Marschall et al. (2016) fit data from the ROSINA/COPS pressure sensor on Rosetta in the period around November 2014 by manipulating the eaf on regional scales. It was concluded that the neck region (Hapi—see Fig. 2.50) needed to have an eaf up to six times greater than the rest of the nucleus in order to fit the observations. Broadly similar conclusions were reached by Bieler et al. (2015a) and Fougere et al. (2016). An example of the effect of stronger emission from the neck is shown in Fig. 3.33 (right). Although the basic appearance of the flow field is similar to the insolation-driven case, there are subtle differences. For example, the gas speed (bottom right) is appreciably faster close to the nucleus in the region above the neck. This illustrates that the flow field can be used to establish variations of the eaf over the nucleus but it is necessary to make precise measurements of the density, speed and temperature of the gas at many points in the coma as close as possible to the nucleus surface. Hoang et al. (2017) also concluded that, during northern summer through to equinox, the $H_2O$ emission was illumination-driven with maxima in the neck. This conclusion was reached assuming free molecular flow from the source to ROSINA which is a less sophisticated model of the outflow than used by, for example, Fougere et al., Bieler et al. and Marschall et al. although a crude comparison with the model of Fougere et al. suggested that the different approaches to determining the spatial distribution of the source can produce broadly similar results if one avoids looking at the details and limiting the physical constraints.

### 3.4.5.4   Degeneracy in Surface Activity Distributions

We have seen in Fig. 3.32 that when there is large scale inhomogeneity in the initial boundary condition, gas flows to equalise the pressure. But above large inactive regions, the gas flow is less dense and is clearly sub-sonic. This was also seen in somewhat more complex activity distribution models (Crifo and Rodionov 1999). In Fig. 3.33, we have seen that regional variations in the eaf can be identified but these distributions can be considered to be continuous. Are models with strong sub-regional variations in emission strength consistent with observation?

The most detailed nucleus shape model we have is that of 67P. In the high fidelity model, we have more than 40 million facets and, in principle, each of these facets can (and probably has) its own eaf ranging from totally inactive to being fully active at the level of free sublimation of water ice (or indeed at the level of an even more volatile species such as $CO_2$). This provides a challenge in that we end up with an additional 40 million+ free parameters for the initial boundary condition. But is it really necessary to consider each facet individually? And can measurements of gas densities in the inner coma be inverted to give source strengths at the surface at facet dimension scales?

The key property of the flow here is the Knudsen penetration number and, through that, the mean free path. The variation with time and position plays an important role. Inhomogeneities in the boundary condition on scales smaller than the mean free path will not be distinguishable and they should be irrelevant to the flow field. This is particularly obvious for very low production rate cases when the Knudsen penetration number is large. Gas emitted in the same direction but from different parts of the nucleus cannot be distinguished because there is no interaction. This has been studied by Liao (2017) who looked at changing the activity distribution from insolation-driven to a distribution where only 2% of the surface was active with these active areas being randomly distributed over the nucleus. The activity was scaled to produce the same total production rate. Several intermediate cases were also tested. The results showed that for cases where only 20% of the surface was actively emitting, the flow field was essentially indistinguishable from the insolation-driven case. Only when the total area contributing to activity was reduced to the minimum (2% of the surface) did the flow field become clearly different.

These tests were performed for a low production case (1.25 kg s$^{-1}$) and obviously as the production rate is increased the mean free path is reduced (Fig. 3.19). For 67P close to perihelion, values in the range 1–5 m would be expected. For these higher production rates, the reduction in the Knudsen penetration number leads to a more radial outflow but, even here, the strong variations in production close to the terminators (where the mean free path rises sharply) and possible deviations from insolation-driven production at scales larger than the mean free path, will result in complex flow geometries leading to degeneracy (Marschall et al. 2020).

It is to be noted that several authors have mapped gas emissions using ROSINA/COPS and ROSINA/DFMS data under the assumption that the activity distribution is solely a function of the solar insolation so that observations in four dimensional space and time can be related by simple functions (e.g. Combi et al. 2020). However, the evidence of transient ice patches on the surface (e.g. Fig. 2.106) and large transient dust jets (see below) implies that this assumption must break down at some level. At present, however, that level is not well defined.

### 3.4.6  Re-Condensation and Surface Reflection of Gas Molecules

As we have seen, gas molecules can be transported from their source on the dayside to the nightside. Rubin et al. (2014a) also showed this and concluded that condensation onto the nightside surface could occur producing surface frosts because the nightside acts as a regional cold trap. This would result in bright frosts being evident on the surface at the morning terminator for example.

Condensation can also occur in shadowed areas on the dayside. The temperature of shadowed surfaces will be low because of the low thermal inertia and the absence of any warming atmosphere. As we have seen in Fig. 2.66, decreases in the surface temperature upon transition from a surface being illuminated to entering shadow can be rapid and large. Hence, molecules returning to the surface will have a good chance of sticking to the surface on contact once the temperature is significantly below 200 K. Even under fairly rarefied conditions, collisions of molecules in the source regions are sufficient to generate a backflux of molecules as can be seen from the velocity distribution functions (Fig. 3.21; Table 3.7). Hence, shadowed areas can also act as local cold traps for emitted gas and the production of visible surface frosts.

Evidence for condensation has been found in OSIRIS images of 67P. Two examples are given in Figs. 3.34 and 3.35. Figure 3.34 shows an extensive condensate close to a shadowed area in the Hapi region. Hapi was the most active region on

**Fig. 3.34** Oblique view of the Hapi region on 67P acquired on 30 Dec. 2014 07:42:59. There is shadow covering much of Hapi in the foreground. At the edge of the illuminated region, the surface is appreciably brighter (**A**). The shadow is retreating towards the lower left corner. This is an indication of a surface condensate. Two bright circular patches can be seen (**B**; cf. Fig. 2.105). Evidence of emission from the surface can also be seen (**C**). (Image number: N20141230T074259892ID10F22)

the nucleus at the time of these observations with a high eaf (Marschall et al. 2016; Fougere et al. 2016). Figure 3.35 was acquired closer to perihelion and shows rougher terrain. Once again, at shadow boundaries brighter deposits are evident. Fornasier et al. (2016) used a time series showing the retreat and disappearance of a condensate to estimate its thickness and concluded that the covering was equivalent to 10–15 μm of solid ice.

Liao et al. (2018) used a DSMC code and determined the returning flux of molecules impacting facets that were set to being inactive (and thus equivalent to being in shadow). It was shown that condensation at the rates needed to match the condensate thickness estimated by Fornasier et al. (2016) were indeed plausible.

An alternative explanation for the observations has been presented by De Sanctis et al. (2015) who proposed a model whereby thermal inertia maintains sub-surface sublimation for a short time after local sunset. The surface then is colder than the sub-surface and sub-surface generated molecules condense at the surface as a frost before being exposed to sunlight the following day. This was supported by a numerical simulation (see Huebner et al. 2006) similar to that of the Marboeuf et al. (2012) model discussed above.

It is difficult to distinguish between these two explanations. Both mechanisms have been modelled numerically and shown to be feasible on the basis of current knowledge and indeed both may be active depending upon the precise geometry and the structure of the sub-surface. Some effort to use the existing data sets to study the

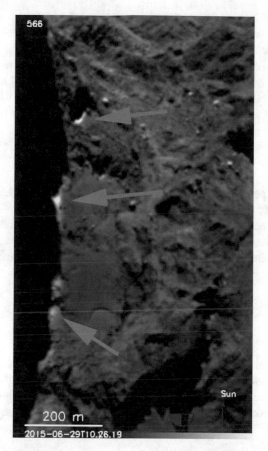

**Fig. 3.35** Image of 67P acquired on 29 June 2015 at 10:26:19. Notice at the edges of the shadows (marked by arrows), the surface is relatively bright. (Image number: N20150629T102619565ID10F24)

brightness of the surface immediately before sunset compared to that at sunrise may provide some additional constraints.

### 3.4.7  The Initial Gas Temperature

In a previous section, the porosity of the surface layer was discussed and the influence of a desiccated, partially insulating, porous layer on the gas production rate was calculated from the perspective of the thermal conductivity. The effect of radiation and gas flow through the layer on the thermal conductivity of the layer was also shown to be rather complex. However, the flow through a porous surface layer can influence the gas properties in other ways.

The first consideration is that the gas velocity at the surface is modified by this process. The flow rate through a layer must match the production rate at the sub-surface source in a pressure driven flow. The sub-surface pressure can then be estimated by application of modifications to Darcy's law. The permeability provides an indication of how rapidly a fluid can flow through rock and has an SI unit of [m$^2$]. In its simplest form, the fluid flow velocity through a medium can be calculated using

$$v_g = \frac{\kappa_p}{\mu_v} \frac{dp}{dz} \tag{3.102}$$

where $\kappa_p$ is the permeability ($\sim 10^{-9}$ m$^2$ for highly fractured rocks) and $\mu_v$ is the dynamic viscosity (typical value for gases at STP $\sim 10^{-5}$ Pa s).

On the other hand, Benkhoff and Boice (1996) noted that there is an analytical solution for an ideal gas flowing through a porous medium in the Knudsen regime where here the Knudsen number is defined through the ratio of the mean free path to the pore size. They quote

$$v_g = \frac{-8\Psi}{3\xi_T{}^2} a_p \sqrt{\frac{R_g}{2\pi m_g}} \left( \sqrt{T} \frac{d \ln \rho_g}{dx} + \frac{d\sqrt{T}}{dx} \right) \tag{3.103}$$

where the tortuosity, $\xi_T$, is defined as the ratio of the length of the tubes to the thickness of the porous layer, $a_p$ is the pore radius, $\rho_g$ is the density of the gas and $m_g$ is the molecular mass for the gas species. One can see from this equation that there is strong dependence on the porosity and tortuosity which is balanced against the density gradient. This leads to pressure build-up within the porous layer even in the Knudsen case.

A lot of work in the oil and gas industries has been performed on this subject to provide better estimates of flow rates but their applicability to highly diffuse flows is questionable. As an example, a semi-empirical equation of possible use is from Carrigy et al. (2013) who found that the gas flux through a porous layer, $q_f$, can be found from

$$q_f = - \left( \frac{\kappa_p}{\mu_v} \frac{p_a + p_b}{2} + D_K^{eff} \right) \frac{1}{kT} \frac{p_b - p_a}{L_{th}} \tag{3.104}$$

where $p_a$ and $p_b$ are the pressures on either side of a layer of thickness, $L_{th}$, and $D_K^{eff}$ is an effective diffusivity for the layer. But, as can be seen, there are numerous parameters here that are difficult to determine in a cometary case and can take a wide range of values although the relationship to Eq. (3.102) is quite apparent.

A further aspect is that the temperature of the gas can be modified by the interaction with the inert surface in the porous layer. In a previous section, we looked at how a porous matrix can be heated or cooled by a gas flow. But the gas itself will either lose or gain energy through this process. The absence of sublimation

**Fig. 3.36** The gas temperature 1 cm above an inert surface layer of 5 cm thickness. Water vapour from a sub-surface source at 200 K passes through the layer (which is at 300 K). The gas is heated to varying degrees by collisions with the inert surface. (Model calculation of Christou et al. 2018)

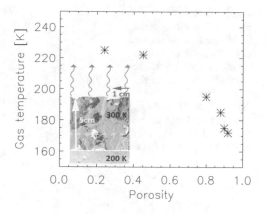

allows the temperature of an inert surface layer to rise to close to the black-body temperature. Gas molecules emitted from below the surface will strike the inert layer which may result in the molecules receiving additional energy and hence increase their temperature. This can be a significant source of additional energy for the subsequent outflow and can lead to increases in the final velocity of the gas. Christou et al. (2018) demonstrated this by using coherence tomography (CT)-scans of real rocks with varying porosity to initiate a DSMC simulation with water vapour emitting at 200 K from 5 cm below the surface. The rock temperature was set to 300 K and the gas temperature was then sampled at 1 cm above the surface in the outstreaming flow. An increase in the gas temperature at the surface with decreasing porosity was found as shown in Fig. 3.36. There are other unknowns that were not investigated in this work (e.g. the influence of pore size) but the basic ideas seems to have substance and provides an explanation for the MIRO observations discussed with respect to Fig. 3.6.

Steiner (1990) provides the equation (apparently traceable back to Knudsen's work) for the mass flux through a tube in the Knudsen regime where both density and temperature gradients are present along the tube axis. He gave the equation as

$$q_f = \sqrt{\frac{32 m_g}{9\pi k}} \frac{a_{tu}}{L_{tu}} \left( \frac{p_a}{\sqrt{T_a}} - \frac{p_b}{\sqrt{T_b}} \right) \tag{3.105}$$

where $a_{tu}$ is the tube radius, $L_{tu}$ is the length, and the temperatures and pressures at either end of the tube are denoted by the subscripts $a$ and $b$. He also pointed out that this equation is applicable for single long tubes and that Clausing had derived a semi-empirical approximation for the flow through short tubes and gave the equation

$$q_f = \frac{20 + 8 L_{tu}/a_{tu}}{20 + \frac{19 L_{tu}}{a_{tu}} + 3\left(\frac{L_{tu}}{a_{tu}}\right)^2} \sqrt{\frac{m_g}{2\pi k}} \left( \frac{p_a}{\sqrt{T_a}} - \frac{p_b}{\sqrt{T_b}} \right) \tag{3.106}$$

**Fig. 3.37** Plot of
Eq. (3.105) for three cases.
Solid line: Internal pressure
of 0.1 Pa going to vacuum
for an internal temperature
of 200 K and an external
temperature of
300 K. Dashed line: As the
solid line but for an internal
pressure of 0.001 Pa. The
external temperature has no
effect

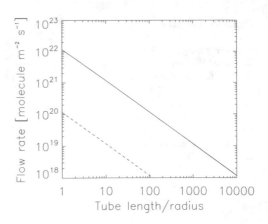

which he showed to have improved properties over the Knudsen equation. Skorov et al. (2011) subsequently used this to investigate, numerically, artificially generated layer structures. This equation shows that a range of solutions for the same mass flux, $q_f$, can be arrived at by balancing the internal pressure and the ratio of the length to the pore radius. In other words, the equation shows the increase in internal pressure required to maintain an identical surface production rate when the subliming source is moved deeper into the nucleus.

Equation (3.105) is shown in Fig. 3.37 and the proportionalities seen in the equation are evident. Temperature has no effect if the external boundary is vacuum while the flow rate is almost directly proportional to the internal pressure and inversely proportional to the ratio of the tube length to its radius. The mass flow rate has been converted to molecular flow rate to allow comparison with Fig. 2.28.

The use of this equation is nonetheless complex for several reasons. For example, the surface boundary pressure may not be strictly vacuum. The temperature will then play a role. The temperature itself within the surface layer is not linear with distance and the temperature of the gas at the surface will depend upon how well the gas accommodates to the layer temperature. The layer itself loses heat to the gas in this process which also needs to be accounted for. Hence, conserving energy in a computation of this sort requires significant care. The key point here is that the temperature of the gas at the surface is not necessarily that of the free sublimation temperature although how far it can deviate from this temperature in a real case has not been established.

### 3.4.8  Effects of Porosity on Small Scales

At local scales, variations in porosity may have quite important effects. Christou et al. (2020) have demonstrated, again using tomographic scans of real highly porous materials to initialise their calculations, that if the porosity of a layer changes along its length, this can lead to significant non-radial flow. If the porosity is lower on one side of a uniform slab, the gas outflow is reduced and significant pressure gradients across the slab arise. In the case of 67P, the surface morphology that was observed varies substantially across the surface (e.g. Figs. 2.69 and 2.73). Sublimation from these areas would probably need to be through a porous layer in order to reduce the detectability of any icy source by remote-sensing. If the morphology differences also represent different porosities then lateral (non-radial) expansion occurs. This might be a means of generating near-surface localised lateral flow to explain observations such as the movement of boulders (El-Maarry et al. 2017b) or some of the smaller formations that appear similar to wind-driven features (Fig. 2.100) on Earth and Mars (Giacomini et al. 2016).

### 3.4.9   Nightside Outgassing

Most gas dynamics calculations of cometary outgassing include some form of uniform outgassing from the nightside at production rates equivalent to 2%–7% of the total production rate. Bieler et al. (2015a), for example, used 7% to match ROSINA/COPS data at 67P while Marschall et al. (2016) deliberately did not use a nightside gas emission at all but, in the latter case, the fitting of the gas to the ROSINA/COPS measurements above the nightside of the nucleus was clearly inadequate and additional gas emission was required.

We shall see below that dust emission indicates that nightside emission of dust is necessary to explain the observations of the dayside to nightside dust column density ratio. This dust emission must be driven by gas emission but it is not obvious which volatile is responsible and how the process works.

Outgassing giving 7% of the total gas production rate from the nightside would seem improbable if the driving volatile is water but ROSINA/COPS cannot distinguish between the species and hence it is conceivable that the major species producing nightside emission is either CO or $CO_2$. This should be evident in plots of the mixing ratio as a function of the phase angle that spacecraft was measuring at. However, the spacecraft trajectory that was flown was not well suited to making this type of assessment because most measurements were acquired at the terminator. Figure 3.38 shows a plot of the DFMS measurements of the ratio of $CO_2$ to $H_2O$ binned in 5° bins of phase angle and it indicates that $CO_2$ production increases rapidly relative to $H_2O$ as one nears the terminator. This is effectively another expression of the initial plots made by Hässig et al. (2015) which showed abundant $CO_2$ production over the southern (unilluminated) hemisphere in the early phases of the Rosetta mission. While this provides some evidence, it is not unambiguous because the clear trend at low phase angles is not evident at the very highest phase angles. Hence, this is by no means conclusive. For comparison, a curve is given for the same ratio but plotted against the absolute difference between the sub-observer latitude and the sub-solar latitude. This shows a similar (but possibly better)

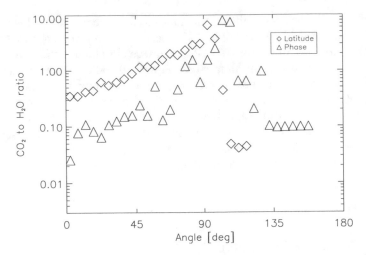

**Fig. 3.38** The ratio of $CO_2$ to $H_2O$ derived from ROSINA/DFMS data by averaging all measurements acquired over the mission within $5°$ bins of phase angle (triangles) and the angle in latitude away the sub-solar latitude (diamonds). Measurements were limited beyond $130°$ of phase angle. No measurements were obtained above $155°$

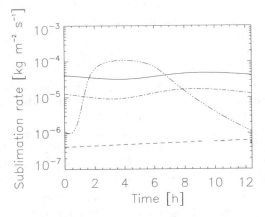

**Fig. 3.39** The sublimation rate of $CO_2$ at the equator over a rotation of a comet with an internal temperature of 50 K. Solid line: thermal inertia $= 126$ TIU and sublimation front at 30 cm. Dashed line: thermal inertia changed to 40 TIU. Dot-dash: thermal inertia $= 40$ TIU and sublimation front at 10 cm. Dash-dot-dot-dot: thermal inertia $= 40$ TIU and sublimation front at 3 cm. The 0 time on the x-axis corresponds to sunrise. The period of 67P has been used for a heliocentric distance of 2 AU

correlation. On the other hand, the rather clear trends through to $105°$ suggest that using Eq. (2.91) with $CO_2$ as the subliming volatile should be considered.

Bockelée-Morvan et al. (2015) suggested that VIRTIS-H observations of the varying $CO_2$ to $H_2O$ ratios when comparing spectra acquired above the dayside and nightside limbs might be explained by $CO_2$ emission from below the thermal skin depth. We can use the thermal model in Sect. 2.9.3.1 to explore this in a little more detail.

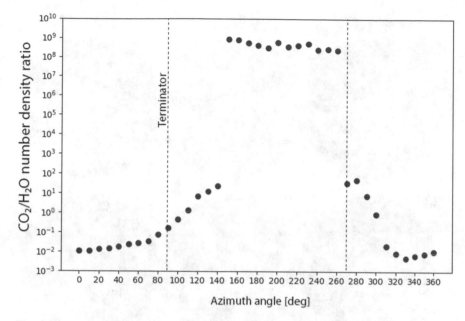

**Fig. 3.40** Number density mixing ratio using a DSMC calculation with $CO_2$ and $H_2O$ sources initialized using a thermal inertia of 40 TIU for a spherical nucleus. Note the increase in $CO_2/H_2O$ as one moves from the sub-solar point (0°) to the terminators. The nightside gas coma is totally dominated by $CO_2$

In Fig. 3.39, a simple calculation of the $CO_2$ sublimation rate at the equator for different idealised surface conditions is shown. Here, the position of a $CO_2$ sublimation front relative to the surface and the thermal inertia of the inert layer above this sublimation front have been modified to look at their influence on the $CO_2$ sublimation rate. It can be seen that low thermal inertia combined with sublimation from a depth of about 10 cm can smooth out the $CO_2$ production rate with rotational phase so that nightside $CO_2$ production is comparable to that seen on the dayside and at a relatively high level. Consequently, $CO_2$ emission through a porous inert layer may be a significant contributor to nightside gas production and modifying the $CO_2$ to $H_2O$ density ratio locally in the coma.

Calculations using idealised geometries show that this can be extreme as illustrated in Fig. 3.40. Here, the dayside is dominated by $H_2O$ while the nightside is totally dominated by $CO_2$. Moving from the sub-solar point towards the terminator results in the $CO_2/H_2O$ ratio increasing by several orders of magnitude with a gradient that is even greater than observed in Fig. 3.38. While detailed fitting has not yet been completed, this shows that $CO_2$-driven nightside emission must be carefully considered.

Modelling using DSMC also illustrated that gas flow velocities are likely to be strongly position dependent with nightside $CO_2$ outgassing slowing the flow significantly above the nightside hemisphere. This is illustrated well in Fig. 3.41. The rightmost column shows the flow speeds. The case with $CO_2$ (bottom two rows)

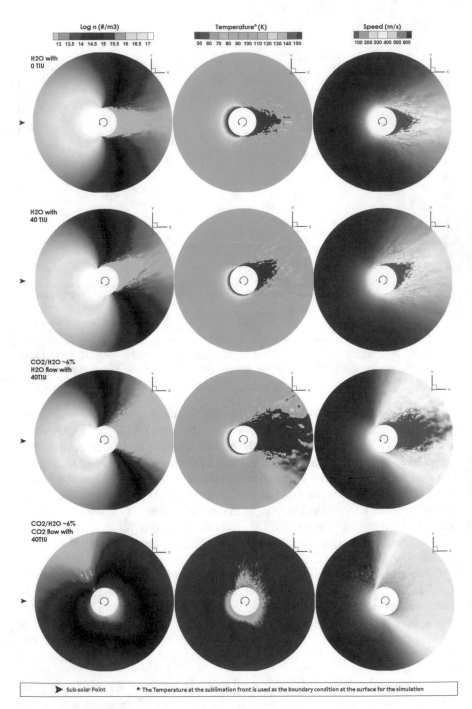

**Fig. 3.41** DSMC calculation for a sphere with three different cases. From top to bottom: $H_2O$ only with no thermal inertia, $H_2O$ only with a thermal inertia of 40 TIU, the $H_2O$ flow with a 6% $CO_2$ contamination from a sub-surface source with 40 TIU, the same case but showing the $CO_2$ distribution. The column show (left to right) the number density, the temperature and the gas speed. The Sun is to the left

shows that the gas velocities over the whole nightside can be factors of several below those on the dayside over most of the hemisphere.

Finally, we can make a computation of the gas flow field for 67P incorporating both $H_2O$ and $CO_2$ using a more physically realistic thermal model to initialise the source. An example of such a computation is shown in Fig. 3.42. These diagrams show 2D cuts through a 3D domain. We can see on the left a spherical nucleus as a test case and on the right, a model based on 67P. In both models, we can see how the $CO_2$ dominates the nightside emission and reduces the dayside to nightside gas density gradient. In the 67P model, we can also see the influence of geometry. There are strong afternoon jets in the $CO_2$ distribution resulting from extended high solar incidence. The plots also indicate that the $CO_2$ flow field can be extremely complex when the shape of the nucleus is irregular.

## 3.5   Reaction Chemistry and the Extended Coma

### 3.5.1   Daughter Products and the Haser Model

Molecules emitted from the nucleus into the inner coma immediately begin to undergo reactions through numerous pathways (Table 3.9). Photo-ionization, photo-dissociation, charge-exchange, and electron impact ionization all influence gas densities even very close to the nucleus and hence the density equations need modification when the influence on parent species begins to become significant. This was well-known at an early stage in comet research. These reactions influence the coma composition and can provide molecules and newly created radicals with additional energy.

The major species are influenced by photo-dissociation reactions for which the rates are given in Table 3.10. The difference in rates between the major species spans two orders of magnitude and consequently the relative abundances of the major species in the coma change rapidly beyond a few thousand kilometres from the nucleus.

Spectroscopy at visible wavelengths in the 1950s revealed that radicals such as CN and OH were present in gas comae. A more recent example from the bright comet C/1996 B2 Hyakutake that came relatively close to the Earth in 1996 is shown in Fig. 3.44. The absence of strong emissions in the visible from putative parents of these radicals led to increased emphasis on the study of these daughter products. The Haser (1957) model was developed for computations of densities arising through these processes and allows one to avoid discussion and modelling of reaction kinetics by treating the reaction chemistry in terms of scalelengths arising from the expansion of the gas.

As the daughter products are produced, the parent is lost, and in the simplest case, an exponential decay function is used so that, the total number of parent molecules crossing a spherical surface, $q_p$, at a distance, $r$, from a point source is given by

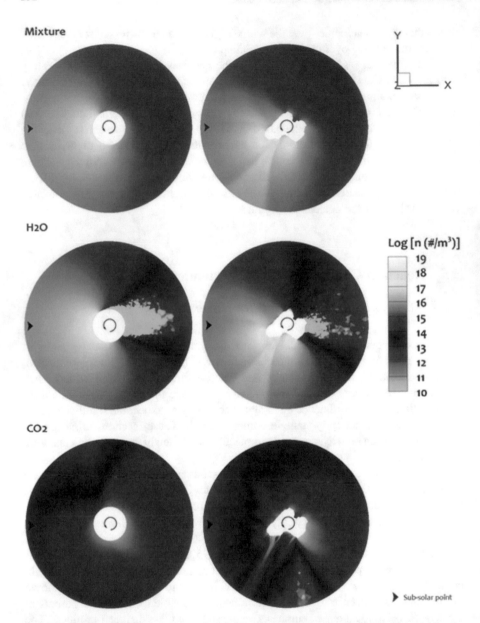

**Fig. 3.42** Flow field for a combined $H_2O$ and $CO_2$ model initialized using a thermal model for the two species with $CO_2$ coming from depth. A spherical case is shown on the left. The shape of 67P has been used in the model on the right. The Sun is to the left in both cases. The individual species are shown separately for comparison. Note the strong jet-like $CO_2$ outgassing on the afternoon side and the more homogeneous outgassing on the nightside

**Table 3.9**  Types of reaction possible within the inner coma of a comet (from Schmidt et al. 1988)

| Reaction number | Reaction type | Example |
|---|---|---|
| R1 | Photodissociation | $h\vartheta + H_2O \rightarrow H + OH$ |
| R2 | Photoionization | $h\vartheta + CO \rightarrow CO^+ + e$ |
| R3 | Photodissociative ionization | $h\vartheta + CO_2 \rightarrow O + CO^+ + e$ |
| R4 | Electron impact dissociation | $e + N_2 \rightarrow N + N + e$ |
| R5 | Electron impact ionization | $e + CO \rightarrow CO^+ + 2e$ |
| R6 | Electron impact dissociative ionization | $e + CO_2 \rightarrow O + CO^+ + 2e$ |
| R7 | Positive ion-atom interchange | $CO^+ + H_2O \rightarrow HCO^+ + OH$ |
| R8 | Positive ion charge transfer (charge exchange) | $CO^+ + H_2O \rightarrow H_2O^+ + CO$ |
| R9 | Electron dissociative recombination | $C_2H^+ + e \rightarrow C_2 + H$ |
| R10 | 3-body positive ion-neutral association | $C_2H_2^+ + H_2 + M \rightarrow C_2H_4^+ + M$ |
| R11 | Neutral rearrangement | $N + CH \rightarrow CN + H$ |
| R12 | 3-body neutral recombination | $C_2H_2 + H + M \rightarrow C_2H_3 + M$ |
| R13 | Radiative electronic state deexcitation | $O(^1D) \rightarrow O(^3P) + h\vartheta$ |
| R14 | Radiative recombination | $e + H^+ \rightarrow H + h\vartheta$ |
| R15 | Radiation stabilised positive ion-neutral association | $C^+ + H \rightarrow CH^+ + h\vartheta$ |
| R16 | Radiation stabilised neutral recombination | $C + C \rightarrow C_2 + h\vartheta$ |
| R17 | Neutral-neutral associative ionization | $CH + O \rightarrow HCO^+ + e$ |
| R18 | Neutral impact electronic state quenching | $O(^1D) + CO_2 \rightarrow O$ $(^3P) + CO_2 + h\vartheta$ |
| R19 | Electron impact electronic state excitation | $CO(^1\Sigma) + e \rightarrow CO(^1\prod) + e$ |

**Table 3.10**  Photo-dissociation rates of the major species at 1 AU (Crovisier and Encrenaz 1983)

| Species | Photo-dissociation rate [$s^{-1}$] |
|---|---|
| $H_2O$ | $6.5 \ 10^{-5}$ |
| $CO_2$ | $2.0 \ 10^{-6}$ |
| $CO$ | $6.5 \ 10^{-7}$ |

$$q_p(r) = Q_{gp}e^{-\beta_1 r} \qquad (3.107)$$

in units of [molecule s$^{-1}$] where $Q_{gp}$ indicates the parent production at source and $\beta_1$ is a reciprocal scalelength. $\beta_1$ can be converted into a lifetime for the parent species by

$$\beta_1 = \frac{1}{v_{gp}\tau_p} \qquad (3.108)$$

where $\tau_p$ is the lifetime of the parent and $v_{gp}$ indicates the velocity of the parent. Close to the nucleus, one can replace $r$ with $r-R_N$ where $R_N$ is again the radius of the nucleus thereby eliminating the point source approximation. However, this is usually an unnecessary finesse and is, in any case, inconsistent if the acceleration of the gas close to the nucleus surface (and hence the change in $v_{gp}$) is not also taken into account.

$q_p(r)$ corresponds to an integral

$$q_p(r) = \int_0^{2\pi} \int_{-\pi}^{\pi} n_{gp}(r) v_{gp} \, r \, d\theta \, r \, \cos \theta \, d\varphi \qquad (3.109)$$

(where $\theta$, $\varphi$ are angles around the source) so that by assuming force-free radial outflow we have

$$q_p(r) = 4\pi r^2 v_{gp} \, n_{gp}(r) \qquad (3.110)$$

and

$$n_{gp}(r) = \frac{Q_g}{4\pi r^2 v_{gp}} e^{-\beta_1 r} \qquad (3.111)$$

thereby giving the local parent density, $n_{gp}$ (cf. Eq. 3.1).

In the general case, the daughter product will also have a finite lifetime which can be included using a second scalelength,

$$q_d(r) = Q_{gp} \frac{\beta_1}{\beta_2 - \beta_1} \left( e^{-\beta_1 r} - e^{-\beta_2 r} \right) \qquad (3.112)$$

in units of [molecule s$^{-1}$] where $\beta_2$ is the reciprocal scalelength of the daughter species. This is the Haser model equation (Cochran 1985).

In Table 3.11, values for the parent and daughter scalelengths for three radicals with strong emissions at optical wavelengths are given and $q_d(r)$ is plotted in Fig. 3.43 with the parent production rate at source normalised to 1. This shows clearly how the mixing ratios of daughter products can vary substantially in the coma. However, the Haser model, as given by Eq. (3.112), makes a large number of assumptions (e.g. A'Hearn 1982; Cochran 1985). These include assuming that

- the coma is collisionless with the free-molecular flow approximation being adequate right down to the source,
- the coma is spherically symmetric,
- each daughter species has only one parent

**Table 3.11** Haser model parent and daughter scalelengths at 1 AU for radicals with strong emissions at optical wavelengths (from Schleicher and Farnham 2004)

| Radical | Parent scalelength [km] | Daughter scalelength [km] |
|---------|-------------------------|---------------------------|
| CN      | $1.3 \ 10^4$            | $2.1 \ 10^5$              |
| $C_2$   | $2.2 \ 10^4$            | $6.6 \ 10^4$              |
| $C_3$   | $2.8 \ 10^3$            | $2.7 \ 10^4$              |

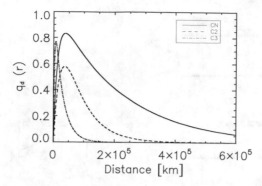

**Fig. 3.43**  Haser model results for CN, $C_3$ and $C_2$ at 1 AU from the Sun based on the scalelengths in Schleicher and Farnham (2004)

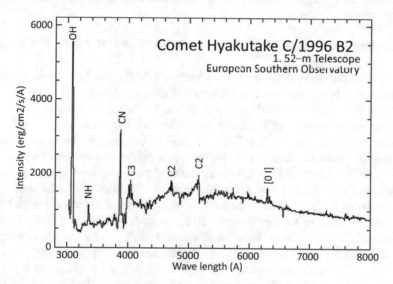

**Fig. 3.44**  A complete, optical spectrum of C/1996 B2 Hyakutake was obtained with the ESO 1.52-m telescope (La Silla Observatory) by Hilmar Duerbeck (ESO) on UT March 8.3, 1996. The principal species seen in the spectrum are indicated. (Image credit: ESO)

- the daughter species has the same velocity vector as the parent (while the velocity of parent species can be estimated from hydrodynamics relations or computed directly, as shown below, that of the daughter species is influenced by the dissociation process), and
- the parent velocity is independent of direction.

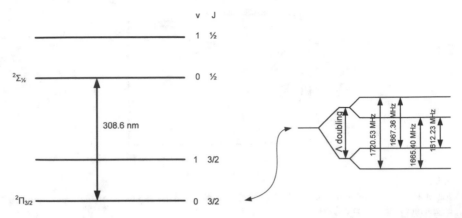

**Fig. 3.45** Energy level diagram of the OH radical showing the main vibrational transition that is observed and the hyperfine structure of the lowest rotational level that leads to lines observable at radio wavelengths

There are also 3-generation Haser models (e.g. O'Dell et al. 1988) but these suffer from the same deficiencies.

The observations of daughter species such as CN, $C_3$, and $C_2$ are interesting for comparison between comets. The most significant radical, however, is OH because its production rate can be used as a tracer for the parent water molecule from which the water production rate of the comet can be derived. The strong $A^2\Sigma - X^2\Pi$ (0, 0) emission band in the near-UV at 308.6 nm arises from resonant fluorescence and it provides an immediate tool to monitor the water production rate and its variations with the comet's heliocentric distance, rotational period and activity changes. An energy level diagram showing the main observed vibrational and rotational transitions is shown in Fig. 3.45. The frequencies of the hyperfine levels are from Maeda et al. (2015) (see also Krishna Swamy 2010 and the Splatalogue database[9]).

The individual transitions between vibrational energy states can cover a significant wavelength range as indicated by Schleicher and A'Hearn (1988) for OH (Fig. 3.46). The bands can overlap. This implies that separation of the bands requires high resolution spectroscopy to extract the individual lines. The situation is further complicated by the presence of other radicals which put their own characteristic lines into the same wavelength range. The CN (0–0) band, for example, can be found at 388 nm while there are NH lines at 335 nm.

In general, a high dry site is needed to observe this far into the UV from ground but has been performed regularly for many years (e.g. Fitzsimmons et al. 1990; Hyland et al. 2019). Comparisons can also be made with radio observations. When the $\Lambda$ value for the transition is not equal to 0, then the rotational energy levels split in an effect referred to as $\Lambda$-doubling. This gives rise to transitions of the ground state $\Lambda$ doublet $X^2\Pi_{3/2}$, J = 3/2 that appear in the radio frequency range at a

---

[9]https://splatalogue.online/sp_basic.html

**Fig. 3.46** Positions and wavelength range for the UV emissions of OH. Note that some bands overlap

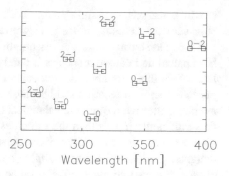

wavelength of 18 cm that are observationally important for $H_2O$ gas production rate determinations, a recent example being Wang et al. (2017) for C/2013 US10 (Catalina). The transitions for OH at 1665 and 1667 MHz are also shown in Fig. 3.45. Examples of observations using these transitions are Crovisier et al. (2013), which were performed in support of the EPOXI programme, and Howell et al. (2007) which were performed in support of Deep Impact.

### 3.5.2   Detailed Reaction Kinetics

Schemes to investigate the reaction kinetics and thereby study the physical chemistry of the coma in detail have been developed and published since the 1980s (Schmidt et al. 1988). A more recent example was produced by Rodgers and Charnley (2002). Both this paper and that of Schmidt et al. (1988) contain descriptions of the numerical approach. A review of the subject was given by Rodgers et al. (2004).

The starting point is the reduction of the Euler equations (Eqs. 3.50, 3.55, and 3.58) to 1D steady state equations for the bulk fluid. The time dependencies can be eliminated and the equations for mass, momentum and energy can be reduced to

$$\frac{1}{r^2}\frac{d}{dr}r^2\rho_g v_g = Q_s \tag{3.113}$$

$$\frac{1}{r^2}\frac{d}{dr}r^2\rho_g v_g^2 + \frac{dp}{dr} = \rho_g F \tag{3.114}$$

and

$$\frac{1}{r^2}\frac{d}{dr}r^2\left[\rho_g\frac{v_g^3}{2} + \frac{\gamma}{\gamma-1}pv_g\right] = \rho_g E \tag{3.115}$$

respectively, under the assumption of spherical symmetry (Schmidt et al. 1988). A multi-fluid system can be constructed. Rodgers and Charnley (2002) used a system in which neutrals, ions and electrons were treated as separate fluids but with the same bulk velocity. (This is, in itself, a significant simplication.)

One of the basic simplifying assumptions in the Haser model is that the outflow of both parent and daughter species is purely radial. However, it was recognized in the late 1970s that suprathermal velocities of daughter products would arise from the photodissociation reaction (R1) because there is excess energy of the order of 2 eV available after overcoming the chemical bond energy and that these velocities could be significantly higher than the velocities of the parent species. Combi et al. (2004) give the exothermic velocities of dissociation products of $H_2O$ which shows that the daughter product velocities strongly depend upon the wavelength of the incoming photon. For the two most probable outcomes (arising from the reaction $H_2O + h\vartheta \rightarrow$ H + OH which occurs in around 70% of cases), the velocity of the OH product will be around 1.05 km s$^{-1}$ whereas the velocity of the H product will be around 18 km s$^{-1}$. For higher energy photons initiating the dissociation, the hydrogen product velocity can reach 37 km s$^{-1}$. The high velocity of the neutral hydrogen arising from photodissociation combined with the relatively long lifetime of hydrogen before ionization occurs, leads to the hydrogen coma of comets being extremely large covering a vast volume of space. The excess energy has the effect of increasing the velocity of the daughter products far above the outflow speed of the parent and the direction of motion is clearly no longer necessarily radial with respect to the nucleus. Hence, this is no longer a "free molecular flow".

Festou (1981) introduced the "vectorial model" to account for the excess energy and thereby correct OH (and thereby $H_2O$) production rates from ground-based observations. At the time of writing a web version of this model (the Web Vectorial Model) is still available[10] and an example calculation is shown in Fig. 3.47. This shows a predicted OH column density along the Sun-comet line for 67P at 1.3 AU using an $H_2O$ production rate of 3.3 $10^{28}$ molecule s$^{-1}$ in steady-state with isotropic emission.

An alternative is to use a Monte Carlo approach (Combi and Delsemme 1980; Combi et al. 2004) which offers a great deal more flexibility in treating specific cases such as non-isotropic outgassing and a larger reaction network.

The hydrogen coma has been studied through observations of Lyman-$\alpha$ as far back as 1972. Observations of 1P/Halley using sounding rockets (McCoy et al. 1992) showed that multiple populations of hydrogen were needed to fit the data corresponding to the different velocities arising from different parents (OH and $H_2O$). However, another slow component was also needed which was attributed to the thermalization of fast hydrogen atoms arising from collisions near the nucleus. Obviously, the presence of this component would strongly depend upon the total comet $H_2O$ production rate.

Measurements of Lyman-$\alpha$ have been performed frequently by the Solar Wind ANisotropies (SWAN) instrument onboard the Solar and Heliospheric Observatory spacecraft, SOHO, from which water production rates and their variation with

---

[10]http://www.boulder.swri.edu/wvm/

**Fig. 3.47** Output from the Web Vectorial Model for OH under conditions approximating those expected for 67P at perihelion (1.3 AU with an $H_2O$ production rate of 3.3 $10^{28}$ molecule/s.) The output shows the OH column density along the Sun-comet line away from the nucleus

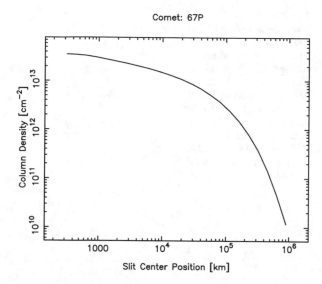

heliocentric distance can be produced. Combi et al. (2018) have studied the production from nine long-period comets, for example, and show that the heliocentric distance dependence of the production rate varies between these objects over a wide range with large differences between inbound and outbound slopes. One example, C/2014 Q1 (PanSTARRS), showed a very steep increase and decrease in activity pre- and post-perihelion respectively, indicating significant water production only within 25 days of perihelion when the comet was <0.7 AU from the Sun.

An example for the water production rates of 67P determined using SWAN data (Combi 2017) is shown in Fig. 3.48. The maximum production is shortly after perihelion with a production rate of 1.2 $10^{28}$ molecule $s^{-1}$. This is from the 2009 apparition of 67P and should be compared with Fig. 1.18 for the 2015 apparition.

Rodgers and Charnley (2002) accounted for the fast by-products of photodissociation reactions in their reaction network calculations by adding additional fluids to their simulations accounting, not merely for reactions such as $H_2O + h\vartheta \rightarrow H + OH$, but also those reactions involving $H_2$, HD and D. These light but fast by-products are the outlet for excess energy that becomes available through reactions and offer a means of providing additional heat to the coma.

The temperature of the fluids in the coma have great significance. For example, we shall see later (Fig. 5.6) that the electron temperature strongly affects the ionization frequency of the $H_2O$ molecule which in turn influences the source distribution of cometary plasma. Neutrals, ions and electrons undergo elastic scattering with each other and the local energy transfer from one type to another can be computed for each combination. The dominant neutral is water vapour although more sophisticated computations can incorporate species of lower abundance.

The collision frequency for ion-neutral collisions is given by

**Fig. 3.48** The water
production rate derived from
SWAN observations of
Lyman-$\alpha$ for 67P acquired
during the 2009 apparition.
The x-axis gives the time in
days with respect to the time
of perihelion. The maximum
seen post-perihelion was
confirmed by ROSINA/
COPS as seen in Fig. 1.18.
(Data credit: Combi 2017)

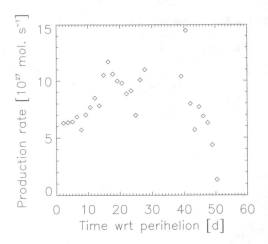

$$f_{col} = \sigma_{in} v_{rel} n_i n_n \tag{3.116}$$

where $v_{rel}$ is the relative velocity, $\sigma_{in}$ is the specific ion-neutral cross-section which is
of the order of $10^{-19}$ m$^2$ and $n_i$ and $n_n$ define the ion and neutral densities
respectively. The energy transfer to the neutral is then calculated as

$$E_n = 3\,k\,\frac{m_i m_n}{\left(m_i + m_n\right)^2}\left(T_i - T_n\right) \tag{3.117}$$

which is a special case of the more general equation for ion-neutral reactions. A
similar equation can be used for electron-neutral elastic collisions by substituting the
terms relating to ions with the analogous terms for electrons. Specific cross-sections
for electron-H$_2$O collisions can be found in Itikawa and Mason (2005).

We have seen that fast neutrals can be produced by photodissociation and we
shall see later that ions are accelerated by interaction with the interplanetary magnetic field. These fast species can collide elastically with the bulk fluid to provide
additional heat for the coma and thus modify the reaction chemistry. It is this
phenomenon that drove Rodgers and Charnley, for example, to model the fast
species as independent fluids.

There are also inelastic collisions that need to be accounted for. Collisions may
result in a neutral or ion being lifted into an excited state. That energy can be released
by photon emission which is an energy loss for the species. For the water molecule,
inelastic neutral-neutral collisions remove energy from those neutrals whereas
inelastic electron-neutral collisions result in the removal of energy from the electrons. For H$_2$O-H$_2$O collisions, Schmidt et al. (1988) have computed the energy
losses (incorporating optical depth effects) whereas the electron energy loss from
electron-H$_2$O collisions has been computed by Cravens and Körösmezey (1986).

From Table 3.9, it can be seen that reactions R13–R17 all have a radiated element leading to energy loss from the system.

Altogether, 181 species were computed in the model of Rodgers and Charnley (2002) with over 3500 individual reactions. One of the resulting figures is shown as Fig. 3.49 and shows the abundances of different species as a function of cometocentric distance. It should be noted that the values have been multiplied by $4\pi r^2$ to give a flux through a spherical shell at each distance from the nucleus. This linearises the function if the species is unaffected by reactions and expands as $1/r^2$. Note also that the abscissa is logarithmic.

The top panel shows the parent molecules. Initially, the fluxes for these molecules show the chosen starting conditions at the nucleus but at distances of $>10^4$ km from the nucleus, the main reactions (e.g. photodissociation) begin to remove these species. The central panel shows the daughter neutrals. In the case of the larger daughter product, $HC_3N$, its increase in local density with distance is slowed and reversed as it undergoes reactions itself. Finally, the bottom panel shows ion species and the electron density. This rises very rapidly with cometocentric distance.

There are several assumptions in the approach used to construct the coma chemistry network described above. Firstly, the coma is assumed to be spherically symmetric. The dayside-nightside asymmetry at the nucleus itself clear violates this assumption. However, gas-gas collisions in the innermost coma where the gas is still collisional will tend to move the spatial distribution to a more symmetric structure and obviously the higher the local density near the nucleus, the more symmetric the spatial distribution should become. Secondly, the coma is assumed to be compositionally uniform. Here again, the evidence suggests that compositional variations in the innermost coma (e.g. the $CO_2/H_2O$ ratio in Fig. 3.2) are actually significant and this could easily challenge 1D models. This has not really been tested in a rigorous manner to this point. Third, the calculations assume that there are no magnetic fields acting on the ions. This is a significant issue that we will address when we discuss the spatial distribution of ions below. However, this does indicate that a coupling between ion-neutral reactions and effects resulting from ion pickup by the solar wind are necessary to describe the spatial distributions more exactly. On the other hand, this breaks the symmetry completely and increases the complexity enormously. Schmidt et al. (1988) were the first to tackle this problem and simplified systems have been studied in detail by, for example, Rubin et al. (2014c). But analysis of the complexities resulting from a detailed chemical network as used by Rodgers and Charnley in a full 3D system has still not been attempted. Finally, as the density falls off, the use of the fluid equations to simulate a system that is close to the free molecular flow regime can introduce significant errors. For example, fast products of dissociation reactions will not be able to transfer their energy to the coma through collision if the mean free path becomes comparable to the size of the coma itself. Hence, temperatures in the outer coma can be substantially overestimated (Ip 1983).

One of main reasons for studying these chemical networks is to derive the nucleus composition of parent species from coma in situ measurements (using mass spectrometers for example) by accounting for known reactions that can influence the

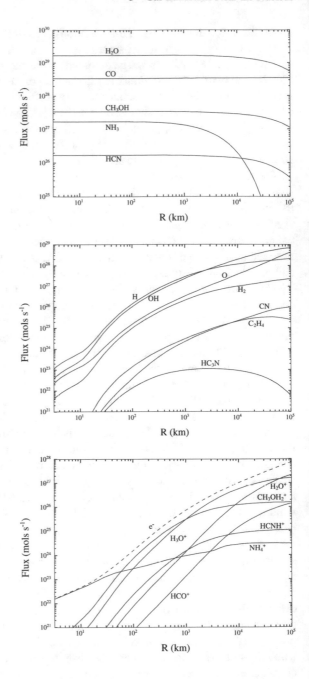

**Fig. 3.49** The distribution of chemical species in the coma computed using a 1D steady-state spherically symmetric model (from Rodgers and Charnley 2002). Note that the authors multiplied the local densities of the species by $4\pi r^2$ to give integrated fluxes and allows one to identify clearly sources and sinks with respect to a $1/r^2$ distribution

local coma densities. A less ambitious goal might be to correct for the influence of the reaction chemistry on relative densities of specific species. Hence, despite the associated limitations, the study of these chemical networks continues to be important and especially for future fast fly-by missions such as Comet Interceptor.

### 3.5.3 The Swings and Greenstein Effects

CN, $C_3$ and $C_2$ all have emissions at visible wavelengths (Fig. 3.44). As in the case of infrared emissions, beyond a few hundred kilometres from the nucleus, collisions are rare and so resonant fluorescence of sunlight by these species at discrete wavelengths dominates.

As with the infrared vibrational bands (Eq. 3.38), the observed intensity scattered by an optically thin column illuminated by the Sun is given by

$$I = \frac{h\vartheta}{4\pi} \frac{g_f}{r_h^2} N_g \, p(\alpha) \tag{3.118}$$

where a phase function for single scattering by gas at a phase angle, $\alpha$, has now been included. $p(\alpha)$ is normalized to 1 and is usually taken to be independent of $\alpha$. However, there may be some effect on $p(\alpha)$ as a result of polarization (see Chamberlain and Hunten 1987). As before, $g_f$ is referred to as the "g-factor" and is given by

$$g_f(J', J_i) = F_\odot \frac{\pi q_c^2}{mc} f_{osc}(J', J_i) \frac{A(J', J_i)}{\sum_i (J', J_i)} \tag{3.119}$$

where $f_{osc}$ is the oscillator strength for the particular transition and $A(J',J_i)$ is the transition probability. (Caution should be exercised with this equation as most authors define the solar flux to be equal to $\pi F_\odot$ rather than simply $F_\odot$ as we have done here.) This value (normalized to 1 AU from the Sun) can be tabulated for important transitions. However, for cometary radicals with near-UV and visible transitions, the Doppler shift with respect to the Sun, caused primarily by the orbital speed, is of significance. The expected wavelength shift is given by

$$\frac{\Delta\lambda}{\lambda} = \frac{v_h}{c} \tag{3.120}$$

where $v_h$ is the heliocentric velocity and hence species moving with respect to the Sun "see" the solar flux at a Doppler-shifted wavelength. A simple example of this with just two transitions within a wavelength range is sodium D-line emission. Figure 3.50 shows the solar spectrum around the 589.0 and 589.6 nm lines.

Absorptions by elements in the solar photosphere are sometimes referred to as the Fraunhofer absorptions and, in the case of sodium, are very deep. A sodium atom with a zero radial velocity component with respect to the Sun, will see a low solar flux in the bottom of this absorption line. However, radial velocities of the order of 15–20 km s$^{-1}$ are sufficient to almost double the incident flux and hence the same column density will appear brighter to an observer. In the case of sodium, this effect has been seen prominently in the Jupiter system associated with sodium emitted from the innermost Galilean moon, Io (e.g. Goldberg et al. 1984; Thomas 1992).

**Fig. 3.50** The sodium
D-lines in the solar spectrum
(data from Kurucz 2005).
The flux is given in units of
[W m$^{-2}$ nm$^{-1}$] at 1 AU
from the Sun. The bar on the
left hand side shows the
range of wavelengths that
can excite a sodium atom if
that atoms moves radially
relative to the Sun by
$\pm 25$ km s$^{-1}$

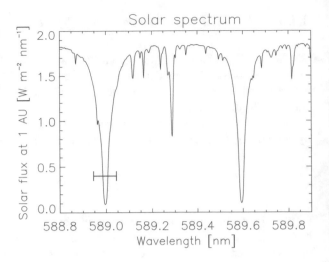

**Fig. 3.51** The solar flux
around 388 nm showing the
strong variability associated
with the UV flux from the
Sun. The bar at the centre of
the plot shows the range
resulting from radial
velocities of $\pm 25$ km s$^{-1}$
(Data from Kurucz 2005)

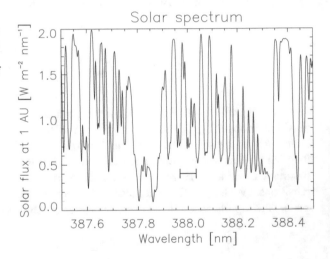

Swings identified this effect in observations of the violet (0,0) band of CN in
cometary spectra and, in cometary sciences, it now bears his name. Unlike the
sodium lines (which constitute a simple doublet), CN, $C_3$, and $C_2$ emissions are
from bands and the Swings effect results in irregular variation of the individual line
strengths. The CN (0,0) band system at 388.3 nm is a particular good example
because of the large number of solar absorption lines in this range of the spectrum
(Fig. 3.51). Schleicher (2010) computed the fluorescence efficiencies over a range of
velocities with respect to the Sun and showed that variations in the (0–0) band can
result in a factor of 1.9 variation while the variation in the adjacent (1–1) band is
significantly less.

An even more extreme example is OH (0–0) at 308.5 nm. Schleicher and A'Hearn
(1988) computed that the fluorescence efficiencies can vary by up to a factor of

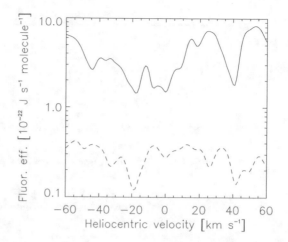

**Fig. 3.52** The fluorescence efficiencies of the OH (0–0) band at 308.5 nm (solid line) and the OH (1–1) band at 313.8 nm (dashed line) as computed by Schleicher and A'Hearn (1988) showing the variation in efficiency with heliocentric velocity of the radical. Note that the y-axis is logarithmic

4 (Fig. 3.52). These efficiencies are still used in modern works (e.g. McKay et al. 2018).

The Greenstein effect can be thought of as a second-order Swings effect. The gas outflow velocity is evident in our fundamental equations with values typically of the order of 1–2 km s$^{-1}$ for parent molecules and potentially higher for daughter species such as CN. Hence, a sunward/anti-sunward velocity asymmetry of order 2–4 km s$^{-1}$ can arise which is again significant when compared to the width of the Fraunhofer lines and affects the observed brightness.

### 3.5.4   Prompt Emission

Prompt emission arises when a dissociative reaction results in a daughter product being in an excited state that subsequently relaxes by a radiative transition to the ground state. In cometary comae, the best known example is that arising from the photodissociative excitation mechanism

$$H_2O + h\vartheta \rightarrow H_2O^* \rightarrow OH^* + H \tag{3.121}$$

where photolysis of water gives rise to hydroxyl (OH) fragments in the first electronically excited state (OH*). The branching ratio is non-negligible but this state is very unstable, with a lifetime of about $10^{-6}$ s (Becker and Haaks 1973). Hence OH* ($A^2\Sigma^+$) radicals promptly decay to the ground state ($X^2\Pi$)—a prompt emission. This results in an emission band in the near-UV ranging approximately from 306 to 325 nm. Normally, in Earth-based observations, the lines are extremely weak compared to the fluorescence of OH in sunlight. Since the prompt emission rate is directly proportional to the column density of water, whereas the fluorescent emission of OH is proportional to the column density of OH, the lines resulting from

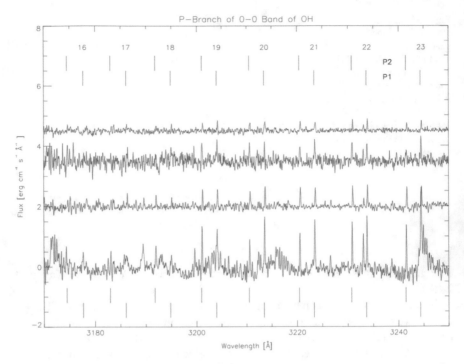

**Fig. 3.53** Spectrum of prompt emission by OH in the P branch of the 0–0 band of the $A_2\Sigma^+ - X^2\Pi$ transition from C/1996 B2 (Hyakutake). From bottom to top are spectra on the nucleus (assuming this to be at the brightest point) and at impact parameters of 155 km, 550 km, and 780 km, plotted with vertical displacements for clarity. Vertical ticks across the top of the figure indicate the features identified as being caused by prompt emission, labeled with the lower rotational quantum number, $N''$. (From A'Hearn et al. 2015 © AAS. Reproduced with permission. Courtesy of Roland Meier)

prompt emission are strongest close to the nucleus, a region that is rarely spatially resolvable from Earth (A'Hearn et al. 2015). Spectrally resolved observations of prompt emission in the near-UV (0–0) band of C/1996 B2 (Hyakutake) were made by A'Hearn et al. (2015) and an example is shown in Fig. 3.53.

La Forgia et al. (2017) studied data from the Deep Impact imaging system and concluded that the OH emission distribution in the inner coma was very different from that expected for a fragment species. Instead, it was well correlated with the spatial distribution of water vapour derived by the imager. Radial profiles of the OH column density and derived water production rates show an excess of OH emission through closest approach that could not be explained with pure fluorescence and they attributed this to the prompt emission process (see also Table 5.2 later in the text).

Bonev et al. (2006) pointed out that the dissociative process also produces ro-vibrationally excited (or "rotationally hot") states leading to infrared emission in the (1–0) and (2–1) bands between 2.9 and 3.6 μm. They also presented observations of C/2000 WM$_1$ (LINEAR) and C/2004 Q2 (Machholz). This is of interest because it implies that OH* and $H_2O$ can be observed in the same spectral range allowing direct comparison of OH and $H_2O$ in the inner coma. Prompt emission from OH* should trace the $H_2O$ spatial distribution unless optical depth effects are important.

Photodissociation of $H_2O$ can also occur through the path

$$H_2O + h\vartheta \rightarrow H_2O^* \rightarrow O^* + H_2 \qquad (3.122)$$

where the atomic oxygen can be in the $^1D$ state, which subsequently decays to the $^3P$ ground state with the release of a photon at either 630.0 nm or 636.4 nm (the red doublet), or the $^1S$ state which can decay to the $^1D$ state with emission of a photon at 557.7 nm (the green line). An example of observations of these lines can be found in Capria et al. (2005) for 153P/2002 C1 (Ikeya-Zhang). The lifetimes of these states is short (the $OI(^1D)$ lifetime is around 130 s while for $OI(^1S)$ it is ~1 s; Feldman et al. 2004) and hence observing these emissions can also trace parent molecules. Interpretation is however far less straightforward in this case because excited oxygen atoms can be produced from dissociation of many species including OH, CO, $CO_2$, and $O_2$. In addition, in the innermost coma, collisional quenching (de-excitation without radiative loss) is probably important (Cochran 2008) and the branching ratios of the reactions producing the $OI(^1D)$ state are not well determined (Budzien et al. 1994). Nonetheless, Decock et al. (2015) concluded that it may be possible to constrain the production rates of parent species by looking at the spatial distribution of these forbidden transitions close to the nucleus.

Another prompt emission of importance arises from the dissociation of $CO_2$. The Cameron bands ($a^3\Pi - X^1\Sigma^+$) of CO were first identified in the laboratory in 1926 (Narahari Rao 1949) and in cometary spectra in 1994 (Weaver et al. 1994). They are in the wavelength range 1900–2500 Å and can result from photodissociative excitation. Feldman and Brune (1976) first detected emission from the Fourth Positive system of CO ($A^1\Pi$-$X^1\Sigma^+$) and this has been used in the past to determine the CO production rate (e.g. Feldman et al. 1997) in the wavelength range 1400–1750 Å. Consequently, the ratio of CO to $CO_2$ can, in principle, be obtained by observing in the far UV with the same UV spectrometer or spectral imager. The Fourth Positive system accounts for most spectral features seen in Hubble Space Telescope observations and Lupu et al. (2007) and Feldman et al. (2018a) were able to model these data to provide production rates of CO relative to water. The spread in values for this ratio (Table 3.12) is notable covering nearly two orders of magnitude with no obvious systematic variation with heliocentric distance.

### 3.5.5   Other Notable UV Line Emissions

In addition to the transitions of CO, one of the other simple molecules to have been observed in the UV is $H_2$. Feldman et al. (2002) reported detection of 2 lines of $H_2$ in the 1070–1170 Å range in C/2001 A2 (LINEAR) using the Far Ultraviolet Spectroscopic Explorer (FUSE) while Lupu et al. (2007) detected $H_2$ in 153P/Ikeya-Zhang

**Table 3.12** CO/$H_2$O production rate ratios derived from HST observations by Lupu et al. (2007) and Feldman et al. (2018a) compared to the ROSINA observations in October 2014 for 67P

| Comet | Source | CO/$H_2$O production rate [%] | $R_h$ [AU] |
|---|---|---|---|
| C/2000 WM1 (LINEAR) | HST | $0.44 \pm 0.03$ | 1.08 |
| 153P/Ikeya-Zhang | | $7.2 \pm 0.4$ | 0.90 |
| C/2001 Q4 (NEAT) | | $8.8 \pm 0.8$ | 1.02 |
| C/1996 B2 (Hyakutake) | | $20.9 \pm 0.3$ | 0.88 |
| 103P/Hartley 2 | | 0.3 | 1.06 |
| C/2009 P1 (Garradd) | | 22 | 1.59 |
| C/2012 S1 (ISON) | | 1.3 | 0.99 |
| C/2014 Q2 (Lovejoy) | | 5.0 | 1.29 |
| 67P (Northern summer hemisphere) | ROSINA | 2.7 | 3.2 |
| 67P (Southern winter hemisphere) | | 20 | 3.2 |

through the (6, 13) P1 line of the Lyman band system ($B^1\Sigma_u^+ - X^1\Sigma_g^+$) at 1607.5 Å. Feldman et al. (2018b) presented evidence that part of the $H_2$ observed in C/2014 Q2 (Lovejoy) came from photodissociation of formaldehyde ($H_2CO$).

The (0,0) band of the $A^1\Pi - X^1\Sigma^+$ system of CS is at 2576 Å and has been detected in nearly all comet spectra acquired by IUE and HST (Feldman et al. 2004). This led to the inclusion of a narrow band filter in the OSIRIS WAC instrument for Rosetta although the final filter was of inadequate quality to provide useful results. This species was detected in 19P/Borrelly by Bockelée-Morvan et al. (2004a) in the sub-mm range (specifically the J(3–2) and J(5–4) lines at 146.9690 and 244.9356 GHz, respectively (Table 3.2)) and has also been detected (using the J = 5–4 line) in C/2001 Q4 (NEAT) and C/2002 T7 (LINEAR) by de Val-Borro et al. (2013) with derived production rates relative to $H_2O$ of 0.08% and 0.2% respectively. Bogelund and Hogerheijde (2017) detected CS in C/2012 S1 (ISON) with the ALMA array and derived a parent scale length for CS of ~200 km (two orders of magnitude smaller than that used in the analysis by De Val-Borro et al. 2013) with a production rate of 0.2% relative to $H_2O$. These types of measurements remain relevant because CS cannot be distinguished from $CO_2$ in a ROSINA-type mass spectrometer (cf. Table 3.1). However, there remain uncertainties about the molecule's parent species (e.g. $CS_2$ and OCS are both plausible).

The region of the spectrum shortward of 2000 Å (the far-ultraviolet, FUV) also contains the resonance transitions of common elements such as carbon, oxygen and sulphur. The principal excitation mechanism in the ultraviolet is resonant fluorescence of solar radiation but the low photon fluxes have limited observations from orbiting observatories, such as the International Ultraviolet Explorer, the Hubble Space Telescope (e.g. Weaver et al. 2011) and the Far Ultraviolet Spectroscopic Explorer, to moderately active comets within ~1.5 AU of the Sun (Feldman et al. 2015)

Observations by the FUV ALICE experiment on Rosetta at 67P showed expected emissions but it was noted that the relative intensities of the neutral hydrogen and oxygen multiplets were indicative of the electron impact dissociative excitation of

$H_2O$ (Feldman et al. 2015). Neutral carbon (CI) emissions could also be attributed to dissociation of $CO_2$ via a similar mechanism. Subsequently, Feldman et al. (2018b) could state that distinct spectral signatures arising from multiple dissociative excitation processes were present in ALICE spectra. Electron impact processes are sensitive to the electron temperature, $T_e$, and, as we shall see below, $T_e$ varies strongly with cometocentric distance so that results are very sensitive to the extrapolation of plasma properties measured at the spacecraft to the immediate vicinity of the nucleus. This is especially true when the cometary activity is low and the contact surface (see below) is close to the nucleus. Evidence for this has also been presented by Bodewits et al. (2016) who used narrow-band filter observations with the OSIRIS Wide Angle Camera. In the WAC, the filters were centred on the emission lines and bands of various daughter species in the optical and near-UV (Keller et al. 2007). The observations at 67P were also interpreted as indicating that the emissions observed in the OH, $OI(^1D)$, CN, NH, and $NH_2$ filters were mostly produced by dissociative electron impact excitation of parent species. A sudden decrease in intensity levels was observed after equinox in March 2015, which was attributed to decreased $T_e$ in the first few kilometers above the surface as the activity of the nucleus increased and the size of the diamagetic cavity increased (see Chap. 5).

### 3.5.6 Spatial and Temporal Variations of Parent and Daughter Species

We have seen in Sect. 3.4.9 that the temperature distribution across the surface can lead to compositional changes in the outgassing. But this was insolation-driven. Is there any evidence for chemical inhomogeneity in the coma driven by nucleus compositional inhomogeneity?

A'Hearn et al. (2012) reviewed observations of the ratios of CO, $CO_2$ and $H_2O$ production rates from various sources. Notable is that measurements of the same comet under similar conditions can produce markedly different results. For example, the $CO_2/H_2O$ ratio of 22P/Kopff was measured five times by Ootsubo et al. (2012) resulting in values ranging from 0.041 to 0.201 which suggests high temporal variability.

Evidence for different spatial distributions of $CO_2$ and $H_2O$ species at the nucleus of 9P/Tempel 1 was presented by Feaga et al. (2007) from infrared imaging spectroscopy data acquired by Deep Impact. Finklenburg et al. (2014) attempted to model the observations of $H_2O$ using a DSMC code and showed that there was some sensitivity to the source distribution on the nucleus (Fig. 3.54). The $CO_2$ distribution shown by Feaga et al. was markedly different suggesting a different source distribution. The situation at 103P/Hartley 2 (Fig. 3.2) appears to have been even more extreme (A'Hearn et al. 2011) with $CO_2$ and $H_2O$ ice being emitted from the end of the small lobe of the bi-lobate structure and $H_2O$ vapour being emitted from the longer side.

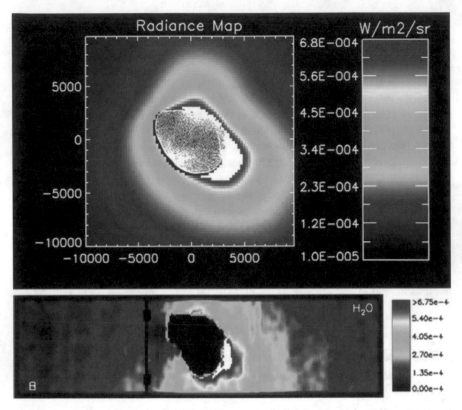

**Fig. 3.54** Water distribution in the innermost coma of 9P/Tempel 1. Below: The Deep Impact observation from Feaga et al. (2007). Above: The model of Finklenburg et al. (2014) for a similar geometry in units of $[W\ m^{-2}\ sr^{-1}]$. The Sun is to the right and ecliptic north is up. The comet nucleus is marked with white dots, the black squared underneath it, are the regions without data. The distances are given in metres. The model required a spatially inhomogeneous source distribution. (Reprinted from Finklenburg et al. 2014 with permission from Elsevier)

We saw in Fig. 3.28, that local CO, $CO_2$ and $H_2O$ densities in the innermost coma vary with respect to each other in Rosetta/ROSINA data. Hässig et al. (2015) extrapolated ROSINA in situ observations from late 2014 back to the nucleus assuming free molecular flow and concluded that the $CO_2/H_2O$ production rate ratio was at a maximum in or near the Imhotep region (Fig. 2.88). Prior to equinox pre-perihelion, there was a broadly illumination-driven source for $H_2O$ but with a maximum in emission from the neck (Hapi) region. However, $CO_2$ emission was more diffuse. A similar conclusion was reached by Migliorini et al. (2016) who used Rosetta/VIRTIS data for 67P in April 2015. The ROSINA measurements suggested that, in this early phase, $CO_2$ could dominate $H_2O$ production from the poorly illuminated southern hemisphere. Hoang et al. (2017), using data from the period between November 2014 and February 2015 showed that the $CO/H_2O$ ratio was highest (>0.3–0.4) in the Imhotep, Khonsu, Wosret, Neith, and Sobek regions, whereas the $CO_2/H_2O$ was highest ($>0.5$) in the same regions as $CO/H_2O$ but also in the whole latitude band between $30°$ and $60°$ S (which includes parts of the Anhur, Bes, Atum, and Geb regions). They suggested that their results supported the concept of $CO_2$ and CO coming from below the surface where thermal variations produced by insolation would be damped (as illustrated in Fig. 3.39) (cf Bockelée-Morvan et al. 2015). This is somewhat at odds with, for example, the Deep Impact observations of 9P/Tempel 1 and 103P/Hartley 2 which would suggest that inhomogeneity of emission of the more volatile species is required.

Herny et al. (2020) tried to fit the ROSINA measurements with the physico-chemical model of Marboeuf et al. (2012) and concluded that the nucleus is dominated by $H_2O$ (91.5% $\pm$ 4.5%), then $CO_2$ (6.7% $\pm$ 3.5%) and CO in small amounts (1.9% $\pm$ 1.2%). However, they also suggest that the data indicated a dichotomy in composition between the northern and southern plains that manifested itself in the form of distinct $CO/CO_2$ ratios. This ratio was found to be about $0.6 \pm 0.1$ on average to reproduce the measurements from the northern hemisphere but only $0.2 \pm 0.1$ for the southern hemisphere. Although these values were not markedly different from previous estimates, the link to a model with physical parameters was a new element. The model results are illustrated in Fig. 3.55 where the optimum composition fits to ROSINA data are mapped into a space with three end-members. The diagram also indicates the need for a dust mantle prior to the first equinox which is subsequently removed (cf Fornasier et al. 2016).

Luspay-Kuti et al. (2015) concluded from ROSINA/DFMS observations that there are inhomogeneities in the minor species as well. Minor species show correlation with either $H_2O$ or $CO_2$.

Ground-based observations of daughter species in the coma have frequently shown jet-like or shell-like anisotropies. However, it is not clear that these are related directly to volatile inhomogeneities in the nucleus source. A good example here is CN. Observations of 1P/Halley indicated repeatable shell-like emissions from

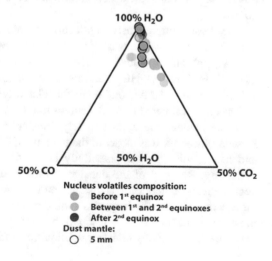

**Fig. 3.55** Representation of the compositions required to fit the observations of ROSINA with the physico-chemical model of Marboeuf et al. (2012). (From Herny et al. 2020, submitted)

the nucleus of a parent. It was suggested that the repeatability of the observed shells reflected the rotation period of the nucleus (Schlosser et al. 1986; Sect. 2.4). More recently, Schleicher et al. (2019) have observed jet-like activity in CN in 41P/Tuttle–Giacobini–Kresák with the curvature of the jet probably being related to the rotation of the nucleus. The spatial distribution of the observed CN was significantly different from that of the OH daughter product of $H_2O$ which showed little evidence of jets. $C_2$ and $C_3$ emissions broadly followed the emissions of CN suggesting source inhomogeneity with respect to the main driving volatile.

While CN, $C_2$, and $C_3$ are commonly observed and frequently show jet-like structures in narrow-band imagery, interpretation in terms of surface volatile inhomogeneity is not straightforward because the parents of these species are not well known and contributions from an extended dust source are also conceivable.

## 3.6   Compositional Variation with Heliocentric Distance

A potential diagnostic of the physico-chemical structure of the surface layer is the variation of the outgassing rates of species with time as the comet orbits the Sun. The changes in insolation ought to be reflected in the outgassing rates with the most volatile species dominant at greater heliocentric distances and less volatile species becoming increasingly evident as the comet approaches perihelion. Additional complexity can arise from non-linear heat input as a consequence of the obliquity of the nucleus (Fig. 2.13). Furthermore, if less volatile species are mixed with more volatile ices then higher mixing ratios of the less volatile species might arise if the more volatile species drags the less volatile species away from the surface in ice form (as seen at 103P/Hartley 2 in Fig. 3.2, for example). The presence of clathrates, where a "guest molecule" is initially held within a "cage" formed by a host molecule or a lattice of host molecules, may also produce less straightforward variations in the production rates of some species (e.g. Gautier and Hersant 2005; Luspay-Kuti et al. 2016).

The most comprehensive collection of observations of a single comet using Earth-based techniques was completed by Biver et al. (2002b) on C/1995 O1 (Hale-Bopp) and included measurements of the production rates of OH, CO, $CH_3OH$, $H_2S$, $H_2CO$, HCN (and its minor tautomer, HNC), CS, and $CH_3CN$ from 7 AU inbound to 10 AU outbound for relatively large fields of view.

The chemical network model shows that the relative densities of neutrals emitted from the nucleus are approximately constant in the coma out to $\sim10^3$ km. The Rosetta spacecraft was closer to the nucleus of 67P than this for much of its mission and hence, this forms a potentially better data set for studies of parent species variability. Ratioing of the density of one species to a standard species removes dependencies on heliocentric distance and changes in the global production rates but the key question is which species should provide the "standard"? While water is the most abundant species, it is also exhibits strong diurnal and heliocentric variations because of its relatively high free sublimation temperature. In Fig. 3.56, CO has been

**Fig. 3.56** ROSINA/DFMS measurements of the relative abundance of 9 species relative to CO for 67P through its perihelion passage in 2015. Top panel: Methanol, ethanol, and methane. Middle panel: Ammonia, HCN, and OCS. Lower panel: water vapour, carbon dioxide and molecular oxygen. The vertical lines mark the times of the equinoxes pre- and post-perihelion. (Data in advance of archiving courtesy of Martin Rubin)

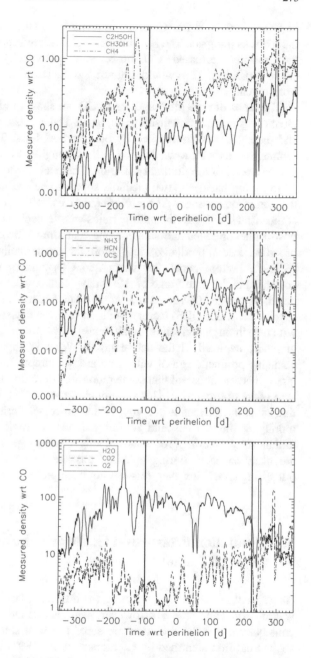

selected as the standard because of its low free sublimation temperature while still having a fairly high mixing ratio. Based on observations at other comets, extended sources of CO are likely to have only a small effect on densities with the spacecraft close to the nucleus and hence it seems safe to ignore this possibility. On the other hand, Feaga et al. (2014) reported that unexpected behavior was seen in the release of CO from C/2009 P1 (Garradd) as observed by the infrared spectrometer instrument on Deep Impact. The $H_2O$ outgassing, increased and peaked pre-perihelion and then steadily decreased. CO, however, monotonically increased throughout the entire apparition. This would suggest the choice of a "standard" molecule is far from trivial.

The ratios of nine other species to CO are shown in Fig. 3.56 and are split into three categories for visualization purposes. The abscissa is the time from perihelion and the equinoxes are marked by bold vertical lines. The ratios are displayed as running means with a window of 7 days (roughly 14 rotations of the nucleus) to reduce effects of longitudinal variation. There are several points to note.

In the top panel, the two alcohols appear to be continuously increasing through the perihelion passage. At the two equinoxes, with the Sun above the equator, a comparison shows that the mixing ratios with respect to CO are factors of ~5 greater post-perihelion. Methane shows somewhat similar behaviour apart from a rather broad increase in production just before the pre-perihelion equinox.

In the bottom panel, the water and $CO_2$ mixing ratios appear almost anti-correlated in the time frame of $\pm 200$ days with respect to perihelion. In the middle panel, we can see that ammonia shows a behaviour similar to water while the other species (HCN and OCS) increase with time. Given that several species are increasing with time through perihelion, an alternative explanation may be that CO is becoming gradually depleted in the active layers. What is apparent, however, is that the changing production rates with time are not simple functions of heliocentric distance. It is also apparent that some ratios are not returning to the values found at the start of the observations. This is particularly evident in the top panel. This suggests that either the nucleus is evolving or that processes during the aphelion passage are modifying the nucleus and returning it to its previous state. The latter is not unthinkable. Sintering processes and longer term outgassing of more volatile species are likely to occur during aphelion passage which might act to modify future outgassing rates. However, this has not been proven.

## 3.7   Radiation Pressure on Gas Molecules and Radicals: The Neutral Tail(s)

As we shall see in the next chapters, the dust and plasma tails of comets have been well observed and studied. However, neutrals and radicals can also form tails as a consequence of solar radiation pressure. The best studied example of this is the sodium tail first identified by Cremonese et al. (1997) in C/1995 O1 (Hale-Bopp). The acceleration of sodium in the anti-sunward direction is the result of the resonant fluorescence mechanism operating at 589.0 and 589.6 nm (Fig. 3.50). The sodium atoms absorb photons coming from one direction but re-emit isotropically. (There has been some discussion in the literature in the 1980s as to whether the emission of the photon is truly isotropic at the 10% level but this has not been proven and is usually ignored although we have made provision for such an effect in Eq. 3.118). The difference in direction between the absorbed and emitted photons gives a change in momentum. The acceleration is given by (Hunten et al. 1988; Thomas 1992)

$$a_n = \left(\frac{\pi q_c^{\,2}}{m_e c}\right) \frac{h}{m_n \lambda} F_\odot(\lambda) \frac{f_{osc}}{r_h^2} \tag{3.123}$$

where $m_e$ and $m_n$ are the masses of the electron and the neutral, respectively, $F_\odot(\lambda)$ is the solar spectral flux at 1 AU at the resonant frequency in the rest frame of the atom (in [photon $cm^{-2}$ $s^{-1}$ $Hz^{-1}$] to obtain an acceleration in [cm $s^{-2}$]), $f_{osc}$ is the oscillator strength of the transition and $q_c$ is the electron charge.

The solar flux that the atoms see strongly depends upon the relative velocity of the atom with respect to the Sun because of the deep solar absorption lines at these wavelengths combined with the Doppler shift as we have seen in connection with the Swings effect (Fig. 3.50). Depending upon the Doppler shift, the flux seen by the sodium atoms can vary by more than a factor of 5. Interestingly, this also indicates that the anti-sunward acceleration of the sodium atoms emitted from the nucleus pre-perihelion will have a different time dependence when compared to the acceleration for atoms emitted post-perihelion. Pre-perihelion, atoms will have their velocities towards the Sun decelerated to lower radiation force whereas post-perihelion atoms will have their velocities away from the Sun increased to even higher radiative forces.

Sodium is the most deeply studied of the atoms and radicals showing this effect because it is extremely bright at optical wavelengths. However, many species should show similar types of behaviour. By analogy with other objects in the Solar System such as Io and Mercury, one can expect atoms such as potassium, magnesium and calcium to show similar effects although the anti-sunward accelerations will be strongly dependent upon the solar flux at the resonant (Doppler shifted) wavelength and the oscillator strength of the transition(s). Other species such as CN may also experience some anti-sunward acceleration.

Brown et al. (1998) concluded from a simple model that the sodium tail observations indicated that around half the sodium observed came directly from the nucleus with the remainder coming from an extended source. Dissociation of salts in the coma or release from dust particles are obvious candidates for the nature of the material producing the extended source. Cremonese et al. (2002) concluded that, while numerous release mechanisms are possible, the data are as yet inadequate to provide strong constraints. It was clear, however, that photolysis of simple molecules (NaOH, NaH, NaCl, and $Na_2$) is extremely rapid so that release of these molecules at the surface would not lead to an identifiable extended source. Evidence for ammonium salts has been presented by Altwegg et al. (2020) and Poch et al. (2020). Stability of salts when in clusters may be longer and they could provide an extended gas source. On the other hand, release from dust still seems the most plausible explanation.

## 3.8   Isotopic Ratios

Was the current inventory of water on Earth present at the time of its formation or has it been transported to our planet by the impact of comet-like bodies? This subject has provoked considerable disagreement and the answer also has implications for the Earth's organic inventory. As we shall see, roughly 25% of a comet's mass is organic. If the water has been brought in, then organics almost certainly came with it. One of the best ways of studying this problem is by looking at isotopic ratios.

Using $CO_2$ and the carbon-13 isotope ($^{13}C$) as an example (O'Leary 1988), the ratio of different isotopic compositions can be expressed as

$$R_{CO_2} = \frac{^{13}CO_2}{^{12}CO_2}$$
(3.124)

which can be converted to standard "per mil" values by

$$\delta^{13}C = \left[ \frac{R_{sample}}{R_{standard}} - 1 \right] \times 1000$$
(3.125)

Chemical reactions (and physical processes of possible relevance such as diffusion) can discriminate between isotopic compounds leading to fractionation. For conversion of a compound A (again using a carbon species as an example) into a compound B, the isotope fractionation can be by defined as

$$\Delta\delta = \frac{\left[ \delta^{13}C(A) - \delta^{13}C(B) \right]}{1 + \delta^{13}C(A)/1000}$$
(3.126)

Isotopic abundances can be determined in situ through mass spectrometry in a fairly straightforward manner but obviously requires a spacecraft. However, they can also be determined at optical wavelengths using high resolution spectroscopy if the bands are sufficient separated. An example here is the $C_2$ (1,0) Swan band. Danks et al. (1974) identified the $^{12}C^{13}C$ (1–0) band at 4745 Å separated from the $^{12}C^{12}C$ feature. However, there is frequently contamination in the spectra from other species and, in this particular case, $NH_2$ forms a significant contaminant of the weak, isotopic band. Observations at infrared and radio wavelengths can also be performed (see Bockelée-Morvan et al. 2004a) with measurements of $H_2O$ isotopes being possible from Earth-orbiting spacecraft such as Herschel (Hartogh et al. 2011). The oxygen isotopic ratio $^{18}O/^{16}O$ is also accessible from the ground using the OH lines as illustrated in Fig. 3.57 and reviewed by Jehin et al. (2009).

Determination of deuterium-to-hydrogen (D/H) ratios is of primary interest because there was a gradient in D/H in the solar nebula resulting from a temperature gradient (Hosseini et al. 2018). It has been assumed that JFCs are formed further away from the Sun, close to or beyond Neptune, whereas Oort cloud comets have

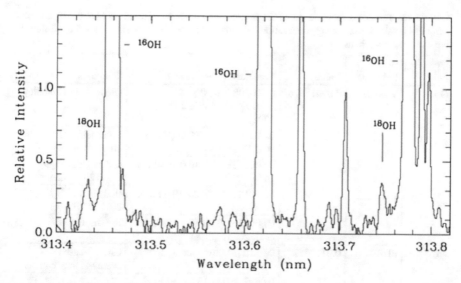

**Fig. 3.57** The $^{18}$OH and $^{16}$OH lines are separated in the UV by around 0.2 Å (0.02 nm) allowing determination of the isotope ratio (from Jehin et al. 2009)

formed closer in. If correct, then a trend of higher D/H ratio in JFCs than in Oort cloud comets should be seen. However, cometary D/H measurements suggest a wide range of D/H ratios in the water within Jupiter family objects although, as noted earlier, the exact place of origin of a specific comet is no longer given if planetary migration was significant in our Solar System.

D/H in the interstellar medium using the hydrogen molecule is estimated to be $2.0–2.3 \times 10^{-5}$ (Geiss and Gloeckler 1998). For the proto-planetary nebula which produced our Solar System, a value of $2.6 \pm 0.7 \times 10^{-5}$ has been determined based on measurements of $H_2$ in the atmosphere of Jupiter (Mahaffy et al. 1998). This is considerably less than that of the Earth as can be seen from the standards given in Table 3.13 (Hoppe et al. 2018). The D/H ratio in comets appears to cover a range of values. Hartogh et al. (2011), using the Heterodyne Instrument for the Far Infrared (HIFI) instrument on the Herschel Space Observatory to measure HDO and $H_2{}^{18}O$ at 509.292 and 547.676 GHz, obtained a value of $1.61 \pm 0.24 \times 10^{-4}$ for 103P/Hartley 2 which would be consistent with the Earth's oceans. On the other hand, the in situ measurements at 1P/Halley and 67P both indicate enrichment compared to the Earth. The similar approach of Biver et al. (2016) for measurements of two different comets giving two very different results (Table 3.13) argues that comets do indeed come from different source regions but it does not really clarify the issue of the source of Earth's water. Altwegg et al. (2019) conclude that the notion that different comet families formed at different distances from the Sun is probably not correct. The large differences in D/H values in cometary water, independent of the comet family, point more to a scenario in which comets formed over a relatively large region of the protoplanetary nebula and were later scattered as we have seen in Fig. 1.12.

**Table 3.13** Terrestrial isotopic ratios from standard references (Hoppe et al. 2018) and some comparison values from comet observations of different types. The table is intended to be illustrative and not comprehensive

| Isotopic ratio | Reference | Value | Cometary values | Source |
|---|---|---|---|---|
| D/H | VSMOW | 0.00015576 | 0.000308 + 0.000038/ −0.000053 (Balsiger et al. 1995) | Giotto $H_2O$ 1P/Halley |
| | | | 0.000161 ± 0.000024 (Hartogh et al. 2011) | Herschel $H_2O$ 103P/Hartley 2 |
| | | | 0.00065 ± 0.00016 (Biver et al. 2016) | IRAM/Odin $H_2O$ C/2012 F6 (Lemmon) |
| | | | 0.00014 ± 0.00004 (Biver et al. 2016) | IRAM/Odin $H_2O$ C/ 2014 Q2 (Lovejoy) |
| | | | 0.00053 ± 0.00007 (Altwegg et al. 2015) | Rosetta $H_2O$ 67P |
| $^{13}C/^{12}C$ | PDB | 0.0112372 | 0.0110 (Manfroid et al. 2009) | Ground-based CN average many comets |
| | | | 0.0092 ± 0.0013 (Biver et al. 2016) | IRAM/Odin HCN C/ 2014 Q2 (Lovejoy) |
| | | | 0.0119 ± 0.0006 (Hässig et al. 2017) | Rosetta $CO_2$ 67P |
| | | | 0.0116 ± 0.0012 (Rubin et al. 2017) | Rosetta CO 67P |
| $^{15}N/^{14}N$ | Air | 0.0036765 | 0.00677 (Manfroid et al. 2009) | Ground-based CN average many comets |
| | | | 0.00737 ± 0.00031 (Shinnaka et al. 2016) | Ground-based $NH_3$ average many comets |
| $^{17}O/^{16}O$ | VSMOW | 0.0003799 | 0.00037 ± 0.00009 (Altwegg et al. 2015) | Rosetta $H_2O$ 67P |
| $^{18}O/^{16}O$ | VSMOW | 0.0020052 | 0.0019 (Balsiger et al. 1995) | Giotto $H_2O$ 1P/Halley |
| | | | 0.00200 ± 0.00010 (Biver et al. 2016) | IRAM/Odin H2O C/ 2014 Q2 (Lovejoy) |
| | | | 0.00202 ± 0.003 (Hässig et al. 2017) | Rosetta $CO_2$ 67P |
| | | | 0.0018 ± 0.0002 (Altwegg et al. 2015) | Rosetta $H_2O$ 67P |
| $^{29}Si/^{28}Si$ | NIST NBS 28 | 0.050804 | 0.0434 ± 0.0050 (Rubin et al. 2017) | Rosetta Si 67P |
| $^{30}Si/^{28}Si$ | NIST NBS 28 | 0.033532 | 0.0263 ± 0.0038 (Rubin et al. 2017) | Rosetta Si 67P |
| $^{33}S/^{32}S$ | VCDT | 0.0078773 | 0.00658 (Calmonte et al. 2017) | Rosetta $H_2S$, OCS, $CS_2$ weight mean, 67P |
| $^{34}S/^{32}S$ | VCDT | 0.0441626 | 0.043 (Altwegg 1996) | Giotto $S^+$ 1P/Halley |
| | | | 0.0420 (Calmonte et al. 2017) | Rosetta $H_2S$, OCS, $CS_2$ weight mean, 67P |
| | | | 0.0463 ± 0.0057 (Paquette et al. 2017) | Rosetta/COSIMA S 67P |

Lis et al. (2019) have proposed an explanation based on their observation that "hyperactive" comets, such as 46P/Wirtanen, require an additional source of water vapour in their comae, explained by the presence of subliming icy grains expelled from the nucleus and that these particular objects have D/H ratios in water consistent with the terrestrial value. They propose that the isotopic properties of water outgassed from the nucleus and that of icy grains may be different because of fractionation effects during the sublimation process. There clearly remain issues of interpretation here that need to be resolved.

As an aside, Oba et al. (2017) have also shown that photolysis of icy interstellar analogues produced a compound (hexamethylenetetramine—HMT) in which the deuteration level of the product far exceeded that of the reactants. They suggest that HMT can play a role as an organic source of interstellar deuterium which may be distributed into other chemical species through molecular evolution. Consequently chemical enhancement of the D/H ratio also needs to be considered.

Ground-based measurements of CN have shown an enrichment of $^{15}N$ while Rosetta data have shown enrichment of some silicon isotopes. The oxygen isotope measurements by Rosetta are consistent with VSMOW (Vienna Standard Mean Ocean Water). Calmonte et al. (2017) have argued that the fractionation observed in Rosetta/ROSINA measurements of sulphur, together with the silicon enrichments, suggests a non-homogeneously mixed proto-planetary nebula. The values for the $^{34}S/^{32}S$ ratio from measurements of dust at 67P (Paquette et al. 2017) by Rosetta/ COSIMA agree within error with the ROSINA measurements and are themselves consistent with the Earth standard, Vienna Canyon Diablo Troilite (VCDT).

## 3.9 Ortho to Para Ratios

Molecular hydrogen can occur in two isomeric forms often referred to as spin isomers. One isomer is with its two proton nuclear spins aligned parallel (ortho-$H_2$), the other with its two proton spins aligned antiparallel (para-$H_2$). Para-$H_2$ is at an energy level equivalent to 23.8 cm$^{-1}$ lower than ortho-$H_2$. At room temperature and thermal equilibrium, thermal excitation causes hydrogen to consist of approximately 75% ortho-$H_2$ and 25% para-$H_2$. When hydrogen is liquified at low temperature, there is a slow transition to predominantly para-$H_2$. In thermal equilibrium, the temperature dependence of the ratio of ortho-$H_2$ to para-$H_2$ (OPR) can be found from the equation (Krishna Swamy 2010)

$$OPR = \frac{(2I_o + 1)\sum(2J + 1)e^{-\frac{E_o}{kT}}}{(2I_p + 1)\sum(2J + 1)e^{-\frac{E_p}{kT}}} \qquad (3.127)$$

where $J$ and $E$ refer to the rotational quantum number and the energy levels respectively and $I$ is the spin angular momentum. The subscripts refer to ortho and para. The transition from exclusively para-$H_2$ to an OPR value of three occurs between 5 K and 50 K (Bonev et al. 2007). Conversion between the two forms in the gas phase by radiative transitions or by collisions is strictly forbidden (Bockelée-Morvan et al. 2004a) with a radiative transition rate calculated to be $6 \; 10^{-14}$ yr.$^{-1}$ (Pachucki and Komasa 2008). The OPR is thus believed to remain constant throughout the coma. The slow rate of conversion of OPR is consequently thought to form a probe of the temperature of the particular species at the time of its formation. On the other hand, Fillion et al. (2012) have noted that the interpretation of the spin temperatures of molecules formed or condensed on grains is perhaps premature, because very little is known concerning the relaxation of spin isomers in cold solids, on surfaces of astrophysical interest, and after thermal/non-thermal desorption. Recent theoretical work by Chapovsky (2019) now suggests that lifetimes are shorter than the lifetime of the Solar System. Nuclear spin conversion also appears to be more efficient when there are molecular impurities, for example.

If the interpretation is correct, however, the results may be extremely useful. The observations by Bonev et al. (2007) at 2.9 μm in the infrared at a resolution of around 25,000 ($\lambda/\Delta\lambda$) suggest differences in OPR between comets from 1.8 to 3 and formation temperatures between 20 and 40 K. Repeated measurements of 103P/Hartley 2 at two different apparitions gave the same value of OPR within error (Bonev et al. 2013) suggesting no variation in OPR with depth in the nucleus.

Other species of interest can exhibit ortho-para states including $NH_3$ and $CH_4$. Shinnaka et al. (2011), for example, determined OPR for $NH_3$ based on studies of the dissociation product, $NH_2$, for 15 comets and concluded that formation temperatures around 30 K were consistent with the observations. $CH_4$ has been studied by Kawakita et al. (2005) from which the spin temperature was derived to be 33 K. These results are significant as they may be an indicator of the internal temperature of the nucleus—a value which is needed to compute the thermal balance (Sect. 2.9.3) and describe possible amorphous to crystalline transitions (Sect. 2.9.3.4).

# Chapter 4
# Dust Emission from the Surface

It is dust that makes bright comets visible to the naked eye through scattering of sunlight. The source of the dust is obviously the nucleus. But the details of the ejection mechanism are not well known. The simplest concept is that the dust is embedded in the ice and that when the ice sublimes, the dust is ejected by the subliming ice that surrounds it. While there are issues with this concept, it is useful as a starting point.

Some of the energy acquired by the gas through the sublimation process is transferred to the dust via gas drag. Hence, as the gas expands to form the inner coma, dust is accelerated away from the nucleus surface. Eventually, the gas density becomes sufficiently low that drag on the dust becomes insignificant, the acceleration drops to near zero and the dust reaches a terminal velocity with respect to the nucleus. Even for relatively active comets, this gas drag acceleration region is quite small in radial extent being typically $\ll$50 km. However, as the distance from the nucleus increases, the influence of more subtle forces begins to become apparent resulting in, for example, the production of the cometary dust tail. By following the forces acting, we are able to describe several phenomena of interest.

## 4.1 The Point Source Approximation

As with the gas, a point source approximation can be used for the dust in the first instance. In analogy with Eq. (3.1), the dust density can be written as

$$n_d(r) = \frac{Q_d}{4\pi r^2 v_d} \tag{4.1}$$

where $v_d$ is a bulk velocity for the dust. This approximation has many of the deficiencies already discussed for the gas including the inadequate assumption of isotropic emission. Furthermore, the gas flow does tend to homogenize more readily

© Springer Nature Switzerland AG 2020
N. Thomas, *An Introduction to Comets*, Astronomy and Astrophysics Library,
https://doi.org/10.1007/978-3-030-50574-5_4

through collisions whereas the decoupling of the dust from the gas leads to a larger residual inhomogeneity in the coma. As with the gas, the integration of the force free outflow equation along the line of sight produces a column density,

$$N_d(b) = \frac{Q_d}{4bv_d} \tag{4.2}$$

that clearly shows that imaging observations of dust emission from a point source should produce radial profiles that vary in brightness with $1/b$. (In the literature, this is usually referred to as a $1/r$ dependence but we are using $r$ here for distances to the nucleus in 3D whereas in the above equation, $b$ is the impact parameter in the image plane.) We can also obtain a further useful equation by analogy namely

$$G_d = \frac{\pi}{2} \frac{Q_d}{v_d} \tag{4.3}$$

which is the integral of the column density on a circle surrounding the nucleus at a constant impact parameter. Hence, we have numerous relations similar to those seen in Sect. 3.1. But differences become evident as we discuss how dust is observed through remote-sensing.

## 4.2 Scattering of Light by Dust

### 4.2.1 Introduction and Rayleigh Scattering

We first discussed the concepts of radiance and irradiance (flux) in Sect. 2.6. Figure 2.15 provides an overview of the nomenclature we use with respect to surface photometry. To look at particle scattering, we need to go a little deeper into the physics. There are numerous textbooks describing the general problem of how an incident wave is scattered by a particle. Bohren and Huffman (1983) and Mishchenko et al. (2002) are examples while Hovenier et al. (2004) look specifically at polarization within planetary atmospheres including scattering by particles that are large with respect to the irradiating wavelength. These books are rigorous and we need not repeat their texts here. However, we do need to summarize one or two aspects of specific relevance to comets.

The radiant energy interacts with the medium it is in. On a clear day on Earth, scattering of sunlight by gas molecules dominates leading to the blue colour of the sky. The physical phenomenon is known as Rayleigh scattering. The wavelength dependence of the observed intensity of the scattering is governed by the proportionality

$$I \propto \frac{1}{\lambda^4} \qquad (4.4)$$

so that scattering at short wavelengths is dominant. The angular dependence is

$$I \propto \left(1 + \cos^2\theta\right) \qquad (4.5)$$

so that the sky brightness at visible wavelengths varies by only about a factor of two as one surveys the horizon on a clear day. Gas molecules are factors of $>1000$ smaller than the wavelength of visible light and scattering by particles that are up to 100 times larger than gas molecules can usually be treated with a Rayleigh scattering approach. However, when the scatterers have a size approaching the wavelength, the result becomes very different. Hence, the sky in an atmosphere where micron-sized dust dominates the scattering of sunlight (e.g. on Mars) has a very different appearance (e.g. Markiewicz et al. 1999).

For comets, Rayleigh scattering by gas molecules is negligible while scattering by dust or ice particles very much smaller than the wavelength is minimal although the contribution of the latter to the study of the cometary outflow should not be immediately ignored. On the other hand, dust in the coma is visible because of larger particles scattering sunlight. A range of particle sizes is evident in the coma and, as we shall see, the most effective scatterers are those with sizes close to the wavelength. This results in strong angular and wavelength dependence of the scattered light from the coma. There are several approaches to modelling the light scattered by individual particles. We begin by looking at the Stokes parameters and Mie theory.

### 4.2.2   Scattering by Particles Close to the Wavelength

#### 4.2.2.1   Preliminaries

Hovenier et al. (2004) present the definition of the Stokes parameters that were introduced to describe a beam of polarized radiation. If these parameters are exactly specified then one can derive the radiance and the polarization state of the beam. Although polarization of light in cometary comae is still a fairly minor sub-topic, it will be discussed later and for convenience we will introduce it here in the context of scattering.

The Stokes parameters are usually combined into a vector called the Stokes vector as

$$S = \begin{pmatrix} I \\ Q \\ U \\ V \end{pmatrix} \qquad (4.6)$$

which is frequently written for convenience as $S = \{I,Q,U,V\}$. The Stokes parameters are functions of the time-averaged amplitudes and initial phases of the electric field vectors of a quasi-monochromatic beam (Hovenier et al. 2004). There are several special cases that illustrate the use of the Stokes formulation.

The $I$ term is of particular interest because it is proportional to the total intensity. For a completely unpolarized beam

$$S_{unp} = \{I, 0, 0, 0\} \qquad (4.7)$$

and represents what is often referred to as natural light. One can assume that sunlight illuminating a dust coma has this property. Light that is completely polarized fulfils the condition

$$I^2 = Q^2 + U^2 + V^2 \qquad (4.8)$$

Radiances are additive and this can be useful in treating partial polarization of a beam as the sum of an unpolarized component and a completely polarized component such that

$$S_{tot} = S_{unp} + S_{pol} \qquad (4.9)$$

and therefore

$$S_{tot} = \left\{ I - \sqrt{Q^2 + U^2 + V^2}, 0, 0, 0 \right\} + \left\{ \sqrt{Q^2 + U^2 + V^2}, Q, U, V \right\} \qquad (4.10)$$

The Stokes parameters, Q, U, and V describe the handedness and ellipticity of the electric vector of a polarised wave in a coordinate system in which $r$ and $l$ are directions perpendicular to the direction of propagation of the wave and orthogonal to each other (Fig. 4.1).

$Q$, $U$, and $V$ are then defined using spherical coordinates as

**Fig. 4.1** The vibration (or polarization) ellipse for an electric vector at a point O of a polarized wave. The direction of propagation is into the paper. (Adapted by permission from Springer Nature from Hovenier et al. 2004)

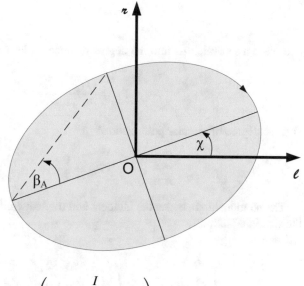

$$S = \begin{pmatrix} I \\ I\mathbb{P}\cos 2\beta_A \cos 2\chi \\ I\mathbb{P}\cos 2\beta_A \sin 2\chi \\ I\mathbb{P}\sin 2\beta_A \end{pmatrix} \qquad (4.11)$$

where $\mathbb{P}$ is the degree of polarization. The influence of linear and circular polarization on the Stokes vector is shown in Fig. 4.2 in which the Stokes parameters are normalized by the intensity, $I$.

The degree of polarization can be expressed (Hovenier et al. 2004) as

$$S_{LHP} = I\begin{pmatrix} 1 \\ 1 \\ 0 \\ 0 \end{pmatrix} \qquad \longleftrightarrow \qquad S_{LVP} = I\begin{pmatrix} 1 \\ -1 \\ 0 \\ 0 \end{pmatrix} \quad \updownarrow$$

$$S_{L+45P} = I\begin{pmatrix} 1 \\ 0 \\ 1 \\ 0 \end{pmatrix} \qquad \nearrow \qquad S_{L-45P} = I\begin{pmatrix} 1 \\ 0 \\ -1 \\ 0 \end{pmatrix} \quad \nwarrow$$

$$S_{RCP} = I\begin{pmatrix} 1 \\ 0 \\ 0 \\ 1 \end{pmatrix} \qquad \circlearrowleft \qquad S_{LCP} = I\begin{pmatrix} 1 \\ 0 \\ 0 \\ -1 \end{pmatrix} \quad \circlearrowright$$

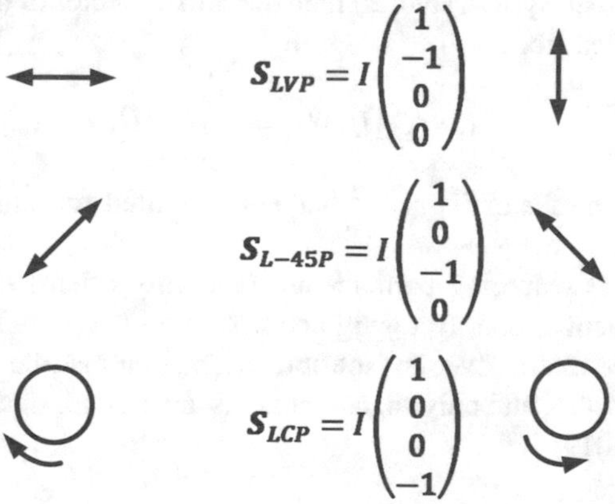

**Fig. 4.2** The pattern of Stokes parameters for various states of polarization. The upper four diagrams indicate linear polarization with $V = 0$. The bottom row indicates circular polarization with $Q = U = 0$. The middle row indicates 45° polarization when $Q = 0$

$$P = \frac{\sqrt{Q^2 + U^2 + V^2}}{I} \tag{4.12}$$

and we can separate this into the degree of linear polarization

$$P_L = \frac{\sqrt{Q^2 + U^2}}{I} \tag{4.13}$$

and the degree of circular polarization

$$P_C = \frac{V}{I} \tag{4.14}$$

The relationship between the incident and the scattered wave can be described by the matrix equation

$$\begin{pmatrix} I_s \\ Q_s \\ U_s \\ V_s \end{pmatrix} = \frac{1}{k_N^2 r_{op}^2} \begin{pmatrix} S_{11} & S_{12} & S_{13} & S_{14} \\ S_{21} & S_{22} & S_{23} & S_{24} \\ S_{31} & S_{32} & S_{33} & S_{34} \\ S_{41} & S_{42} & S_{43} & S_{44} \end{pmatrix} \begin{pmatrix} I_i \\ Q_i \\ U_i \\ V_i \end{pmatrix} \tag{4.15}$$

where the $1 \times 4$ matrices describe the scattered (subscript $s$) and incident (subscript $i$) waves and the $4 \times 4$ matrix is the Müller matrix (or sometimes, phase matrix). $k_N$ is the wavenumber of the incident radiation ($=2\pi/\lambda$) and $r_{op}$ is the observer-particle distance.

The components of the Müller matrix are fully specified in Bohren and Huffman (1983). If the incident illumination is unpolarized (as would usually be assumed for most Solar System studies) then the Stokes vector of the scattered component can be computed from

$$I_s = S_{11}I_i \quad Q_s = S_{21}I_i \quad U_s = S_{31}I_i \quad V_s = S_{41}I_i \tag{4.16}$$

where the factor $(k_N r_{op})^{-2}$ has been omitted for simplicity (Bohren and Huffman 1983).

If the scattering particles are randomly oriented then all scattering planes are equivalent. Upon averaging and assuming the particles present mirror symmetries, then the matrix over an isotropic orientation distribution has eight components that are nonzero, and only six components are unique. Equation (4.15) reduces to (Frattin et al. 2019)

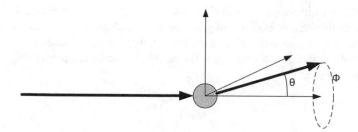

**Fig. 4.3** Definition of the scattering angle for radiation incident upon a scattering particle

$$
\begin{pmatrix} I_s \\ Q_s \\ U_s \\ V_s \end{pmatrix} = \frac{1}{k_N^2 r_{op}^2} \begin{pmatrix} S_{11} & S_{12} & 0 & 0 \\ S_{12} & S_{22} & 0 & 0 \\ 0 & 0 & S_{33} & S_{34} \\ 0 & 0 & -S_{34} & S_{44} \end{pmatrix} \begin{pmatrix} I_i \\ Q_i \\ U_i \\ V_i \end{pmatrix}
\tag{4.17}
$$

Furthermore, if the incident light is unpolarized (as in the case of solar illumination) then the Stokes vector of the scattered light reduces to

$$
\begin{aligned}
I_s &= S_{11} I_i / k_N^2 r_{op}^2 \\
Q_s &= S_{12} Q_i / k_N^2 r_{op}^2 \\
U_s &= 0 \\
V_s &= 0
\end{aligned}
\tag{4.18}
$$

We can now define the scattering angles, $\theta$ and $\phi$ as in Fig. 4.3. The angle $\theta$ (frequently referred to as the scattering angle) can be recognized as $\pi-\alpha$ (where $\alpha$ is the phase angle in planetary photometry) while the angle $\phi$ is a rotation about the vector defining the direction of the incident light.

The linear polarization of the scattered light for unpolarized incident light also has a simple form being

$$
P_L(\theta, \phi) = -\frac{S_{12}(\theta, \phi)}{S_{11}(\theta, \phi)}
\tag{4.19}
$$

where $P_L$ is a function of the scattering angles in a polar coordinate system with $\theta$ being the scattering angle (e.g. Frattin et al. 2019). If $P_L$ is positive, the scattered light is, to some degree, polarized perpendicular to the scattering plane. A negative value for $P_L$ implies that the scattered light is partially polarized parallel to the scattering plane.

Within a cometary coma, dependencies on $\phi$, are often ignored as it is assumed that the number of particles involved in scattering combined with their rotation would lead to randomization and smoothing out of any dependencies. However, this may not be a valid assumption in all cases. One might envisage elongated particles all aligning themselves because of gas drag or electrostatic effects which could, in principle, generate dependencies on $\phi$ although there is no evidence to support this.

The scattering cross section $\sigma_{sca}$ is the ratio of the scattered power, $W_s$, to that of the incident irradiance

$$\sigma_{sca} = \frac{W_s}{I_i} \tag{4.20}$$

while the differential scattering cross-section, $d\sigma_{sca}/d\Omega_s$ specifies the angular distribution of the scattered light. ($\Omega_s$ is the solid angle.) This latter quantity is related to the Stokes parameters by

$$\frac{d\sigma_{sca}}{d\Omega_s} = \frac{S_{11}}{k_N{}^2} \tag{4.21}$$

If we now follow the assumption that there is no dependency on $\phi$, then we can define the single particle angular scattering function as

$$\Phi_S = \frac{1}{\sigma_{sca}} \frac{d\sigma_{sca}}{d\Omega_s} \tag{4.22}$$

where $\Phi_S$ is normalized such that

$$\int_{4\pi} \Phi_S \, d\Omega_s = 1 \tag{4.23}$$

and is in units of $[\text{sr}^{-1}]$. A more convenient form for the normalization, useful for computation purposes, is to substitute for the solid angle so that

$$2\pi \int_0^\pi \Phi_S(\theta) \sin \theta \, d\theta = 1 \tag{4.24}$$

Some authors may choose to normalize to $4\pi$ (e.g. Fink and Rubin 2012) and therefore care is required.

The choice of scattering theory to show the dependence on $\theta$ depends upon the size parameter as can be seen from Fig. 4.4.

### 4.2.3  Mie Theory

We can now introduce Mie theory as the simplest approach to dealing with scattering by dust particles with sizes close to the wavelength of light. Mie theory assumes that particles are spherical and uniform. The theory (which is fully explained in Bohren

**Fig. 4.4** Scattering regimes identified in a plot of particle radius against the wavelength of the incident light. Visible wavelengths are identified to the top left

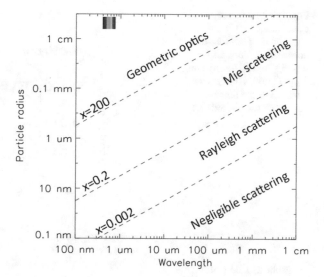

and Huffman who also include a computer code for its computation) describes the way in which the electromagnetic wave passes through the particle and thus requires the setting of a refractive index, $m_{ref}$. The theory uses a dimensionless parameter, the size parameter, $x$, defined as

$$x = \frac{2\pi a}{\lambda} = k_N a \qquad (4.25)$$

where $a$ is the particle radius and $\lambda$ is the wavelength which can be used to scale results for identical particle size to wavelength ratios.

In Fig. 4.5, Mie theory results are shown for three values of $x$ for $m_{ref} = (1.60, +0.01)$. The refractive index is complex and the imaginary part, $m_{ref,i}$, plays a role in the absorption by the particle. For the three cases, the curves have been normalized such that the integral over the $4\pi$ solid angle is one.

$m_{ref,i}$ is related to the absorption coefficient by the equation

$$\kappa_\lambda = \frac{4\pi m_{ref,i}}{\lambda} \qquad (4.26)$$

and also we can relate this to the real ($\varepsilon_r$) and imaginary ($\varepsilon_i$) parts of the dielectric constant using

$$\varepsilon_r = m_{ref,r}^2 - m_{ref,i}^2 \qquad (4.27)$$

and

**Fig. 4.5** Mie theory
calculations for size
parameters of 0.5 (dot-dot-
dot-dash), 2 (dot-dash),
5 (solid) and 10 (dash) using
$m_{ref} = (1.60, +0.01)$. The
integrals of the functions
have been normalised to 1

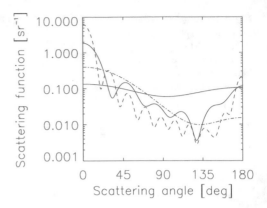

$$\varepsilon_i = 2\, m_{ref,r}\, m_{ref,i} \tag{4.28}$$

It should be noted that

$$|\varepsilon| = \sqrt{\varepsilon_r^2 + \varepsilon_i^2} \tag{4.29}$$

which is the complex modulus of the dielectric constant.

Figure 4.5 illustrates several things. Firstly, the angular dependence of the scattered intensity can vary by three orders of magnitude which is very different from the Rayleigh scattering case. Secondly, for individual particles with values of $x > 2$, there are oscillations of the scattering function. This would be evident in a cometary coma if all the particles had exactly the same size but this is not physically realistic. When a size distribution is used, even over a relatively narrow range of size parameter (e.g. $4 < x < 6$), these oscillations smooth out. Thirdly, we can see that as $x$ increases, the peak at zero scattering angle increases. In other words, the particles become more forward scattering. The size parameter can increase in one of two ways. Either the size of the particle increases or the wavelength decreases. Hence, for the same particle or particle distribution, the ratio of forward scattered light to the total scattered increases as one moves towards the blue.

The refractive index used for Fig. 4.5 was selected to produce a back scattering peak (at 180°). In general, higher values of the imaginary part of $m_{ref}$ lead to more back scattering. This, in turn, leads to a rough generalisation that the phase function has three regimes. The forward scattering regime provides information on the size of the particles, the back scattering regime is influenced by the composition, while the intermediate scattering angles are influenced by the particle shape. Mie theory assumes spherical particles and hence there are no degrees of freedom to modify the intermediate scattering angles of the phase function in Mie theory.

The geometric cross-section of the particles is given by the trivial equation

**Fig. 4.6** $Q_{sca}$ (solid line) and $Q_{ext}$ (dashed) as a function of the size parameter for particles with a refractive index of (1.60, 0.01) computed from Mie theory. $Q_{abs}$ can be derived from this curve. Note the steep rise in the efficiencies between $x = 0$ and $x = 2$

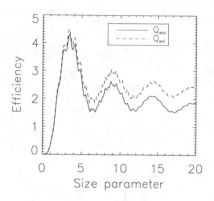

$$\sigma_x = \pi a^2 \tag{4.30}$$

On the other hand, the efficiency of the scattering by particles is defined through the scattering cross-section, $\sigma_{sca}$. The scattering efficiency is then

$$Q_{sca} = \frac{\sigma_{sca}}{\sigma_x}. \tag{4.31}$$

The extinction cross-section, $\sigma_{ext}$, defines the total amount of light removed from the beam by the particle and is the sum of the light removed by scattering and the light removed by absorption. The extinction efficiency is given by

$$Q_{ext} = \frac{\sigma_{ext}}{\sigma_x} = \frac{\sigma_{sca} + \sigma_{abs}}{\sigma_x} \tag{4.32}$$

and

$$Q_{ext} = Q_{sca} + Q_{abs} \tag{4.33}$$

where $Q_{abs}$ is the absorption efficiency. The single-scattering albedo of the particles is then given by

$$\omega = \frac{Q_{sca}}{Q_{ext}} \tag{4.34}$$

Mie theory can be used to compute the magnitudes of $Q_{abs}$ and $Q_{sca}$. An example is shown in Fig. 4.6.

One can see in this plot that at low values of the size parameter, $Q_{abs}$ and $Q_{sca}$ are identical but that they deviate. This can be seen more easily by plotting the single scattering albedo against the size parameter (Fig. 4.7). It is important to recognize here that the single scattering albedo can vary between 0 and 1 depending upon the particle size—something which is not our everyday experience. There is a sharp rise

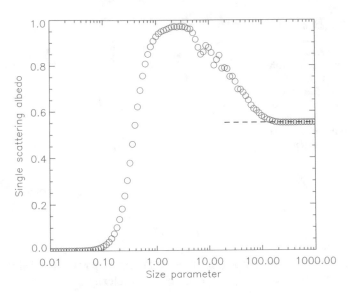

**Fig. 4.7** The single scattering albedo as a function of the size parameter for $m_{ref} = (1.55, 0.007)$ (following Moosmüller and Arnott 2009)

in $\omega$ as the size parameter increases from zero, peaking between $x = 2$ and $x = 3$. It then drops to a large particle limit. In Fig. 4.7, this is slightly above 0.5 because of diffraction and reflection combined with the imaginary part of the refractive index being greater than zero. This partially explains why the small dust particles which might be influenced by the plasma environment for example, are not easily visible at optical wavelengths.

Authors interested in the scattering of aerosols for Earth atmosphere applications discuss separate regimes for the single scattering albedo as shown in Fig. 4.8 where we have used $\log_{10} \omega$ on the ordinate to illustrate the almost linear dependence of the log of the single scattering albedo on the log of the particle size in the Rayleigh regime.

The single scattering albedo is sensitive to the particle size and the refractive index. In Fig. 4.9, the dependence of $\omega$ on $x$ is shown for four different refractive indices. Here, only the imaginary part of the index has been changed. However, one can see clearly that a variation of a factor of two in $\omega$ can arise from this property alone (cf discussion of $\omega$ in Sect. 2.6).

The use of a constant refractive index with wavelength in the above plots is a significant simplification. This optical constant (the complex refractive index) is in fact not constant with wavelength. This is evident in Fig. 4.10 which shows the real and imaginary refractive indices of water ice, amorphous magnesium silicate (Jäger et al. 2003) and carbon at 670 K (Jäger et al. 1998).[1] The optical constants for water

---

[1]https://www.astro.uni-jena.de/Laboratory/OCDB/

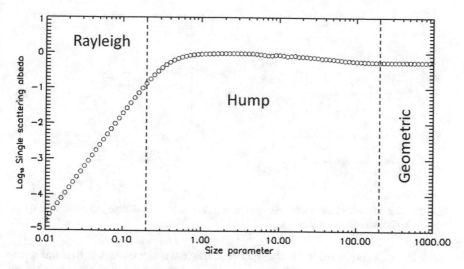

**Fig. 4.8**  Definitions of regimes for the single scattering albedo as a function of the size parameter (following Moosmüller and Sorensen 2018)

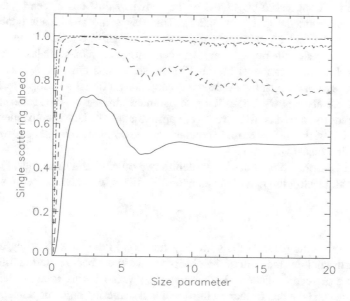

**Fig. 4.9**  The single scattering albedo for four different values of the imaginary part of the refractive index. Solid: (1.60, 0.1); Dashed: (1.60, 0.01); Dot-dash: (1.60, 0.001); Dot-dot-dash: (1.60, 0.0001)

ice from the ultraviolet to the microwave have been given in Warren and Brandt (2008). The change in the $m_{ref,i}$ with wavelength for water ice at visible wavelengths is particularly striking but should not be surprising because, as we have noted before

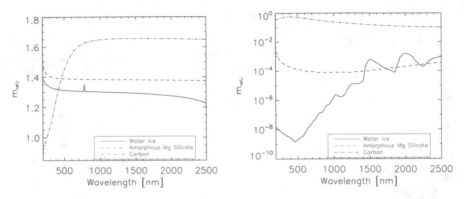

**Fig. 4.10** The refractive indices of water ice, amorphous magnesium silicate and carbon. Left: the real part. Right: the imaginary part

(Eq. 4.26), $m_{ref,i}$ is related to the absorption coefficient and we see here that water ice is not absorbing at visible wavelengths.

Figure 4.6 shows that the scattering efficiency of particles drops steeply as $x$ decreases below about 2. This illustrates that particles smaller than $\lambda/\pi$ contribute very little to the observed brightness of the cometary dust coma seen by the naked eye (as was alluded to in Sect. 4.2.1). This has some significant implications. Ground-based CCD observations of cometary comae are made down to about 400 nm in wavelength but this lack of scattering by small particles implies that there could be large amounts of sub-100 nm-sized dust emitted by a comet that would simply not be seen by an observer using conventional telescopic techniques.

Figure 4.7 also shows that as the size parameter increases, the scattering cross-section tends towards a constant value of approximately 2. This is convenient for the investigation of the radiance from cometary comae dominated by large particles and is a frequently used approximation.

One can use the above relations to define two other useful quantities. Following Divine et al. (1986), the geometric albedo of particles, $p_d$, can be determined from

$$p_d = \pi Q_{sca} \Phi_s(\pi) \tag{4.35}$$

where $\Phi_S(\pi)$ is the value of the scattering function in the backscattering direction. This quantity defines the ratio of the observed flux in the backscattering direction to that from a perfectly reflecting Lambertian disc with a radius equal to the particle radius. One can take this further by introducing a scattering angle dependent albedo, $p_s$, which is normalized to the geometric albedo in the backscattering geometry but follows the behaviour of the scattering function, i.e.

$$p_s = p_d \frac{\Phi_s(\theta)}{\Phi_s(\pi)} \tag{4.36}$$

**Fig. 4.11** A Mie theory
phase function compared to
two T-matrix calculations
from the database of Meng
et al. (2010). Solid: Mie
theory $x = 10.026$,
$m_{ref} = (1.6, i0.1)$. Dashed:
T-matrix for a spherical
particle with identical
properties. Dot-dash:
T-matrix for an ellipsoidal
particle with dimensions in
the ratio 1:1:3.3 but with
similar refractive index

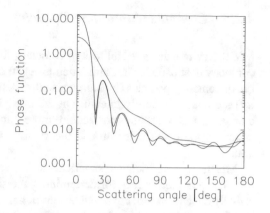

### 4.2.4   The T-Matrix Method

Mie theory assumes spherical particles. A technique for computing light scattering
by nonspherical particles is the T-matrix method. It was originally formulated by
P. C. Waterman and is, effectively, an extension to Mie theory. The approach is
based on the Huygens principle which, when applied to the propagation of light
waves, states that every point on a wavefront may be considered a source of
secondary spherical wavelets which spread out in the forward direction at the
speed of light. The new wavefront is the tangential surface to all of these secondary
wavelets. The conceptual idea is that the incident and scattered waves can be
expanded into appropriate vector spherical wave functions and then related using a
transition (or T) matrix. It also has the advantage that it reduces exactly to the Mie
theory solution when the scattering particle is spherical and homogeneous. As noted
by Mishchenko et al. (2002), the T-matrix approach can, in many applications,
surpass other techniques (such as the discrete dipole approximation; Sect. 4.2.7) in
terms of efficiency. It is also possible to control the numerical accuracy so that it can
form a benchmark for particles lacking spherical symmetry.

Codes are available for use[2] with simplified geometries allowing computation of
scattering functions and the relevant cross-sections needed for cometary dust particle
scattering computations. A simplified Java interface is also available (Halder et al.
2014). Meng et al. (2010) compiled a database of T-matrix solutions for ellipsoidal
particles with different axial ratios and an example is shown in Fig. 4.11. The Mie
and T-matrix calculations agree well for spherical particles with the same properties
but the ellipsoidal particle phase function is smoother with less of a dip at intermediate scattering angles.

---

[2]https://www.giss.nasa.gov/staff/mmishchenko/t_matrix.html

## 4.2.5  *Computer Simulated Particles*

Mie theory remains a useful tool in cometary research because, despite there being cometary dust particles in Earth laboratories, the scattering properties of the ensemble of cometary particles (including size distribution, shape, composition and porosity) are relatively poorly known. On the other hand, this situation is beginning to change as spacecraft missions acquire more information and, as more knowledge is acquired, the deficiencies in Mie theory for modelling work are becoming more apparent.

Mie theory is restricted to spherical particles of constant refractive index throughout. There are techniques for determining the scattering of spherical particles with radially dependent refractive indices but in general it is the spherical approximation that is most questionable when discussing cometary dust particles. Evidence for the fluffy, porous nature of cometary dust particles has been presented on numerous occasions (Dollfus 1989; Hadamcik et al. 2007) although various shapes and degrees of porosity might be expected when studying particles in detail. Simplified approaches such as using equivalent spheres (i.e. using Mie theory for particles with the same cross-sectional area) are becoming obsolete as computing power has increased.

The determination of the scattering properties of irregular-shaped particles has been the subject of a lot of research over the past 20 years because of applications in Earth atmosphere remote sensing and the literature in that field is a rich source of information. An important element is the computational construction of particles. There are two commonly used approaches that result in significantly different particle structures.

The Ballistic Particle-Cluster Aggregation (BPCA) procedure randomly shoots monomers of a given size at a particle. The monomer hits a cluster of monomers and sticks where it hits. The dust agglomerate then grows until a given number of monomers have been accumulated. In the Ballistic Cluster-Cluster Agglomeration (BCCA) process, hit-and-stick collisions occur between dust agglomerates of equal size.

These two mechanisms lead to remarkably different particle structures. BPCA particles are more compact with BCCA particles being more fluffy (Fig. 4.12) and this results in differences in the scattering properties. The degree of fluffiness can be described in terms of a fractal mass dimension such that

$$m_f \sim a^{D_f} \tag{4.37}$$

where $m_f$ is the particle mass and $a$ is the radius of a single dust particle. BCCA is characterized by a fractal mass dimension, $D_f$, of 2 whereas BPCA, being more compact have $D_f \sim 3$ (Wurm and Blum 1998; Mukai et al. 1992). Long chain type structures with very low values of $D_f$ are also possible and have been observed to form in microgravity experiments (Krause and Blum 2004).

**Fig. 4.12** A BCCA particle (left) in comparison to a BPCA particle (right). Both of them consist of 1024 spheres made of "Halley-like" dust (silicates, carbon, organics, a hint of Fe and S). BPCA has the gyration radius of 1.92 µm and porosity 85.5%. For BCCA, those numbers are 3.83 µm and porosity 98.6%. The plot shows the scattering functions of the two particles at 650 nm. (Courtesy of Ludmilla Kolokolova)

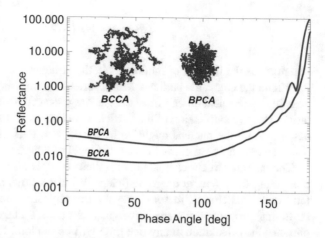

The appearance of the particles in Fig. 4.12 is anything but spherical so the definition of a "radius" is not straightforward. A "radius of gyration" was introduced by Wurm and Blum which is related to the moment of inertia as one can see in the equation for the general case

$$R_{gyr} = \sqrt{\frac{\sum_{i=1}^{N} m_0 r_i^2}{\sum_{i=1}^{N} m_0}} \tag{4.38}$$

where $m_0$ is the mass of the individual monomer located at a distance $r_i$ from the centre of mass of the whole particle. The aggregate mass and its radius are then related to $D_f$ through

$$\frac{m_d}{m_0} = \beta_c \left(\frac{R_{gyr}}{R_m}\right)^{D_f} \tag{4.39}$$

where $m_d$ is the total particle mass, $R_m$ is the radius of the individual monomers and $\beta_c$ is a constant of proportionality and, for example, takes a value of about 0.5 for BPCA.

The porosity relates the density of the constituent material to the bulk density of the particle itself. It is a moot point whether long chain structures can be described as having "porosity" but one can calculate a characteristic radius, $R_c$, if the monomers are of the same type, such that

$$R_c = \sqrt{\frac{5}{3}} R_{gyr} \tag{4.40}$$

and the porosity, $\Psi$, is then (Kozasa et al. 1992)

$$\Psi = 1 - N_m \left(\frac{R_m}{R_c}\right)^3 \tag{4.41}$$

where $N_m$ is the number of monomers in the structure.

Within the coma, the particles will experience significant stress, both thermal and possibly mechanical arising from the gas drag, as well as sublimation and fragmentation. Hence modification of the particle structure and its scattering properties with distance from the nucleus might be expected. However, light scattering investigations have rarely been carried out at this level of detail.

The particle structures arising from these models also illustrate the difficulty in computing the effective drag coefficient. Within a flow, asymmetric particles will tend to rotate. This is, in some ways, a simplification because we can determine scattering properties averaged over all orientations and assume that over many particles the orientation at any one time will be random. However, as noted above, it is conceivable that particles could align themselves in the gas flow to minimize the cross-sectional area orthogonal to the flow. (Think of a weathercock on a church tower for example.) This would result in lower drag and consequently a lower terminal velocity.

In Fig. 4.12, we can see the phase functions for the two types of particle (BPCA and BCCA) shown in the inset, computed using the T-matrix approach. Note that the plots show the scattered brightness against phase angle so that the forward scattering peak is to the right. The plots have not been normalized and show the higher value of $Q_{abs}$ for the BCCA structure clearly with the red line being below the blue line over the entire angular range. Note in particular the shallow minimum near 90° for the BPCA particle for a radius of 1.92 µm.

### 4.2.6   Observed Particle Structures

Observing the structure of real cometary particles is challenging. Interplanetary dust particles determined from sampling the Earth's atmosphere will be affected by the entry process. Particles sampled by fast fly-bys, such as by NASA's Stardust, experience a rapid deceleration during capture that, given the assumed fragility of these particles, influences the particle properties. The only way to guarantee accurate results is in situ sampling and analysis but even here the collection process may influence the results.

The atomic force microscope, MIDAS, onboard Rosetta has provided the best attempt to study individual particle shapes in situ. Two examples of particles seen by the MIDAS experiment are shown in Fig. 4.13. The difference in size between the two examples is around a factor of 10. They are both irregular but are roughly circular on the detector and roughness at smaller scales is evident.

Mannel et al. (2016) studied the fractal dimensions of MIDAS particles and obtained a value for $D_f$ (Eq. 4.37) of $1.7 \pm 0.1$ for one particle indicating more

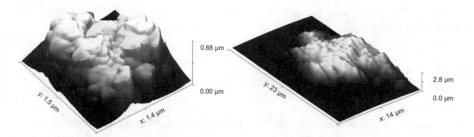

**Fig. 4.13** MIDAS observations of two dust particles. Left: A small particle (a ~ 0.8 μm) which would be the typical size of particles that are most optically active (scan_md_m029_s108_2016-05-11t120928z_tgt03). Right: A larger (a ~ 10 μm) particle. Both particles exhibit roughness at scales comparable to 1/tenth of the radius (scan_md_m021_s078_2015-10-14t080823z _tgt10). (Courtesy of Mark Bentley)

chain-like properties. Nanometre-scale analysis by Mannel et al. (2019) resulted in the description of three types of particles

- ~1 μm sized particles with surface features at the 100 nm scale
- larger ~10 μm agglomerates with 1 μm subunits in a fragile arrangement with moderate packing density
- other ~10 μm agglomerates with chain-like ($D_f < 2$) properties.

MIDAS data do not show evidence for solid particles but this could have been caused by instrumental biases.

### 4.2.7   The Discrete Dipole Approximation

The properties of the BPCA and BCCA particles were computed using the T-matrix method. Another well-known method to compute the scattering behaviour of complex particle shapes is the discrete dipole approximation (DDA). DDA (Draine 1988; Draine and Flatau 2012) addresses the deficiencies of Mie theory but at the cost of greater computational time. It was first proposed by Purcell and Pennypacker (1973) and approximates individual dust particles as a combination of dipoles. Maxwell's equations are then solved precisely to obtain the scattering function.

The approach has many advantages over Mie Theory. Dust particles can be accurately modelled through the particle construction procedures outlined above and run through the DDA algorithm. One thereby obtains an accurate result even for complex particles. Thus the spherical assumption of Mie theory is removed and particles of mixed composition can also be addressed.

The applicability is only limited by the need to have a separation between the dipoles, $d_{sep}$, that is small compared to both the size of the particle and the size of the wavelength being considered. The applicability with respect to wavelength can be tested using the criterion $|m_{ref}|k_N d_{sep} < 0.5$. On the other hand, the limit of its applicability is generally given by the computing power available as the method is computationally highly intensive.

### 4.2.8  The Observed Phase Function at 67P

The absence of observations by the Rosetta imagers in forward scattering geometries limits the range over which the phase function of dust emitted by 67P can be derived. Fink and Doose (2018) carried out an analysis of low phase angle (i.e. back scattering) geometries and matched these to other observations analysed by Bertini et al. (2017). They concluded that a simple Mie scattering model with a particle size distribution can reproduce the general characteristics of the observed phase curve, but it could not match its details. This is illustrated in Fig. 4.14 which compares Fink and Doose's fit to the observations with a Mie scattering calculation using particles in the size range 0.2 to 0.6 $\mu$m and $m_{ref} = (1.70, i0.015)$. Because of the absence of high phase angle measurements, normalization of the observed phase function is subject to assumptions and can be crudely scaled to match the model.

Bertini et al. (2017) provided phase functions derived from Rosetta/OSIRIS data covering intermediate scattering angles. An example is shown in Fig. 4.15 as the triangles with error bars. A characteristic of all the curves is a minimum at scattering angles of 80–100°. As shown by comparison with the Mie theory and T-matrix calculations in Fig. 4.15, this behaviour is seen in the scattering functions of particles with relatively small size parameters. This also illustrates an important property of phase functions arising from size distributions. For similar total cross-sections, smaller particles in the distribution contribute relatively more brightness at intermediate scattering angles than larger particles. At low and high scattering angles, the reverse is true. Using this property, a fairly reasonable fit to the phase curve in Fig. 4.15 can be obtained by combining scattering from different particle sizes.

**Fig. 4.14** Mie scattering curve (solid line) compared to the Fink and Doose (2018) scattering curve (dashed line). The Mie theory curve was produced by summing the phase curves of 200 particles linearly distributed between 0.2 and 0.6 $\mu$m with $m_{ref} = (1.7, i0.015)$. The error on the observed curve is at least 50%

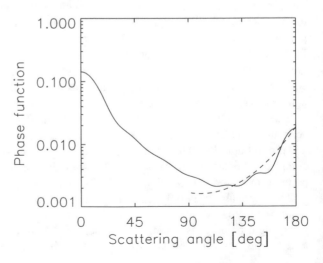

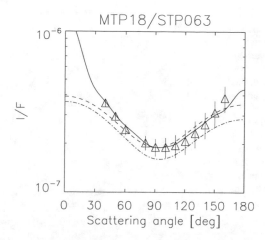

**Fig. 4.15** Phase curve derived by Bertini et al. (2017) from Rosetta/OSIRIS data (triangles with error bars). Dash: T-matrix calculation with ellipsoidal particles (1:1:3.3), $x = 0.62$, and $m_{ref} = (1.60, i0.1)$. Dot-dash: A Mie theory calculation with similar composition but $x = 0.5$. Solid: T-matrix solution added to the Mie theory solution found from the Fink and Doose paper to produce a composite phase function

Muñoz et al. (2020) have used laboratory measurements to demonstrate that highly porous, large (millimetre-sized) particles fit the Bertini et al. phase function. On the other hand, we have just demonstrated an almost identical fit using models with sub-micron particles. The BPCA curve from Kolokolova in Fig. 4.12 also shows similar behaviour with phase angle for small values of $R_g$. This might be interpreted as indicating a major problem. However, the monomer size of larger aggregates plays a significant part in defining the scattering properties as indicated by T-matrix calculations of computer generated aggregates that have been used to model ground-based observations (Dlugach et al. 2018). This suggests that the bulk sizes of the particles in the cometary coma cannot be well constrained through this technique particularly because the forward scattering peak was not measured. Furthermore, if the size distribution is undefined (and there was no instrument to measure the size distribution below 30 μm), there are suddenly large numbers of free parameters to fit the observations.

Finally, we note that there are now techniques available to allow computation of Mie theory for very large particles ($x > 3000$). However, this is not physically realistic because the deviations from sphericity become increasingly important. For large "natural" particles, the computed phase functions from Mie theory are strongly misleading. On the other hand, there is no consensus on how to treat large particle scattering. This is particularly problematic because we have an ensemble of particles with different shapes. One possibility is to use the Hapke parameters derived by Fornasier et al. (2016) to compute the phase dependence of a regular sphere with the same reflectance properties (as discussed in relation to Fig. 2.19). The idea is rather obvious in that it assumes that the particle is just an element of the surface. However, one should also be aware of the limitations. The surface phase function at 180°, for

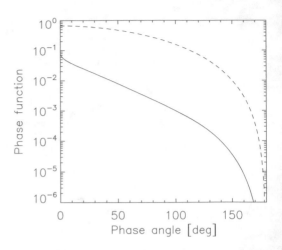

**Fig. 4.16** The phase function of a large spherical particle (>1 mm radius) derived from the scattering properties of the surface of 67P (solid line). The dashed line gives the corresponding function for a Lambertian surface with a directional-hemispherical albedo of 1. Note that at zero phase angle (the geometric albedo geometry), the function is 0.6667 as explained by Eq. (2.64)

example, is zero and large particles certainly do have non-zero forward scattering. Hence, this "phase function" is potentially unreliable at high phase (low scattering) angles. The particle shape also influences the backscattering. On the other hand, if one can make this assumption, conversion of column densities to brightnesses becomes straightforward. The result of this approach is shown in Fig. 4.16. However, as Fink and Doose (2018) point out, surface scattering methodologies which use purely geometrical optics and neglect diffraction are probably inappropriate for dispersed scattering particles in the coma of a comet and it is arguable whether the approach accurately conserves energy.

### 4.2.9   The Observed Radiance

#### 4.2.9.1   The Optical Thin Case

For the single scattering approximation in the absence of optical depth effects, the scattered radiance from the dust along a line of sight from $s_1$ to $s_2$ is given by

$$I = \int_{s_1}^{s_2} \int_0^\infty n_d(a)\sigma_{ext}(a) \frac{Q_{sca}}{Q_{ext}} \Phi_s \frac{F_\odot}{r_h^2} \, da \, ds \qquad (4.42)$$

for a single particle type at one wavelength and this can be generalised further for all particle types (e.g. water ice particles, silicate particles, etc.), $k_T$,

$$I = \int_{s_1}^{s_2} \sum_{k_T} \int_0^\infty n_d(a, k_T)\sigma_{ext}(a, k_T) \frac{Q_{sca}(k_T)}{Q_{ext}(k_T)} \Phi_s(k_T) \frac{F_\odot}{r_h^2} \, da \, ds \qquad (4.43)$$

where we have indicated the quantities that are affected by the choice of particle type (Divine et al. 1986). Once the scattered radiance has been found, the reflectance is computed according to Eq. (2.62).

There are obviously ways in which this equation can be simplified if assumptions can be made. For example, assuming a single particle type and size, we can remove the integrals and, by substitution for $Q_{sca}$ we can obtain

$$I = N_d \sigma_{ext} \frac{1}{Q_{ext}} \frac{p_d}{\pi} \frac{\Phi_s(\theta)}{\Phi_s(\pi)} \frac{F_\odot}{r_h^2} \tag{4.44}$$

where $N_d$ is the column density and

$$\rho_F = N_d \sigma_x p_d \frac{\Phi_s(\theta)}{\Phi_s(\pi)} . \tag{4.45}$$

These equations can be useful as long as the importance of the assumptions is appreciated.

#### 4.2.9.2 Optical Thickness Effects

For an active comet such as 1P/Halley or comets that make close approaches to the Sun (e.g. 96P/Machholz), the dust production rate can be sufficiently large that optical depth effects need to be accounted for. The optical depth through a part of the dust coma can be expressed as

$$\tau_d = \int_{s_1}^{s_2} \int_0^\infty n_d(a) \, \sigma_{ext}(a) \, da \, ds \tag{4.46}$$

where $n_d(a)$ is a local dust density of particles of radius, $a$, and $s$ is a distance along a line of sight. This equation is valid for a single particle type (i.e. no variation in $m_{ref}$) but can be summed over all particle types as shown in Divine et al. (1986) to give

$$\tau_d = \int_{s_1}^{s_2} \sum_{k_T} \int_0^\infty n_{k_T}(a) \, \sigma_{ext}(a, k_T) \, da \, ds \tag{4.47}$$

where $n_k$ is the density of each particle type. It should be clear here that the optical depth seen by an observer viewing the dust coma is not the only optical depth of importance in the computation of radiances. The solar irradiance of the dust coma can also be affected by optical depth effects. This is particularly true along the Sun-nucleus line because the peak dust production is usually related to the maximum insolation and thus is often assumed to be from the sub-solar point. Hence, the maximum optical depth towards the nucleus is, to first order, from the direction of the Sun. We have seen that, in the general case, $\sigma_{ext}$ is also a function of wavelength and thus, for dust, $\tau_d$ is wavelength dependent.

The direct solar irradiance at the sub-solar point of the nucleus is given by

$$F = \frac{F_\odot}{r_h^2} e^{-\tau_d} \tag{4.48}$$

where the integration limits for $\tau_d$ go from the sub-solar point to the Sun.

It is often argued that there is a natural limit to the activity of cometary nuclei because dust emission provides a throttle. As the dust emission increases, the optical depth increases up to the point when the energy reaching the nucleus surface is limited as can be seen in the exponential term in Eq. (4.48). However, there is also diffuse illumination of the nucleus arising from multiple scattering by the coma. As we have seen, particles larger than the wavelength are forward scattering with the efficiency increasing for particles close to the wavelength and especially in the blue. Hence, the energy input to the surface is not choked by increased optical depth totally and the wavelength dependence of the irradiance of the surface becomes non-solar. Hellmich (1981) appears to have been the first to look at this particular problem but Salo (1988) provided an assessment of the importance of multiple scattering for the total energy input and concluded that it is only weakly dependent on the coma opacity, $\tau_d$.

In the specific case of 67P, the relative brightness of the dust seen close to the limb when compared to the nucleus was rarely greater than 0.1. One can interpret this as implying that the dust particle filling factor in any column along a line of sight was <10% and hence that $\tau_d < 0.1$. An example from 7 July 2015 (5 weeks before perihelion) at a phase angle of 89.5° is shown in Fig. 4.17. The ratio of the brightness of the surface to that of the dust close to the limb is around a factor of 20 ($\tau_d \sim 0.05$) even in the core of the brightest jet.

### 4.2.10  Inhomogeneous Particles and Maxwell Garnett Theory

Mie theory describes extinction by homogeneous spheres. However, all particles are inhomogeneous at some level and cometary dust particles are assumed to be fluffy in nature with significant voids. Several theories have been developed to treat these types of particles but all require approximations and application to a specific problem requires assumptions on the applicability of one solution to an ensemble of particles. Probably the two best-known theories are Bruggeman's theory (often called effective medium theory) and Maxwell Garnett theory. Bruggeman's theory applies to a randomly inhomogeneous medium (Bohren and Huffman 1983) and results in an expression for an average dielectric function that can then be used to describe the interaction of the incoming wave with the particle. This theory applies to a

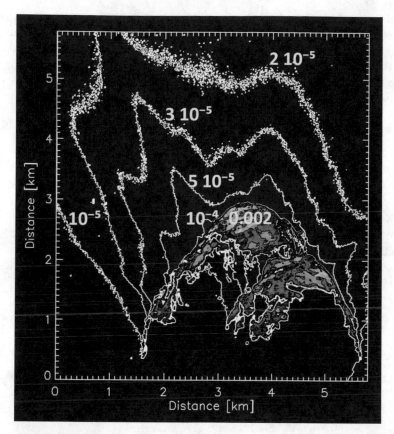

**Fig. 4.17** Contour map of an image acquired on 7 July 2015 14.00.38 (W20150707T140038894ID30F18). The contour levels are given in units of reflectance. The minimum (black) contour level on the nucleus is at 0.002. The highest contour level for the dust is a factor of 20 below this

two-component mixture in which there are no distinguishable inclusions of one component within the other. As such it is probably not a reasonable description of fluffy particles.

Maxwell Garnett (1904) (the name "Maxwell" was actually a given name to the son of the physicist William Garnett because of the latter's admiration of his friend, James Clerk Maxwell) derived an average dielectric function under the assumption of small inclusions in vacuo, i.e.

$$\varepsilon_{MG} = \frac{1 + 2(1 - \Psi)\frac{\varepsilon_p - 1}{\varepsilon_p + 2}}{1 - (1 - \Psi)\frac{\varepsilon_p - 1}{\varepsilon_p + 2}} \tag{4.49}$$

(Bohren and Huffman 1983; Markel 2016) where $\Psi$ is the porosity as previously defined for larger scales and so $1-\Psi$ is the volume fraction of the inclusions. $\varepsilon_p$ is the permittivity of the material. A more general expression can be obtained by removing the assumption that the background medium is vacuum and defining the permittivity of the host medium as $\varepsilon_h$ and that of the inclusion as $\varepsilon_i$. This results in

$$\varepsilon_{MG} = \varepsilon_h \frac{1 + 2(1 - \Psi)\frac{\varepsilon_i - \varepsilon_h}{\varepsilon_i + 2\varepsilon_h}}{1 - (1 - \Psi)\frac{\varepsilon_i - \varepsilon_h}{\varepsilon_i + 2\varepsilon_h}} \tag{4.50}$$

Given the dielectric constant, the refractive index can be computed from

$$m_{ref} = m_{ref,r} + i\, m_{ref,i} = \sqrt{\varepsilon \mu_p} \tag{4.51}$$

where the real part of the dielectric constant $\varepsilon' = m_{\text{ref},r}^2 - m_{\text{ref},i}^2$ is called the relative permittivity and describes how the electric field is stored. The imaginary part, $\varepsilon'' = 2\, m_{\text{ref},r}\, m_{\text{ref,I}}$ is the loss factor. $\mu_p$ is the relative permeability, which for most materials is very close to 1 at optical frequencies. We can also write

$$m_{ref,r} = \sqrt{\frac{\sqrt{\varepsilon'^2 + \varepsilon''^2} + \varepsilon'}{2}} \tag{4.52}$$

and

$$m_{ref,i} = \sqrt{\frac{\sqrt{\varepsilon'^2 + \varepsilon''^2} - \varepsilon'}{2}} \quad . \tag{4.53}$$

Bohren and Huffman (1983) have written quite extensively about whether the resulting values of $m_{\text{ref}}$ can be used in combination with Mie theory to produce the differential angular scattering function and the associated scattering and extinction cross-sections. Essentially, one treats $m_{\text{ref}}$ as an "effective" optical constant. However, Bohren and Huffman note that, in the general case, this is incorrect and that treating an inhomogeneous particle as if it were homogeneous does not necessarily result in consistency between measured and predicted optical parameters. One can easily imagine crossings from one material to the other within the particle and the local refractive index change will modify the light path—something that simply does not occur in a homogeneous particle. Hence, caution needs to be exercised.

## 4.3   Afρ ("Afrho")

A'Hearn et al. (1984) defined the *Afρ* quantity (or sometimes "Afrho") that is now used frequently to quantify the dust emission from comets and is related to the dependence of dust brightness on the impact parameter, *b*. *Afρ* is an equation where ρ is the radial size of a circular aperture on the sky in units of [km] and is thus equivalent to the impact parameter, $b$, in the nomenclature used here. $f$ is a filling factor which represents the ratio of the geometric cross-section of dust grains within the aperture to the area of the aperture itself. $A$ is an "albedo", which we place in inverted commas to denote that it is sometimes described as such but is actually the product of the scattering efficiency, $Q_{sca}$, and the scattering function $\Phi_s$, i.e.

$$A = 4\pi\, \Phi_s(\alpha)Q_{sca} \tag{4.54}$$

where $\Phi_S$ is the single scattering particle phase function and is again normalized such that

$$\int_{4\pi} \Phi_S\, d\Omega = 1 \tag{4.55}$$

$\Phi_S$ is in units of $[\text{sr}^{-1}]$. (We repeat that caution should be exercised because there are alternate normalizations for $\Phi_S$. Here we follow Divine et al. 1986.)

The key point of this definition is that while $A$ and $f$ cannot be determined independently, the product, $Af$, is an observable by comparing the observed flux from the comet to the flux from the Sun at the comet's distance. Mathematically, this is

$$Af = \left(\frac{2\Delta r_h}{b}\right)^2 \frac{F_{coma}}{F_\odot} \tag{4.56}$$

in which we have substituted our symbol definitions. The resulting $Af\rho$ quantity has units of length. Free radial outflow would suggest that $f$ decreases with $1/b$ and hence $Af\rho$ should be independent of aperture size for distances where other effects (such as radiation pressure) can be ignored.

The convenience of $Af\rho$ lies in the relatively straightforward comparison of dust production rates with time and between comets. However, the dependence of $Af$ on several physical parameters should not be forgotten.

The filling factor can be related to the production rate through the column density, $N_d(b)$,

$$f = \frac{\int_0^{b_{max}} N_d(b)\sigma_x\, 2\pi b\, db}{\int_0^{b_{max}} 2\pi b\, db} . \tag{4.57}$$

The geometric cross-section, $\sigma_x$, can be used for the particles in this equation if other effects (such as the scattering efficiency and the phase function) are taken into account in the computation of $A$ as has been performed in Eq. (4.54). Substituting for $N_d(b)$ in the equation for $f$ gives

$$f = \frac{Q_d \sigma_x}{2 v_d b_{max}}$$

which shows the relationship of the filling factor to fundamental quantities under the assumption that the optical depth, $\tau << 1$. A full derivation of $Af\rho$ is given by Fink and Rubin (2012) who also formulated the integral required to account for the dependence of the scattering properties on particle size

$$Af\rho = \int_{a=0}^{a=a_{max}} 2\pi \, \Phi_s(\alpha, a) Q_{sca}(a) \frac{dQ_d(a)}{da} \frac{\sigma_x(a)}{v_d(a)} da \qquad (4.58)$$

where $dQ_d/da$ is the differential particle size production rate for particle radius, $a$. Any dependence on the material properties of the particles could be included by summation over particle types using, for example, Mie theory to compute $Q_{sca}$ from the relevant refractive indices. A convenient way to compute this is by discretizing this equation (Fink and Rubin 2012) so that

$$Af\rho = \sum_a 2\pi \, \Phi_s(\alpha, a) Q_{sca}(a) \frac{Q_d(a)}{2 v_d(a)} \qquad (4.59)$$

$Af\rho$ was defined to provide a meaningful way to characterize the brightness of cometary comae as seen from the ground. $Af\rho$ can also be determined for inner comae observations from spacecraft and is related to the average azimuthal dust reflectance. The dust column density in the force-free radial outflow approximation was shown in Eq. (4.2). Assuming $\tau_d < <1$ then the reflectance from the dust is proportional to the column density. If we now average the product of the reflectance and the impact parameter on a circle surrounding the nucleus then this should be a constant independent of the impact parameter. This is actually just another way of expressing what is shown by Eq. (4.3). We can write this as

$$\bar{A} = \langle \rho \rangle b \qquad (4.60)$$

where $<\rho>$ implies that we are averaging the reflectance on a circle at constant impact parameter, $b$. Fink and Rubin et al. (2012) use the equation

$$F_{coma} = \int_0^{b_{max}} \frac{2\pi I b}{\Delta^2} db \qquad (4.61)$$

for the flux from a comet when using a circular aperture centred on the comet nucleus that is equivalent to a projected radius $b_{max}$ at an observer-comet distance of $\Delta$. Multiplying the observed radiance, $I$, by $\pi/(F_\odot/r_h^2)$ allows us to replace the flux with the reflectance factor, $\rho_F$, and we obtain

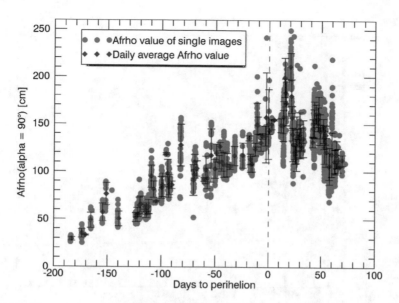

**Fig. 4.18** *Af$\rho$* values calculated from the constant value derived from fits to the azimuthal average with impact parameter, *b*. A phase angle normalization to $\alpha = 90°$ has been applied to the data shown in this plot. The green data points mark *Af$\rho$* values calculated for single images and the red data points indicate averaged *Af$\rho$* values over one comet day. Variation with heliocentric distance is clearly evident as is the lag in the maximum with respect to perihelion. (Reprinted from Gerig et al. 2018, with permission from Elsevier)

$$\frac{F_{coma}}{F_{\odot}} = \frac{2}{\Delta^2 r_h^2} \int_0^{b_{max}} \rho_F b \, db \qquad (4.62)$$

and we can substitute this back into Eq. (4.56). This gives

$$Af = \frac{8}{b_{max}^2} \int_0^{b_{max}} \rho_F \, b \, db \qquad (4.63)$$

Using $\rho$ now for $b_{max}$ and substituting using Eq. (4.60), we see that

$$Af\rho = 8\overline{A} \qquad (4.64)$$

In other words, averaging the reflectance on a circle about the nucleus and multiplying that by impact parameter of that circle gives a very simple relation to *Af$\rho$*. ($\overline{A}$ has been occasionally referred to as the "azimuthal average" of the reflectance although this is actually a misnomer.)

This approach was taken by Gerig et al. (2018) in the construction of Fig. 4.18 which shows *Af$\rho$* in the innermost coma of 67P derived from Rosetta/OSIRIS data. This gives one of the most complete descriptions of the evolution of *Af$\rho$* for any comet. It should be noted that the measurements have been normalised to a phase

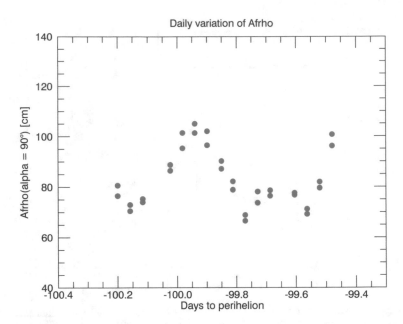

**Fig. 4.19**   The variation of *Afρ* over one nucleus rotation for 67P near equinox obtained by Gerig et al. (2018). (Reprinted from Gerig et al. 2018, with permission from Elsevier)

angle of 90° using one of the phase functions determined by Bertini et al. (2017). The phase function is a critical component of *A* in the equation for *Afρ* and comparisons between ground-based observations require careful consideration of this, especially at low scattering angles. Figure 4.18 shows that the lag in the production rate of 67P relative to perihelion seen in the gas (Figs. 1.18 and 3.48) is also evident in the dust. The maximum here is about 20 days after perihelion.

Figure 4.19 shows the variation in *Afρ* with the rotational phase of 67P over one nucleus rotation near equinox and indicates a 30% variation with a double-peak behaviour as might be expected from the rotation of an aspherical nucleus with roughly uniform dust emission on large scales.

## 4.4   Radiation Pressure

We will discuss acceleration of the dust by gas drag in Sect. 4.7 but here we look at the main force acting on the particles on larger scales—radiation pressure.

## 4.4.1   Radiation Pressure Efficiency

After the dust has de-coupled from the gas, the equation of motion needs to include more slowly acting forces such as solar gravity and solar radiation pressure. The gravitational force can be described in its simplest form as

$$F_g = \frac{GM_\odot m_d}{r_h^2} \tag{4.65}$$

where $GM_\odot$ is the geopotential of the Sun and $m_d$ is the particle mass. It should be noted that the Newtonian approximation for studies of dust particle motion will break down close to the Sun in, for example, investigations of the outgassing of the Kreutz family comets.

The dust particle mass can be expressed in terms of its radius under the assumption of spherical particles according to the trivial equation

$$m_d = \frac{4}{3} \pi a^3 \rho_a \tag{4.66}$$

where $\rho_a$ is the particle bulk density. Solar radiation pressure is an opposing force and, by use of de Broglie's hypothesis in combination with the solar flux and assuming spherical particles, we can obtain

$$F_{pr} = \frac{\pi a^2 L_\odot Q_{pr}}{4\pi r_h^2 c} \tag{4.67}$$

where $Q_{pr}$ is a radiation pressure efficiency factor that is related to the scattering properties of the particles. The radiation pressure cross-section, $Q_{pr}$, can be defined as

$$Q_{pr} = Q_{abs} + Q_{sca}(1 - \langle \cos \Phi \rangle) \tag{4.68}$$

where $<\cos \Phi>$ is the average of the cosine of the scattering angle weighted with the scattering function. This can be expressed as

$$\langle \cos \Phi \rangle \equiv 2\pi \int_0^\pi \Phi_s(\theta) \sin \theta \cos \theta d\theta \tag{4.69}$$

where $\Phi_S$ has been normalised as shown above. For an individual particle, $Q_{pr}$ is a function of the wavelength of the incoming light through the changing size parameter and hence to compute the radiation pressure force, an integration over the illuminating flux must be performed, i.e.,

$$F_{pr} = \int_0^\infty \frac{\pi a^2 Q_{pr}(\lambda)}{c} \frac{F_\odot(\lambda)}{r_h^2} d\lambda \; . \tag{4.70}$$

Similarly, an effective radiation pressure efficiency over the solar spectrum can be computed by "averaging" over the solar spectrum. This can be calculated analytically by assuming the solar spectrum is a black-body at 5770 K and can be written as

$$\langle Q_{pr} \rangle = \frac{\int_0^\infty Q_{pr}(\lambda) F_\odot(\lambda) d\lambda}{\int_0^\infty F_\odot(\lambda) d\lambda} = \frac{\pi}{\sigma T^4} \int_0^\infty Q_{pr}(\lambda) B_\lambda d\lambda \tag{4.71}$$

where $B_\lambda$ is the non-normalised Planck function at a temperature of $T$ (Silsbee and Draine 2016). For crude calculations, $<Q_{pr}>$ can be assumed to be 2. Once again this equation becomes invalid for Sun-grazing comets when the Sun can no longer be considered as a point source.

### 4.4.2   The Fountain Model

The ratio of $F_{pr}/F_g$ is known as $\beta$ and is given by

$$\beta = \frac{3L_\odot Q_{pr}}{16\pi c G M_\odot \rho_d a} \tag{4.72}$$

where we have written the equation in terms of particle bulk density, $\rho_d$, rather than mass and have dropped the averaging brackets for $Q_{pr}$. $\beta$ indicates how effective solar radiation pressure is compared to gravity. Values of between 0.5 and 2 have been presented in the literature from analyses of cometary dust tails.

Physically, particles experience an initial acceleration from the gas drag. This acceleration is, in general, towards the Sun so that particles acquire a sunward velocity relative to the nucleus close to the source. Deceleration relative to the nucleus then occurs as a result of radiation pressure, the acceleration in the cometocentric frame being given by

$$\frac{dv_d}{dt} = -\frac{GM_\odot}{r_h^2} \beta \tag{4.73}$$

Eventually, the sunward component of the initial velocity is reduced to zero and begins to be reversed in the cometocentric reference frame. The relative motion of the particles with respect to the nucleus then becomes anti-sunward and the particles are further accelerated to form the dust tail in the anti-sunward direction.

A particle emitted from the nucleus in the sunward direction and rapidly accelerated to a velocity, $v_{d0}$, along the Sun-comet line will reach a maximum distance, $E_{ap}$, from the nucleus in the sunward direction given by the equation

$$E_{ap} = \frac{v_{d0}^2 r_h^2}{2\beta GM_\odot} \tag{4.74}$$

which can be derived from simple mechanics in combination with the equation for $\beta$ (see also Grün and Jessberger 1990).

The equations above form the basis for the fountain model which is attributed to Eddington (1910) (see e.g. Mendis and Ip 1976). In the fountain model, particle emission is studied in a cometocentric system. If the motion of the comet around the Sun is ignored, then the density at each point in the coma can be calculated analytically as shown by Divine et al. (1986). In this simplified case, the model predicts that at any point there are zero, one or two solutions for particles trajectories to reach that point. In the case of two solutions, one trajectory is a "direct" trajectory from the nucleus, whereas the other solution is "reflected" in that the particle reaches the point after having the sunward component of its velocity reversed.

The analytical solution is very convenient and can also be used for distributions of particles by combining results for several values of $v_{d0}$ and $\beta$. However, the situation does become more complex for time-dependent solutions including variable production rates with rotation and motion about the Sun when the direction of the radiation pressure force changes with time in the inertial cometocentric coordinate frame. As a result of this and the increase in computing speed, numerical integrations of the trajectories are of interest. Examples are shown in Fig. 4.20. The non-linear colour table should be noted. Solar radiation pressure is applied from the right. Case a (lower left) is the simplest case and shows the result of isotropic emission from a point source at a constant velocity. The viewing direction is from a remote point orthogonal to the comet-Sun line. The unevenness of the colour contours is purely statistical resulting from the use of only 200,000 test particles.

A direct comparison can be made with case c. Here, the only difference is an increase of the $\beta$ value by a factor of 4. The acceleration from solar radiation pressure is much higher thereby compressing the dust envelope in the sunward direction.

Case b is similar to case a but here only the dayside is active with a distribution corresponding to the cosine of the solar incidence angle—an insolation-driven case. Finally, case d shows the same model but with the observer being only 60° from the Sun-comet line. Note that there is significant dust column density above the nightside seen in projection.

The contribution of reflected particles to the dust column density on the nightside of the nucleus is small but can be significant for some applications. We have seen earlier how the gas flow field leads to significant gas densities above the nightside. Entrainment of the dust particles in the flow field leads to the dust being present on the nightside as well. The evidence of lateral flow of dust from the dayside to the nightside was apparent in Giotto images of 1P/Halley, for example, and a dayside to

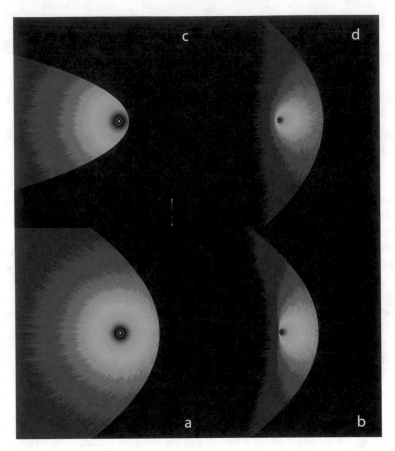

**Fig. 4.20** Fountain models for different input parameters drawn at the same scale. (**a**) $\beta = 0.5$, isotropic emission, viewed orthogonal to the Sun-comet line. (**b**) As a. but with only dayside emission. (**c**) As a. but with $\beta = 2.0$. (**d**) As b. but viewed at 60 degrees from the Sun-comet line

nightside dust brightness ratio of 3.2:1 was observed in this case. But how much of that dust came directly from the nucleus when compared to the dust on reflected trajectories? We can use the fountain model to indicate when this effect becomes important.

In Fig. 4.21 (left) the dust column density along the Sun-comet axis for case b in Fig. 4.20 is shown. The column density above the nightside is clearly evident and drops relatively slowly with distance. The decrease in column density from the nucleus towards the Sun is much steeper. This of course leads to the ratio of reflected particles to those on direct trajectories changing with distance from the nucleus. An example is shown in Fig. 4.21 (right) where we can see the relative contribution of reflected particles to the total column density in the anti-sunward direction for the isotropic case (case a).

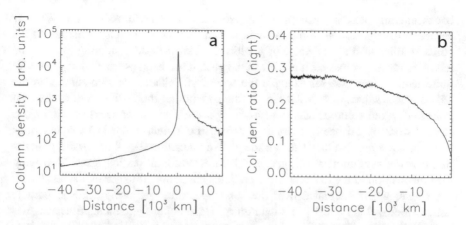

**Fig. 4.21** Left: The dust column density for a fountain model where only the dayside is active. The profile is along the Sun-comet line with the line of sight orthogonal to that line. The positive x-axis is towards the Sun. The dust column density drops to zero at around 16,000 km from the nucleus. The column density in the anti-Sun direction arising from radiation pressure is small but significant. Right: The ratio of the column density arising from reflected particles to the total for a numerically-computed isotropic fountain model. Only the ratio in the anti-solar direction is shown. Close to the nucleus, direct particles dominate. However, the further one gets from the nucleus, the more reflected particles are seen

Clearly, the contribution of reflected particles is significant. However, very close to the nucleus the particles on direct trajectories are dominant. In our case a (isotropic emission), 100 km from the nucleus, the contribution from reflected particles is <3% of the column density and is smaller as the direct contribution increases with $1/r$ towards the nucleus. Even if direct emission in the nightside direction is strongly reduced with respect to an isotropic emission distribution (e.g. a factor of 10) the contribution of reflected particles to the column density within 10 km of the nucleus is only of the order of a percent.

Equation (4.72) shows how $\beta$ depends on $a$ and $Q_{pr}$ and this in turn influences the distance to the apex of the fountain, $E_{ap}$. This indicates that the average particle properties within a column changes as one moves towards the Sun. Changes in $\beta$ are often thought of as being changes in particle size but this is not completely true because while $Q_{pr}$ is size dependent, the radiation pressure efficiency also depends on composition through the refractive index of the particles. This can then lead to modification of the bulk scattering properties within the extended coma.

### 4.4.3   The Dynamics of Fluffy Particles

We have seen that the nature of the particle structure influences the scattering properties. It also effects the dynamics. For example, particles that are strongly forward scattering are less influenced by radiation pressure while a strong backscattering peak suggests a higher radiation pressure cross-section. We saw earlier in the discussion of the fountain model that the large scale structure of the coma is influenced by the value of $\beta$ which is related to the scattering through the radiation pressure cross-section, $Q_{pr}$. In the past, it has been assumed that $\beta$ can take quite large values (see for example the works of Sekanina and co-authors in the 1980s where values of $\beta > 2.0$ can be found) and that this is the result of the dust grains being more porous than solid spheres producing a larger backscattering peak. Recent work by Silsbee and Draine (2016) using T-matrix and DDA calculations suggest, however, that this is not an adequate explanation. For fluffy grains with the properties of astronomical silicates, $\beta < 1.0$ is seen for all particle sizes. Modifying the composition of the material in the particles has some influence on the results but even here, when integrated over the solar spectrum and using realistic particle compositions, $\beta$ is typically less than unity. Hence, there remain significant issues in resolving inconsistencies between measurement and theory.

The fluffy nature of particles is also important for the dust-gas interaction during the initial ejection and acceleration of particles because of its relationship to the drag coefficient, $C_D$ (Eq. 4.87) as we shall see in Sects. 4.6.1 and 4.7.1.

## 4.5   Dust Size Distributions

Within the equations for radiation pressure, we see the particle size, $a$, appearing. It will also appear shortly in the equations for gas drag close to the nucleus and we have already seen it in the equations for the radiance from the dust coma (Eq. 4.43). But what is the particle size? Incorporation of some form of continuous size-frequency distribution (SFD) of particle sizes into cometary studies was foreseen well before the Halley fly-bys and various mathematical forms have been proposed. In general, these equations are some variant on a power law distribution so that

$$n_d(a) = k_p \, a^{-b_d} \tag{4.75}$$

where $n_d$ is the number density of particles with a size, $a$, and $k_p$ is a proportionality constant. Note that the power law description can only hold above a certain size threshold because the density diverges as $a \to 0$ if $b_d > 1$. In our physical case this threshold ($a_{min}$) must be larger than, for example, molecules. Thus to be accurate we must say that for this model "the tail of the distribution follows a power law" (Virkar and Clauset 2014). The power law can also be thought of as a probability distribution and $k_p$ then becomes a normalization constant and can be computed if $a_{min}$ is defined using the equation

$$k_p = (b_d - 1) \, a_{min}{}^{b_d - 1} \tag{4.76}$$

It is usual to express the number of particles, $dn_d$, within a size interval, $da$, so that

$$\frac{dn_d(a)}{da} = k_p{}'' a^{-b_d-1} \tag{4.77}$$

where $k_p{}''$ is another proportionality constant. This is the differential particle size distribution within a linear bin, $da$.

The sensitivity of experiments to particle masses typically results in these experiments being able to determine particle number densities over a very large range of masses (sizes) often covering many orders of magnitude. Consequently, log plots are most appropriate. We can see that taking the log of $n_d$ in Eq. (4.77) results in

$$\frac{d \log n(a)}{d \log a} = -b_d \tag{4.78}$$

so that the slope of the SFD can be found by linear regression of the log of the particle number density against the log of the particle size.

With dust particle counters usually producing discrete events, binning of the data into size bins is performed. If these size bins are logarithmic then plotting the log of the number per bin against the particle size in each (logarithmic) bin gives a slope of $1-b_d$.

Finally, cumulative SFDs are often used so that we require an integral of the particle densities for sizes above a certain value. We can express this as

$$n_c(a > a_c) = \int_{a_c}^{\infty} k_p\, a^{-b_d}\, da = k_p' a_c{}^{1-b_d} \tag{4.79}$$

under the physically realistic assumption that $b$ is positive and greater than 1. Hence, if bins are linear, i.e., $a = 1$–$2$, $2$–$3$, $3$–$4$, ..., and data have a power-law dependence, $n_d(a) \sim a^{-bd}$ then the resulting cumulative distribution will be $n_c \sim a^{-bd+1}$ and the exponents differ by 1. The change in the exponent (from $-b_d$ to $1-b_d$) can lead to considerable confusion unless it is clear what quantity is actually being presented. On the other hand, if bins are logarithmic, i.e., $a = 1$–$2$, $2$–$4$, $4$–$8$, ..., and data have a power-law dependence, $n_d(a) \sim a^{-bd}$, then the resulting cumulative distribution will be $n_c \sim a^{-bd}$. The exponents are the same.

In the general case, the detection and characterization of power laws is complicated by the large fluctuations that occur in the tail of the distribution, i.e. the part of the distribution representing large but rare particle sizes, and also by the difficulty of identifying the range over which the power law behaviour is valid. Commonly used methods for analysing power-law data, such as least-squares fitting, can produce substantially inaccurate estimates of parameters for power-law distributions because reliance on identifying a minimum gives no indication of whether the data obey a power law at all (Clauset et al. 2009). Furthermore, least-squares fitting algorithms are often used to determine the slope of the SFD which gives substantial weight to bins with low frequencies and can be bin size dependent where bin "placement" is ambiguous.

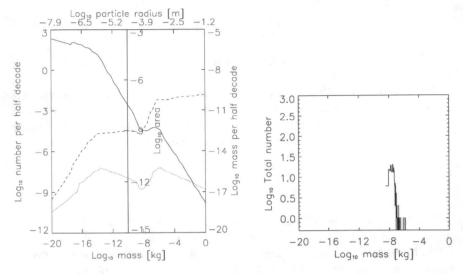

**Fig. 4.22** Left: The dust size distribution in the coma derived from the PIA and DIDSY observations made from the Giotto spacecraft and normalised to 600 km (re-drawn after McDonnell et al. 1991). Right: The 269 GIADA measurements of particle masses made during the Rosetta mission. Note that the axes have been deliberately chosen to make a comparison with the McDonnell work. Although the x axis is the same, the y axis only covers three decades

On the other hand, the SFD is important because the value of the exponent, $b_d$, implies certain properties of the dust coma. The cross-sectional area of the particles and their mass is of course a function of their size and we can crudely use $\pi a^2$ and $4\pi a^3 \rho_d/3$ for cross-sectional area and mass respectively for illustration purposes. The total cross-sectional area at one particle size is then

$$\sigma_p(a) = k_p \pi \, a^{-b_d+2} \tag{4.80}$$

and the total mass at one size is then

$$m(a) = k_p \frac{4}{3} \pi \, \rho_d a^{-b_d+3} \tag{4.81}$$

with $k_p$ being a proportionality constant. If these equations specify the emitted size distribution and if $b_d > 3$ then the mass loss is dominated by the largest particles in the distribution and if $b_d > 2$ then both the mass and the cross-sectional area, which is related to the scattering area, are dominated by large particles. In Fig. 4.22 left, we can see the dust size distribution (and associated area and mass plots) from measurements made by the DIDSY and PIA experiments on Giotto (McDonnell et al. 1991) which remains the most comprehensive data set for this purpose. The plot shows that the quantities as the total (number density, cross-section and mass) within a size interval equal to half a decade as a function of the logarithm of the mass

**Table 4.1** Power law exponents for particles in the range 30 μm – 1 mm measured at 67P by Merouane et al. (2017). The bold indicates when the Sun was over the southern hemisphere of the comet; the italics indicate when over the northern hemisphere. These are compared to the values of $b_d$ for 1P/Halley, 26P/Grigg-Skjellerup, and 81P/Wild 2 compiled by Price et al. (2010)

| Comet | Times | Orbital position | Power law exponent, $b_d$ |
|---|---|---|---|
| 67P | *Aug. 2014 –May 2015* | *Encounter to pre-perihelion equinox* | *1.8 ± 0.4* |
|  | **May 2015–Aug. 2015** | **Pre-perihelion equinox to perihelion** | **2.8 ± 0.9** |
|  | **Aug. 2015–Apr. 2016** | **Perihelion to post-perihelion equinox** | **2.1 ± 0.5** |
|  | *Apr. 2016–Sept. 2016* | *Post-perihelion equinox to end of mission* | *1.6 ± 0.5* |
| 1P | See Table 2 in the Preface | | 2.6 ± 0.2 |
| 26P | | | 0.93 |
| 81P | | | 1.89 |

bin normalised to 600 km from the nucleus. Integration over the size distribution then shows the importance of determining $b_d$. It should, however, be remembered that these data are for one comet and its applicability to other comets and at other heliocentric distances is not given.

A variation in the size distribution with heliocentric distance and when pre- and post-perihelion observations were compared, was noted by Merouane et al. (2017) in measurements by Rosetta/COSIMA at 67P (Table 4.1). These data were acquired over a narrower particle size range (30 μm–1 mm) and in this case the particle diameters were measured and not the particle masses as in McDonnell et al. At sizes smaller than 30 μm, the detection of particles is biased as their size reaches the limit of detection of COSIMA (Merouane et al. 2017). In comparing these values with McDonnell et al.'s observations, assumptions have to be made regarding the density of the particles which is probably a strong function of particle radius. Hornung et al. (2016) have estimated a dependence proportional to $a^{-2.5}$, for example.

The GIADA experiment on Rosetta measured the momentum of particles impacting a target (Colangeli et al. 2007). A time-of-flight system was used sporadically to determine the velocity of the particles pre-impact allowing determination of the mass to an estimated accuracy of 20%. Over the course of the mission, 269 particles could be measured (~3–4 orders of magnitude less than predicted). A histogram is shown in Fig. 4.22 right with an axis scaling designed to make a comparison with the results from the 1P/Halley encounters. The restricted mass range is a consequence of the difficulty in constructing a low mass instrument to determine particle masses at very low relative velocities. However, it is interesting (but perhaps co-incidental) that the peak seen in the distribution is at a mass similar to the local maximum seen in the 1P/Halley data.

The limitations in mass range of the GIADA data are well illustrated in Fig. 4.23 (left) which shows the masses and velocities of the particles detected (for all particles that passed through the Grain Detection System giving the velocity and were also

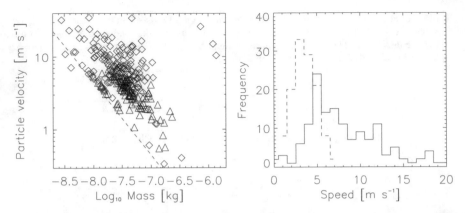

**Fig. 4.23** Left: GIADA measurements of mass and velocity for individual particles using the Grain Detection System (a laser curtain) for the velocity and the Impact Sensor for the momentum. Particles detected beyond and closer than $r_h = 2.5$ AU are shown as triangles and diamonds respectively. Right: Histograms of the detected particle speeds. The solid line shows particles detected at $r_h < 2.5$ AU. The dashed line particles detected at $r_h > 2.5$ AU

detected by the Impact Sensor giving the momentum). The dashed line is a line of constant momentum and one can see the detection threshold clearly. However, the histograms in Fig. 4.23 right show that one can use the data to investigate several dependencies. In this example, particles seen inside a heliocentric distance of 2.5 AU have a broader, faster velocity distribution when compared to particles seen beyond 2.5 AU which probably indicates the detection of newly produced particles leaving the nucleus roughly radially. The narrower lower velocity distribution for particles detected at $r_h > 2.5$ AU may be older residuals (cf Della Corte et al. 2015). These also have higher masses when looking at Fig. 4.23 left.

The SFD in the GIADA case cannot be described by a power law but appears rather as a peaked distribution with a maximum at around 200–300 μm in radius (depending upon the chosen density). This has led to significant controversy about the size distribution and whether or not the inner coma dust size distribution can be assumed to be solely made up of particles of this size. There is clearly inconsistency between the GIADA result and the power law distributions derived from COSISCOPE but there are obviously observational selection effects resulting from the dynamic range of the instruments. As can be seen from Fig. 4.22 (left), the peak optical scattering efficiency (radius ~ 1 μm) corresponds to a peak in the cross-sectional area function derived from 1P/Halley measurements. There was no instrument on Rosetta capable of providing a size distribution for particles <30 μm in diameter. However, Mannel et al. (2019) detected particles with the MIDAS atomic force microscope that were ≲10 μm in diameter and therefore at least a factor 2–3 smaller than those seen by COSISCOPE and more than an order of magnitude smaller than seen by GIADA. Hence, particle sizes down to at least $a = 5$ μm were present in the coma.

Studies of crater statistics also use SFDs and it is interesting to note that this field is moving away from use of least-squares fits to more stringent statistical methods (Robbins et al. 2018). Maximum likelihood estimators would appear to be a more robust approach because there is no dependence upon how the data are binned. Robbins et al. (2018) recommend use of the truncated Pareto distribution which has the form

$$f(D) = \frac{b_d \, D_{min}^{b_d} \, D^{-b_d-1}}{1 - (D_{min}/D_{max})^{b_d}} \quad \text{for } D \in [D_{min}, D_{max}] \qquad (4.82)$$

(this equation can be found in many texts including Wikipedia) where $D_{min}$ and $D_{max}$ are the minimum and maximum sizes possible in the SFD, respectively. For a distribution, the maximum likelihood estimate for $b_d$ can be found by iteration on the equation

$$\frac{N}{\widehat{b}} + \frac{N(D_{min}/D_{max})^{\widehat{b}} \ln (D_{min}/D_{max})}{1 - (D_{min}/D_{max})^{\widehat{b}}} - \sum_{i=1}^{N} (\ln D_i - \ln D_{min}) = 0 \qquad (4.83)$$

where the circumflex over the parameter, $b$, indicates that this variable should be varied until the equation becomes valid to a specifiable tolerance.

Robbins et al. (2018) also provide the uncertainty on this parameter, $\lambda_b$, as

$$\lambda_b = \sqrt{2} \, \text{erf}^{-1}(\text{CI}) \, \frac{A_{var}}{N} \, S_b^{\,2} \qquad (4.84)$$

where CI specifies the confidence interval that is required, $A_{var}$ is defined through

$$A_{var} = \left( \frac{1}{b^2} - \frac{(D_{min}/D_{max})^{b_d} (\ln (D_{min}/D_{max}))^2}{\left(1 - (D_{min}/D_{max})^{b_d}\right)^2} \right)^{-1} \qquad (4.85)$$

and $S_b$ is given by

$$S_b = \frac{D_{min}^{b_d} \ln\left(D_{min}\right)\left(d^{-b_d} - D_{max}^{-b_d}\right)}{1 - \left(\frac{D_{min}}{D_{max}}\right)^2}$$

$$+ \frac{D_{min}^{b_d}\left(D_{min}/D_{max}\right)^{b_d}\left(d^{-b_d} - D_{max}^{-b_d}\right)\ln\left(D_{min}/D_{max}\right)}{\left(1 - \left(\frac{D_{min}}{D_{max}}\right)^{b_d}\right)^2}$$

$$+ \frac{D_{min}^{b_d}\left(D_{max}^{-b_d}\ln\left(D_{max}\right) - d^{-b_d}\ln d\right)}{1 - \left(\frac{D_{min}}{D_{max}}\right)^2} \tag{4.86}$$

which is derived in full in an appendix of Robbins et al.

Although this approach has not been used to this point on dust SFDs for comets, it clearly warrants attention.

## 4.6   The Lifting Dust Ejection Process

### 4.6.1   Drag Force at the Nucleus Surface

So far we have looked at force-free radial outflow and the effects of radiation pressure on particles far from the nucleus. We also have a size frequency distribution but we have not looked at the emission process or the mechanism by which particles are accelerated from the nucleus surface.

It is clear that there must be a minimum particle size for the dust size distribution. But what is the maximum dust particle size? Do larger and larger particles leave the nucleus—just not very frequently? The usual approach to addressing this problem has been to look at the largest particle size gas drag can lift from the nucleus surface by first calculating the drag force on a particle.

In its simplest form, the drag force on an object in a fluid can be written generally as the standard drag equation

$$F_D = \frac{1}{2}C_D\rho_d\pi a^2\left|v_g - v_a\right|\left(v_g - v_a\right) \tag{4.87}$$

where $F_D$ is the drag force, $\rho_d$ is the gas mass density, $v_g$-$v_a$ is the relative velocity of the dust particle of size, $a$, in the gas flow, $C_D$ is the drag coefficient and $\pi a^2$ is the cross-sectional area of a spherical particle. Note that we use $v_a$ for the velocity of a single dust particle. The drag coefficient is an important element but can be set to 2 (as is appropriate for smooth spheres) if further simplification is required. However, in reality, the complexity is higher because $C_D$ is not a constant for a particular body but depends upon the Reynolds number (Re). For laminar flow (Re $< 10^6$), the drag coefficient is markedly lower than if the flow is turbulent. $C_D$ is therefore often set through the equation

**Fig. 4.24** Values for the drag coefficient, $C_D$, as a function of s for three values (0.1—solid, 1.0—dashed, 10.0—dot-dash) of the ratio of $T_d/T_g$. After Zakharov et al. (2018)

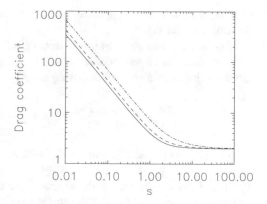

$$C_D = \frac{2s^2 + 1}{\sqrt{\pi}s^3}e^{-s^2} + \frac{4s^4 + 4s^2 - 1}{2s^4}\operatorname{erf}(s) + \frac{2(1 - \varepsilon_s)\sqrt{\pi}}{3s}\sqrt{\frac{T_d}{T_g}} \qquad (4.88)$$

where

$$s = \frac{|v_g - v_a|}{\sqrt{\frac{2kT_g}{m_g}}} \qquad (4.89)$$

(Baines et al. 1965) where the parameter $\varepsilon_s$ accounts for specular reflection from the surface of a sphere. Solutions to this equation show (Fig. 4.24) that if $s < 1$, $C_D$ rises exponentially. Hence if the gas is warm or if the differential velocity of the dust is low, $C_D$ can be appreciably higher than 2. A further complexity arises from the possibility that emitted particles are rotating fluffy aggregates which leads to a requirement that an "average" $C_D$ is defined. The larger geometric cross-section to mass ratio implies that the drag acceleration of the particles will be higher for these particles leading to higher terminal velocities.

The drag force, $F_D$, can be used to solve the equation of motion of a dust particle within the flow field numerically under the assumptions that dust-dust collisions are negligible and that non-LTE effects within the innermost coma (i.e. in the Knudsen layer) can also be neglected. These simplifying assumptions are usually adequate. For example, we can see from the optical depth that, in weakly active comets, dust-dust collisions should be negligible. The assumption of LTE is less straightforward.

In Fig. 3.21, we can see that the gas velocity distribution function (VDF) is non-Maxwellian close to the source and can remain so if the mean free path is sufficiently long. This gives rise to the question of whether the drag force, $F_D$, is influenced by the VDF as the gas velocity goes directly into the equation for $F_D$ and also into the equation for the drag coefficient, $C_D$. Finklenburg et al. (2014) made some initial studies of this problem and showed that a conical VDF does increase the drag coefficient but the magnitude of the effect is (perhaps surprisingly) small. She concluded that it can be ignored given the size of the other uncertainties in gas dynamics modelling.

If we assume that gas drag in a uniform gas flow field is responsible for lifting gas particles then the largest liftable radius, $a_m$, can be computed from the equation of motion using drag force and opposing gravity. This can be written as

$$\frac{4\pi}{3}\rho_d a^3 \frac{dv_a}{dt} = \frac{1}{2}a^2 \pi C_D \rho_g \left(v_g - v_a\right)^2 - \frac{4\pi}{3}\rho_d a^3 \frac{GM_N}{r^2} \tag{4.90}$$

where $\rho_d$ is the dust bulk density, $GM_N$ is the standard gravitational parameter for the nucleus, $v_g$ and $\rho_g$ are the gas velocity and density, respectively (Gombosi et al. 1986) and $r$ is the distance to the centre of the nucleus. A simple solution exists by balancing the gas drag force at the surface and the surface gravity. $a_m$ is then given by

$$a_m = \frac{3\,C_D\,Z\,v_o R_N^2}{8 G M_N\,\rho_d} \tag{4.91}$$

where $R_N$ is the nucleus radius, $v_o$ is the velocity of the gas at the surface and $Z$ is the gas flux (Eq. 2.35). This is identical to the equation given by Harmon et al. (2004) where it is expressed as

$$a_m = \frac{9\,C_D\,Z\,v_{th}}{32\pi G R_N \rho_N \rho_g} \tag{4.92}$$

where $v_{th}$ is the thermal expansion velocity of the gas at the surface multiplied by a correction factor to allow for expansion effects. This factor is 9/4 in Finson and Probstein (1968). $\rho_N$ is the density of the nucleus.

Huebner (1970) gives for the fluid dynamic limit

$$a_m = \sqrt{\frac{27\,\mu_v\,v_{th}}{8\pi G R_N d_n d_g}} \tag{4.93}$$

where $\mu_v$ is the viscosity given by

$$\mu_v = \frac{1.85\ 10^{-6}\ T^{1/2}}{1 + 680/T} \tag{4.94}$$

which is traceable back to Sutherland's law for the relationship between the dynamic viscosity and the temperature for an ideal gas. A derivation including the centripetal acceleration is included in Huebner et al. (2006) page 49.

It should be recalled that the acceleration derived here ignores local effects. As discussed earlier (Sect. 2.9.3.2), the Fanale and Salvail (1984) expression for the acceleration included a term accounting for the ratio of tortuosity and porosity—effectively providing locally modified accelerations so that $a_m$ is not a rigorous upper limit. However, Eq. (4.92) does provide the notable (and unsurprising) result that the largest liftable mass is proportional to the gas flux at the surface. As this varies strongly with heliocentric distance, it implies that the particle distribution function in the coma will change significantly as the comet approaches and recedes from the Sun. Given that the gas production rate of 67P increased by two orders of magnitude from 3 AU to perihelion, Eq. (4.92) indicates that the largest liftable mass should increase proportionally. This has also been clearly illustrated in the context of numerical model calculations by, for example, Marschall (2017) (see also Thomas et al. 2015b; Fig. 11).

### 4.6.2   Cohesive Forces

The equations for the maximum liftable mass/radius given above exclude any cohesive forces between the masses being lifted and the surface they are initially attached to. The significance of cohesive forces was recognized more than 20 years ago (Kührt and Keller 1996) but it remains a difficult subject because the surface structure is essentially unknown. One can imagine two extremes. Either the particle to be emitted is coupled directly to the non-volatile surface so that cohesive forces are at a maximum or the particle is coupled to the surface via a sort of ice "bridge" which releases the particle from the surface by sublimation without any cohesive force. These two possibilities are illustrated schematically in Fig. 4.25.

In the case of the subliming ice bridge, the equation for maximum liftable mass holds. However, the cohesive forces, if the particle is in contact with non-volatile material, can be of high magnitude. Representative values are shown in Table 4.2.

**Fig. 4.25**  On the left, a particle is in direct contact with a flat smooth surface. Cohesive forces will provide a strong link between the particle and the surface. On the right, an ice bridge between the surface and the particle exists. If this ice bridge sublimes, the particle may only experience the weaker gravitational force allowing it to be ejected more easily by gas drag

**Table 4.2** Comparison of values for van der Waals (vdW) forces to the gravitational force on a typical cometary nucleus

| Acceleration | Value | Comment/Notes |
|---|---|---|
| Surface gravitational force | $8.12 \ 10^{-14}$ N | $GM_N = 620$ m$^3$ s$^{-2}$ <br> $R_N = 2$ km <br> 100 micron particles |
| Opposing vdW force | $1.56 \ 10^{-6}$ N | With a Hamaker constant of $3 \ 10^{-20}$ J (Eq. 4.95). |
| Opposing vdW force | $1.80 \ 10^{-8}$ N | Formulation of Scheeres et al. (Eq. 4.96). |

The simplest computation of the van der Waals force is given by

$$F_{vdW} = \frac{Ha}{6z_0^2} \tag{4.95}$$

where $H$ is the Hamaker constant (typically $3 \ 10^{-20}$ J), $a$ is the grain size (radius) and $z_0$ is the particle to surface distance and often assumed to be around 0.4 nm. This equation gives values for the cohesive force that are seven orders of magnitude larger than the gravitational force at a comet such as 67P and only comparable to the gravitational force if the particle-surface separation exceeds 1 micron. It also far exceeds the drag force for reasonable assumptions about the local gas production rate. However, this equation applies to a dust particle on a flat smooth surface—which is clearly not applicable at a comet. The key question, though, is how much this force is reduced by the specific conditions. There are several items that are poorly understood at this point. For example:

- The cross-sectional area of the contact points between surface particles is unknown. This issue is similar to the problem of the thermal conductivity of porous, fragile structures.
- The influence of torque on the (probably) highly fragile particles is unknown.
- Saffman lift force is caused by the sharp gradient in the fluid velocity above a particle bed, which creates a lower pressure above the particle than below it as a consequence of the Bernoulli effect (Kok et al. 2012). This can lower the effective cohesive force.
- The effects of local turbulence may be strong.

Scheeres et al. (2010) suggested use of the equation

$$F_C = 0.9 \ 10^{-2} S_c^2 a \tag{4.96}$$

where $S_c$ is a numerical constant approximately equal to 0.1, to compute the cohesive forces in lunar regolith and argued that this will underestimate the van der Waals force for particles on asteroids or in micro-gravity. But even here the cohesive force can far exceed the drag force at the surface. Given the magnitude of the forces involved, it might be considered surprising that there is any dust loss at all!

### 4.6.3   Advanced Dust Ejection Concepts

The difficulty in overcoming cohesive forces leads one to consider if there are other mechanisms at work. There are several ideas that have not really been explored. I give three examples. Firstly, there is a widespread belief that pressure cannot build up underneath particles because the material is porous. This is probably incorrect as illustrated in Fig. 2.87. Even if the surface layer is highly porous, there can be a pressure gradient across it. It is merely a question of how far below the actual surface the subliming front has to be to produce sufficient pressure to rupture the surface. If sub-surface pressure can build up in this way then small-scale quasi-explosive events may eject material. Agarwal et al. (2017) have invoked this idea, without substantiating the exact mechanism, in their explanation for an outburst recorded at 67P in July 2016. Attainable pressures have been calculated in an idealised system by Skorov et al. (2017) using a numerical simulation. They showed that water vapour pressures close to 20 Pa (and thus comparable to the tensile strengths estimated for the bulk material at 67P) can be reached at 5 mm depth if the particle size of the non-volatile material is sufficiently small.

Secondly, the importance of the "super-volatiles", CO and $CO_2$, may be underestimated. Sub-surface pressure build-up may be driven by the super-volatiles in some way rather than relying solely on $H_2O$. In this respect, 103P/Hartley 2 may be an extreme example where the super-volatile is so dominant that the water ice has no chance to sublime during the ejection process. Skorov et al. (2017) also demonstrated that this may be of significance by studying $CO_2$ and CO. They showed (Fig. 4.26) that even higher internal pressures than those seen for water vapour may be reached with these super-volatiles.

Finally, ejected particles may just be large enough so that coherent forces are low. However, this requires explanation of the measured particle size distribution, which for 1P/Halley contained particles right down to the 10–100 nm size range (Fig. 4.22), and an understanding of whether these large particles can be accelerated sufficiently to produce the observed coma.

What the above discussion shows is that we are rather ignorant of the whole ejection process and that additional effort and experiment appears to be necessary (Thomas et al. 2019). This has been recognized in the past few years and some progress has been made although we remain far from having clear answers. Indeed, there may be processes at work that we have not yet identified. An example of the challenges is given in Fig. 4.27 which is from the work of Poch et al. (2016a, b) and is a schematic diagram of an observation made in the laboratory.

Water ice and a non-volatile organic tholin were mixed in two different ways. In the first method, liquid water was mixed with the tholin before being sprayed through a nebuliser and immediately frozen. The tholins were contained within the ice particles in an "intra-mixture" (Fig. 4.27 left). In the second method, the ice particles were produced first from pure liquid water and then simply mixed with the tholins. This was referred to as an "inter-mixture". The two types of sample were then allowed to evolve by illuminating their surfaces in a vacuum chamber. The

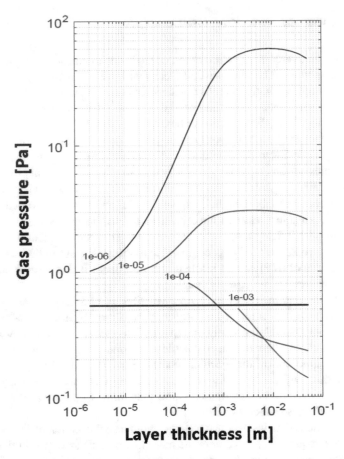

**Fig. 4.26** The $CO_2$ gas pressure below thicknesses of an ice-free porous layer composed of different particle sizes. Each particle size is given by a different line with its size in [m] marked next to it (From Skorov et al. 2017)

resulting phenomena were remarkably different. The intra-mixture sample produced a cohesive foam-like mantle. Foamy large (cm-sized) chunks were then ejected from the surface against the retarding force of Earth's gravity. The inter-mixture produced a less cohesive and more compact mantle on the surface. The water ice sintered (Fig. 2.37) in the sub-surface generating a harder layer. This shows that the physical properties of layers produced by heating water ice and organic materials can be very different even if the bulk composition is held constant and depends upon the way the substances are arranged at microscopic scales.

These tests have also shown how large amounts of material, including larger particle sizes, might leave a nucleus surface. In Fig. 4.28 (bottom), we can see two images of a test where the later image (to the right) shows that a chunk of a tholin-water ice mixture has been ejected from the surface. The upper panels of Fig. 4.28 show images taken just a few seconds apart. This mixture was more dynamic in its

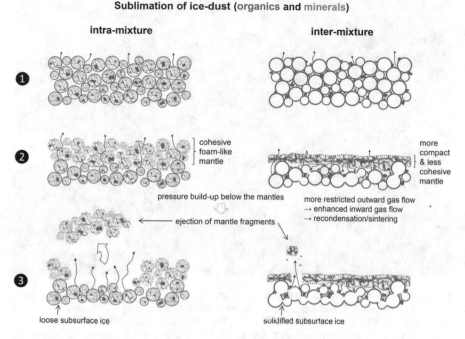

**Fig. 4.27** Schematic diagram of the behaviour of water ice-tholin mixtures. Left: The tholins are contained within the ice particles. Right: The tholins are mixed with the ice particles after the ice has been produced. The evolution of the two types of mixture under heating by illumination is markedly different. (Reprinted from Poch et al. 2016b, with permission from Elsevier)

response to illumination. The material took on a fluffy texture and could be seen moving in the gas flow arising from the sublimation of the water ice. In the image to the right, pieces of material of the order of 0.5 cm in diameter were seen to be ejected from the surface. It is to be noted that this was against the retarding force of Earth's gravity—i.e. several orders of magnitude greater than the retarding force on 67P. Consequently, the ejected size was well above that expected from calculation using Eq. (4.92).

## 4.7   The Influence of Drag on the Equations of Motion for the Dust

### 4.7.1   Analytical Solutions

The point source approximation and the fountain model have their uses but the two approaches require us to assume a dust particle velocity, $v_d$. Consequently, it is necessary not merely to look at how drag lifts the particles but also to look in detail at

**Fig. 4.28** Views of two laboratory tests using water ice and tholin mixtures. Bottom: Two images taken 1.39 hours apart. The later image (bottom right) shows that a large chunk of material has been lost from the surface (marked by the blue square). Top: Two images taken just a few seconds apart of a similar test. On the right, a chunk of material can be seen flying from the surface towards the observation port (red arrow). The surface has been disrupted at the position indicated by the yellow arrow

how gas drag influences the acceleration of the dust within 10 km of the nucleus surface.

A semi-analytical approach has been developed by Zakharov et al. (2018) which includes the drag, gravity and solar radiation pressure forces that have already been introduced. We can take Eq. (4.90) and write

$$m_d \frac{dv_a}{dt} = \frac{1}{2}\pi a^2 C_D \rho_g |v_g - v_a|(v_g - v_a) - m_d \frac{GM_N}{r^2} - \frac{\pi a^2 L_\odot Q_{pr}}{4\pi r_h^2 c} \qquad (4.97)$$

where again we are referring to a single particle size. Note that we have substituted for the particle mass and added the solar radiation pressure term. The radiation pressure term can be neglected close to the nucleus in most cases. The particle velocity is then

$$\frac{dr}{dt} = v_a \qquad (4.98)$$

Zakharov et al. have shown that, if the nucleus is assumed to be homogeneous, spherical, and uniformly active, then the equation of motion can be re-written in a dimensionless form which is extremely useful for studying gas outflow from a range of objects of different size and position in the Solar System. The basic equation is

$$\frac{d\bar{v}_a}{d\bar{t}} = \bar{\rho}_g\left(\bar{v}_g - \bar{v}_a\right)^2 C_D \, \boldsymbol{Iv} - \boldsymbol{Fu}\,\frac{1}{\bar{r}^2} - \boldsymbol{Ro}\,\cos i \qquad (4.99)$$

where the coefficients, $\boldsymbol{Iv}$, $\boldsymbol{Fu}$, and $\boldsymbol{Ro}$ are related to the physics of the gas-dust motion. $\boldsymbol{Iv}$ characterizes the efficiency of the entrainment of the particle within the gas flow, $\boldsymbol{Fu}$ characterizes the efficiency of the gravitational interaction, and $\boldsymbol{Ro}$ characterizes the radiation pressure effects. They are defined through the equations

$$\boldsymbol{Iv} = \sqrt{\frac{\gamma+1}{2(\gamma-1)}}\,\frac{3ZR_N}{\sqrt{32}\,v_g^{max}a\,\rho_d}$$

$$\boldsymbol{Fu} = \frac{GM_N}{R_N}\,\frac{1}{\left(v_g^{max}\right)^2} \qquad\qquad (4.100)$$

$$\boldsymbol{Ro} = \frac{1}{m_d\left(v_g^{max}\right)^2}R_N\,\frac{\pi a^2 Q_{pr}S}{cr_h^2}$$

where $v_g^{max}$ has already been given by Eq. (3.73) and $Z$ is the gas production rate per unit area in [kg m$^{-2}$ s$^{-1}$]. The barred quantities in Eq. (4.99) (e.g. $\bar{v}_g$) give characteristic dimensionless parameters such that $\bar{v}_g = v_g/v_g^{max}$ (Eq. 3.72), $\bar{v}_a = v_a/v_g^{max}$, $\bar{r} = r/R_N$ (see Eq. 3.69), $\bar{t} = t\,v_g^{max}/R_N$, and $\bar{\rho}_g = \rho_g/\rho^*$ (cf. Eq. 3.70 which is for number density) where $\rho^*$ and $T^*$ indicate the density and temperature at the sonic surface (see Zakharov et al. for details).

### 4.7.2   Numerical Solutions

For high resolution observations of the gas flow in the immediate vicinity of irregular nuclei, the analytical solutions become inadequate and numerical solutions are required. However, analytical solutions are fundamental in checking numerical solutions of simple cases while the numerical solutions provide some additional physical insight.

Examples of numerical solutions to the dust distribution for an insolation-driven 3D spherical nucleus are shown in Fig. 4.29. Two-D slices through the centre of the domain are shown. The gas production rate in all cases was 200 kg s$^{-1}$ (cf Fig. 3.32). It should be noted that the panels are individually scaled and differing by, for example, a factor of 30 in dust velocity, in order to use the colour table to the full extent.

The gas flow from the active dayside to the inactive nightside entrains small particles and brings them into the nightside hemisphere but the larger particles are

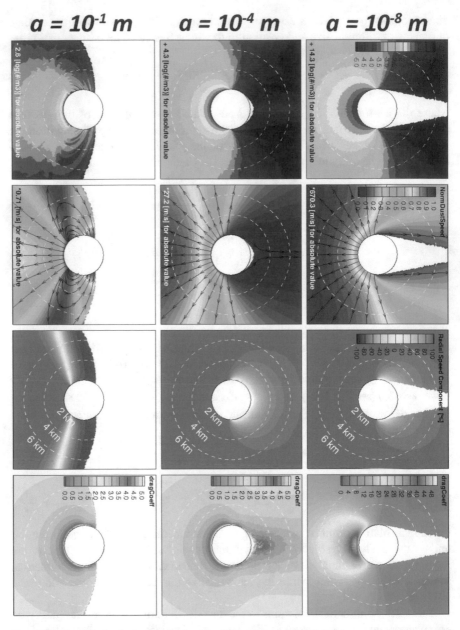

**Fig. 4.29** Numerical solutions to the dust distribution for an insolation-driven spherical nucleus of 2 km in diameter for three different particle sizes. The rows indicate (top to bottom) the local dust density, the dust speed, the radial component of the dust speed, and the drag coefficient (from Marschall 2017)

decoupled quickly so the resulting dust distribution is strongly dependent upon the particle radius. Streamlines are superposed on the dust speed panels and show that large particles can return to the nucleus under the influence of gravity to produce airfall. This verifies that the concept of airfall to produce the smooth surfaces seen in Ma'at and Ash on 67P is feasible. It is interesting to note that the dust speed of the re-impacting particles increases as one moves towards the nightside. This is in some ways obvious—particles have to be thrown out further to get as far as possible into the nightside hemisphere and thus the impact speed is maximized and approaches the escape velocity as they experience the gravitational pull for the longest duration. The plot also shows the non-radial component of the dust outflow at the terminator. Dust flow is almost parallel to the surface at the terminator for intermediate-sized and smaller particles. The variation in the drag coefficient should also be noted. The terminal velocity of the dust is clearly dependent upon the gas production rate but Marschall (2017) has shown that the dependence is $\sim 1/\sqrt{a}$ over a large range of particle sizes and that scaling according to

$$v_a \propto \sqrt{\frac{Q_g}{a}} \qquad (4.101)$$

can be employed for much of the typical parameter space. It should be clear from this that the dust size distribution at the nucleus is not the same as that in a unit volume at some point in the coma with the distribution in the coma being skewed towards larger sizes such that the exponent in the coma, $b_c$, is

$$b_c = b_d - 0.5 \qquad (4.102)$$

if $b_d$ defines the power law distribution at the surface. McDonnell et al. (1991) provided a computation for the size distribution at the surface of 1P/Halley based on semi-empirical dust velocities from Divine (1981) using their observations (Fig. 4.22). Della Corte et al. (2016) using the GIADA data determined power law relationships between the mass and velocity of detected particles and noted a phase angle dependence. Using all particles detected inside 2.5 AU with the phase angle of $<75°$ (i.e. within the dayside outflow), a least squares fit gives

$$v_a \propto m_d^{-0.21 \pm 0.06} \qquad (4.103)$$

showing a slightly steeper dependence of velocity on particle size than implied by Eq. (4.102) assuming density is independent of size over the fairly narrow range of particle masses sampled by GIADA. However, two low velocity outliers in the dataset have a very strong influence on the slope of the least squares fit so that this is not necessarily inconsistent with Eq. (4.102).

It should also be clear from this discussion that the size distribution measured in the coma will vary with heliocentric distance for large particles because of the relationship between the efficiency of gas drag and particle size. Given that

1P/Halley was very active at the time of the Giotto encounter, other comets with a similar size distribution at the surface may appear relatively depleted in these large particles when their inner comae are investigated. Hence, Eq. 4.102 will not hold for particle sizes close to the tail of the distribution.

Finally, in the acceleration region close to the nucleus (within the first 10–12 km), where small particles are being accelerated to higher velocities than larger particles, it is apparent that there will be a rapid change in the size distribution as the cometocentric distance increases.

### 4.7.3  Gas-Dust Energy Exchange within the Coma

The dust in the inner coma can influence the radiation field close to the nucleus and the gas flow. However, the dust density and distribution close to the nucleus varies with heliocentric distance, position, and level of activity. It should be apparent that when there is very little dust emission, the influence on the gas flow and the radiation field is negligible. But when does it become important?

A useful approach here is to use the dust optical depth, $\tau_d$, as a proxy. The optical depth through a part of the dust coma was given in Eq. (4.46). If the brightness of the dust close to the limb is much smaller than the brightness of the nucleus, then this indicates that the optical depth must be small. We assume here that the scattering properties of the dust are similar to those of the nucleus itself. This is clearly not true, particularly if the particle size distribution is dominated by particles close to the wavelength of the observation, but this at least allows us to assess when more detailed calculations might be needed. For the case of 67P, we have seen that $\tau_d < 0.1$ in almost all cases and in images close to perihelion (Fig. 4.17), $\tau_d$ is typically 0.05. If the gas velocity is much greater than the dust velocity, this suggests that, in this case, <5% of gas molecules have momentum exchange with the dust and hence the gas flow field is only marginally affected by the presence of dust.

This conclusion is a huge simplification and one that has usually been made for 67P even well before the Rosetta rendezvous (see, for example, Tenishev et al. 2011). Without this simplification, the back reaction on the gas arising from the dust must be calculated which results in an iterative procedure for DSMC calculations or the treatment of dust as an additional fluid in the Navier-Stokes approach (Rodionov et al. 2002). The consequences of high dust density are fairly clear. Primarily, the mass loading of the gas flow results in a slowing of the flow. In addition, the collisions should result in a more homogeneous coma and an increased gas velocity component parallel to the surface.

For comets such as 1P/Halley near perihelion, the effects of dust on the gas probably need to be accounted for. The dust reflectance above the limb can be seen in Fig. 2.20 and shows it to be comparable to the reflectance of the nucleus itself and thus $\tau_d \approx 1$. Other calculations based on integrating the results from several Giotto experiments suggested that $\tau_d \approx 0.4$ (Thomas and Keller 1991) although there are grounds to believe that the value may have been higher because of inaccuracies.

Rodionov et al. (2002) addressed the multi-fluid approach for the dust and showed that the mean free path for dust-dust collisions can be assumed to be infinite. Hence, the dust distribution function follows the Collisionless Boltzmann Equation (Eq. 3.91) which is a significant simplification. But the individual dust particles sizes cannot be grouped together because different particles sizes emitted from the same position, at the same time will follow different streamlines and this is not the behaviour of a single species fluid. Hence, a separate set of fluid equations are needed for each particle size (and clearly some grouping is still required to discretize the problem). In addition, the interaction between the gas and dust fluids results in additional source and sink terms for the momentum and energy budgets. Hence, we arrive at

$$\frac{\partial \rho_{d,i}}{\partial t} + \nabla \left( \rho_{d,i} \boldsymbol{u}_{d,i} \right) = Q_{d,i} \qquad (4.104)$$

for the continuity equation for the dust with similar modifications for the momentum and energy equations where the external forces and energy changes should now include those arising from the gas-dust interaction. These adaptations can be made to both the Euler system of equations (Rauer 2010) and the Navier-Stokes system (Rodionov et al. 2002).

The main issue with this approach is in defining the details of the interactions. There are several uncertainties. For example, the gas is not merely slowed by the momentum transfer to the dust but it can also be heated by the dust. The expansion of the gas implies that it is always colder than the dust but during a collision it is in contact with the dust and can therefore receive thermal energy through that contact. This is not an entirely moot point because the gas obtains the energy to expand from its internal degrees of freedom and that energy might be replenished by interaction with the dust giving the gas more energy to increase its outflow (terminal) velocity. The dust also emits thermal radiation in the infrared which the gas can absorb.

There is also the question of whether the dust provides an additional gas source which needs to be included in the fluid equations. Rodionov et al. (2002) included this in their model based on work by Crifo (1995). There are a vast number of possibilities and relatively little reliable data to constrain the problem.

We have seen that column densities of dust follow a $1/r$ law in the case of force-free radial outflow. If the scattering cross-section within a column is unmodified by sublimation then the observed scattered brightness from the dust should also follow $1/r$ and it was indeed shown at 1P/Halley that, for distances from 100 km to ~2000 km, this was an excellent approximation. In other words, the particle cross-sections remained constant in this distance range. But nothing could be said at distances closer than 100 km because of the influence of other effects (e.g. non-point source geometry, see below). The observations of the unusual 103P/Hartley 2 by EPOXI revealed water ice particles close to the nucleus. This suggests that in some cases, sublimation from ice particles can provide a significant additional gas source. Protopapa et al. (2014) concluded that within one nucleus radius of the surface, dust dominated water ice by a factor of ten but that the water ice and the dust were not uniformly distributed. The lifetime of pure water ice particles is long compared to typical outflow timescales (see below) but impurities may modify this. Hence, there is a case for including extended sources from dust in gas dynamics schemes. However, other claims of evidence of sublimation effects in the innermost coma are, at this point, less quantitative or dubious.

## 4.8   Converting Afρ to a Dust Loss Rate

Combining the size distribution, the particle size-dependent velocity, and the scattering properties allows us to convert *Afρ* to a dust mass loss rate. It is instructive to look at the contribution of each particle size to the observed intensity assuming Mie theory. An example is shown in Fig. 4.30 for size distribution power-law exponents of 2.1, 2.6, and 3.0 (cf Table 4.1) using a complex refractive index of $m_{ref} = (1.6, i0.001)$ viewed at a phase angle of 90°. The curves have been normalised to 1 at the maximum of the distributions. The main point of this plot is to show that at these intermediate phase angles, the observed intensity is totally dominated by particles in the 0.1–1 µm range for exponents $\geq 2.6$ (i.e. 1P/Halley-like distributions) and that orders of magnitude more particles of larger sizes are needed to influence the observed brightness under the assumption that Mie theory is appropriate. Only when the exponent is decreased to values approaching $b_d = 2.1$ do we see the larger particles beginning to dominate (as indicated by the size at which the curve for $b_d = 2.1$ is normalized in Fig. 4.30).

It is perhaps surprising that the contribution to the intensity as a function of particle size drops so quickly from sub-micron sizes to 10 µm-sized particles in Fig. 4.30 because the geometric cross-section is only dropping with $a$ and this is

**Fig. 4.30** The relative contribution to the observed intensity at 650 nm as a function of the particle size for power law exponents of 2.1, 2.6, and 3.0, using a refractive index of (1.6, i0.001), viewed at a phase angle of 90°. In each case, normalization is made with respect to the maximum of the intensity distribution

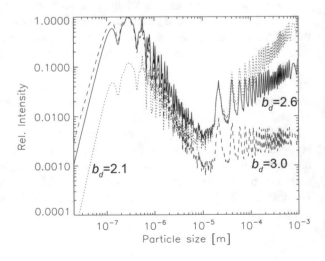

**Fig. 4.31** The value of the scattering function computed from Mie theory for $m_{ref} = (1.6, i0.001)$ at two different scattering angles. Solid line: Scattering angle of 3° Dashed line 90°

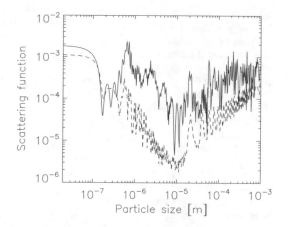

partially compensated for by the decrease in velocity with particle size which increases the local density of larger particles. However, the effect of the scattering function is highly non-linear and depends strongly on the scattering angle. This is illustrated in Fig. 4.31 which shows the scattering function at two different scattering angles for a range of particle size. The absolute values of the scattering function in this plot are somewhat less important than the variation with size. The low scattering angle corresponds to forward scattering which becomes increasingly dominant for large particle sizes. Although the scattering function at a scattering angle of 90° is, in general, lower, smaller particles scatter more efficiently to these intermediate scattering angles. Hence, the result in Fig. 4.30 is a complex function of the particle size distribution, the dust particle velocity, AND the scattering function and the form is dependent upon the Sun-comet-observer angle.

Combining the various properties, we can sum over particle sizes to get an *Afρ* using

$$Af\rho = 2\pi \sum_a \frac{\sigma_x\, n_d(a) Q_{sca} \Phi_s(a)}{v_d(a)} \tag{4.105}$$

where $n_d(a)$ is defined in Eq. (4.75). For $v_d(a)$ an approximation to results in Marschall (2017) roughly corresponding to Eq. 4.101 can be made. This can be rearranged to provide a production rate per unit *Afρ* as a function of exponent for the size frequency distribution as shown in Fig. 4.32 for three different values of the refractive index.

Two important points are evident in Fig. 4.32. Firstly, there is a minimum for a constant refractive index which, in this case, is around $b_d \sim 3.5$. If the exponent becomes larger than this, the smaller particle sizes become more numerous and hence the mass of optically inactive particles increases. Consequently, the emitted mass must increase to produce the same value of *Afρ*. On the other hand, as the exponent decreases, the number of large particles increases but the scattering area

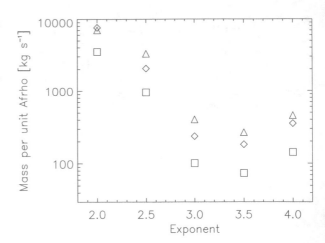

**Fig. 4.32** The mass implied by an *Afρ* value of 1 m for different values of the dust size frequency distribution exponent, $b_d$ assuming a gas production rate of 200 kg s$^{-1}$ to provide the dust velocity distribution and Mie theory for a scattering model. Diamonds: $m_{ref} = (1.6, i0.001)$ Triangles: $m_{ref} = (1.6, i0.1)$. Squares: $m_{ref} = (2.0, i0.01)$. A scattering angle of 90° has been used

does not keep pace with the mass increase. So the mass needed to produce the same *Afρ* again increases.

A second point of note is the strong dependence on the real part of the refractive index which is a parameter that is poorly known and therefore ill-constrained.

A large number of assumptions have been made in reaching this result. Notably, the dust velocity has been assumed to be a function of particle size only and not a function of emission direction. This assumption is clearly violated upon inspection of Fig. 4.29 although it should be recalled that most of the emission is from the dayside and here the velocities are more uniform. More critical however is that the dust velocity is strong function of the assumed gas production rate. This calculation was performed using the velocities appropriate for $Q_g = 200$ kg s$^{-1}$. The values would be roughly 1 order of magnitude less for $Q_g = 2$ kg s$^{-1}$ (Eq. 4.101). The assumption of a power-law distribution for the dust size distribution is also not clearly demonstrable.

The above also assumes that Mie scattering is dominant. However, Mie theory is physically unrealistic because, as we saw in Sect. 4.2, the particles are not smooth spheres and this becomes an increasingly critical assumption as the size parameter increases. It should be evident from this that there are multiple solutions to the production rate for a given *Afρ* and that the scheme outlined above is an approximation.

## 4.9  Observation of Non-uniform Dust Emission

### 4.9.1  Large-Scale Structures

Non-symmetric comae of comets had been observed in the past but it was Larson and Sekanina (1984) who developed digital image processing techniques to bring out

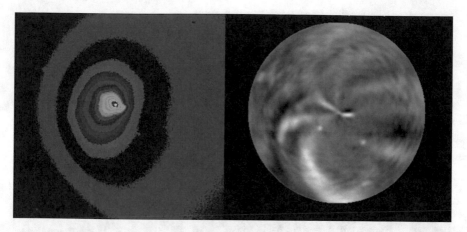

**Fig. 4.33** Left: Image of 1P/Halley acquired at the Anglo Australian Telescope on 11 March 1986. False colour has been used to show the non-circular isophotes in the innermost coma. Right: A ring-masked version of the same image revealing large jets (in white) with significant curvature (adapted from Keller et al. 1995)

and enhance fine structures in the comae of comets observed with ground-based telescopes so that they were clearly visible. The basic idea was to subtract from an image, a rotated version of itself leading to the name of "shift-differencing". This technique revealed curved jet-like structures emanating from the nucleus that were interpreted as non-uniform emission from the surface combined with the effects of nucleus rotation. Subsequent works by Sekanina and co-workers attempted to derive the number of jets (enhancements in the dust emission) and the rotational properties of several comets using this approach. However, it was not quantitative and the angles used to generate the "best" result were arbitrary. On the other hand, this showed that the nucleus was a source of non-uniform dust emission and that large scale jet structures were to be expected in the inner coma.

More sophisticated image processing studies used a technique referred to as "ring-masking" (e.g. Hoban et al. 1988). Here, the observed intensities on circles of constant distance, $b$, from the nucleus in projection were multiplied by $b$ and averaged (note the similarity here to Eq. 4.60). These values were subtracted from the original image. This is effectively producing an image of deviations from a coma averaged over all directions at every distance from the nucleus. An early example of this is shown in Fig. 4.33. The image on the left is an original image of 1P/Halley acquired at the Anglo-Australian Telescope shortly before Giotto's encounter with the comet. False colour has been used to show the non-circular isophotes surrounding the nucleus at the centre. Ring-masking (Fig. 4.33; right) reveals the curvature and strength of the jet-like structures emanating from the nucleus. Similar techniques have been used by observers of gas emissions to study non-isotropic emissions of radicals (e.g. Hoban et al. 1988).

Observations of the innermost comae of 1P/Halley and 19P/Borrelly both showed evidence for non-uniform dust emission. In Fig. 4.34, we use 19P/Borrelly as an

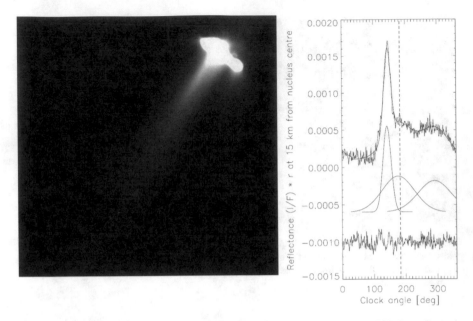

**Fig. 4.34** Left: 19P/Borrelly observed with the MICAS instrument onboard NASA's Deep Space 1 spacecraft. The image has been stretched to show the dust coma surrounding the nucleus. Right: An earlier image was used to determine the brightness on a circle 15 km from the nucleus centre. The distribution was then fit with 3 Gaussians to estimate the angular direction, strength and width of discrete jet-like structures. The non-uniformity of the dust emission from the nucleus is clearly evident in both representations. The clock angle towards the Sun is at 177° and is marked (Reprinted from Ho et al. 2003, with permission from Elsevier)

example. The image to the left of the figure already shows non-uniform emission with a jet-like structure emitted from the nucleus towards the bottom left of the image. The nucleus in this image has been deliberately saturated in post-processing. On the right, an earlier image that gave a more comprehensive overview of the emission into all directions has been analysed to determine the directions and strengths of the emission. Three Gaussians have been fit to the distribution. It can be seen that what appears to be the strongest jet is actually rather narrow with a full width half maximum of only 18 degrees and contributes only 19% of the total emission from the comet (assuming no angular dependence of velocity). Hence, what might appear at first sight to be the dominant structure in the coma may not necessarily contain the most mass loss.

For objects with non-zero obliquity and little or no precession, there is always a constantly illuminated pole which swaps from one pole to the other at the equinox. In the case of comets, this may have added significance because the surface layer at the constantly illuminated pole does not have the chance to cool off over night and the penetration of the thermal wave into the sub-surface may provide a source for higher activity. This concept was first promoted by Keller et al. (1987) as an explanation for the Giotto/HMC observations of 1P/Halley (Fig. 4.35) which showed the maximum

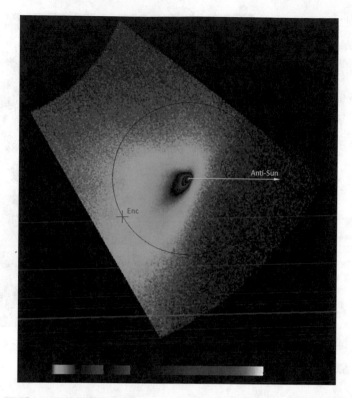

**Fig. 4.35** HMC image of 1P/Halley showing the dust coma. The anti-sunward direction is marked. The encounter point 596 km from the centre of the nucleus is also marked. The maximum of the dust emission is not directed towards the Sun but was between 45° and 60° from the Sun-comet line. The dust brightness integrated over the sunward facing hemisphere in projection to that above the projected nightside is around 3.2 to 1

dust emission in a direction at an angle of 60° to the Sun-comet line but almost co-incident with the rotation pole derived by Wilhelm (1987) from an initial interpretation of the Giotto and Vega observations of the nucleus. This pole solution was subsequently questioned but the idea of a constantly illuminated pole providing an enhanced source of emission remains. For example, the Hapi region of 67P, which was identified as a maximum in terms of $H_2O$ production during northern summer (Fougere et al. 2016; Marschall et al. 2017) contains the north pole. Schleicher et al. (2003) suggested that the absence of curvature of a jet from 19P/Borrelly could be explained by enhanced emission from a constantly illuminated pole. Thus, this concept may still have validity.

**Fig. 4.36** Image of the
inner coma of 1P/Halley
post-processed to reveal fine
structures extending from
the nucleus using a shift
differencing method.
(Following the approach of
Thomas and Keller 1987)

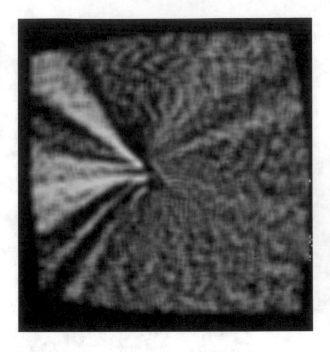

## 4.9.2   Small-Scale Structures

Larger scale structures similar to those seen in Figs. 4.35 and 4.36 are properly called
jets and are related to either the non-spherical shape of the nucleus and/or variations
in the dust emission rate that are not solely arising from changes in illumination
(energy input). The latter are deviations from a purely insolation-driven dust emission from the nucleus.

By looking even more closely at the angular distribution of dust emission, finer
structures can be identified in the innermost coma. The first example of this was at
1P/Halley (Fig. 4.36) and was fairly crude at around 500 metres resolution. The
visual observations of 67P at sub-decametre resolutions have taken studies of these
structures to a completely different level.

Figure 4.37 shows an image acquired about a month before the perihelion of 67P.
Four main characteristics of the observations are marked. Relatively diffuse dust
emission is present over most of the nucleus which is supplemented by narrow, well-
defined structures with widths at least an order of magnitude below those seen at
1P/Halley. Enhancements in the dust emission can appear to cross each other but
care must be taken here when considering line-of-sight effects. In the neck region in
this particular image there is some slight evidence of curvature of the jets emanating.

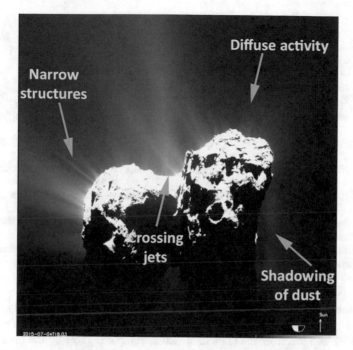

**Fig. 4.37** Slightly enhanced version of an OSIRIS image acquired on 4 July 2015 close to perihelion. (Image number: N20150704T180343702ID30F22). The image shows diffuse activity from the head of the nucleus. On the body, there are narrow, well-defined structures emanating from the nucleus. Jets appear to be crossing, particularly in the neck region, but line of sight effects here need to be carefully considered. Evidence of the nucleus casting a shadow on dust above the nightside hemisphere of the nucleus is also evident

There are better examples in the Rosetta data set but again one must be aware of optical illusions arising from crossing of diffuse structures along the line of sight. There are also sharp boundaries that arise from dust being shadowed above the nightside hemisphere of the nucleus.

In Fig. 4.38 some of the effects are shown more quantitatively. The image on the right is from the OSIRIS Wide-Angle Camera. The Sun is vertically upwards. In the centre is a plot extracted from a polar coordinate transform of the image about the centre of the nucleus showing the brightness distribution at three impact parameters from the nucleus centre. The curves have been linearised for the distance to the nucleus by multiplying by the impact parameter, $b$. Several things are evident. Firstly, the overall level of each curve decreases slightly as the distance to the nucleus increases. This is evidence of particle acceleration (although we note that other processes may also be active as will be discussed below). Secondly, the nucleus casts a clearly visible shadow on the dust indicating that there is dust above the nightside hemisphere as a consequence of either nightside activity or dayside to nightside flow of material (either through gas flow or gravitation). Thirdly, the spatial distribution suggests that some of the fine structures are curved

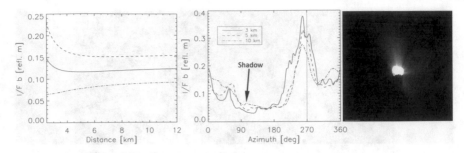

**Fig. 4.38** The centre panel shows the distribution of dust aroud the nucleus at three different distances from the nucleus centre on 29 April 2015 at 16:04:09 (W20150429T160409605ID30F18). The original image is shown to the right. The nucleus casts a shadow on dust in the coma which is evident also in the centre panel. The left panel shows the change in the azimuthal average (solid line) with distance from the centre of the nucleus and also shows azimuthal average values restricted to the projected dayside (dashed line) and the nightside (dot-dash) hemispheres. The plot shows the dust brightness multiplied by the distance from the centre of the nucleus in units of reflectance x metres. The Sun direction is marked by the vertical line in the centre panel

close to the source implying non-radial hydrodynamic expansion. Finally, the finer structures are actually very weak producing a variation in the brightness of only around 10% at 3 km from the nucleus which is completely lost by the time the flow reaches 10 km from the nucleus. This loss of structure is presumably caused by non-radial emission at source or by non-radial gas flow (the acceleration indicates that the dust is still coupled to the gas).

Originally, the fine structures were not referred to as "jets", in part, because the mechanism by which they arise was (and remains) unclear. We consider four of the possibilities here,

- local enhancements of the production rate,
- local changes in the dust size distribution,
- local minima in the production rate,
- topographic focussing.

Localised enhancements of the dust production rate can of course occur. This would be a "jet". However, there are subtle issues with this mechanism. If the dust to gas production rate ratio remains constant, then the gas production rate must also increase. The expansion of the gas into the lower density medium surrounding the jet would dilute the observed enhancement in the dust and reduce its collimation. The increase in brightness within the finer structure is only of the order of 10% with respect to the surroundings but the structure is relatively narrow and we are looking through a column of dust. Hence, the actual increase in production in the structure relative to its immediate surroundings must be much higher. Although no published test of the parameter space has been made, this suggests that there are limits to how bright these structures can become with this mechanism. We shall see extreme examples of this in the next sub-section.

A way around this is either to enhance the dust only or to modify the local dust size distribution so that the scattering properties are changed. This implies that the source region is fundamentally different in terms of physical properties—either the dust to gas production rate ratio is larger or the emitted sizes are different when compared to adjacent areas. Vincent et al. (2016a) traced back the structures from the coma back to the nucleus to try to determine their footpoints and, by comparing the distribution to the surface appearance, suggested that active cliffs on the nucleus may be responsible for these structures. It was subsequently shown by Marschall et al. (2017) by fitting of measured gas densities in the inner coma that Rosetta data could not distinguish between a model with insolation-driven sublimation and a comparable model with only cliffs being active. Consequently, the idea is consistent with available data. The only challenge to this hypothesis is that cliffs are emitting roughly tangentially to the main body of the nucleus. In order to become (roughly) radial as observed the emitted material must be forced into that direction presumably by gas flow.

A third possibility is that the fine structure represents the result of local minima in the production rate. We saw in Figs. 3.26 and 3.27 that if the $Kn_p$ is small ($<1$) then interaction between adjacent active areas can occur. The interaction region can then have locally higher densities. If the particles are well coupled to the gas, then an enhancement in the dust density along the line of the interaction region can result. This, rather counter-intuitive, idea was studied by Knollenberg (1994) (see also Rauer 2010). A simple case was that of an inactive cap surrounded by uniformly active regions. The flow of gas into the region above the inactive surface led to enhancements in the local dust density above the *inactive* part of the nucleus. This is instructive in that it illustrates that density enhancements in the dust coma do not necessarily require local increases in production and indeed the opposite can be the case. A similar concept was also investigated in 2-D by Finklenburg (2009) using Bird's DSMC code. Figure 4.39 shows how interaction between gas emitted from separate sources can lead to maxima in the gas density above the inactive area between them (see also Fig. 3.27). This concept requires that the dust is well coupled to the gas and that the Knudsen penetration number is less than 1.

Another possibility is that we are seeing effects of topography. This was also investigated by Knollenberg (1994) (see also Keller et al. 1994) using an Euler equation solver and trace particles. The result of a similar calculation, using DSMC for the gas in this case, is shown in Fig. 4.40. The calculation is based on a spherical nucleus with a concave depression (or cavity) directly facing the Sun. The gas production rate is set at 100 kg s$^{-1}$ with an insolation-driven distribution. The key point is that the concave cavity focusses the outflow and fine structure is produced. This illustrates that topography can strongly control the appearance of the innermost coma.

It should be apparent that cliffs and larger scale roughness can generate these types of structures. This raises the possibility that by tracing coma structures back to cliffs, Vincent et al. (2016a) may have been identifying focussing by topography rather than locally increased dust production.

Although this has not yet been investigated in detail, the dust brightness distribution along the jet axis should allow one to distinguish between the jet production

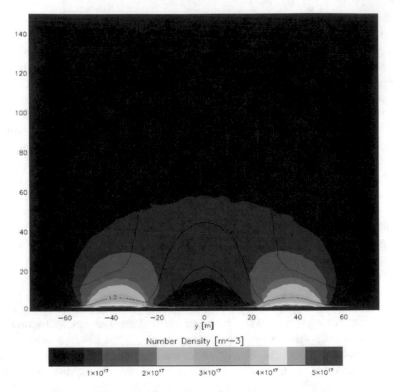

**Fig. 4.39** Two sources, 25 m wide and separated by 75 m emit water vapour. The reservoir has a density of $10^{18}$ m$^{-3}$ and a temperature of 200 K. The colour scheme shows the local density and the contours indicate the Mach number. The density distribution rapidly takes on the appearance of a single jet with the maximum above the inactive area between the two sources. (From Finklenburg 2009)

mechanisms. It remains to be seen to what extent deviations from idealised modelling cases blur this distinction.

In the above, we have touched on the question of whether the dust to gas production rate ratio from a specific source area is a constant over a cometary nucleus. It is almost self-evident that at some scale this assumption cannot hold. Looking at Fig. 2.105, it is hard to imagine that the dust to gas production rate ratio from the bright, icy, area is equal to the ratio for the surroundings. The appearance of the surface is clearly different. However, these are relatively small patches compared to the surface area of the whole nucleus. Tenishev et al. (2016) have argued that the dust and gas may have distinctly different production rate distributions on the surface of 67P and this may indeed have substance. On the other hand, the introduction of a variable dust to gas production rate ratio introduces further free parameters to fits of the gas and dust distribution in a problem that is already under-constrained.

**Fig. 4.40** A curved depression carved into a spherical nucleus has been modelled. A DSMC calculation was then performed assuming insolation-driven sublimation from the surface with a total production of $100 \, \text{kg s}^{-1}$. A test particle approach was used to compute the dust distribution and column density assuming $b = 3$ for the dust size distribution. Above the concave cavity, fine structure can be seen

It can also be asked whether the dust to gas production rate ratio is constant with time (and therefore heliocentric distance). In this case, if gas drag alone is responsible for dust loss then the answer is fairly straightforward in that it cannot be because the maximum liftable mass is a function of the gas production rate which is itself a function of heliocentric distance. However, the degree to which this simple relation holds depends on the exact dust ejection mechanism which is still poorly understood.

### 4.9.3 Transient Jet/Filament Structures

The imaging systems on Rosetta observed numerous short-lived emission events. Events seen near perihelion were catalogued by Vincent et al. (2016b). The event that recorded the highest reflectance is shown in Fig. 4.41. An image acquired 30 min earlier is shown on the top right for comparison. It has been processed in an identical way. An image 30 min later showed that activity had ceased (at least at this high level). The maximum reflectance close to the surface was approaching 0.012 in the visible and near-infrared suggesting that the optical depth was still below one. The curve of growth along the axis of the brightest streamline of the jet (bottom right) also suggests that the dust was not optically thick.

The opening angle of the transient structure is around 50° full cone angle. This is similar to results obtained by Kitamura (1986) for axisymmetric dusty gas jets using sub-micron (a single size of 0.65 μm) dust particles entrained in the gas flow. The

**Fig. 4.41** The highest reflectance transient in the dust emission recorded by OSIRIS near perihe-lion. Top left: A slightly enhanced version of an image on 12 August 2015 at 17:20:02. (Image number: N20150812T172002741ID30F22). Bottom left shows a larger format version of the same image showing the source of the activity. Top right: The appearance of the nucleus 30 min earlier (Image number: N20150812T165002770ID30F22). Bottom right: The reflectance along the bright-ness element of the jet in the orange (solid line) and 880 nm (dashed line) filters of OSIRIS

radial velocities of the gas and dust are enhanced inside the jet, compared with those in the background. This type of effect can be seen by close inspection of Fig. 3.33, for example. The gas also expands laterally. Dragged by this laterally expanding gas outflow, the dust particles are swept away from the central axis to produce a cone shaped enhancement. The non-uniformity of the observed conical structure in Fig. 4.41 may be caused by inhomogeneities of the source.

Vincent et al. (2016b) catagorized observed transients into three groups. Type A transients produce a very collimated jet. Type B transients form broad plumes or wide dust fans with type C being complex. As we have seen, collimated structures can be generated in several ways but the broader plumes and fans with large opening

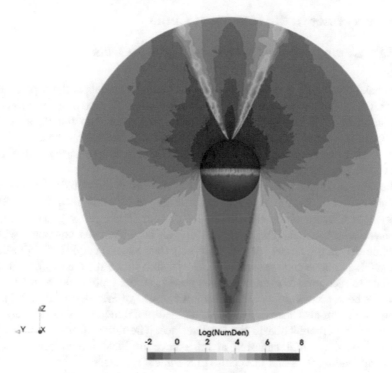

**Fig. 4.42** Log particle number density of a simulation of a dusty gas jet emanating from the sub-solar point and illustrating the dust distribution. Note the rarefaction forming a conical shape. On the outside of this is a narrow region where the density is enhanced over the background. (Courtesy of S.-B. Gerig)

angles can be generated by large differences between the gas density in the jet and the surroundings leading to significant lateral expansion. Kitamura (1986) used Gaussian distributions of enhancement to model a jet with the maximum being a factor of ten higher than the background. The width of the Gaussian was varied. The opening angle is, unsurprisingly, a strong function of the jet strength with respect to the background. There is a weaker dependence upon the dust particle size. An example showing the basic features of the behaviour in simplified geometry is shown in Fig. 4.42 (from S.-B. Gerig, PhD thesis in preparation) and its similarity to the contour plots of Kitamura is qualitatively quite striking.

## 4.10    Processes in the Innermost Coma

### 4.10.1    Comparisons of Models of 67P with Data

Crifo et al. (2002b) made the first comparisons between models of the appearance of the innermost dust coma with spacecraft observation by studying the brightness distribution at 1P/Halley recorded by the imaging system on the Russian Vega spacecraft. The detail shown by the observations was limited by the de-focussing of the Vega 1 camera but this was the first attempt at a self-consistent model of an observed dust outflow.

The scheme needed to establish a model of the dust coma of 67P was illustrated by Marschall et al. (2016). A model result is shown in comparison to an OSIRIS Wide-Angle Camera observation in Fig. 4.43 and shows fairly good qualitative agreement with the data. The model calculation here has been computed with an insolation-driven gas model but with regionally (large-scale) varying effective active fractions. It is noticeable that several linear jet-like structures are evident and it is important to recognize that these features arise from the topography of the nucleus which has been accurately described using the results of Preusker et al. (2017). The fine structures are also sharper and better defined in the model. The model assumes dust emission perpendicular to the local surface. The rather diffuse coma may be indicative of non-radial emission at the source or local inhomogeneity in the outgassing leading to lateral flow close to the nucleus surface.

### 4.10.2    Deviations from Force-Free Radial Outflow

In the force-free radial outflow approximation, the column density at any point in the coma is inversely proportional to the product of the impact parameter and the velocity (Eq. 4.2). The velocity of the particles when still part of the surface is

**Fig. 4.43** Comparison of an observation obtained with the OSIRIS wide-angle camera at 2015-05-04 T21.10.46 (centre) with two models of the dust outflow. Left: A solely insolation-driven model. Right: A model with regional variations in emission strength (adapted from Gerig et al. 2020, submitted). The coloured arrows identify common structures in the data and the model

zero. The ejection mechanism may give the particles an initial velocity but, even if not, the drag force from the gas (Eq. 4.87) accelerates the particles. Hence, the product of $N_d b$ is no longer a constant—we have what is referred to as "a deviation from 1/r".

While it is clear that $N_d b$ must decrease as one moves away from the nucleus because of the drag force, it is not the only issue close to the nucleus. This is not solely of academic interest. Such deviations might also include the signatures of other processes influencing the dusty-gas outflow which, in turn, may have causes relating to the properties of emitted material. For example, the acceleration is related to the particle structure through the drag equation. The initial velocity prior to acceleration is related to the ejection process itself. Any non-radial component of the motion also influences the coma brightness distribution and is related to the gas dynamics directly above the nucleus and its inhomogeneity.

In addition to the breakdown of the force-free radial outflow approximation because of drag, the point source approximation must also breakdown when close to the nucleus. How close is "close" in this case is unclear as it depends upon the shape of the nucleus and the activity distribution but the diameter of the nucleus would be a reasonable approximation for the length scale. There are also other effects such as fragmentation, sublimation, and effects on particles from the gravity of the nucleus that can also produce deviations from 1/r. Viewing and interpreting the dust column density is one of the few ways of constraining the effectiveness of each process although it was recognized during analyses of the data from Giotto that there are difficulties in interpretation because the processes involved are not well known and natural assumptions lead to significant ambiguity.

We begin to illustrate the problem by showing an example from Rosetta/OSIRIS observations of 67P. In Fig. 4.44, an example intensity profile from the coma of 67P is shown. The zero position for the distance is placed at the projected surface of the nucleus—in this case at the limb. The reflectance from the dust has been multiplied by the distance from the surface. Force-free radial outflow from a point source would predict this value to be constant and clearly it is not.

This can easily be explained by the finite dimensions of the source. One can assume that each position on the object emits dust in a narrow cone which alone would produce a 1/r function in column density. Multiple cones dotted across the finite source make up the total emission but, as one approaches the nucleus along a fixed direction, a decreasing number of cones contribute to the column and hence the column density decreases relative to a 1/r distribution and we obtain a result similar to that seen in Fig. 4.44. Boice et al. (2002) used this type of approach for the initial interpretation of data from Deep Space 1 at 19P/Borrelly, for example, following ideas presented in Reitsema et al. (1989).

It is also clear from the dust distribution around the comet that the production rate, $Q_d$, has a strong angular dependence. For the force free approximation, this would be of no relevance if the nucleus were a point source. Radial profiles would show a 1/r distribution but the final value of $N_d b$ would be dependent on the angular direction of the profile from the nucleus. However, non-point source geometry, non-radial

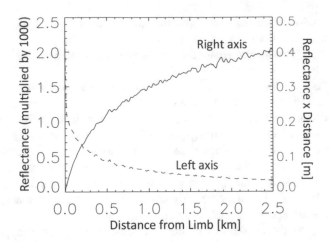

**Fig. 4.44** Dashed line: Intensity profile along a jet structure. Solid line (right axis): The reflectance multiplied by the projected distance from the surface. Physical interpretation of these profiles is compromised by numerous processes (e.g. non-radial flow) that are hard to quantify in specific cases. Nonetheless fits to these curves can be made by making assumptions about the finite dimensions of the source. (From image number: N20150824T081620627ID3BF22)

flow arising from the gas dynamics and unusual (non-spherical) surface geometries complicate the interpretation severely.

One approach of attacking this problem with a view to understanding the physics of the initial outflow is to construct a surface about the nucleus and applying conservation "laws". In the force-free approximation using a point source, the proportionality

$$\langle \rho \rangle \propto \frac{1}{2\pi} \int_0^{2\pi} N_d(b, \theta) d\theta = \frac{1}{2\pi} \int_0^{2\pi} \frac{Q_d(\theta)}{4bv_d(\theta)} d\theta \qquad (4.106)$$

where $<\rho>$ has been previously defined through Eq. (4.60) and is related to *Afρ*. In other words, the dust flux through a surface defined by a cylinder, with its long axis along the line of sight, and with the nucleus at its centre is a constant and can be related to the observed intensity if the scattering properties of the dust particles are neither a function of the impact parameter, $b$, nor a function of angle.

There is a large list of assumptions made in adopting this equation. The following processes are neglected,

- Non-point source geometry
- Non-radial flow
- Unusual source geometries
- Acceleration by gas drag
- Optical depth effects
- Excessive numbers of gravitationally bound particles close to the nucleus
- Particle fragmentation

• Particle sublimation/condensation effects

It might appear that we have made no progress, however, the influences of the above processes on the behaviour of Eq. 4.106 are different allowing us to use the observations to constrain which processes are dominant.

For example, we have seen that gas drag accelerates the particles from zero velocity at the surface to a terminal velocity and therefore $v_d$ is not constant with $b$ leading to a decrease in the "conserved" quantity, $<\rho>b$ ($=\overline{A}$; see Eq. 4.60) as $b$ increases until the particles reach terminal velocity. This can be illustrated more precisely using simple models with spherical geometries. The effect of acceleration on the column density of dust emitted isotropically from a 2 km radius nucleus can be clearly seen in Fig. 4.45 and is almost independent of the particle size when normalized.

Figure 4.46 shows $\overline{A}$ observed using the Giotto/HMC instrument at 1P/Halley (following Keller et al. 1994). As we have just seen in Fig. 4.45, gas drag on the dust should lead to a decrease $\overline{A}$ with impact parameter, $b$. Here, this is clearly not the case and hence another process must be dominant.

The rise with impact parameter can be interpreted in terms of particle fragmentation. Fragmentation of large particles into smaller particles that are still optical active will increase the effective scattering cross-section in the coma and $\overline{A}$ should rise with distance from the nucleus. A very simple way to simulate this would be to use an exponential function for the splitting of a large particle into two smaller particles of equal volume and compute the increase in cross-sectional area. An example is seen in Fig. 4.47. This illustrates that the magnitude of the increase in

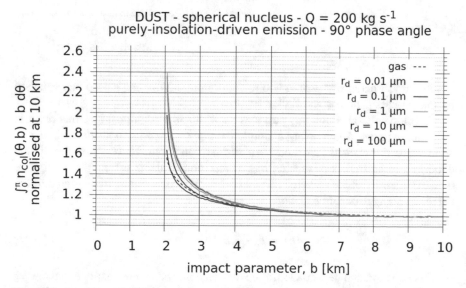

**Fig. 4.45** The column density of dust multiplied by the distance from the centre of a spherical nucleus driven by the outflow of a spherically symmetric gas emission. Different dust sizes (rd is the particle radius here) are shown for comparison. The curves have been normalized at 10 km

**Fig. 4.46** The azimuthal
average, $\overline{A}$, as a function of
impact parameter with
respect to the nucleus seen at
1P/Halley by Giotto/HMC
(Image number 3416). The
mean radius and the radius
of the long-axis are marked
as is the azimuthal average
found at 120 km on previous
images

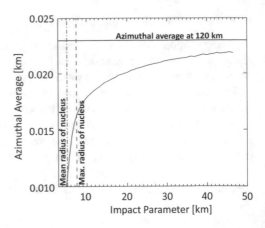

**Fig. 4.47** A simple
calculation of how particle
fragmentation can lead to
increased cross-
sectional area

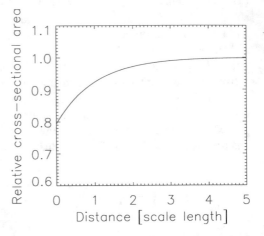

$\overline{A}$ with impact parameter seen in Giotto data (around 20%) can indeed be explained
by this process (Thomas and Keller 1990b).

Unfortunately, although this provides us with some additional information, it is
not unambiguous. Again, using a simple example, increasing optical depth as one
approaches the nucleus can produce a similar behaviour. If we know the optical
depth, $\tau_d$, at an impact parameter, $b_{far}$, where optical depth effects are negligible,
then the intensity of the scattered flux can be calculated assuming free-radial outflow
using the straightforward equation.

$$\rho(b) = \rho_0 \frac{b_{far}}{b} e^{-\tau_d\left(b_{far}\right)b_{far}/b} \tag{4.107}$$

This equation and the particle fragmentation concept are both exponential curves
and, with the various free parameters, they can produce identical behaviour. Therefore, the processes cannot be distinguished without additional information.

However, it is probable that one of these processes must have been influencing the inner coma of 1P/Halley as all the other processes that are thought to be of possible significance produce decreasing values of $\overline{A}$ as the impact parameter increases. A schematic summary of this is shown in Fig. 4.48.

In Fig. 4.49, a plot is shown from data acquired at 67P close to perihelion. The data set was selected because the phase angle is 90° and therefore dust above the dayside and nightside hemispheres are clearly separable. It is to be noted that the image from which this plot was extracted shows only weak evidence of shadowing of dust by the nucleus. The left panel of Fig. 4.38 shows the same type of plot but there the shadowing is more prominent (see the image in Fig. 4.35 right). Shadowing can lead to $\overline{A}$ initially decreasing and then slightly increasing with impact parameter before stabilizing at the expected constant value. The evidence of the importance of shadowing can be seen when comparing the contribution to $\overline{A}$ from the dayside (dashed line) and the nightside (dot-dashed line) (compare with Fig. 4.38). On the dayside, unlike the observation at 1P/Halley, $\overline{A}$ reaches an almost constant value at <5 nucleus radii from the surface. This suggests that the processes itemized above have ceased having any influence on the dust flow <10 km from the surface. (The alternative is that the effects of the processes are cancelling each other out which would be contrary to Ockham's razor) Furthermore, the product rises as the nucleus is approached indicating a steeper profile than $1/r$ close to the nucleus. Several of the itemized effects could be responsible for this behaviour. The nightside shows a steady increase in both our examples which is partially attributable to the decrease in the solid angle subtended by the nucleus shadowing the nightside dust as the impact parameter increases. Further effects of non-radial outflow are also likely to be present.

Dust acceleration as illustrated in Fig. 4.45 must occur and hence this is the prime candidate to explain the observation steeper profile than $1/r$ close to the nucleus. However, non-radial flow from and close to the nucleus produces a similar effect. Some combination of these two processes is highly probable. Particle fragmentation can also be responsible. If particles split into sub-particles that are no longer optically active (i.e. < 100 nm in the case of observations at visible wavelengths; see Fig. 4.30) then you lose scattering efficiency and $\overline{A}$ drops with distance from the nucleus. Sublimation of particles produces a similar effect. The cross-section may reduce as the particles lose volatiles.

As we shall see below, there is strong evidence that 67P emits a significant number of large, slow moving, possibly non-escaping large particles. Large particles in gravitationally-bound orbits are clearly not undergoing force-free radial outflow and would appear as a change in $\overline{A}$ as one approaches the nucleus as observed. An increase or a decrease in $\overline{A}$ can occur depending upon the dominant particle population (escaping or orbiting) as we shall discuss later. If the innermost coma region cannot be sampled by, for example, multi-point in situ investigation, the only way to tackle the problem is through detailed modelling. However, we can already see by comparison of Fig. 4.45 with Fig. 4.49 that the percentage increase in $\overline{A}$ from 10 km inwards to 3.5 km from the centre of the nucleus is almost identical in the

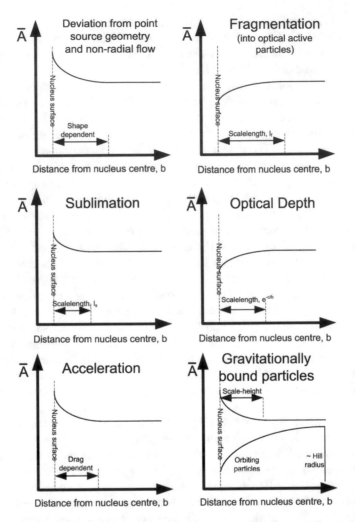

**Fig. 4.48** Schematic diagram of how various near-nucleus processes can affect the azimuthal average. Top left: Deviations from point source geometry and non-radial outflow produce a decrease with increasing distance. Middle left: Sublimation reduces the scattering area but the scale length will be different to that of the non-radial outflow and dependent on particle properties. Bottom left: Acceleration shows a similar effect but here the scale length is a function of the drag coefficient and the production rate. Top right: Rapid fragmentation close to the nucleus increases the effective scattering area so that the azimuthal average rises with distance. The form of the curve is unclear but an exponential function is probably reasonable. Middle right: Optical depth effects close to the nucleus mask scattering particles. This reduces with distance as the dust expands, increasing the visible scattering area. Bottom right: Gravitationally bound particles remain close to the nucleus and obviously do not conform to the free-radial outflow approximation. The exact form of the azimuthal average will be dependent on the initial velocity distribution but a Boltzmann-type distribution is a plausible starting point

**Fig. 4.49** Left: Solid line: The value of the azimuthal average against the impact parameter for 67P at 2015-07-31 T06.23.04 at a phase angle of 90.0° (N20150731T062304549ID30F22). Dashed line: The azimuthal average but restricted to the dayside hemisphere only. Dot-dash: The azimuthal average but restricted to the nightside hemisphere only. The nightside hemisphere curve is partially influenced by shadowing of coma dust by the nucleus itself. Centre: The original image (enhanced to show the coma). Right: The product of the reflectance and the impact parameter on concentric circles of fixed distance from the centre of the nucleus. Note how the fine structure smoothes out with distance from the nucleus. The vertical line is the sunward direction

spherically symmetric model and the observation, suggesting that acceleration only may be sufficient to explain this particular example. On the other hand, details are probably of considerable importance here and so it is necessary to look at the other potential processes and their effects. We will look at the process for which we have the most evidence—slow-moving, non-escaping particles—in a subsequent sub-section.

### 4.10.3 Dust Above the Nightside Hemisphere

Another noticeable feature of Fig. 4.49 is the ratio of the dayside to nightside brightness. The observation here was obtained at a phase angle of 90° and hence the observer was directly above the terminator so that there are no projection effects (i.e. material above the dayside surface but projected to be above the nightside in a 2D image and vice versa). The ratio therefore provides an exact value of the dayside to nightside brightness asymmetry. What is remarkable is that this value is only 3.3 and thus very low compared to what one might expect from an outflow dominated by dayside emission. This essentially confirms a conclusion reached from observations of 1P/Halley's dust coma. Thomas and Keller (1989) found a dayside to nightside brightness asymmetry of 3.2 at 1P/Halley at a phase angle of 107.2° and noted that this was a factor of three lower than for an emission model that is isotropic on the dayside and viewed in projection. Thomas and Keller (1989) suggested possible explanations connected to solar radiation pressure driving particles from the dayside to the nightside but this mechanism is not really adequate so close to the nucleus as can be deduced from Fig. 4.21. Using observations from the MICAS camera on Deep Space 1, Ho et al. (2003) found for 19P/Borrelly a dayside to nightside coma brightness ratio of just 1.7 at a phase angle of 88° and $r_h = 1.36$ AU and subsequently compared this to the results from 1P/Halley (Ho et al. 2007). Other evidence of substantial dust densities above the nightside hemisphere can be seen in, for example, Fig. 4.38 where we can see an increase in observable brightness at the position of the penumbra behind the nucleus.

On the basis of what we now know, there are two alternatives that need to be investigated to explain these observations. Firstly, is nightside dust emission significant? Alternatively, are we seeing here evidence of slow moving particles that are gravitationally influenced or even bound by the nucleus? We shall see that the first explanation provides more consistent agreement with the observations.

A purely insolation-driven source, as seen in the simple, spherically symmetric, models shown in Fig. 4.29, results in no gas or dust emission from the nightside hemisphere. Furthermore, even if the thermal inertia is non-zero (albeit very low as indicated by observations) and water ice is the subliming volatile close to the surface, Fig. 2.30 shows that the surface temperature drops rapidly after sunset and this chokes off any water sublimation extremely quickly as indicated in Fig. 2.30. If sublimation is from the surface then gas emission drops by more than four orders of magnitude in less than an hour for a nucleus with a 12 hour rotation period and even if sublimation is from the deep sub-surface, a decrease of three orders of magnitude occurs between sunset and sunrise.

Nightside dust emission arising from the effects of thermal inertia was used by Shi et al. (2016) to estimate that the sublimation of water ice at 67P was from a depth of 6 mm assuming a thermal inertia of 50 TIU although, as the authors themselves point out, the parameter space of degenerate solutions was not fully explored. In addition, these calculations referred to "sunset" jets (jet-like emission seen after sunset). There were observations of sudden jet-like emissions from the nightside on several occasions (e.g. Fig. 4.50) but it is not obvious that the bulk emission from the nightside is in the form of (water) jets nor does it appear restricted to the sunset terminator. Hence, there are physically good reasons to question nightside emission as a means of producing the observed dayside to nightside asymmetry.

The solution may be evident in Figs. 3.39 and 3.42. Sub-surface sublimation of $CO_2$ can result in a nightside outgassing at rates that are fairly constant with rotational phase, as postulated by Bockelée-Morvan et al. (2015), although variable with latitude. Dust entrainment within this flow field should conform roughly to the $1/r$ law as indicated by the observations in Fig. 4.49. Hence, this qualitiatively fits the data.

One point to note in addition is that the dayside to nightside dust column density ratio may be only ~3 but this does not imply that the production rate has the same ratio. The outflow velocity may be appreciably slower on the nightside leading to higher local densities (Eq. 3.1; Fig. 3.41). Consequently, the gas production rate ratio (dayside to nightside) inferred from gas dynamics modelling (12:1–20:1 e.g. Bieler et al. 2015a) could be appreciably higher than the dust production rate ratio. However, this is somewhat speculative because we have no published measurements of dust flow speeds on the nightside.

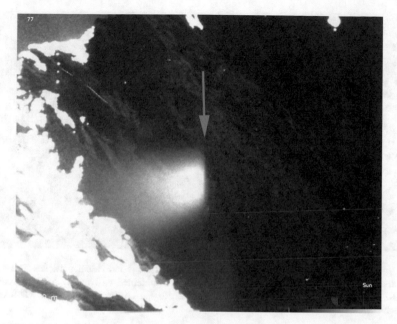

**Fig. 4.50** Observation of a dust jet emanating from 67P on 2 Jan 2016 made by Rosetta/OSIRIS (Image number: N20160102T120150600ID10F22). Note that the jet appears to enter sunlight suddenly from a shadowed/nightside source on the right

To address the alternative explanation, we must look first at the evidence for slow moving dust particles in the coma.


## 4.11   Slow (Large) Moving Particles in the Coma

### 4.11.1   Observations and Significance of Slow-Moving Particles

Prior to the EPOXI mission at 103P/Hartley 2, it was widely assumed that gas drag accelerated most particles and that most of these particles escaped the gravitational influence of the nucleus. The discovery of $>10^4$ chunks of material in the inner coma where 10 to 20% of the chunks were moving at less than their local escape velocity (A'Hearn et al. 2011) started to change this thinking (Fig. 4.51). When Rosetta reached 67P, the evidence of airfall deposits on the nucleus prompted a complete re-evaluation. It was shown (Thomas et al. 2015a) that individual particles could be seen in single images (see Figs. 2.80 and 4.52). The lack of smearing in the images of these particles within the exposure times used for the camera system led to the conclusion that many particles near the nucleus were travelling at speeds close or

**Fig. 4.51** Image of comet 103P/Hartley 2 showing numerous individual particles. The sun is illuminating the nucleus from the right. This image was obtained on Nov. 4, 2010, the day the EPOXI mission spacecraft made its closest approach to the comet. Image Credit: NASA/JPL-Caltech/UMD

even below the escape velocity. Particles with speeds less than the escape velocity do not follow radial trajectories and populate the Hill sphere of the nucleus.

The Hill sphere of a body is the region in which the body dominates the gravitational force acting on particles within it. The outer shell of the Hill sphere constitutes a zero-velocity surface. To be retained by the nucleus, any particle must have an orbit that lies within the Hill sphere. The radius of the Hill sphere can be approximated by the equation

$$r_{Hill} \approx a(1 - e)\sqrt[3]{\frac{M_N}{3M_\odot}} \tag{4.108}$$

The actual surface within which closed orbits about the nucleus can be found also depends upon the solar gravitational field and, if the nucleus makes a close fly-by of a planet for example, other gravitational perturbations.

Figure 4.52 shows a nice example of the emission of slow moving particles from the Bes region. The image on the left shows that the emission here was from one of the cliffs when it was close to the terminator. Processing the image reveals the jet and the particles emerging into sunlight (position A). A large number of individual particles can be seen. (As a curiosity, there is also an image of a large particle very close to the camera. It is clearly out of focus and moving with respect to the spacecraft.) Using a rough estimate of the distance to the particles and the exposure time for this image, we can place an upper limit on the velocity component of the particles in the image plane of <10.5 m/s. In this particular case, it is not a strong

**Fig. 4.52** Small jet structure emanating from the nucleus close to the terminator seen after the image left has been stretched (Image number: N20160130T190323815ID10F22). The dust from the jet is illuminated (**A**) and seen against the nightside of the nucleus. Numerous individual particles can be seen. Note the large dust particle (**B**) that is out of focus and therefore near the camera

indicator of bound particles but other less spectacular images have revealed tighter constraints (Thomas et al. 2015a).

Not all jet emissions show this level of large particle ejection. Figure 4.53 is an illustration of this. The strong jet is smooth indicating that any large particles in the flow, if they exist, are moving quickly compared to the exposure time for the image. The arrow points to a series of dark traces in the image that may be large particles between the camera and the jet. These particles are either dark compared to the background or not illuminated and therefore occult a small part of the jet in the far-field.

Large particles become increasingly evident when looking off the nucleus (Fig. 4.54). The brightness of the smaller particles that form the diffuse (smeared) brightness drops as approximately $1/b$ (where $b$ is the impact parameter) while the brightness of individual large particles remains constant. Long exposures reveal the presence of large individuals and provides information about their motion. Repeat imaging is normally desirable to ensure that observations are not affected by cosmic ray events. An example of a short sequence to view individuals is shown in Fig. 4.54. Three images are displayed. They were obtained 24 seconds apart with exposure times of 12.5 seconds. The spacecraft was 145 km from the centre of the nucleus and the viewing direction was pointed around 73 km off the nucleus. Individual particles seen in all three images are identified by the same coloured arrow. Not all individuals evident have been identified. Static bright points in the field are likely to be stars with the SIMBAD database indicating more than 350 stars brighter than magnitude 15 should be present. Hence motion is needed to identify a particle. Nonetheless there are clearly large numbers of individuals just in this one observation.

Figure 4.55 shows an OSIRIS NAC image in which more than 1750 individual particles have been identified using an edge detection algorithm. The orientation of the tracks is shown in the centre panel of Fig. 4.55 while the apparent velocity is

**Fig. 4.53** A jet emitted from the nucleus showing very little slow large particle emission. The jet structure is fairly smooth. Interestingly, at the position of the arrow, some dark tracks can be seen. One possible explanation is that these are large particles between the camera and the smooth jet. The particles are either weaker scatters than the jet material or are in shadow. (Image number: N20151207T062446727ID10F22)

shown in the right panel. The camera pointed 30° off the nightside limb with the spacecraft 142 km from the nucleus. The spacecraft moves during the exposure producing smear. The vertical dot-dashed line here shows the angle one would expect if a particle were stationary and the track were solely from this motion of the spacecraft. (This angle actually shows slight dependence upon the distance of the spacecraft to the particle. Stars would also appear close to this angle.) One can plot the brightness of the streaks against their length (bottom left) and this shows a strong concentration at low speed and a particular I/F. But of course, the faster a particle is,

**Fig. 4.54** Three 12.5 second exposures off the nucleus of 67P. A number of individual dust grains can be seen. Their presence is revealed by their motion during the exposure and multiple exposures indicate that they are not random cosmic ray events. Several individuals are indicated by arrows in the three images. The same coloured arrow indicates the same particle across the three images. Not all individuals have been identified in this way. (Multiple images including N20151120T210807063ID30F22)

the fainter the streak which leads to detection limits for faint fast particles and observational selection effects. It is to be noted that the individuals identified by the algorithm in this image contribute only 10% to the total brightness in the frame. The dashed line in Fig. 4.55 (bottom right) shows how the brightness of a particle at rest with an I/F of $2 \ 10^{-7}$ would need to increase to have the same apparent brightness per pixel as its speed increases in the image plane. This represents a crude detection threshold for individuals in this image and corresponds roughly to stationary 2 mm diameter particles at 10 km distance. Figure 4.55 (bottom left) shows the speeds in 2D (the velocity component along the line of sight cannot be determined) that 100 particles in the sample would need to have if the observed streak were a combination of spacecraft motion and particle motion directly away from the comet. This speed is however a strong function of the distance between the particle and the camera and solutions show a large degree of degeneracy. However, if these assumptions are valid then the observed particles will in general be at speeds of 0.5–10 m s$^{-1}$ and 1–20 mm in diameter.

The number of individual large particles in the vicinity of 67P has become the subject of considerable controversy. It is an important quantity for two major reasons. Firstly, slow moving, large particles are probably the main contributors to the sedimentation mechanism (airfall) that influences the surface morphology. The rate of deposition is, however, not well established. Secondly, large particles may be responsible for most of the cometary mass loss. While large particle loss cannot be doubted, the subject remains controversial because the total mass of large particles (both that ejected from the system and that falling back onto the nucleus) and the possible errors on the numbers are highly disputed (Choukroun et al. 2020).

The main problem in analysing images such as Fig. 4.54 is that the distance to the particles is not known and hence velocities can only be estimated. In addition, the brightness of a particle decreases with $1/a^2$ so that derived column densities are ultimately lower limits as the particle brightness drops below the detection threshold

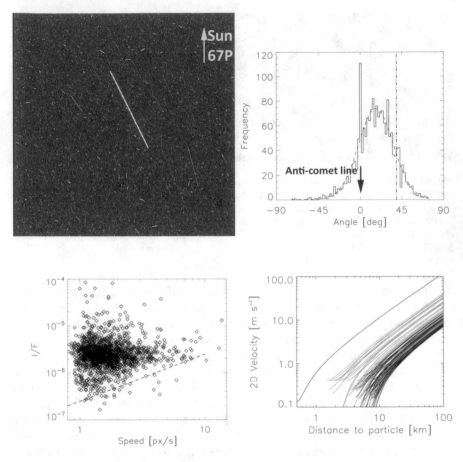

**Fig. 4.55** An edge detection algorithm has been used to extract 1769 individual dust tracks in the image (top left) acquired on 20 Nov. 2015 (Image number: N20151120T182742402ID30F32). The top right panel shows a histogram of the angles of the tracks with respect to the anti-comet direction. The dot-dash line gives the angle expected if spacecraft smear were solely responsible. The I/F against apparent velocity is shown to the bottom left while the results of a model calculation showing the ambiguity between distance to the dust particle and its speed in the plane of sky is shown to the bottom right. (Algorithm courtesy of Valentin Bickel)

with increased distance from the observer. And clearly a specific size or size distribution cannot be assumed without large uncertainty.

Drolshagen et al. (2017) attempted to overcome this issue by using the fact that the narrow angle and wide angle cameras on Rosetta were deliberately set to the greatest possible distance apart to allow for stereo acquisitions. In an example, the parallax was used to determine the distance to a specific particle resulting in an error of ~4% over 1.5 km. This approach is therefore accurate but limited in the distance over which it is effective by the baseline distance between the two cameras. Ott et al. (2017) subsequently used this approach for a data set with 262 large particles acquired with the spacecraft around 200 km from the nucleus. It should be noted that even here the velocity component along the line of sight cannot be derived. Ott et al. assumed velocities and derived a total large particle dust production rate assuming force-free radial outflow of more than 8 tonne s$^{-1}$ in August 2015. Fig. 1.18 shows the gas production rate to be around 900 kg s$^{-1}$ at this time indicating a dust to gas mass production rate ratio of ~9. This seems to be remarkably high compared to observations of the dense interstellar medium and protostellar envelopes which suggest an ice/rock ratio of ~1.5 (Pontoppidan et al. 2014) although it is likely that this ratio in planetesimals is a function of the heliocentric distance of their formation.

There are two main difficulties with the Ott et al. result. Firstly, Choukroun et al. (2020) have pointed out that the mass loss rate can only be sustained for 15–30 days before the hard limit on the total mass lost by 67P during the whole perihelion passage based on the RSI experiment (Pätzold et al. 2019) is reached. The second is connected to the trajectories of slow moving particles as shown in Fig. 2.81. The $Af\rho$ value inferred by Ott et al. for the large particles alone is higher than that found by Gerig et al. (2018) suggesting that these particles should dominate the optical flux from the dust coma. However, if velocities are of the order of 1–2 m s$^{-1}$ then the nucleus will rotate 20° to 40° while a particle travels from the nucleus to the edge of the field of view in images such as Fig. 2.81. If the emission is dominated by large slow moving particles then the jet structures must therefore be curved. In Fig. 2.81, only one structure shows this behaviour and it is, by far, the best example. Lin et al. (2016) noted this evidence of curvature resulting from gravitational effects on unbound particles. Subsequently, Marshall et al. (2018), with a slightly more sophisticated analysis, showed that these particles were still being measurably accelerated beyond escape velocity by the gas drag and were moving at velocities ~20 m s$^{-1}$ towards the end of the observed part of the flight. This is again inconsistent with the Ott et al. result.

Models of large particle motion in the vicinity of the nucleus were first discussed by Richter and Keller et al. (1995) at a time when the main issue was whether large particles could be retained in stable orbits for long periods of time and thereby present a danger to the Rosetta spacecraft. An alternative analytical approach would be to assume that the non-escaping dust particle distribution in the innermost coma can be modelled with a Chamberlain-like model for an exosphere (Chamberlain and Hunten 1987; Gerig et al. 2018). This model separates molecules into three different components using a partition function. Molecules may be on ballistic trajectories, escaping trajectories or end up in orbits in the event of collisions. In terms of the trajectories, this bears some similarity to the dust particle case. However, the Chamberlain model is an equilibrium model at source and, in addition, dust particles are brought into orbit not by dust-dust collisions but by other forces including external solar-driven ones that are not accounted for within the scheme.

Because this issue is the subject of detailed debate at the time of writing, we will look at an alternative approach to quantifying the mass production rate of strongly, gravitationally-influenced, particles. A numerical model of ejection and ballistic motion can be easily constructed for the simple case of a 2 km spherical nucleus of 67P's mass with uniform (isotropic) emission from the dayside only. Assuming a

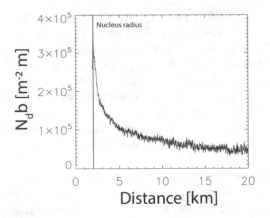

**Fig. 4.56** The column density of modelled dust particles arising from emission of slow particles from the dayside hemisphere of a 2 km nucleus. The radius of the nucleus is marked. Some particles apparently closer to the origin than the surface are along the line of sight in the integrated column and originate from near the terminator. The ordinate is the column density multiplied by the impact parameter, $b$, as measured from the centre of the sphere. For radial force-free, $N_d b$ would be constant. $N_d b$ is given in units of [#m$^{-2}$ m] for a particle production rate of 2500 s$^{-1}$. The scatter on the curve is a consequence of the numerical approach adopted

one-sided Boltzmann distribution for the dust ejection speeds (1 $\sigma = 0.6$ m/s), the column density over the dayside when steady-state is reached can be determined and is shown in Fig. 4.56. The ordinate shows the column density multiplied by the impact parameter, $b$, measured from the centre of the sphere. This quantity, $N_d\, b$, would be a constant for force-free radial outflow and hence the change in its value with distance indicates the effect of the gravitational field. In this case, 82.3% of the particles re-impact because they have initial velocities below escape velocity.

This calculation (Fig. 4.56) is highly simplified but illustrates the decrease in the column density-impact parameter product as the impact parameter increases. This is contrary to the observations seen in OSIRIS images (Fig. 4.49) and already suggests that this model has deficiencies (Gerig et al. 2020). Other forces can be included and local densities and column densities of large particles computed numerically to compare with observation. In combination with determination of the total contributing brightness, it is the most credible approach for attempting to determine the total mass production rate for large slow moving particles.

Nonetheless, we can use this result to arrive at a very simple equation for the relationship of $Af\rho$ to the dust mass loss rate if large particles of this type are dominant. The geometric cross-section of each particle is $\sigma_x(a)$. The reflectance is then

$$\rho_{dl} = N_{d(a)}\pi a^2 \Phi_S(\alpha) \tag{4.109}$$

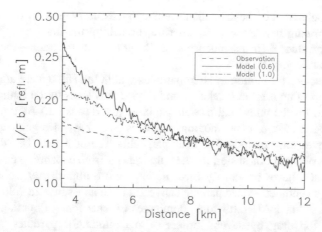

**Fig. 4.57** The azimuthal average for an image of 67P at 2019-07-31 T06.23.04 (dashed line and taken from Fig. 4.49) compared to numerical models of slow-moving particles ejected from the nucleus at two different speeds. The models have been scaled to match the observation at around 8 km in order to estimate production rates. Note that the models do NOT fit the trend with distance of the azimuthal average

where the single particle angular scattering function is computed from the Hapke parameters from Fornasier et al. (2016).

So that the reflectance from a column density of particles, $N_d$, in units of $[\# \, \mathrm{m}^{-2}]$, under the assumption of negligible optical depth and a single size, is given by

$$\rho_F = \rho_{s-d}(\alpha)\sigma_x(a)N_d \tag{4.110}$$

where $\rho_{s-d}$ is the reflectance of a single particle at a phase angle, $\alpha$, for which we can use Fig. 4.16. Combining this equation with Eqs. (4.60) and (4.2) and substituting for the geometric cross-section with $\pi a^2$ we can arrive at

$$Q_d = \frac{2}{3} Af\rho \, \frac{v_d}{\rho_{s-d}(\alpha)} a \, \rho_d \tag{4.111}$$

If we use 67P as an example again and assume $Af\rho$ is 2 m at maximum (Fig. 4.18), $\rho_{s-d}(90°) \approx 1.5 \; 10^{-3}$ (Fig. 4.16), $a = 1$ mm, $\rho_d = 1000$ kg m$^{-3}$, and $v_d = 2$ m/s, then we get $Q_d \sim 2 \; 10^3$ kg s$^{-1}$. This would be at least 2–3 times higher than the gas production rate (Fig. 3.7) and possibly more given the rather low estimate for the outflow velocity, $v_d$ and therefore giving a dust/gas mass loss rate ratio much higher than can be tolerated by the measurements of Pätzold et al. (2019).

For a 3D distribution generated by a numerical model, the azimuthal average, $\overline{A}$, can then be computed and compared to the data. This has been performed for the geometry of the image shown in Fig. 4.49 with the result shown in Fig. 4.57.

It is apparent from this plot that particles moving solely under the influence of gravity with ejection speeds close to the escape velocity do not reproduce the asymptotic behaviour of the azimuthal average in the observation. This should not be terribly surprising because we are, after all, breaching the main condition that is required to produce a $1/r$ dependence of the brightness distribution—namely force-free radial outflow.

The timescales for motion of large particles within the Hill sphere are very long. A single loop of an orbit may take several days. (This leads to the long integration times needed for the numerical models we have just discussed.) As a result, small forces acting on large, slow moving, particles may have a significant effect if integrated over long periods. For example, while the dust may be "de-coupled" from the gas on short timescales, momentum transfer from expanding gas to the dust particles would never be exactly zero. Perhaps more importantly, solar radiation pressure on a centimetre-sized particle can result in an anti-sunward velocity change of 5–10 cm s$^{-1}$ at 1 AU. This force dominates Lorentz force, Poynting-Robertson effect, and ion drag by several orders of magnitude for particles of this size (e.g. Grün 2007). This implies that particles with apoapses approaching the edge of the Hill sphere do not follow purely ballistic trajectories and will not re-impact the nucleus at pericentre. Consequently, particles producing airfall deposits on the nucleus are not those that have ejection velocities just below escape velocity for objects of the size of 67P.

### 4.11.2  Individual Particle Dynamics

A further force of possible importance is the so-called "rocket effect". Here, an anti-sunward force acts on the particle as a consequence of sublimation. Evidence of the influence of this effect on dust particle trajectories has been presented by Agarwal et al. (2016). The influence of rocket effect is constrained by two particle properties—the mass of volatile available for sublimation and the moment of inertia which determines a particle's reaction to applied torque. The maximum momentum transfer rate to the particle is given by the mass sublimation rate multiplied by the effective velocity of the emitted molecules. This can be estimated as

$$\frac{dp_m}{dt} = Z\,\pi a^2\, m_{H2O} v_{H2O} \tag{4.112}$$

where $Z$ is the sublimation rate per unit area, $p_m$ is the momentum, $\pi a^2$ is the cross-sectional area of the particle, $m_{H2O}$ is the mass of the water molecule and $v_{H2O}$ is the effective velocity with which the water molecule leaves the particle. The relative change in mass of a particle per second gives an approximation for the lifetime against sublimation and is of the order of

$$\frac{dm}{m} = \frac{3}{4}\frac{Z\,m_{H2O}}{a\,\rho_d} \tag{4.113}$$

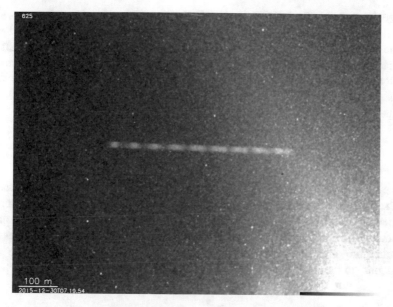

**Fig. 4.58** A rotating dust particle caught during a 6 second OSIRIS exposure. The particle exhibits roughly nine rotations within the exposure. The particle is clearly out of focus and was therefore less than 2 km from the camera at the time of acquisition. (Image number: N20151230T071954750ID10F22)

where $\rho_d$ is the particle density. $dm/m$ takes a value of around $2 \ 10^{-4} \ s^{-1}$ for 1 mm particles giving a lifetime of just over 1 h. We shall see shortly that there are significant issues with this calculation because of the requirement for a high absorption coefficient at optical wavelengths. However, it indicates that sublimation rates from particles at levels comparable to the free sublimation rate cannot be maintained and are required to be orders of magnitude lower to allow there to be significant numbers of large particles with lifetimes long enough to populate the Hill sphere if the large particles are mostly composed of ice.

Unbalanced reaction forces on the particle will also initiate rotation. One of the surprises from Rosetta was that the imaging system could observe the rotation of individual large dust particles. An example is shown in Fig. 4.58. The particle (observed on 30 Dec. 2015) was less than 2 km from the camera and was therefore out of focus but the nine rotations of the particle can be clearly seen in the brightness variation as it moved during the 6 second exposure time. This is equivalent to around $10 \ rad \ s^{-1}$. Ivanovski et al. (2017a, b) looked at how deviations from sphericity can affect the equation of motion and showed that there could be a significant influence on the velocity dispersion of particles. This can be particularly important for slower moving particles in a highly non-uniform gas flow field.

To include this effect in a forward model, however, requires knowledge of individual particle shapes which generates further free parameters making it challenging to produce non-degenerate results.

We have seen (Sect. 2.9.3.1) that if the thermal parameter, $\Theta$, approaches 100, then the surface of an object becomes isothermal with rotation. For a thermal inertia of 50 TIU this implies that rotation faster than 0.7 rad s$^{-1}$ will result in a particle that is thermally a fast rotator and the net rocket effect will tend to zero. We can look at the proportionalities for torque-induced rotation with a simple set of assumptions.

The angular acceleration is

$$\frac{d\Omega_d}{dt} = \frac{d_l \times F_A}{I} \tag{4.114}$$

where $F_A$ is the applied force, $d_l$ is the lever arm with respect to the rotation axis, and $I$ is the moment of inertia. The net imbalancing force can be written as the change in momentum (as above) with sublimation from a surface area that is proportional to the square of the particle radius. For this exercise we assume that a torque arises from a sublimating area that is ¼ of the cross-sectional area and directed in the plane orthogonal to the rotation axis. This gives

$$F_A = Z \frac{\pi a^2}{4} m_{H2O} v_{H2O} \tag{4.115}$$

If we further assume that the lever arm, $d_l$, is also a function of radius (we will use $2a/3$) and if a moment of inertia for a filled sphere is used we can arrive at the equation

$$\frac{d\Omega_d}{dt} = \frac{5}{16} Z \frac{m_{H2O} v_{H2O}}{\rho_d \, a^2} \tag{4.116}$$

For decimetre-sized (0.1 m) particles, the resulting angular acceleration is of the order of $10^{-3}$ rad s$^{-2}$ assuming free sublimation and a water vapour ejection velocity of 100 m s$^{-1}$. This result implies that a decimetre particle would become isothermal in around 1000 seconds. The numerical constant on the right-hand side depends on the numerous assumptions that have been made (in particular the surface area of the imbalancing force and the effective sublimation rate, both of which could an order of magnitude smaller). However, the fundamental points are that the angular acceleration is proportional to $1/a^2$ and that any net rocket effect from sub-decimetre particles will be lost within, at most, a few hours as the surface temperature becomes isothermal.

### 4.11.3   Neck-Lines and Dust Trails

Another phenomenon associated with the presence of slow moving large particles is commonly referred to as a "neck-line structure". Figure 4.59 shows this structure in a

**Fig. 4.59** The neck-line
structure of C/1995 O1
(Hale-Bopp) observed on
5 January 1998. (Courtesy
of G. Cremonese; see also
Fulle et al. 1998 for neck-
line observations
pre-perihelion)

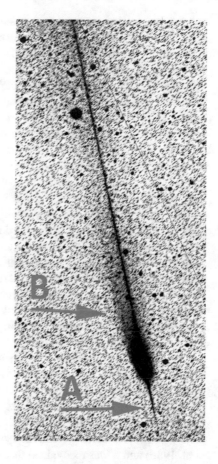

ground-based image of C/1995 O1 Hale-Bopp taken in January 1998. A narrow line-like structure is, in projection, superimposed on the usual dust coma and tail. This observation also shows the narrow structure on both sides of the nucleus. Observations of this phenomenon have led to phrases such as "sunward spike" and "anti-tail" being used to describe the appearance. In the case of this observation of Hale-Bopp, there is a spike extending from the nucleus in the opposite direction to the normal, more diffuse, dust tail as well as the rest of the neck-line going roughly, but not exactly, in the same direction as the tail.

The neck-line structure results from the presence of particles that have escaped the Hill sphere but their velocities are still low relative to the nucleus. They must also be large otherwise radiation pressure effects will dominate their motion through an anti-sunward acceleration. The slow motion leads to the particles staying close to the orbital plane of the nucleus. Consequently, when an observer images the comet from a position in the orbital plane, he sees a sharp rise in particle column density. This indicates the importance of projection effects in studying dust trail phenomena.

We can illustrate this with a simple Monte Carlo calculation using a Keplerian orbit for a nucleus source combined with the difference in the gravitational acceleration arising between the nucleus and a particle as it slows drifts away. With Monte Carlo techniques it is, of course, straightforward to initialise particle velocities within limits. Sunward ejection within a 30° half-angle cone would be a reasonable starting point, for example, at velocities slightly higher than the escape velocity. The particles can then be tracked through time. The influence of radiation pressure can also be included through the $\beta$ parameter (Eq. 4.72). Two example results for two different values of $\beta$ are shown in Fig. 4.60. The test case shown here is rather generic—a Jupiter family comet with a perihelion at 1.3 AU with emission of particles in a 30° cone about the sub-solar point at a velocity of 6 m s$^{-1}$ relative to the nucleus. The plots show distributions after 1.6 orbits of the comet about the Sun with t = 0 at aphelion. In the top panels, we can see the distribution of particles when looking down upon the orbital plane of the comet and the influence of $\beta$ is clearly evident. In the centre panels, the viewing direction is in the orbital plane of the comet from the direction of the Sun. The particles are seen to be distributed about the orbital plane independent of $\beta$ although the structure of the distribution shows dependence upon $\beta$. Note that particles can appear on both sides of the nucleus depending upon viewing geometry and the value of $\beta$. As discussed by Fulle and Sedmak (1988), the photometric analysis of these structures can provide information on large (millimetre-sized) particles ejected by nuclei. However, there are numerous free parameters in the modelling including the distribution of $\beta$ values, the assumed surface emission distribution, and its time dependence.

The neck-line structures are seen in the vicinity of the nucleus itself and are, relatively speaking, quite young. However, the trajectories of these large particles will evolve and, with time, spread out further to produce a cloud of particles along the orbital path. These older particles can be detected in the infrared through their thermal emission and are referred to as cometary dust trails. The significance became clear when observations were made by the IRAS spacecraft (Eaton et al. 1984; Sykes et al. 1986) and it was recognized that these trails were consistent with low velocity emission over timescales of many years and possibly decades.

Spitzer observations were acquired of many dust trails associated with known comets. Two examples can be seen in Fig. 4.61. The upper panel shows 2P/Encke (Kelley 2006). The white lines indicate the cometary dust trail. The lower panel shows a similar observation for 67P (see Kelley et al. 2008). Here the dust trail is far less bright particularly to the left of the nucleus in the image. The orbit of 67P was perturbed by a close approach to Jupiter in 1959 and hence we may be seeing here a reduced dust trail brightness purely because 67P has not been in its present orbit for more than a few orbital periods. The particles in the trails are themselves subject to perturbations and hence orbital stability over long timescales is not always assured.

Dust trails are effectively meteor streams and if they intersect the Earth then this can result in a meteor shower. When (3200) Phaethon was discovered (Davies et al. 1984), it was recognized as being the probable parent of the Geminid meteor stream (Williams and Wu 1993) and on that basis it was proposed as being a "dead" comet meaning that it had been active once but had ceased being active as a result of either volatile loss or an activity choking mechanism. The stream itself has been estimated as being about 1000 years old. While this link hinted at Phaethon's unusual

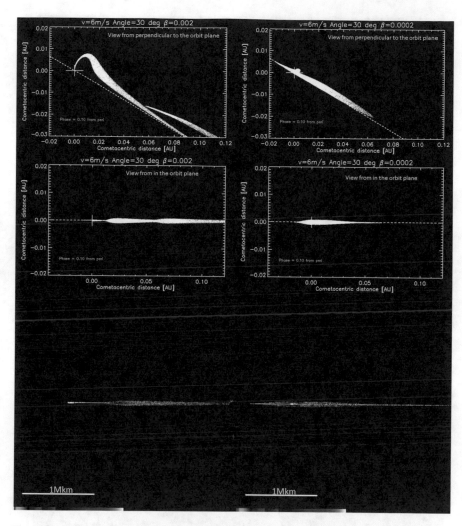

**Fig. 4.60** Monte Carlo model calculation of the dynamics of slow moving particles emitted by comets. Snapshot of particle positions seen after 1.6 orbits of the comet about the Sun ($0 =$ aphelion). Emission is from a $30°$ cone about the sub-solar point. Left: Model for a high beta value. Right: Model for a lower beta value. first row: Particle positions seen from perpendicular to the orbit plane. Middle row: Particle positions seen from in the orbital plane of the comet. Bottom row: How the structure in the middle row would appear using a weighting of the production proportional to $1/r_h^2$. A non-linear colour table has been used to emphasize low column densities

character, subsequent improvements in observational capabilities have now shown Phaethon to be one of a class of objects now generally referred to "active asteroids". A comet-like tail was discovered in two apparitions by Jewitt et al. (2013) (cf. Hsieh and Jewitt 2005). It is now the target of a space mission (JAXA's DESTINY+) to study its exact nature.

**Fig. 4.61** Spitzer observations of two comets showing dust trails at 24 μm wavelength using the MIPS instrument. Top: 2P/Encke. The white lines indicate the dust trail. Credit: NASA/JPL-Caltech/M. Kelley (Univ. of Minnesota). Bottom: 67P/Churyumov-Gerasimenko. Credit: NASA/JPL-Caltech/W. Reach (SSC/Caltech)

The particles in the dust trails are suggested to have diameters of the order of 1 mm although there are numerous assumptions in this value such that it should be taken as indicating an order of magnitude. Trails represent the principal means by which comets contribute dust to the zodiacal light and their total contribution is of the order of $10^{11}$ kg/year (Sykes et al. 2004) but with large uncertainties.

## 4.12   Radiometric Properties of Dust

### 4.12.1   The Colour of Dust

Studies of the colour of dust emitted from nuclei have been made for many years. Lamy et al. (1989), for example, summarized measurements of 1P/Halley made through ground-based and spacecraft observations. Dust colours using the spectral gradient definition from Eq. (2.85) appear to cover a range of values. For 1P/Halley, values of the spectral gradient (Sect. 2.7) of between 6 and 9% $(100 \text{ nm})^{-1}$ were obtained. Kolokolova et al. (2004) give a general range of 5–18% $(100 \text{ nm})^{-1}$ for wavelengths between 350 and 650 nm.

It is widely assumed that the dust colour should be the same as that of the non-volatile material at the source. However, because light is being scattered preferentially by particles close to the wavelength, there may be subtleties to account for. One of the interesting but not yet fully explained results from the observations of 1P/Halley by the Halley Multicolour Camera (HMC) onboard Giotto spacecraft concerns the relative colour difference between the dust and the nucleus. Thomas and Keller (1989) showed difference images that revealed that the nucleus was less red than dust in the coma at a scattering angle of 73°. Although the absolute colour of the dust and the nucleus were the same within error, the relative difference was clearly visible and equal to ~3–4% $(100 \text{ nm})^{-1}$ for the range 440 nm to 650 nm. Given that the visible nucleus surface is the source of the dust, the surface might contain an additional component that, when combined with the dust, produces a more blue surface colour. A mixture with water ice might produce such an effect. Before major activity starts, the surface may be more dusty and redder. But when erosion is rapid, more ice becomes evident and one no longer see the pure, redder colour of the dust. Such a model was suggested by Fornasier et al. (2016) for observed colour changes on 67P. Given that water ice must be present close to the surface in order to drive dust emission and given its relatively blue colour in pure form (Pommerol et al. 2015), this is clearly plausible although it requires a mixture of the components that provides an almost constant colour over much of the illuminated and visible surface of 1P/Halley as seen by HMC. In other words, it requires a large degree of homogeneity over the surface.

An alternative in the case of 1P/Halley is that dust in the coma between the Sun and the nucleus preferentially scatters blue light in the forward direction so that the nucleus is illuminated by a slightly bluer light distribution than a solar spectrum (Fig. 4.5). This effect was seen at Mars where relative variations in the colour of surfaces with time of day could be used to demonstrate clearly the effectiveness of this phenomenon in an atmosphere with an optical depth in the range 0.6–0.8 (Thomas et al. 1999). With the optical depth from the Sun to the surface at 1P/Halley at the time of the Giotto fly-by also being close to this range, this explanation is plausible (cf Salo 1988). In cases where the optical depth is <0.1 (as is the case for most observations of 67P), the influence of this effect on measurements should be minimal.

Some of the ground-based studies have included the determination of spatial variations in the colour of dust comae on large scales by subtraction of images acquired through different filters. A recent example for 67P is given in Rosenbush et al. (2017). The cause of spatial variations in colour remains somewhat speculative with the obvious candidates being variations in size and/or composition at the source. While Rosetta/OSIRIS observations at 67P might be useful in addressing this problem, there are issues with "optical ghosts" caused by having weak but non-negligible internal reflections within the cameras that produce uncertainty with these subtle effects at low optical depths.

### 4.12.2   The Thermal Properties of Dust and Sublimation

We have so far focussed on scattering of sunlight by the dust particles. However, they absorb and re-radiate sunlight if the single scattering albedo is less than 1. As we have alluded to in the previous section, absorbed energy can sublime residual ice in the particle (potentially affecting its cross-section). It can also heat the gas (either through radiation or conduction) and thereby affect the energy available for expansion. The absorbed energy per second is given by (Lien 1990).

$$E_{sol} = \pi a^2 \int_0^\infty Q_{abs}(\lambda) \, \frac{F_\odot}{r_h^2} \, d\lambda \qquad (4.117)$$

The absorption efficiency is related to the size parameter, $x$, and thus it is dependent upon both wavelength and particle size. This can be seen in Fig. 4.62 for four particles sizes with a refractive index of $m_{ref} = (1.60, 0.01)$.

**Fig. 4.62** The wavelength dependence of the absorption efficiency, $Q_{abs}$, for four particle sizes computed using Mie theory. Solid: 0.1 um. Dashed: 0.5 $\mu$m. Dot-dash: 1.0 $\mu$m. Dot-dot-dash: 5 $\mu$m. The refractive index used was $m_{ref} = (1.60, 0.01)$

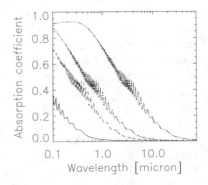

**Fig. 4.63** The size
dependence of the
equilibrium temperature of
dust particles for three
different refractive indices at
1 AU. Solid line:
$m_{ref} = (1.6, i0.01)$, dash:
$m_{ref} = (1.6, i0.1)$, dot-dash:
$m_{ref} = (1.8, i0.01)$

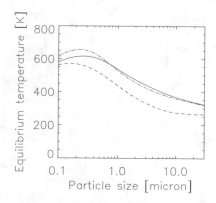

As pointed out by Lien (1990), there are additional heating terms arising from heating by the gas (collision, conduction, and radiation terms), from interaction with the solar wind, and from re-radiation by the nucleus. In most cases, these factors can be ignored. However, the heating of slow moving, possibly bound, large particles by the thermal emission from the nucleus is of potential importance given the large solid angle that the nucleus can subtend at the particle's position.

The power thermally re-radiated by the particle can be expressed as

$$E_{rad} = 4\pi a^2 \int_0^\infty Q_{abs}\,(a, \lambda) B_\lambda(T)\, d\lambda \qquad (4.118)$$

where $B_\lambda(T)$ is the wavelength-dependent Planck radiation function at the particle's temperature (Eq. 2.9). $Q_{abs}$ appears in this equation as a consequence of Kirchhoff's law because, in equilibrium, the absorption efficiency must equal the emission efficiency. One can now find the value of the temperature that leads to an equilibrium between $E_{sol}$ and $E_{rad}$. As Fig. 4.63 shows, the temperature is strongly particle size dependent.

There are two aspects that increase the complexity of the problem dramatically. Firstly, the refractive indices of most materials of interest are wavelength dependent as we saw in Fig. 3.32 for water ice. The consequences are quite profound (Fig. 4.64). The temperature of water ice particles in the 0.1 to 50 μm range are well below 200 K and decrease with particle size. This leads to typical lifetimes for the particles that range from $10^3$ (sub-micron particles) to $10^{10}$ seconds (>50 μm) at 1 AU because of the exponential dependence on temperature as shown in the more detailed calculations of Lien (1990). This implies that optically active water ice particles can take many hours and even days to sublime even at 1 AU and hence they would not normally be a major extended source for water in the innermost coma. Sunshine et al. (2005) noted that ice absorptions were still detectable at 9P/Tempel 1, $2.7\ 10^3$ seconds (45 min) after the impact of the Deep Impact impactor and this is consistent with microscopic ice particles having lifetimes of the order of hours.

This should not be taken as implying that sublimation of particles in the innermost coma ($\lesssim$100 km from the nucleus) cannot occur. The case we have just examined is

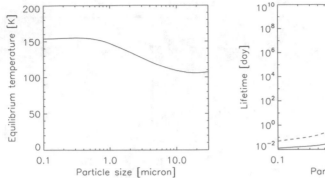
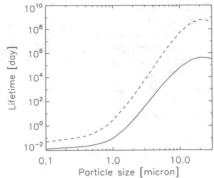

**Fig. 4.64** Left: The equilibrium temperature of pure water ice particles at 1 AU following the approach of Lien (1990). The temperatures are well below the free sublimation temperature of water ice leading to relatively long lifetimes for these particles. Right: The lifetimes at 1 AU (solid line) and 1.3 AU (dashed line—appropriate for 67P near perihelion). Note that the units here are in days

an ideal one. Impurities will affect the absorption coefficient and it is straightforward to imagine that, for example, carbonaceous materials with high absorption at visible wavelengths could be combined with water ice in particles at the source leading to more rapid sublimation once the particles have left the surface. However, at present we have no concrete evidence for such a scenario. On the other hand, we have already seen (Sect. 3.2) that extended sources of gas species have been seen both in situ and from the ground.

VIRTIS-H observations of the thermal and reflected dust continuum have been studied by Bockelée-Morvan et al. (2019). The basic approach is illustrated in Fig. 4.65 which uses a style of presentation similar to that seen in the Bockelée-Morvan paper. The reflected continuum is fit with a Planck function at the temperature of the Sun while the thermal emission from the dust is fit with a Planck function at a temperature to be determined and found in this case to be 295.2 K. However, the reflected and thermal continua are not independent because they are linked through the scattering properties of the particles—the absorbed solar flux needs to be matched self-consistently with the thermal emission. Here, we use a slightly different approach to that given by Bockelée-Morvan et al. (2019) for illustrative purposes.

Taking the Sun as a black-body, the reflected spectral intensity (in [W m$^{-2}$ sr$^{-1}$ nm$^{-1}$] for example) from a particle is given by

$$I_{ref}(\lambda) = \frac{B_\lambda(T_\odot)}{k_N} \frac{S_\odot}{r_h^2} Q_{sca}(\lambda) \, \pi a^2 \, \Phi(\alpha) \qquad (4.119)$$

where $S_\odot$ is the integrated solar flux, and $B_\lambda$ and $k_N$ are defined in Eqs. (2.9) and (2.10) to normalize the Planck function and $B_\lambda$ is computed at the effective temperature of the Sun, $T_\odot$. The phase function is normalized over all solid angles to

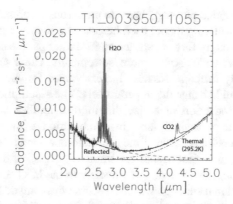

**Fig. 4.65** VIRTIS-H spectrum from off the limb of the nucleus of 67P acquired on 2015-07-08 T21:10 at a phase angle of 90° and a heliocentric distance of 1.31 AU. The continuum reflected and thermal emission from the dust has been fit by a simple algorithm using Mie theory. The gas emissions of $H_2O$ and $CO_2$ are marked. (Diagram construction following Bockelée-Morvan et al. 2019)

1 (Divine et al. 1986). The absorbed power is given by Eq. 4.117 which allows us to compute a particle temperature. Rather than using Eq. 4.118, for the radiated power we can make a simplifying approximation by replacing the integral of $Q_{abs}$ over wavelength with a single emissivity constant, $\varepsilon$, as is done for surfaces and using this as a free parameter. This allows us to write

$$T^4 = \frac{1}{4\varepsilon\sigma} \frac{S_\odot}{r_h^2} \int_0^\infty Q_{abs}(\lambda) \frac{B_\lambda(T_\odot)}{k_N} \, d\lambda \qquad (4.120)$$

so that the thermal spectral intensity is

$$I_{th}(\lambda) = 4 \frac{B_\lambda(T)}{k_N} \varepsilon\sigma T^4 \qquad (4.121)$$

Using Mie theory to compute $Q_{sca}$ and $Q_{abs}$, we can fit the continuum. The values required to make the model curve in Fig. 4.65 use a size parameter, $x$, of 1.0 at a wavelength of 2 μm, $m_{ref} = (1.7, i1.5)$, and $\varepsilon = 0.63$ to give the observed dust temperature. The refractive index indicates strongly absorbing particles and its values are close to those expected for carbonaceous materials (e.g. Fig. 4.10) which probably make up ~50% of the dust composition (see Sect. 4.14). The size parameter is indicative of the dominant contribution of small particles to the scattering cross-section. This might be expected for values of the power law exponent, $b$, greater than 2 (Table 4.1). The dust temperature indicates super-heating of the particles as discussed by Bockelée-Morvan et al. (2019). The similarity of these Mie parameters to those needed to fit the phase function of Fink and Doose (2018) (Sect. 4.2.8) is interesting although possibly co-incidental.

While this might sound pleasing, there are numerous assumptions leading to large uncertainties. The scattering model used here is Mie theory which produces phase functions that are strongly forward scattering for larger particles. This forces the fits to lower particle sizes which scatter more isotropically. Other scattering models can produce significantly different (usually larger) values for the phase curve at 90° phase angle and would change the fit parameters. The assumption of a single size is not tenable but introduction of a size distribution implies at least 1 more free parameter. (The use of a single size in combination with Mie theory actually complicates the inversion because of the oscillations we have seen in, for example, Fig. 4.5 affect the numerical search for minima.) Also, as discussed above and evident in Fig. 4.5, $m_{ref}$ is a function of wavelength but has been assumed constant.

It should be noted that although we have spectra, there are de facto only three data points to fit; the temperature and amplitude of the Planck function of the thermal emission and the amplitude of the Planck function corresponding to the reflected sunlight. As we have at least five free parameters here ($m_{ref,i}$, $m_{ref,r}$, $\varepsilon$, $x$, and a scaling factor corresponding to the column density) as well as a choice of scattering model, some degeneracy in the solutions cannot be avoided.

Bockelée-Morvan et al. (2019) performed a more sophisticated fitting by incorporating dust colour into the fit (although the magnitude of the colour gradient was small as can also be seen by the rather good fit to the reflected continuum seen in Fig. 4.65 which does not include this term) and limiting the free parameters by using a bolometric albedo to avoid specifying a particle size. However, in all cases, one obtains a dust temperature fairly accurately. The derived temperature here is above the free sublimation temperature of water ice suggesting that sublimation cooling of the particles is not a dominant process even in the innermost coma.

The VIRTIS-H data for this analysis were line of sight integrations above the limb of the nucleus. Motion of the spacecraft pointing allowed study of the variation in dust properties with the impact parameter. Bockelée-Morvan et al. (2019) concluded that there were variations in colour temperature with impact parameter with the computed albedo increasing with distance above the limb (from 0 km to ~8 km). It needs to be noted that this region is extremely dynamic with large changes in the dust size distribution with cometocentric distance occurring because of the acceleration gradient between small and large particles. Hence, the interpretation of this result is by no means straightforward.

### 4.12.3   The Polarization of the Scattered Light

In the discussion of the Stokes vector (Eq. 4.6), we saw how the linear polarization is defined. Sunlight scattered from comets is partially polarized. The plane of linear polarization is either perpendicular or parallel to the scattering plane that is defined by the Sun, Earth, and the comet (Zubko et al. 2016). The degree of linear polarization can then be expressed as

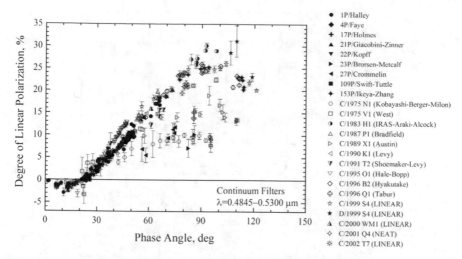

**Fig. 4.66** Degree of linear polarization as a function of phase angle for 23 comets. (Reprinted from Zubko et al. 2016, with permission from Elsevier)

$$\mathcal{P} = \frac{I_\perp - I_\parallel}{I_\perp + I_\parallel} \tag{4.122}$$

where $I_\perp$ and $I_\parallel$ are the intensities of the scattered electromagnetic fields perpendicular and parallel to the scattering plane. The degree of linear polarization is usually plotted as a function of phase angle, $\alpha$ and an excellent example showing $\mathcal{P}(\alpha)$ for 23 comets is shown in Fig. 4.66 (from Zubko et al. 2016). At $\alpha \sim 21°$, the polarization crosses from being negative (the so-called negative branch) at lower phase angles to being positive at higher phase angles. The plot gives the impression of a bi-modal distribution (see also Levasseur-Regourd et al. 2004) such that there appear to be two types of comets—one showing a high polarization at 90° phase angle with the other group showing ~50% lower polarization. However, this may be an artifact caused by low spatial and spectral resolution. Close to the nucleus, the emitted flux in a broad-band optical filter will be dominated by the dust. The contribution from gas emissions to the total radiance increases with the impact parameter and this unpolarized source becomes increasingly important. Hence, higher spatial resolution measurements should see higher polarization in the near-nucleus region. The low $\mathcal{P}_{max}$ comets are also thought to be mostly gas rich and ignoring the contribution of molecular emissions could lead to artificially low polarization values for these comets (Chernova et al. 1993). Zubko et al. (2016), however, conclude that the range of polarizations observed (from 7% up to more than 30%) cannot be explained through this depolarization by gaseous emissions. They conclude that $\mathcal{P}_{max}$ unambiguously measures the relative abundance of refractory materials such as Mg-rich silicates, organics and/or amorphous carbon.

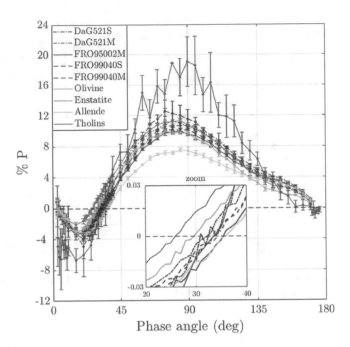

**Fig. 4.67** Laboratory measurements of the degree of linear polarization of a set of cometary analogues. The inset shows that the angle at which the polarization goes to zero may be related to composition. (Reprinted from Frattin et al., MNRAS, Frattin et al. 2019)

For phase angles larger than 30–40°, the polarization at a given phase angle increases with wavelength, a phenomenon sometimes referred to as the polarization colour effect. A convenient way of representing this effect is to use a polarimetric colour gradient

$$\frac{\Delta P}{\Delta \lambda} = \frac{P(\lambda_2) - P(\lambda_1)}{\lambda_2 - \lambda_1} \tag{4.123}$$

which may be adequate over a limited wavelength range.

Frattin et al. (2019) have presented experimental phase functions and values of $P(\alpha)$ for seven cometary dust analogues. The curves (Fig. 4.67) are consistent with those seen in Fig. 4.66. They suggest that the phase angle at which the degree of polarization reaches zero between the negative and the positive branches is indicative of the composition of the particles but this appears to be a rather subtle effect.

Sen et al. (2017) completed a systematic study of linear polarization using particle simulations. They investigated composition and also showed that the depth of the negative branch requires significant porosity but their calculations were restricted to the 0.01–1.00 μm size range. Previously, Ivanova et al. (2015) presented measurements of $P(\alpha)$ of the distant comets C/2010 S1 (LINEAR) and C/2010 R1 (LINEAR) at heliocentric distances of 5.9–7.0 AU. Using T-matrix calculations, they also fit their data with the 1.3 μm radius dust aggregates composed of 1000 spherical monomers of 0.1 μm and refractive indices of $m_{ref} = (1.65, i0.05)$ although the number of measurements obtained was small and exclusively at low phase angle. The similarity to values suggested as fitting the 67P phase function and the infrared spectra of 67P is somewhat remarkable.

## 4.13   Equation of Motion of Charged Dust

Once the dust particles have de-coupled from the gas and particles are outside the Hill sphere (typically at distances >300 km), the gas drag and gravity terms disappear from the equation of motion and the major influence on particle motion results from radiation pressure resulting in the "fountain-like" appearance of the dust coma. However, there are other, more subtle, forces at work. In particular, dust particle charging combined with the inter-planetary magnetic field can influence dust particle motion. The basic equation describing the influence of charging is

$$m_d \frac{dv_a}{dt} = q(t)(v_a - u_{sw}) \times B + F_{pr} + F_c + F_{ig} + F_{gss} \qquad (4.124)$$

where $q(t)$ is the time-dependent charge on the particle, and $F_{pr}$, $F_c$ and $F_{ig}$ are the radiation pressure force (seen in Eq. 4.97), the plasma drag force, and the intergrain Coulomb force respectively (cf Sterken et al. 2012; Mendis and Horányi 2013; Lhotka et al. 2016). We again use $v_a$ here for the velocity of a specific particle size. $F_{gss}$ expresses the perturbations arising from other massive objects in the Solar System if required.

To solve the equation of motion, a model of the plasma environment is required. Simple models will be discussed in Chap. 5. However, the solar wind interaction with the comet is rarely stable for any length in time and this can lead to relatively complex behaviour.

The influence of charging is most evident in the behaviour of the smallest particle sizes. Electromagnetic effects can greatly distort the spatial distribution of small particles, leading to their swift non-symmetrical dispersal (Mendis and Horányi 2013) and it should be noted that small particles also have a low scattering efficiency at optical wavelengths (e.g. Fig. 4.8). Hence, their spatial distributions are not really well characterized by observation.

Dust charging has been proposed as a possible cause for striations observed in some cometary dust tails. Hill and Mendis (1980) suggested that fragmentation of grains could occur because of electrostatic charging by keV electrons. The most recent discussion of this mechanism has been by Price et al. (2019) with relation to the unusual appearance of C/2006 P1 (McNaught) as observed using the SOHO/ LASCO C3 coronograph (Fig. 4.68) and the STEREO-A spacecraft. They concluded that dust-solar wind interaction does occur and that a mechanism, proposed by Nishioka (1998), in which a continuous cascade of fragmentation, occurs, possibly as a result of charging.

## 4.14   The Non-volatile Composition of Dust and the Nucleus

It is perhaps surprising that, while we have discussed ices and volatiles coming from the nucleus extensively, we have appeared to ignore the surface non-volatile composition. This arises because, up to this point in time, direct measurements are limited. Remote-sensing observations have limitations and consequently direct measurements in the inner coma of the dust emitted by the nucleus play a very large role in trying to deduce the surface composition.

Jessberger et al. (1988) described the abundance of elements in dust particles analysed by the PUMA experiment at 1P/Halley during the Vega fly-bys while Bardyn et al. (2017) has provided the average composition of 67P's dust particles as deduced from COSIMA measurements on Rosetta. The 1P/Halley data showed the importance of organics and the dominance of carbon, hydrogen, oxygen and nitrogen in the composition. This is illustrated in Fig. 4.69 which shows the relative abundance of C, N, and O with respect to heavier elements (hydrogen has been excluded). Around ¼ of the element mass in the particles was carbon which led to the general assumption that roughly 50% of the emitted dust from comets is organic. Bardyn et al. (2017) deduced that, at 67P, organics make up ~45% and hence this conclusion seems relatively robust. However, analysis of data from the CONSERT

**Fig. 4.68**  Image from the SOHO/LASCO C3 coronograph of C/2006 P1 (McNaught) in January 2007. The nucleus region was saturated but faint irregular linear structures could be seen in the dust tail

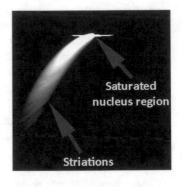

**Fig. 4.69**  Comet 1P/Halley dust composition grouping the elements C, N, and O and comparing the mass to that of heavier elements. Data from Jessberger et al. (1988)

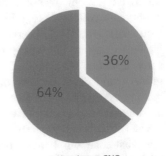

CNO/Heavy Elements by Mass

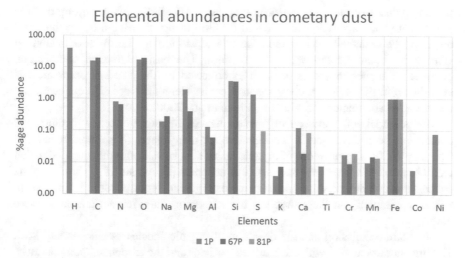

**Fig. 4.70** Comparison of elemental abundances in the dust from in situ measurements at comets 1P/Halley, 67P/C-G, and 81P/Wild 2. (Data from Jessberger et al. 1988 (renormalized to Fe = 1), Bardyn et al. 2017, and Flynn et al. 2006)

experiment led to an organic to silicate ratio of the non-volatile component of at least 75:25 (Hérique et al. 2016). Comparisons are, however, difficult. The mass spectrometers actually derive a C/Si ratio and must assume how the carbon is bound to other elements while the radiowave sounding experiment deduces the ratio through comparison of the permittivity with standards through calibration using density constraints (Hérique et al. 2019).

The comparison in Fig. 4.70 of mass spectrometer data shows that 1P/Halley and 67P were broadly similar. The scale is, however, logarithmic and there are some subtle differences. Notably, 1P/Halley was around five times richer in magnesium and calcium.

The in situ mass spectrometer data are vital but the detailed organic and mineralogical composition is not well defined through these experiments. The Philae lander on Rosetta carried a gas chromatograph coupled to a stepped combustion analysis system specifically to look at the isotopic ratios in organics (Wright et al. 2007) but the problems encountered with the landing system compromised this experiment. The difficulty in adapting Earth-based analytical experiments for space implementation is probably the strongest argument for sample return missions. Ground-based techniques are highly advanced but high performance equipment is rarely portable and often highly specific.

In principle, interplanetary dust particles (IDPs) captured at high altitude in the Earth's atmosphere can be used to study composition in the laboratory. Some of these particles have a cometary origin (Brownlee 1977) as is clearly imaginable from when the Earth passes through a meteor stream such as the Perseids—the stream associated with 109P/Swift–Tuttle (a periodic comet with an orbital period of 133 years). The Stardust mission was designed to take this a step further by collecting dust particles at a comet, in this case, 81P/Wild 2, and returning these to Earth. With such a mission there could be little or no discussion about the exact origin of the collected material and the full capabilities of Earth-based analytical laboratories could be thrown at the samples. The basic elemental composition obtained from measurements of particles collected by the Stardust experiment is also shown in Fig. 4.70 for comparison. Heavier elements were also collected and their relative abundances provided by Flynn et al. (2006).

The potential link between organic molecules in the interstellar medium and comets has been discussed in the literature for many years (e.g. Ehrenfreund and Charnley 2000). The presence of organic material in comets and the knowledge that their ejected dust has entered the Earth's atmosphere (Brownlee 1977) has led to the concept that comets carry and spread complex organics throughout the Solar System. But how complex can these organics become before they impact a Solar System body?

The Stardust mission had the detection of organic species as a secondary goal. The organics associated with the returned particles and the associated particle tracks in the aerogel collection system were found to be a mixture of aromatic, aliphatic, and polycyclic aromatic hydrocarbons (PAHs) (Sandford et al. 2006). Glavin et al. (2008) identified methylamine and ethylamine above background levels in hot-water extracts of returned Stardust materials. Glycine, the simplest amino acid ($NH_2$–$CH_2$–COOH) was also found but was also present in the control sample initially indicating a terrestrial origin. Subsequently, Elsila et al. (2009) reported detection of glycine in the samples in addition to β- and L-alanine (2-aminopropanoic acid—$C_3H_7NO_2$) which is a left-handed molecule that is incorporated into a significant fraction of known proteins.

Glycine was detected in the gas phase of 67P by ROSINA (Altwegg et al. 2016). Hadraoui et al. (2019) suggest that the observations are consistent with an extended source with the glycine being embedded in water ice that sublimes from dust particles after ejection from the surface. On the other hand, despite several attempts, interstellar glycine has not yet been detected (Ohishi et al. 2019) and there are grounds to suggest that its destruction through energetic particle radiation may be quite rapid unless it is "shielded" by a sufficient depth of material such as water ice (Maté et al. 2015).

Some insoluble inorganic matter was also found in the Stardust samples and analysis determined that it contained a consistent functional group chemistry, dominated by carboxyl and ketone bonding, with a variable amount of aromatic C=C bonding (De Gregorio et al. 2011; Westphal et al. 2017).

Carbonates are products of aqueous alteration, so those found in extraterrestrial samples are taken as a signature of the presence of liquid water. Detections in comets imply that reaction pathways such as

$$CO_2 + H_2O \rightarrow OH^+ + CO_3^-$$
$$CO_3^- + Ca^{++} \rightarrow Ca(CO_3)_2$$

or with different cations might have occurred at some point. Curiously, although no large carbonate grains have been identified in Stardust samples from 81P/Wild 2, numerous submicron carbonate grains, frequently intermixed with other mineral grains, have been found (Westphal et al. 2017). The presence of sulphides in 81P/Wild 2 particles (Westphal et al. 2009) in the form of pyrrhotite and troilite would seem to agree with interpretations of the Spitzer data during the Deep Impact experiment (see below).

Carbonaceous chondrite meteorites are among the most pristine material known in the Solar System and hence there may be some relationship to comets. Their classification has been described by Sears and Dodd (1988). From the organic composition point of view, CI and CM are the most interesting classes, in particular because they contain up to 2 wt% of organic carbon. In laboratory studies, more than 80 different amino acids have been identified in the Murchison meteorite alone including some that are rare in terrestrial environments and unquestionably of extraterrestrial origin (Botta et al. 2007).

Remote-sensing observations of the composition of the resolved nucleus rely on infrared spectroscopy. VIRTIS-M measurements of the surface of 67P suggest that it consists of an assemblage of various organic components, minerals and the water ice that could be deduced from the outgassing. Filacchione et al. (2019) suggested that the organic compounds contain COOH and OH-groups and that a refractory macromolecular material bearing aliphatic ($CH_2$ and $CH_3$) and polycyclic aromatic hydrocarbons would fit the infrared reflectance spectra of 67P. They also suggested that this material is mixed with minerals, including silicates, Fe-sulphides (pyrrhotite and/or troilite) and possibly salts with ammonium cations (Altwegg et al. 2020; Poch et al. 2020). However, both the resolution of the instrument and the intimate mixtures of many compounds containing similar chemical groups inevitably leads to ambiguity. The infrared reflectance spectra of 9P/Tempel 1 acquired by Deep Impact show very little structure and ratios between icy and non-icy regions were required to identify surface water ice with very little evidence of other absorptions in the spectra.

The Deep Impact mission to 9P/Tempel 1 was complemented by an extensive Earth-based observational campaign. Of specific interest here are the results from the Spitzer Space Telescope observations at thermal infrared wavelengths (Lisse et al. 2006). A spectrum acquired 45 min after the impact is shown in Fig. 4.71. There are numerous features evident that might be attributable to different minerals.

A factor of ten increase in the flux density at 8–12 µm is indicative of silicates and specifically pyroxenes and olivines. Carbonates show emission features typically appear at 6.5 to 7.2 µm, while the broad emission around 27 µm might be attributable to sulphides. However, the features that one sees in the spectrum have limited contrast and this can lead to ambiguity. Lisse et al. (2006) performed spectral fitting to the data and provided a best fit composition but the result almost certainly has significant degeneracy (e.g. Gicquel et al. 2012).

Harker et al. (2007) performed ground-based mid-infrared spectroscopy of the 9P/Tempel 1 Deep Impact ejecta and concluded that small dust grains (~0.2 µm) of diverse mineralogy (amorphous olivine, amorphous pyroxene, amorphous carbon,

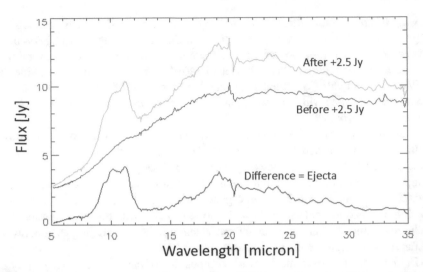

**Fig. 4.71** Spitzer Space Telescope/IRS observation of 9P/Tempel 1 acquired 23 hours before the impact of the Deep Impact impactor and 45 min after impact. The difference is also plotted. (Courtesy of Casey Lisse following Lisse et al. 2006)

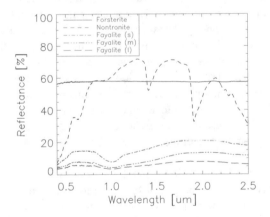

**Fig. 4.72** Laboratory spectra in the visible and reflected IR of three of the minerals identified by Lisse et al. (2006) in Spitzer Space Telescope thermal infrared observations. Data from Grove et al. (1992)

and crystalline olivine) were evident. There were significant changes in the mineralogical composition observed post-impact suggesting heterogeneity with depth.

Ambiguity in spectral fitting is a significant issue in most attempts to derive a detailed mineral composition from spectroscopy. Grain size and intimate mixtures of multiple species can influence the depths of absorption features, for example. The difficulties are clearly evident at optical and infrared wavelengths. Figure 4.72 shows reflectance spectra in the optical and near infrared from the JPL spectral library (Grove et al. 1992) of three of the minerals indicated as contributing to the spectra

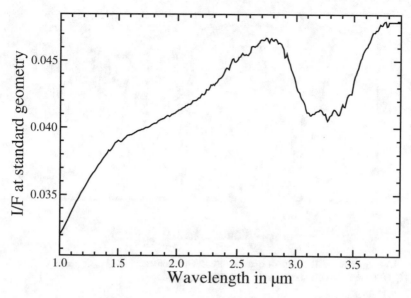

**Fig. 4.73** Composite spectrum of the nucleus of 67P produced by combining five spectra from the VIRTIS experiment converted to a standard geometry (i = 0°, e = 30°, α = 30°). (Courtesy of David Kappel)

observed by Spitzer. Forsterite has a very flat spectrum in the optical and near-IR and this would be consistent with resolved near-IR spectra from Deep Impact which only shows features attributable to water ice absorption and only then when water ice is present on the surface (Sunshine et al. 2006). The phyllosilicate, nontronite, is one of the more surprising components of the mixture derived from Spitzer data and this might be compared to the analysis of Stardust particles where no phyllosilicates were found (Westphal et al. 2017). Furthermore, nontronite has numerous absorption bands that were not seen in the Deep Impact infrared spectrometer data. Fayalite has an absorption at around 1 micron that would not have been evident in the near-IR observations at 9P/Tempel 1. However, Grove et al. (1992) studied three different grain sizes and the effect of grain size can be seen in the three spectra for fayalite which are for small (<45 μm), medium (45–125 μm) and large particles (125–500 μm). Large particles result in a lower reflectance but, in addition and perhaps not so clearly evident, is that the contrast of the absorption bands decreases with particle size making the mineral more difficult to identify in mixtures and/or low signal to noise data. Hence, deduction of detailed composition from remote sensing data is by no means straightforward.

Another example is that of organics. Combes et al. (1988) determined that there were organics on the surface of 1P/Halley in 1988 by detection of the C-H stretch band at 3.4 μm. Figure 4.73 shows a spectrum of the nucleus of 67P acquired by summing five spectra from the Rosetta/VIRTIS experiment. It shows a strong absorption between 3.2 and 3.4 μm attributable to organics.

Despite the clarity of the observations, the determination of the exact molecules contributing is again challenging. The issues are apparent when looking at Fig. 4.74

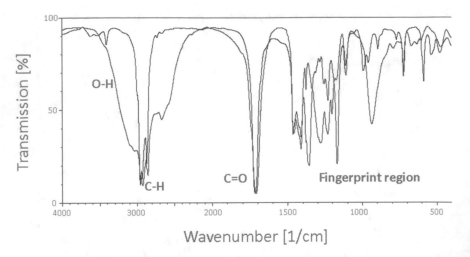

**Fig. 4.74** Spectra of two pure organics. Red: Octanoic acid ($C_8H_{16}O_2$) Black: 2-hexanone ($C_6H_{12}O$) from the SDBS database (https://sdbs.db.aist.go.jp)

which shows transmission spectra of two pure organic species (octoanoic acid and 2-hexanone) from laboratory measurements.[3] 2-hexanone shows the jagged peak at approximately 2900–3000 cm$^{-1}$ (3.33–3.44 microns) that is characteristic of tetrahedral carbon-hydrogen bonds. However, this is not terribly useful, as just about every organic molecule has these bonds. (Indeed, I once heard Carl Sagan suggest that red wine would provide a good fit to the data although his publications (e.g. Chyba and Sagan 1988) fitting the 3.4 μm spectrum of 1P/Halley in an attempt to refute a claim that the band allowed identification of a specific organic, used slightly less controversial substances.) Nevertheless, the 3.4 μm feature can serve as a familiar reference point to orient yourself in a spectrum. The O-H stretching mode of alcohols is around 3400 cm$^{-1}$ (2.95 μm) but the exact wavelength changes depending upon the molecule (e.g. the position of the hydroxyl group with respect to a carbonyl group). In Fig. 4.74, the broad O-H mode is blending with the C-H mode in octanoic acid.

The carbonyl group is seen at 1650–1700 cm$^{-1}$ (~6.0 μm) in both our example molecules. At wavelengths longward of this, the features become more diagnostic and this is referred to as the fingerprint region but there have not been any clear identifications to date using this wavelength range. Indeed, in the event that more complex molecules and mixtures are present, combined with the difficulty of observing in reflection (rather than transmission), it becomes arguable whether the task of identifying individual organics in the solid form through this technique is actually feasible.

---

[3]SDBSWeb: https://sdbs.db.aist.go.jp National Institute of Advanced Industrial Science and Technology, accessed 12 May 2019

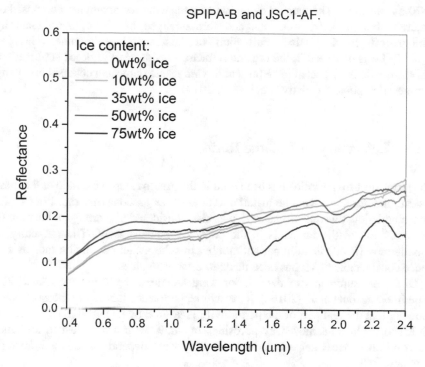

**Fig. 4.75** The spectral reflectance of a mixture of water ice (produced with a reproducible set-up called SPIPA-B) with a lunar simulant (JSC-1-AF) for varying ice contents (from Yoldi et al. 2018). Different colours correspond to different percentages of ice within the sample. The diagram shows that significant amounts of water ice (~35%) can be masked by the simulant in spite of the high signal to noise in this laboratory experiment

Even though pure ices can be more readily identified by remote sensing, the effect of mixing them with dark minerals leads to non-linearities that can further lead to misinterpretation of their mixing ratios. For example, the depth of water ice absorptions shows this non-linear dependency when mixed with darker non-volatiles as is illustrated in Fig. 4.75 (Yoldi et al. 2018). As the ice content of a sample changes from 10% ice to 35% ice, the reflectance actually decreases over the whole visible and near-IR range before increasing with higher ice contents. This behaviour is partially explained by the presence of relatively large ice particles in this test, which, when present at low mixing ratios, increase the optical thickness of the sample. Consequently, determination of mixing ratios of ices in comet-like material from visible and near-infrared spectra is not straightforward.

Laboratory experiments aiming to reproduce the evolution of cosmically abundant molecules have a long history (e.g. Sagan and Khare 1979). The investigation of the evolution of laboratory mixtures of simple ices has provided several interesting developments in recent years. For example, Esmaili et al. (2018) have reported the formation of glycine in a mixture of $CO_2$, $CH_4$ and $NH_3$ ice when bombarded by

0–70 eV electrons. The thermal evolution of interstellar ice analogues composed of water, carbon dioxide, ammonia, and formaldehyde has been investigated by Vinogradoff et al. (2015). Formation of hexamethylenetetramine (HMT – $C_6N_4H_{12}$) was observed in the organic refractory residue left after ice sublimation. This molecule is suspected to be present in various astrophysical contexts involving nitrogen-rich photochemistry (Pirali et al. 2014).

## 4.15  Refractory to Volatile Ratios

The term "dust-to-gas ratio" has been used in the past to refer to the ratio of the mass lost in refractory, non-volatile material to that lost as gaseous material. Prior to the Rosetta mission, this quantity, $Q_d/Q_g$, was also considered as a proxy for the ratio of refractory to volatile materials in the cometary nucleus itself. The refractory to volatile ratio (RVR) is an important number in solar system evolution models and hence considerable effort has been made to establish values.

Direct measurements of the interior were attempted by using the CONSERT experiment. Brouet et al. (2016b), for example, performed laboratory measurements showing that the real part of the permittivity measured by CONSERT ($\varepsilon' = 1.27$ at 90 MHz) could be achieved by porosities of ~80% with a low dust to ice ratio. However, these measurements indicated a fairly weak dependence on the refractory to volatile ratio.

The complexity of determining the RVR without actually sampling the interior of the nucleus and measuring it directly should not be underestimated. Two attempts to combine space-borne measurements of the dust and gas production have been made. McDonnell et al. (1991) provided a number for $Q_d/Q_g$ (the value was two) by combining Giotto data of the gas production with the measured dust production rate but it was clear to all at the time that the error bar on this number was very large. Subsequent work by Thomas and Keller (1991) showed that the available data were sufficiently uncertain that an error bar of at least one order of magnitude could be associated with McDonnell's value. Rotundi et al. (2015) reported a dust to gas ratio of $4 \pm 2$ at 67P but based on sampling a small range of the dust size distribution at high heliocentric distances early in the mission and the authors suggested that this value might change as the comet became more active.

On the other hand, RVR ~ 1 is a number that intuitively feels right although there is actually only limited justification for this. Attempts by ground-based observers have given values of $\lesssim 1$ (e.g. Singh et al. 1992) but the assumptions necessary to reach these results were no less significant than those used in the analysis of Giotto data. We noted above that Pontoppidan et al. (2014) studied dense interstellar medium and protostellar envelopes and gave a rock/ice ratio of ~0.67. The difficulty in arriving at an accurate number is immediately evident by looking at the two parts of the ratio.

The total water vapour production of a comet can be estimated from the hydrogen emission at Lyman-$\alpha$ (e.g. Fig. 3.48) or from the OH emission using, for example, the vectorial model (e.g. Fig. 3.47) and appropriate g-factors for the emission rate (taking into account phenomena such as the Swings effect for near-UV bands of OH). The total volatile production rate also includes the other major species, $CO_2$ and CO. Their contribution is less easily established but assuming that water contributes roughly 90% to the total volatile loss (Sect. 3.2) seems to be a reasonable assumption. Consequently, the volatile emission from the nucleus at any one time can be obtained reasonably well. However, variability is significant. The ROSINA data provide by far the most comprehensive measurements of local gas densities at the spacecraft that can be extrapolated to gas production rates over the mission (Hansen et al. 2016). Integration of Hansen's fit to the Rosetta data sets and assuming water as the dominant species leads a volatile loss of 6.3 $10^9$ kg which in combination with the total mass lost (Pätzold et al. 2019; Sect. 2.3) leads to a $Q_d/Q_g$ (integrated over the orbit) of ~0.7 $\pm$ 0.5 where we have used Pätzold's error estimate alone. There are clearly assumptions needed to produce the gas production rates and Hansen's model is also somewhat artificial but this is probably one of the most reliable approaches to determine the gas mass loss.

Biver et al. (2019) arrived at a total water production for the apparition in the range of 4.0–5.8 $10^9$ kg (see Choukroun et al. 2020 for a review) which would place $Q_d/Q_g$ at between 0.7 and 2.3 using the nominal value of Pätzold et al. (2019).

For the non-volatile component, an independent measurement of the total production is far more difficult to establish because the mass loss seems to be dominated by the largest particles emitted by the nucleus. Potentially, the loss of a few individual particles (perhaps better described as boulders) that are unseen from the ground (because of resolution) or from in situ observation (because of instrumental measurement limits on detection and outflow velocity) dominate the mass loss. Nonetheless, it is instructive to see if the values of **Afρ** are consistent with the estimated total mass loss while keeping in mind the caveats.

Using Table 4.1, we can set the dust size distribution power law exponent to 2.8 which encompasses values for both 1P/Halley and 67P. Figure 4.18 indicates **Afρ** through the Rosetta mission and we have seen in Fig. 4.32 that individual points can be converted to obtain a dust mass production rate depending upon the scattering properties such as the use of Mie theory, the refractive index for the Mie equations, and an assumed particle size dependent velocity distribution, $v_d(a)$, using Eq. 4.105. We have also seen Eq. 4.101 that $v_d(a)$ is a straightforward function of the gas production rate. Hence, we can use the gas production rate of Hansen et al. (2016) to support interpolation of the values of $v_d(a)$ derived by Marschall (2017) from gas dynamics simulations. Inversion of Eq. 4.105 for each measured **Afρ** value of Gerig et al. (2018) then provides a function that can be integrated to give a total dust mass loss. This is novel because all other measurements for dust mass loss are either derived from the difference between the total mass loss and the gas mass lost or they are extrapolations from in situ dust measurements obtained mostly at the terminator in the case of Rosetta. Here, 2D remote-sensing provides a more global estimate.

The result of this procedure gives 12.3 $10^9$ kg. Taken at face value, this is inconsistent with other measurements because it is close to Pätzold et al.'s upper limit for the total mass loss including gas. However, two assumptions have been

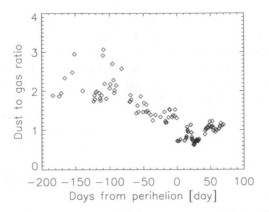

**Fig. 4.76** Dust to gas production rate ratios computed from Hansen et al. (2016) model of the gas production rate and the **Afρ** values of Gerig et al. (2018) assuming Mie theory with a refractive index of $m_{ref} = (2.0, i0.01)$ and gas drag on spherical particles. While the absolute value may have uncertainty, the trend with heliocentric distance is very evident

made in reaching this result that are significant in that they suggest that this value should be an upper limit. Firstly, Mie theory in general predicts a low reflectance at 90° phase angle. Irregular particles, such as those seen in Fig. 4.13, scatter more efficiently at these intermediate scattering angles (see e.g. Fig. 4.11 for ellipsoidal particles and note the logarithmic scale of the y-axis). In other words, **Afρ** becomes larger for the same particle mass. Secondly, the gas production function of Hansen et al. (2016) is higher than the more recent value of Biver et al. (2019) and higher gas production rates drive higher gas velocities which in turn produce higher dust velocities. Both of these effects would reduce the computed total dust mass loss bringing the result closer to consistency. Increasing the dust size distribution power law exponent to three would have a similar effect but here the measurement evidence would argue against. Consequently, with some subtle manipulation, the **Afρ** values can be claimed as consistent although using these values to define RVR would obviously be over-interpretation of the data.

A final point of interest here is the variation in the dust to gas production rate ratio with respect to perihelion. This can be calculated in the same way (from the gas production rate of Hansen et al. and the **Afρ** value of Gerig et al.) and is shown in Fig. 4.76. While the absolute value may be a little high (as discussed above) the trend in $Q_d/Q_g$ is clear and emphasizes Rotundi et al. (2015)'s caution that the dust to gas production rate ratio at high heliocentric distance may not be representative for other value of $r_h$.

The general approach of measuring the mass production rates of the gas species and the dust independently and assuming that the RVR is equivalent may be correct to first order but there are numerous pitfalls here that should be highlighted (see also Choukroun et al. 2020).

We have already seen that the size distribution and specifically the high mass cut-off of the distribution strongly influences the dust mass loss rate if the exponent of the size distribution is smaller than three (as most evidence suggests is the case)

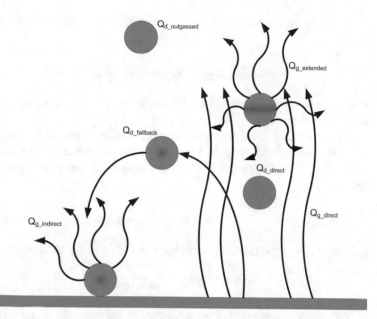

**Fig. 4.77** Definitions of different production rates of volatile (gas) and refractory (dust) components from a cometary nucleus. The production rates of each component is time-dependent which further complicates the picture

and this has a marked effect on the computation of $Q_d/Q_g$. However, it is not obvious that large dust particles are devoid of volatile material that might be subsequently released. The indisputable evidence that some large particles fail to reach escape velocity further complicates the scenario. The complexity is summarized in Fig. 4.77.

Gas is produced and escapes into the inner coma directly through activity ($Q_{g\_direct}$). This emission removes particles from the activity source. These particles can either escape ($Q_{d\_direct}$) or fallback to the surface ($Q_{d\_fallback}$). The escaping particles can lose volatiles thereby acting as an extended source of gas ($Q_{g\_extended}$) in the inner coma. The final refractory mass loss from the comet is then $Q_{d\_outgassed}$. At the surface, the dust that falls back can also outgas. The gas lost from the comet through this process is defined here as $Q_{g\_indirect}$. We have seen bright spots on the surface of the comet in Fig. 1.3 that may be extreme examples of this process.

With these definitions, we can begin to relate measured quantities to the refractory to volatile ratio in the interior if we can assume that we have a steady-state.

Under this assumption, the comet mass production rate ratio of refractory to volatile materials is given by

$$= \frac{Q_{d\_outgassed}}{Q_{g\_direct} + Q_{g\_indirect} + Q_{g\_extended}} \tag{4.125}$$

The mass production rate of volatile-bearing dust leaving the surface and escaping is

$$Q_{d\_direct} = Q_{d\_outgassed} + Q_{g\_extended} \qquad (4.126)$$

where obviously the mass leaving the surface in solid form is greater than the final solid production rate.

If $f_{vol}$ is a factor that defines the escaping volatile content of the dust once the dust falls back then we can write an equation for the gas emitted by the material that has fallen back onto the surface under the assumption that that material completely desiccates, i.e.,

$$Q_{g\_indirect} = f_{vol} Q_{d\_fallback} \qquad (4.127)$$

with the mass of the refractory material remaining on the surface as

$$= (1 - f_{vol}) Q_{d\_fallback} \qquad (4.128)$$

This allows us to see that the refractory to volatile ratio at the activity source is given by

$$= \frac{(1 - f_{vol}) Q_{d_{fallback}} + Q_{d\_outgassed}}{Q_{g\_direct} + Q_{g\_indirect} + Q_{g\_extended}} \qquad (4.129)$$

While the sum of $Q_{g\_direct}$, $Q_{g\_indirect}$, and $Q_{g\_extended}$ can be determined to reasonable accuracy from ground-based or Earth-orbiting observation of OH or similar taking into account less numerous species, none of the terms in the numerator can be accurately determined. Furthermore, the individual production rates are neither steady nor do the ratios between the sources remain constant with time. Every production rate in the equations has a time dependence and that time dependence is not the same for all variables. In some cases, the time dependence might also manifest itself as a spatial variability. Let us consider fallback as an example.

For 67P, the southern hemisphere was the most active at perihelion as a result of the obliquity and the orientation of the rotation axis. Larger dust particles laden with volatiles were emitted from the southern hemisphere and, if not escaping, returned to the nucleus. Many of these would have landed on the constantly unilluminated northern polar region (Keller et al. 2017) where they would remain until equinox. At this time, southern hemisphere outgassing would be slowing but the gas emission from the fallback in the north would now be augmenting the emission from the northern regions that would just be starting. Hence $Q_{d\_fallback}(t)$ would, at least locally, not follow a simple relationship with respect to true anomaly. This is just one example.

Finally, there may be some dependence of $f_{vol}$ on particle properties such as size because of the low thermal inertia—particles larger than several cms will have better insulated interiors.

This implies that the RVR at the activity source defined by Eq. (4.129) is an over-simplification. One way around this is to integrate the production rate over the orbit but it is inevitable that measurement accuracy and frequency must limit the precision of the final result. A further issue is that fallback can result in producing a less volatile surface so that we have evolution of RVR over time. This is probably not of importance to the value of RVR for the whole nucleus but in situ measurements made locally might be highly misleading if the fallback phenomenon is as significant as is currently assumed. This would be important for future in situ studies. Static, immobile, surface elements making measurements of cometary properties may provide highly misleading results if their landing sites are not well chosen.

Quantifying $(1 - f_{vol})Q_{d\_fallback}(t)$ is difficult in part because the thickness of the airfall deposit on the northern hemisphere of 67P is poorly known. The smallest crater structure in Fig. 2.56 implies a total depth of material of ~5 m over many apparitions; the loss of new tiny craters (Fig. 2.82) implies new deposition of <1 m during one apparition; the thickness of the deposit on the Aswan structure after collapse of the cliff is too small to quantify (Fig. 2.51); the Anubis region in the northern hemisphere appears to have lost material (Fig. 2.100). However, we do know from Pätzold et al. (2019) that the average global mass loss is 0.55 m per apparition. If a further 0.55 m equivalent were ejected and re-deposited on the northern hemisphere only, it would provide a 1 m thick layer which is arguably already exceeding the depth that can be supported by visual imaging.

# Chapter 5
# The Plasma Environment

## 5.1 Initial Considerations and the Solar Wind

Ultimately, all gas molecules leaving the nucleus become ionized by solar radiation or electron impact and, after acceleration, assume the velocity of the solar wind. This process leads to the production of a plasma tail. Historically, cometary plasma tails (Fig. 5.1) have played an important role in the study of the interplanetary space environment. It was Ludwig Biermann in 1951 who deduced the existence of an outflow of material from the Sun based upon the motion of cometary ions with respect to the (then assumed) nucleus. When it became clear that Coulomb collisions between the cometary ions and ionized particles in this "solar wind" were not sufficiently effective at providing the required momentum transfer to the cometary ions, Alfvén proposed the presence of an interplanetary magnetic field to produce the necessary acceleration. It subsequently became apparent that the solar wind was exhibiting the properties of a conducting fluid in a magnetic field.

The relative motion of a conducting fluid with respect to a magnetic field produces an electromotive force and electrical currents arise. These currents generate a second, induced magnetic field which adds to the original field. The combination of these two field components interact with the induced current density to produce a $\mathbf{J} \times \mathbf{B}$ (Lorentz) force that acts on the conducting fluid and, usually, inhibits the relative motion of the fluid with respect to the field. The relative motion of the fluid with respect to the magnetic field is therefore reduced. One can imagine this in two ways. Magnetic fields can pull on the fluid. Alternatively, the conducting fluid can drag the magnetic field lines. This is where the concept of a "frozen-in" magnetic field arises when discussing the fields associated with the outflowing plasma of the solar wind.

The fundamental properties of the solar wind were deduced through comprehension of the dynamical properties of cometary ions and from understanding of the physical effects of relative motion between a plasma and a magnetic field through application of Faraday's and Ampere's laws. The basic properties of the solar wind

© Springer Nature Switzerland AG 2020
N. Thomas, *An Introduction to Comets*, Astronomy and Astrophysics Library,
https://doi.org/10.1007/978-3-030-50574-5_5

**Fig. 5.1** Plasma tail of comet 1P/Halley from April 1986. This figure is an enhanced version of data (LSPN 2671) in the International Halley Watch Large Scale Phenomenon Network archive

**Table 5.1** Properties of the solar wind from Hundhausen in Kivelson and Russell (1995)

| Property | Value |
| --- | --- |
| Proton density | $6.6\ \text{cm}^{-3}$ |
| Electron density | $7.1\ \text{cm}^{-3}$ |
| $He^{2+}$ density | $0.25\ \text{cm}^{-3}$ |
| Flow speed | $450\ \text{km s}^{-1}$ |
| Proton temperature | $1.2 \cdot 10^5\ \text{K}$ |
| Electron temperature | $1.4 \cdot 10^5\ \text{K}$ |
| Dynamic pressure | $2.6 \cdot 10^{-9}\ \text{Pa}$ |
| Sound speed | $60\ \text{km s}^{-1}$ |
| Magnetic pressure | $1.9 \cdot 10^{-11}\ \text{Pa}$ |
| Proton gyroradius | $80\ \text{km}$ |
| Proton-proton collision time | $4 \cdot 10^6\ \text{s}$ |
| Electron-electron collision time | $3 \cdot 10^5\ \text{s}$ |
| Magnetic field strength | $5\ \text{nT}$ |

at 1 AU from the Sun, determined by spacecraft observations, are summarized in Table 5.1. The ion density drops as roughly $1/r_h^2$. The dynamic (ram) pressure is given by the equation

$$P_{dyn} = \rho_{sw} u_{sw}^2 \tag{5.1}$$

where $u_{sw}$ is the flow speed and $\rho_{sw}$ is the mass density with most of the momentum being carried by the protons (Cravens 1997) which implies values of 1–3 nPa at 1 AU.

The presence of a magnetic field in an ionized gas can lead to a "magnetic pressure" which is an energy density associated with the field. The magnetic pressure is given by

$$P_{mag} = \frac{B^2}{2\mu_0} \tag{5.2}$$

where $B$ is the magnetic field strength and $\mu_0$ is the permeability of free space. A gradient in $B$ leads to a magnetic pressure force and in our case it is the plasma that experiences this force.

To first order, the major effect of the magnetized solar wind is to exert pressure on obstacles in its path. The relative importance of the magnetic pressure with respect to the thermal pressure is expressed as a value, $\beta_{mag}$, which is

$$\beta_{mag} = \frac{p}{P_{mag}} = \frac{2\mu_0}{B^2} n_e k(T_e + T_i) \tag{5.3}$$

where $T_e$ and $T_i$ are the electron and ion temperatures respectively with the total pressure being the sum of the two components.

The pressure of the solar wind is exerted on obstacles in its path. When the nucleus is inactive, it forms almost no barrier at all to the solar wind because of its small physical size. Solar wind ions can strike the surface of the nucleus directly thereby sputtering its surface. This asteroidal interaction is of little real interest in the case of comets. (It is more interesting at asteroids because it provides a means of sampling the chemical composition of the surface without requiring the complicated process of landing on the surface.) But when activity and gas emission starts, the processes involved rapidly become considerably more interesting as the outgassing exerts forces against the pressure of the solar wind.

While there are many phenomena resulting from the outgassing, the influence on the interplanetary environment arises from two principal physical mechanisms. Firstly, the cometary outgassing, initially mostly comprising neutrals, expands away from the nucleus providing an opposing pressure. The ion-neutral interaction is through collisions (which one can also think of as friction). The neutral outflow exerts a force on the plasma which can balance the inward magnetic pressure gradient (the $\boldsymbol{J} \times \boldsymbol{B}$ force) if the neutral density is sufficient. The position with respect to the nucleus where such a balance occurs is then a function of the gas production rate. Secondly, as they expand, the neutrals are ionized by solar EUV photons. The ions are initially moving at the velocities of their parent neutrals ($\sim$ 0.8 km s$^{-1}$ relative to the nucleus) but this velocity is negligible compared to the relative velocity with respect to the solar wind and the interplanetary magnetic field. Consequently, the ions are "picked up" by the solar wind and accelerated. However, this pick-up requires a momentum transfer from the solar wind to the newly created ions. This process of mass loading the flow, which is not only significant because of the numbers of created ions but also because the masses of species such as O$^+$ are heavier than solar protons, acts to progressively reduce the flow velocity as the cometary ion density increases. It is the combination of the two mechanisms that generates most of the observed phenomena. We start by looking at the ionization process.

## 5.2   The Production of Cometary Ions

We have seen a long list of reaction types that can occur within the coma in Table 3.9. The main sources that are sufficient to explain the observed plasma densities in the vicinity of comets appear to be photo-ionization and electron-impact ionization although the influence of charge-exchange can be seen in some phenomena (see, for example, Heritier et al. 2018 for 67P).

Consequently, the three major reaction types of interest are

- Photoionization (R2)
- Charge exchange (R8)
- Electron impact (dissociative) ionization (R5, R6)

### 5.2.1   Photoionization

Photoionization cross-sections for several cometary species in computer readable form can be found online.[1] Some examples of relevance are shown in Fig. 5.2. The photo-ionization of atomic hydrogen (indicated by H $\rightarrow$ H+ in the diagram) has a threshold at 91.175 nm (Table 5.2). The cross-section is at a maximum at this wavelength and drops rapidly with increasing photon energy so that higher energy photons will penetrate a hydrogen coma further than photons closer to threshold. The other lines in Fig. 5.2 show branches where ionized products from photo-absorption by the water molecule arise. There are, of course, dissociation branches where only neutrals are produced but they are not shown here to limit the details in the diagram.

---

[1] https://phidrates.space.swri.edu/

**Fig. 5.2** Photo-ionization cross-sections for atomic hydrogen and various species arising from photo-absorption by the water molecule. The ionization of atomic hydrogen is marked as H → H⁺. The other lines are marked with the products. E.g. H⁺ + OH indicates the cross-section for the reaction H₂O + hϑ → H⁺ + OH

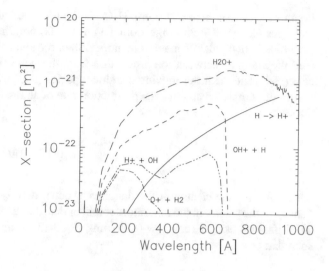

**Table 5.2** Photolytic reactions involving water vapour and its daughter products. The threshold wavelengths are given. Reactions marked in bold would produce pick-up ions in the cometary environment (Data from Huebner et al. 1992)

| Precursors | Product | Threshold wavelength [Å] | Rate coefficient [s⁻¹] | |
|---|---|---|---|---|
| | | | Quiet Sun | Active Sun |
| $H_2O + h\vartheta$ | OH + H | 2424.6 | $1.03\ 10^{-5}$ | $1.76\ 10^{-5}$ |
| | OH (A²Σ⁺) + H | 1357.1 | | |
| | $H_2 + O(^1D)$ | 1770 | $5.97\ 10^{-7}$ | $1.48\ 10^{-6}$ |
| | $H_2 + O(^1S)$ | 1450 | | |
| | $H + H + O(^3P)$ | 1304 | $7.55\ 10^{-7}$ | $1.91\ 10^{-6}$ |
| | **$H_2O^+ + e$** | **984** | **$3.31\ 10^{-7}$** | **$8.28\ 10^{-7}$** |
| | **$H + OH^+ + e$** | **684.4** | **$5.54\ 10^{-8}$** | **$1.51\ 10^{-7}$** |
| | **$H_2 + O^+ + e$** | **664.4** | **$5.85\ 10^{-9}$** | **$2.21\ 10^{-8}$** |
| | **$OH + H^+ + e$** | **662.3** | **$1.31\ 10^{-8}$** | **$4.07\ 10^{-8}$** |
| $OH + h\vartheta$ | O(¹D) + H | 5114 | $7.01\ 10^{-6a}$ | $1.76\ 10^{-5a}$ |
| | O(¹S) + H | 61,320 | $8.33\ 10^{-7a}$ | $2.11\ 10^{-6a}$ |
| | O(³P) + H | 2823 | $1.20\ 10^{-5a}$ | $1.38\ 10^{-5a}$ |
| | **$OH^+ + e$** | **928** | **$2.43\ 10^{-7a}$** | **$6.43\ 10^{-7a}$** |
| $H_2 + h\vartheta$ | H(1 s) + H(1 s) | 2768.85 | $4.80\ 10^{-8}$ | $1.09\ 10^{-7}$ |
| | H(1 s) + H(2 s, 2p) | 844.79 | $3.44\ 10^{-8}$ | $8.21\ 10^{-8}$ |
| | **$H_2^+ + e$** | **803.67** | **$5.41\ 10^{-8}$** | **$1.15\ 10^{-7}$** |
| | **$H + H^+ + e$** | **685.8** | **$9.52\ 10^{-9}$** | **$2.79\ 10^{-8}$** |
| $O(^3P) + h\vartheta$ | **$O^+ + e$** | **910.44** | **$2.12\ 10^{-7}$** | **$5.88\ 10^{-7}$** |
| $O(^1D) + h\vartheta$ | | 827.9 | $1.82\ 10^{-7}$ | $5.04\ 10^{-7}$ |
| $O(^1S) + h\vartheta$ | | 858.3 | $1.96\ 10^{-7}$ | $5.28\ 10^{-7}$ |
| $H + h\vartheta$ | **$H^+ + e$** | **911.75** | **$7.26\ 10^{-8}$** | **$1.72\ 10^{-7}$** |

[a]Indicates based on experimental cross-sections. Theoretical values may differ substantially

It can be seen that, below 100 nm, $H_2O^+$ is the most probable ionized product and that over the wavelength range from 10 nm to 100 nm, the cross-section is larger (and mostly one order of magnitude larger) than for ionization of H. This indicates well the interplay between ion production from the slowly expanding parent molecule ($H_2O$) and the faster moving daughter species (H).

For the simplest reactions, the production rate of an ion locally can be computed through

$$\frac{dn_{dau}}{dt} = n_{par} \int_0^{\lambda_{threshold}} \sigma_x(\lambda) F_\odot(\lambda) d\lambda \tag{5.4}$$

where $dn_{\text{dau}}$ is the change in the daughter product density and $n_{\text{par}}$ is the number density of the parent. In the innermost coma, optical depth must be considered for the EUV photons using Beer's law (following Eq. 3.13) in the illuminating irradiance such that

$$\frac{F_\odot(\lambda)}{r_h^2} e^{-\tau(\lambda)} = \frac{F_\odot(\lambda)}{r_h^2} e^{-\sum_{k_T} \int n_{k_T}(l) \sigma_{k_T,ph}(\lambda) dl} \tag{5.5}$$

where the subscript $k_T$ is indicating summation over all gas species (types) and the product, $n_{k_T} \sigma_{k_T,ph}$, is the extinction coefficient with $\sigma_{k_T,ph}$ being the ionization cross-section (see Avakyan et al. 1998 for a compilation of many species or e.g. Haddad and Samson 1986 for $H_2O$ as a primary source) of species $k_T$. The photo-ionization frequency for each species is given by

$$f_{k_T,ph} = \int_0^{\lambda_{k_T,th}} \sigma_{k_T,ph}(\lambda) F(\lambda) \, d\lambda \tag{5.6}$$

where $\lambda_{k_T,\text{th}}$ is the photoionization threshold for the species. Although integration is over all wavelengths below threshold, there are natural limits arising from the solar flux distribution as one moves toward the X-ray region of the spectrum (Heritier et al. 2018).

In the case of a spherically symmetric uniformly expanding species, the column density from the Sun to a point in the coma sunward of the nucleus would be

$$N = \frac{Q}{4\pi v b} \left( \frac{\pi}{2} - \tan^{-1} \frac{x}{b} \right) \tag{5.7}$$

where $x$ is the distance along the line of sight towards the Sun from the plane orthogonal to the Sun-comet line containing the nucleus and $b$ is the distance from the nucleus at the point where the line of sight intersects that plane (i.e. the impact parameter). (We have not used a subscript for the velocity, $v$, and the column density, $N$, here to avoid specifying the species in any way.) In more realistic calculations, the reduction in $n_{par}$ with the distance from the nucleus through

**Fig. 5.3** Total solar
irradiance at Lyman-alpha
(1216 Å) at 1 AU from the
Sun. The data are from the
SORCE data repository
(http://lasp.colorado.edu/
home/sorce/data/) and show
16 years of measurements.
Note the 50% change in flux
over the solar cycle

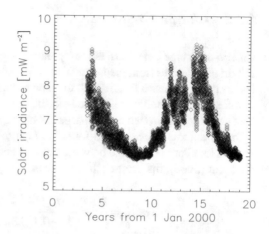

photolysis must also be accounted for requiring an exponential decay term. Therefore, this calculation can quickly become complicated for a network of reactions.

The dominant neutral species in the outflow is usually assumed to be water
vapour although, as we have noted above, other molecules may compete or dominate
under specific conditions. Huebner et al. (1992) have given threshold wavelengths
for the multiple photolytic pathways in which water vapour can be broken down
including those for photoionization. Solar EUV is a strong function of the solar cycle
as illustrated in Fig. 5.3 which shows the solar flux at 1 AU in Lyman-$\alpha$ at 121.6 nm
over a period of 15 years. There is a 50% change in flux over this period. In general,
the amplitude of the variation of the solar flux over the cycle increases as the
wavelength decreases. Huebner et al. have given the rate coefficients
(by integrating over wavelength using Eq. 5.4) for the photolysis reactions of $H_2O$
both under quiet (low EUV flux) and active Sun conditions (see Table 5.2).

One can see here that the dissociation reactions are quick, implying loss of $H_2O$
on length scales of the order of 100,000 km from the nucleus. This is also evident in
the 1D reaction network results shown in Fig. 3.49 (top panel). Care must be taken
with very active comets when computing rates close to the nucleus. The spectral
photodissociation cross-section of $H_2O$ ($\sigma_{H2O}$) can reach values as high as $10^{-20}$ m$^2$
and thus the optical depth, as defined through

$$\tau_{H2O} = \int_{-\infty}^{l} \sigma_{H2O}\, n_{H2O}(s)\, ds \qquad (5.8)$$

where $s$ is a distance along a line of sight from the Sun through the coma, can exceed
1 at tens to hundreds of km from the nucleus reducing photodissociation rates in the
innermost coma (see also Eq. 3.14).

## 5.2.2   Charge-Exchange

Charge-exchange reactions are of greatest significance near the nucleus. Their importance derives from the fact that while the old ion (now transformed to a neutral) leaves the system as a fast neutral, the new ion has almost zero velocity with respect to the solar wind and therefore adds to the mass loading in the system.

The proton-hydrogen atom charge exchange cross-section has been investigated by many authors. Freeman and Jones (1974) provide a simple formula for the cross-section for charge exchange between protons and hydrogen atoms that, according to Gerhardt (2004), fits most of the data reasonably well. It has the form

$$\sigma_{cx,H} = \frac{0.6937 \; 10^{-14} \; (1 - 0.155 \; \log_{10}E)^2}{1 + 0.1112 \; 10^{-14} \; E^{3.3}} \tag{5.9}$$

and is shown in Fig. 5.4. Proton energies in the solar wind are typically around 1 keV. Figure 5.4 also shows the charge exchange cross-sections for proton-$H_2O$ and proton-$CO_2$ reactions.

The main point here is that the cross-sections are relatively large. Combi et al. (2004) summarized this by stating the proton-$H_2O$ charge exchange reaction at 1 AU removes neutral $H_2O$ as fast as photoionization for quiet Sun conditions. It is less competitive for active Sun conditions (see also Huebner et al. 1992) but this indicates that charge exchange is a notable process in any solar wind interaction with either the neutral parents or the hydrogen coma.

It may also play a significant role in the generation of cometary X-ray emissions (Cravens 2000 and refs therein). In March 1996, X-ray (0.1–2 keV) and extreme ultraviolet (0.09–0.2 keV) emission was observed from C/1996 B2 (Hyakutake) by instruments on ROSAT (Lisse et al. 1996) and EUVE (Mumma et al. 1997). Subsequently several other detections of X-ray emissions have been reported (e.g. from C/2007 N3 (Lulin) by Carter et al. 2012, from 103P/Hartley 2 by Lisse

**Fig. 5.4** The charge-exchange cross-sections for three reactions of cometary significance. Solid line: Calculation of the proton-hydrogen neutral cross-section according to Eq. (5.9). Triangles: Measurements of proton-$CO_2$ charge exchange by Greenwood et al. (2000). Diamonds: Measurements of proton-$H_2O$ charge exchange by Greenwood et al. (2000)

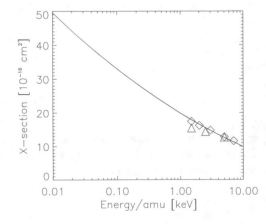

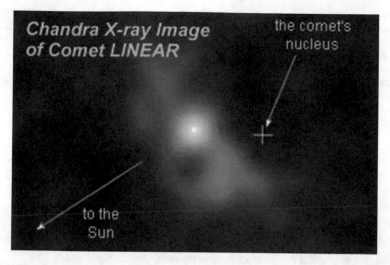

**Fig. 5.5**   Chandra observation of X-ray emission from C/1999 S4 (LINEAR). Credit: NASA

et al. 2013, from C/2012 S1 (ISON) and C/2011 L4 (PanSTARRS) by Snios et al. 2016,; see also Fig. 5.5).

The measurements showed that X-ray emission was being produced over a large volume and that the luminosity decreased with heliocentric distance. It also varied with the gas production rate and was not centred at the nucleus but ~20,000 km sunward.

Although several mechanisms to explain the emission have been presented (see review by Lisse et al. 2004), the most widely accepted is based around multiply-charged, heavy, solar wind ions such as $O^{7+}$ and $O^{6+}$ undergoing charge-exchange with neutral species. With the advent of higher resolution X-ray spectroscopy, cross-sections and models for this process are becoming more sophisticated (e.g. Mullen et al. 2017). Snios et al. (2018) produced a comprehensive model including many species that could undergo charge-exchange. They concluded that dust and ice scattering is also significant in the 1–2 keV range and, by modelling the interaction with small particles, suggested that a power-law distribution for the particle size could be extracted from data acquired by X-ray observatories. Observations of 153P/Ikeya-Zhang resulted in a value for $b_d = 2.5$ (cf Table 4.1). Recent work by Rigby et al. (2018), however, attempts to imply that the X-ray emission is a consequence of electron acceleration by wave turbulence in a strong magnetic field although a quantitative assessment has not yet been presented.

### 5.2.3   Electron Impact Ionization

The importance of electron impact ionization was first discussed quantitatively around the time of the 1P/Halley encounters (see for example, Cravens et al. 1987

**Fig. 5.6** Ionization frequencies for electron impact with water molecules. Solid line: Frequency of $H_2O^+$ production. Broken line: Frequency of $OH^+$ production

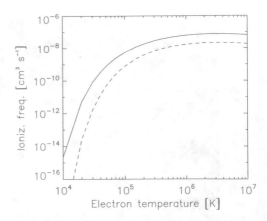

who compared photoionization rates with electron impact ionization in various regions of the plasma environment). The process can be represented by

$$e + M \longrightarrow M^+ + e + e \qquad (5.10)$$

Cravens et al. (1987) computed ionization frequencies, $f_i$, for electron impacts with water molecules, $CO_2$ and atomic oxygen. The values for electron impact onto water to produce $H_2O^+$ and $OH^+$ are shown in Fig. 5.6. Other ions can be produced from this single reaction but these two ionized products cover 80% or more of the branching at a particular electron temperature. The calculations were performed assuming a Maxwellian velocity distribution. The ion production rate can then be calculated from

$$\frac{dn_i}{dt} = f_i \, n_e n_{H2O} = f_i \, [e][H_2O] \qquad (5.11)$$

where $[e]$ and $[H_2O]$ are concentrations expressed in a form familiar in analytical chemistry. This can be compared with the cross-sections for this and other electron impact phenomena given in Vigren and Galand (2013). As shown in Table 5.1, the electron temperature in the solar wind is around $1.4 \ 10^5$ K. One of the key findings at 1P/Halley was the abrupt change in electron temperature between 10,000 and 20,000 km from the nucleus (Fig. 5.7).

This would indicate a sharp rise in ion production through this mechanism as one goes out into the coma. It has also been pointed out that the electron distribution function can be non-Maxwellian so that high energy tails of the distribution may provide higher ionization rates than would be implied by solely using the temperature of the core of the distribution.

Ions can also be lost from the system by ion–electron recombination. This is important in the close vicinity of the nucleus where the electron temperature is low (Fig. 5.7). The sharp increase in electron temperature with cometocentric distance was identified to be responsible for the formation of what is commonly referred to as

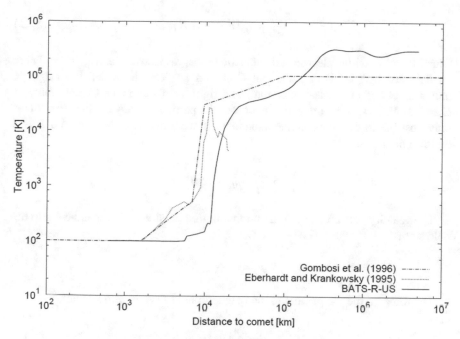

**Fig. 5.7** The electron temperature profile at 1P/Halley. The data are given by the dotted line. The model from Gombosi et al. (1986) is given by the dot-dashed line. The BATS-R-US solid line is a model calculation obtained by Rubin et al. (2014c) from which this diagram is taken. Obtained with the BATS-R-US tool developed at The University of Michigan Center for Space Environment Modeling (CSEM)

an "ion pile-up" region between 10,000 and 20,000 km from the nucleus of 1P/Halley that was identified in Giotto/IMS measurements. The higher electron temperature in this region reduced the effect of ion-electron recombination so that the plasma density increased (Rubin et al. 2014c).

## 5.3 Dynamics of the Interaction

The ion production mechanisms produce a plasma which, on average, has no net charge. The ions move in response to electric and magnetic fields and this motion can be described either through study of single particles or by treating the ions as a fluid, depending upon the application. The basic physics of ion motion has been presented by many authors. The works of Cravens (1997) and Kivelson (1995) can be recommended. Both provide considerably more detail than is possible here. However, there are some important concepts that need to be summarized.

In the single particle description, the equation of motion is given by

$$m\frac{dv}{dt} = q_c(E + v \times B) \tag{5.12}$$

where $E$ and $B$ are the electric and magnetic fields, respectively and $q_c$ is the charge on the particle. (Kivelson notes that $B$ should properly be called the magnetic induction but this is far less common.) The right hand side is the Lorentz force. If $E$ and $B$ can be specified then the motion of the particle is known. However, this is only possible in the most simple situations. If we allow $E$ to be zero and $B$ is a uniform field then

$$m\frac{dv}{dt} = q_c(v \times B) \tag{5.13}$$

The velocity vector can be split into the motion parallel and perpendicular to the field so that

$$v = v_\parallel + v_\perp \tag{5.14}$$

The force on the ion parallel to the magnetic field lines is zero so that

$$m\frac{dv_\parallel}{dt} = 0 \tag{5.15}$$

such that motion along the field line is constant and independent of time. Perpendicular to the field, the velocity can be determined to be

$$m\frac{dv_\perp}{dt} = q_c(v_\perp \times B) = mv_\perp \times \Omega_f \tag{5.16}$$

where $\Omega_f$ is the gyrofrequency (or cyclotron frequency) and has a magnitude of

$$\Omega_f = \frac{q_c B}{m} \tag{5.17}$$

The solution to this equation is uniform circular motion in a plane orthogonal to the magnetic field vector. The radius of the circle, $r_c$, is determined by the particle velocity and the magnitude of the magnetic field and given by

$$r_c = \frac{mv_\perp}{q_c B} \tag{5.18}$$

which is called the gyroradius (or Larmor radius or cyclotron radius). If there is motion parallel to the magnetic field then this implies that motion of the particle will be a helical about the magnetic field direction. In our case, as soon as an ion is created in the interplanetary magnetic field, it starts to gyrate with the given gyroradius. But, because the ambient field is moving with respect to the original

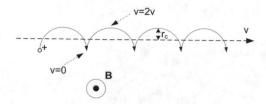

**Fig. 5.8** Illustration of the pick-up process. An ion is created on the left in a magnetic field, $B$. The field moves with respect to the newly created ion at a velocity $v$. The ion begins to gyrate with a radius, $r_c$ and is accelerated such that the mean speed of the particle becomes equal to $v$

neutral at the solar wind speed, the new ion is accelerated or "picked-up" as shown in Fig. 5.8. The speed of the ion in the original rest frame of the neutral oscillates between 0 and twice the velocity of the field in the neutral's reference frame as the ion follows the cycloid path.

Because there is no acceleration parallel to the field line and the speed of the original neutral is fractions of a percent of the solar wind speed, the ratio of the perpendicular to the parallel velocity is large. This is expressed as the pitch angle which is defined through

$$\tan \alpha_{pa} = \frac{v_\perp}{v_\parallel} \tag{5.19}$$

and $\alpha_{pa}$ can be assumed to be 90° in many cases. The effects of ultra-low frequency waves lead to pitch-angle scattering. The particle distribution function is said to transform from a ring distribution to a shell distribution through these phenomena. Subsequent collisions and interactions with other ions, neutrals and/or fields modify this distribution function further.

The acceleration of the particle is a transfer of momentum and energy from the solar wind to the ion whilst increasing the density of the plasma. A single ion would, of course, have no effect on the solar wind, but the effect of mass loading of the flow by an active comet is substantial. The size of the gyroradius is also important. For a newly created proton, $r_c$ will be approximately 1000 km and the gyrofrequency is around 0.5 rad s$^{-1}$ but, for heavy ions, the gyroradius can be tens of thousands of kilometres and the gyrofrequency can be much smaller. This shows that spatial inhomogeneity in ion production on fairly large scales (e.g. several thousand kilometres) can lead to inhomogeneity of the plasma downstream because of the common cycloidal motion of pick-up ions combined with the highly localised source of ions (the near-nucleus region of the comet).

The net effect of the mass loading from an active comet is to slow down the solar wind and produce an interaction similar to that shown schematically in Fig. 5.9. As the solar wind starts to encounter cometary material deceleration of the flow begins. Initially the deceleration process is continuous until a critical mass flux is reached at which point a shock forms in a collisionless plasma.

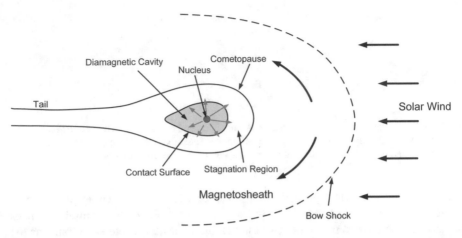

**Fig. 5.9** Schematic diagram of the main regions and features arising from the interaction of the solar wind (coming from the right) with an active comet. (Modified after Cravens 1997)

The criterion defining a shock wave is that the bulk velocity of the plasma drops from "supersonic" to "subsonic", where the speed of sound $c_s$ is defined by

$$c_s = \sqrt{\frac{\gamma p_p}{\rho_p}} \tag{5.20}$$

where $\gamma$ is the ratio of specific heats, $p_p$ is the plasma pressure, and $\rho_p$ is its density. In a plasma, the pressure term must include the partial pressures from both electrons and ions leading to the didactically clearer equation

$$c_s = \sqrt{\frac{\gamma k(T_e + T_i)}{m_i}} \tag{5.21}$$

where $T_e$ and $T_i$ define the electron and ion temperatures. The ideal gas law is here an assumption. The gyrovelocity of the ions around field lines can also be used to assign a thermal velocity to the plasma which is comparable to the sound speed.

The sonic Mach number, $Ma_s$, is then given by

$$Ma_s = \frac{u_{sw}}{c_s} \tag{5.22}$$

and using the values given in Table 5.1 we see that, in the undisturbed solar wind, $Ma_s \approx 10$. In other words, it is highly supersonic. But the mass loading remote from the nucleus already slows the solar wind and, even more importantly, is a significant heat source. The picked-up protons have an initial temperature around 60 times higher than the solar wind species. This leads to a rise in the thermal pressure of the plasma and a reduction of the Mach number to approximately two at the position of the bow shock. 1P/Halley was emitting gas at $>20$ tonne s$^{-1}$ (Krankowsky et al. 1986) at the time of the Halley Armada resulting in a bow shock position roughly $3 \times 10^5$ km from the nucleus in the direction towards the Sun. The presence of cometary pick-up ions was detected by Giotto more than 20 times this distance from the nucleus (at $8 \times 10^6$ km; Neugebauer et al. 1987).

The plasma at this point is still collisionless and so flow continues until charge exchange collisions with cometary neutrals become important. This next region inside the bow shock is called the magnetosheath (Fig. 5.9) and is typically characterized by plasma turbulence (Verigin et al. 1987). The flow speed within the magnetosheath continues to drop as the distance to the nucleus reduces until it approaches zero in the case of an active comet.

It is informative here to look at the flow in one dimension with a simplified set of equations. The single fluid continuity equation is given by

$$\frac{\partial \rho_p}{\partial t} + \nabla \cdot (\rho_p \boldsymbol{u_p}) = m_i S_i \qquad (5.23)$$

where the electron contribution to the mass has been neglected. We use the symbol $u_p$ for the plasma velocity. The right hand side is the source term. The mass density is given by the sum of the solar wind ions and the cometary ions, viz.,

$$\rho_p(r) = n_{sw}(r)\, m_{sw} + n_i(r)\, m_i \qquad (5.24)$$

with the source term arising from the ionization frequency, $R_i$, of cometary species. Using force-free radial outflow allowing us to compute the neutral density with Eq. (3.1) leads to the equation

$$-\frac{d}{dr}(\rho_p u_p) = \frac{m_i R_i Q_g}{4\pi v_g r^2} \qquad (5.25)$$

where $v_g$ is the neutral outflow speed and $Q_g$ is the gas production rate. The flow speed, $u_p(r)$, comes from the 1D single-fluid momentum equation under the assumption that the neutral flow speed is much smaller than plasma speed. This results in the equation

$$-\frac{d}{dr}\left(\rho_p u_p^2 + p_p + \frac{B^2}{2\mu_0}\right) = 0 \qquad (5.26)$$

where we can recognize the dynamic pressure and the magnetic pressure as two of the terms and $p_p$ is the plasma pressure incorporating all ionized species. Cravens (1997) shows that this equation can be solved for the flow speed relative to the upstream solar wind speed and gives the result

**Fig. 5.10** Simplified 1D solution for the plasma flow speed as the solar wind interacts with the ionizing cometary gas. Typical solar wind parameters have been used. The lines show results for two different cometary gas production rates. Solid line: $10^{30}$ molecule s$^{-1}$. Dashed line: $10^{28}$ molecule s$^{-1}$

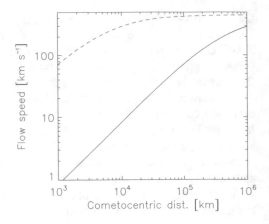

$$\frac{u_p(r)}{u_\infty} = \frac{1}{1 + \frac{r_{ml}}{r}} \qquad (5.27)$$

where

$$r_{ml} = \frac{1}{3} \frac{m_i R_i Q_g}{\pi v_g \rho_\infty u_\infty} \qquad (5.28)$$

This equation is shown for two values of $Q_g$ in Fig. 5.10. The higher production rate would be appropriate for 1P/Halley at 1 AU. The lower production rate corresponds to 300 kg s$^{-1}$ and is equivalent to a comet such as 67P shortly before perihelion. For the lower activity case, the plasma is slowed but still has appreciable speed as it reaches the vicinity of the nucleus. We can see here that for a Halley-type interaction the flow speed drops to close to zero well before the nucleus is reached. Near-stagnation of the flow was indeed observed at 1P/Halley (Lammerzahl et al. 1987).

The cometopause (Fig. 5.9) is essentially where the composition changes from being predominantly solar to being dominated by ions of cometary origin and was found between 1 and 2 $10^5$ km from the nucleus in the case of 1P/Halley (e.g. Vaisberg et al. 1987). Already here, the flow towards the nucleus along the Sun-comet line is beginning to stagnate as the ion density builds up. Even within the magnetosheath, the flow can no longer be considered one-dimensional as plasma is now forced to flow around the stagnating material. However, as one moves away from the Sun-comet line, the deceleration of the solar wind becomes less pronounced until eventually, when one is far enough away from the Sun-comet line, the solar wind flow reasserts itself. The frozen-in magnetic field of the solar wind therefore drapes around the near-nucleus region.

Inside the cometopause, the pressure from the inflowing material becomes balanced by the outflow of material from the nucleus and a stagnation region develops (Fig. 5.9). The magnetic field lines have slowed to close to zero velocity relative to

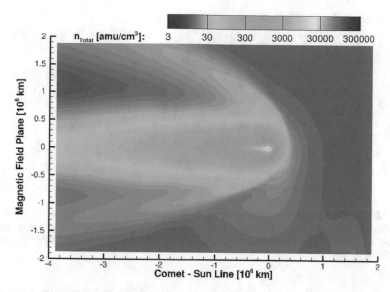

**Fig. 5.11** The total ion density in the model of 1P/Halley during the Giotto flyby by Rubin et al. (2014c). The Sun is to the right as in Fig. 5.9. Derived from the work in Rubin et al. (2014c) obtained with the BATS-R-US tool developed at The University of Michigan Center for Space Environment Modeling (CSEM)

the cometary neutrals. Consequently, the ion and electron temperatures drop rapidly because the energy given to new plasma by the pickup process is now far lower. Finally, one crosses a boundary into a field-free region. This boundary is sometimes called the contact surface and the field-free region is called the diamagnetic cavity.

A diamagnetic cavity arises when solar wind ions are unable to penetrate the inner coma because of collisions with outflowing neutrals (Haerendel 1987). The solar wind ions carry the magnetic field and this absence of penetration also excludes the field from this region. Consequently, within this cavity, newly created ions do not "see" any magnetic field and are instead coupled to the outflowing neutrals and flow away from the nucleus through ion-neutral drag (friction).

This description of the interaction is evident in Rubin et al. (2014c) who modelled the flow using a multifluid magnetohydrodynamics (MHD) model tuned to fit the Giotto fly-by observations. In Fig. 5.11, the total ion density is shown. The start of the increase in ion density along the Sun-comet line is around $10^6$ km from the nucleus. At around $2.5 \ 10^5$ km, the density rises rapidly as solar wind ions encounter the cometary ions. The density of solar wind ions increases slightly as the flow towards the nucleus slows but then drops quickly as cometary ions begin to dominate. This is seen in Fig. 5.12 (left) in which the cometary outgassing forms a barrier to the solar wind flow and solar wind ions are less and less able to penetrate the outflowing cometary material.

In Fig. 5.12 right, we focus on the near-nucleus part of the simulation domain. Here, within a few thousand kilometres of the nucleus, the contribution of the solar wind to the ion density drops to almost zero and the plasma is dominated by

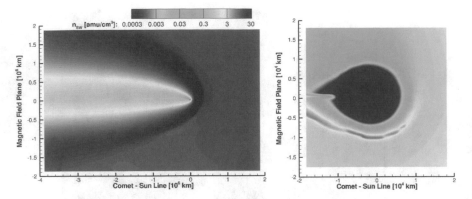

**Fig. 5.12** The density of solar wind ions in the simulation domain. Left: Large scale. Right: Near the nucleus at 100 times higher resolution. Derived from the work in Rubin et al. (2014c) obtained with the BATS-R-US tool developed at The University of Michigan Center for Space Environment Modeling (CSEM)

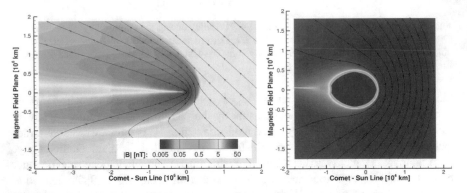

**Fig. 5.13** The magnetic field strength and field lines from the model of Rubin et al. (2014c) for a simulation of 1P/Halley. Left: Large scale. Right: The near-nucleus region at 100 times higher resolution. Derived from the work in Rubin et al. (2014c) obtained with the BATS-R-US tool developed at The University of Michigan Center for Space Environment Modeling (CSEM)

cometary ions. The asymmetries should be noted. As Rubin et al. (2014c) showed, solar wind ions are deflected in the opposite direction from cometary pick-up ions because of the difference in their velocities with respect to the charge-averaged ion velocity. This results in opposite signs in the Lorentz force term and thus gyration starts in the opposite direction (cf Fig. 5.8).

Within the stagnation region on the sunward side of the nucleus the solar wind is slowed to a standstill and the frozen-in magnetic field carried by the solar wind drapes around the stagnation region. In the case of Rubin et al. (2014c), the magnetic field was oriented at 45 degrees to the Sun-comet line leading to the field line orientation shown in Fig. 5.13 (left). However, the physics of the draping of the field lines about the obstacle remains unchanged. Thus simulations of the interaction between the solar wind and a strongly active comet show that a bilobed magnetic field configuration arises. The International Cometary Explorer (ICE) spacecraft flew ~7800 km down the plasma tail of 21P/Giacobini-Zinner making magnetic field observations. It confirmed the presence of a well developed magnetotail (Slavin et al. 1986). The tail was indeed composed of two lobes of opposite magnetic polarity (note the direction of the arrows on the magnetic field lines in Fig. 5.13, right) separated by a 1500 km thick plasma sheet.

In Fig. 5.13 (right), we can see the diamagnetic cavity and the high field strengths surrounding it in the stagnation region.

The figures illustrate how MHD can be used to treat the mutual interaction between flows of electrically conducting, non-magnetic fluids and magnetic fields. We have already seen the Navier-Stokes equations for fluid flow. The basic MHD equation incorporates the Lorentz force into the Navier-Stokes equations. Previously we saw the equation

$$\frac{\partial \rho_p u_p}{\partial t} + \left(u_p \cdot \nabla\right)\rho_p u_p + \rho_p u_p\left(\nabla \cdot u_p\right) + \nabla p_p - \upsilon\rho_p \nabla^2 u_p = \rho_p F \qquad (5.29)$$

and incorporating the Lorentz force, we obtain

$$\frac{\partial \rho_p u_p}{\partial t} + \left(u_p \cdot \nabla\right)\rho_p u_p + \rho_p u_p\left(\nabla \cdot u_p\right) + \nabla p_p - \upsilon\rho_p \nabla^2 u_p + (\mathbf{J} \times \mathbf{B})$$
$$= \rho_p F \qquad (5.30)$$

By using Ampere's law, the Lorentz force can be re-written (see Davidson 2001) so that

$$\mathbf{J} \times \mathbf{B} = (\mathbf{B} \cdot \nabla)\left(\frac{\mathbf{B}}{\mu_0}\right) - \nabla\left(\frac{\mathbf{B}^2}{2\mu_0}\right) \qquad (5.31)$$

from which we can again recognize the magnetic pressure as the second term on the right hand side. The momentum equation still contains the $\rho_p F$ term on the right side which can be considered as a means to include additional, potentially minor, forces (e.g. gravity).

Figure 5.12 is a result from a multifluid simulation and follows work by Najib et al. (2011) in which there are separate mass, momentum, and energy equations for each ion species in the total fluid. These are coupled through ensuring charge neutrality by defining the electron density as

$$n_e = \sum_{k_T} q_{k_T} n_{k_T} \qquad (5.32)$$

where $q_{kT}$ is the charge of species, $k_T$, and $n_{kT}$ is the number density of that species and through source terms which include ion-ion elastic collisions that couple the ion fluids, ion-electron elastic collisions that couple the ion and electron fluids, and electron-neutral elastic collisions that couple the electrons to the neutrals.

## 5.4  Processes within the Diamagnetic Cavity

The position of the boundary of the diamagnetic cavity is usually assumed to be the result of pressure balance between ion-neutral drag force inside the cavity and the magnetic pressure outside it (e.g. Ip and Axford 1987). This can be written as (Goetz et al. 2016a)

$$-\frac{d}{dr}\left(\frac{B^2}{2\mu_0}\right) = n_g n_i m_i k_{in}\left(v_i - v_g\right) \tag{5.33}$$

where $B$ is again the magnetic field strength, $v_i$, $n_i$, and $m_i$ are the velocity, number density and mass of the ion species (i.e. $H_2O^+$ or $H_3O^+$), $n_g$ and $v_g$ are the density and velocity of the neutrals, and $k_{in}$ is a ion-neutral momentum transfer collision rate and given as $1.85 \times 10^{-9}$ cm$^3$ s$^{-1}$ by Gombosi et al. (1996).

At 1P/Halley, the boundary to the cavity was detected around 3840 km from the nucleus (Neubauer et al. 1986) but this was expected to be much closer to the nucleus of the relatively weakly active comet, 67P, and even non-existant at large heliocentric distances. Nonetheless, the orbital distance of the Rosetta spacecraft was sufficiently close that passage into the diamagnetic cavity was recorded on 665 occasions as catalogued by Goetz et al. (2016a). The boundary, however, appears to be dynamic. Engelhardt et al. (2018) have reported large but short-lived plasma density and energy enhancements close to the contact surface. This is possibly arising from a Kelvin-Helmholtz instability (KHI) at the diamagnetic cavity boundary (Ershkovich and Mendis 1986) as predicted in MHD simulations by Rubin et al. (2012) (see also Eriksson et al. 2017). The KHI is an instability at an interface between two streams with different velocities and/or densities. It is seen when there is velocity shear in a single continuous fluid, or where there is a velocity difference across an interface between two fluids. The most commonly used examples in planetary sciences are from giant planet atmospheres. Goetz et al. (2016b) have reported that the position of the diamagnetic cavity boundary at 67P is not well described by Eq. (5.33) but the KHI and the fact that many of the measurements were made at the flanks of the interaction region (as a consequence of the terminator orbits flown by Rosetta) may be significant.

As noted in Sect. 5.2, the ions generated from cometary neutrals are produced mainly by photoionization, charge-exchange with solar wind ions and by high-energy electron impact ionization. Obviously, there are also electrons produced by these processes. The typical energy of fresh photoelectrons is expected to be 12–15 eV (Huebner et al. 1992) which might be considered a little surprising. For photoionization, the threshold for $H_2O$ is around, 12.619 $\pm$ 0.006 eV

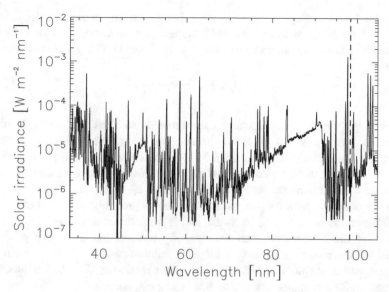

**Fig. 5.14** The solar spectrum in the EUV at 1 AU from July 9, 2018 at 03:23 UT. The photoionization threshold of $H_2O$ is marked by the vertical dashed line. Taken from the Extreme ultraviolet Variability Experiment (EVE) database (http://lasp.colorado.edu/home/eve/)

(982.47 $\pm$ 0.5 Å; Table 5.2; Katayama et al. 1973) in the extreme ultraviolet. With the daughter electron energy arising from the excess energy above threshold, photoionization by photons of around 500 Å wavelength are required to give the electrons that amount of energy. The relative amount of energy that the newborn electron receives is a stochastic variable which can vary between 0 and 100% of the excess energy but the energy distribution function between these two extremes is not well known (Lummerzheim and Lilensten 1994; Heritier et al. 2018). The slightly surprising point can be seen in Fig. 5.14 which shows that the solar flux does not drop off in the EUV like a Planck function. On the contrary, the solar flux is fairly constant (within an order of magnitude) with wavelength between 330 Å and threshold. Hence, fresh photoelectrons will get a broad distribution of energies from at least 40 eV downwards.

The 1D continuity equation for the ions in spherical symmetry and under steady-state conditions is similar in form to that seen in Eq. (3.113) for the gas. Galand et al. (2016) write it as

$$\frac{1}{r^2}\frac{d}{dr}r^2 n_i v_i = n_g\left(f_e + f_{ph}\right) = Q_i \tag{5.34}$$

where $n_i$, $v_i$, $f_e$ and $f_{ph}$ are all functions of the cometocentric distance, $r$. Here, $n_i$ is the local ion density, $v_i$ is the bulk ion velocity, $f_e$ is the electron impact ionization frequency, and $f_{ph}$ is the photo-ionization frequency. The bulk ion velocity can be taken as being equal to the neutral gas velocity (i.e. $v_i = v_g$) where ion-neutral drag is dominant although there is now evidence that, in the diamagnetic cavity, ion velocities are on average 2–4 times higher than that of the neutrals (Odelstad et al. 2018). If both the frequencies are assumed constant with distance and the velocity is assumed constant, then an analytical solution is available (Galand et al. 2016), i.e.

$$n_i = \left( f_e + f_{ph} \right) n_g \, \frac{r - r_0}{v_g}. \tag{5.35}$$

When gas production rates and therefore neutral densities are high, photo-ionization dominates but at lower production rates both terms are required. Johansson et al. (2017) have considered the influence of dust absorption. This will also be dependent upon the dust optical depth and hence its importance will depend on the dust production rate. As illustrated in Fig. 4.17, the optical depth at 67P was quite low at optical wavelengths and so the effect is probably of limited importance there. However, one needs to be aware that if particles smaller than optical wave-lengths are more numerous than we currently believe then these can produce a significant absorption cross-section for EUV photons and for 67P we have little knowledge of their column densities. For comets as active as 1P/Halley at the time of the spacecraft encounters, dust would need to be considered.

For typical electron densities within a cometary coma, the electron flux from the local plasma overcomes the photoemission from the spacecraft itself and the space-craft potential becomes negative. This was commonly observed at Rosetta. In the limit where the plasma electron flux greatly exceeds the flux of electrons from the spacecraft, the spacecraft potential is expected to have a negative value several times the electron thermal energy equivalent (Eriksson et al. 2017). An example at 67P from 10 Jan 2015 presented by Eriksson et al. (2017) shows the spacecraft potential oscillating between about $-5$ and $-15$ V over a single rotation of the nucleus with the electron density varying between 100 and 400 cm$^{-3}$ at a cometocentric distance of 27.5 km from the nucleus above the terminator (phase angle of 91°). However, a cold electron population in the innermost coma of 67P with an electron temperature, $T_e$, $\lesssim 0.1$ eV was also identified. This population was insufficient to modify the spacecraft potential but was clear evidence of electron cooling by collisions with the cold (e.g. Fig. 3.4), neutral gas molecules (Gan and Cravens 1990). When the neutral coma and ionosphere becomes dense enough, the neutrals and local ions act as a large sink for the electron energy input produced by the photoionization resulting in the cold electron population. As the cometocentric distance increases, the neutral and ion densities drop faster than the decrease in the electron heating rate and the cold electron population disappears. Mandt et al. (2016) defined a limit for electron collisionality (which might be called an electron collisionopause or exobase and given the symbol, $r_{ce}$) as the cometocentric distance at which the mean free path is equal to that distance. This can be related to the neutral production rate in an isotropic model. The mean free path is

$$\lambda_{ce} = \frac{1}{n_g \sigma_{en}} \qquad (5.36)$$

where $\sigma_{en}$ is the electron-neutral collisional cross-section and equal to $1.5 \ 10^{-19} \ m^2$ (Itikawa and Mason 2005). Substitution using Eq. (3.1) results in

$$r_{ce} = \frac{Q_g \sigma_{en}}{4 \pi v_g} \qquad (5.37)$$

which indicates that production rates in excess of about $1.3 \ 10^{26} \ s^{-1}$ are needed from a 2 km sized nucleus for collisional cooling to be of any significance. The influence of bi-modal electron distributions on the plasma flow at the boundary of the diamagnetic cavity has not yet been investigated (cf. Fig. 5.7).

The low electron temperatures imply that reaction R9 (electron-impact dissociative recombination) in Table 3.9 can play a significant role in modifying the ion densities close to the nucleus with models indicating that ion densities 380 km from the nucleus of 67P near perihelion can be overestimated by 50% (Heritier et al. 2018). The cross-section for the reaction is a function of the electron temperature, i.e.

$$\sigma_{DR} \propto \frac{1}{\sqrt{T_e}} \qquad (5.38)$$

and is therefore sensitive to the coupling of the electron temperature to the strongly decreasing neutral gas temperature (e.g. Fig. 3.30). The exact details of the variation in ion densities in the innermost coma, particularly near perihelion, are still the subject of research (e.g. Vigren et al. 2019).

The calculations of Rubin et al. (2014c) for 1P/Halley shown above were for a very active comet ($Q_{H2O} \approx 6.9 \times 10^{29}$ molecule $s^{-1}$; Krankowsky et al. 1986) which was near perihelion at the time. The flight through the tail of 21P/Giacobini-Zinner on 11 Sept. 1985 by the International Cometary Explorer (ICE) spacecraft (von Rosenvinge et al. 1986) occurred when the water production rate of the comet was between 3.2 and $4.3 \times 10^{28}$ molecule $s^{-1}$ (Combi and Feldman 1992). Schmidt-Voigt (1989) used MHD calculations to study the differences between the two comets at the times of the spacecraft encounters. While most of the main elements of the interaction are similar, it was shown that the larger the gas production rate, the more the tail is flattened by the effects of magnetic stresses on the plasma. On the other hand, draping of the magnetic field around the "obstacle" would still occur.

The situation is appreciably different for the weak interaction occurring at 67P— particularly at larger heliocentric distances—producing effects that can be difficult to model (Rubin et al. 2014b). Near perihelion with production rates of $\sim 10^{27} - 10^{28}$ molecule $s^{-1}$ (e.g. Fig. 3.48), Goetz et al. (2017) showed that draping patterns were still evident although the opening angle of the tail between the lines of opposite polarity (i.e. between the two lobes) was much larger than seen at 21P/Giacobini-Zimmer. The diamagnetic cavity was also more dynamic in its response to changing

source and solar wind conditions. This is now being addressed with more sophisticated time-dependent models (e.g. Nemeth 2020).

In the case of very weak outgassing ($10^{25}$–$10^{26}$ molecule s$^{-1}$), the coma can be
penetrated by the solar wind particles through to the nucleus surface itself. Volwerk
et al. (2018) showed that the interaction is very different in cases similar to these and
does not show the bilobed induced magnetotail structure found at higher production
rates as one would qualitatively expect from consideration of the pressure balance.

One of the key issues that arises in the case of weak outgassing is that the gyroradii of newly picked-up ions become much larger than the interaction region and the
MHD approach to modelling used above is no longer tenable. Hybrid models
describe the plasma environment in a semi-kinetic manner. In these models, electrons are assumed to be a massless charge-neutralizing fluid and are modeled using
typical fluid methods (conservation equations). The ions, which have much larger
gyroradii, are treated as particles. A hybrid code therefore solves individual equations of motion to construct local densities (Hansen et al. 2007). This method
resolves physical structures down to the required length scales and also allows
non-Maxwellian distribution functions. Crudely, one can imagine this as the kinetic
counterpart to the fluid approach we also saw in the description of the gas dynamics.

Delamere (2006) used a hybrid code to simulate the solar wind interaction with
19P/Borrelly at the time of the Deep Space 1 encounter. The DS1/PEPE experiment
had shown a north-south asymmetry in the plasma boundaries which were initially
unexpected because the neutral coma was observed to be symmetric. Delamere
showed that this asymmetry was a natural consequence of the large pickup ion
gyroradius (Fig. 5.8) compared to the size of the obstacle. Test particle methods have
also been used to study effects upstream of the bow shock for more active comets
(e.g. McKenzie et al. 1994). Koenders et al. (2013, 2015) provided pre-Rosetta
models of 67P using a hybrid code.

## 5.5   Spectroscopy and Imaging of Plasma Tails

### 5.5.1   Main Ion Species

As can be seen in Fig. 5.1, plasma tails can be imaged by ground-based observers out
to several tens of millions kilometres from the nucleus. Both broad-band and narrow-
band imaging as well as spectroscopy have been used to study the composition and
dynamics of tails. A compilation of large-scale images for 1P/Halley has been
produced by Brandt et al. (1993) and various phenomena associated with plasma
tails can be seen therein. The plasma tails are bright as a result of emission from $CO^+$
(4260 Å and 5077 Å) and $H_2O^+$ (typically 6148 Å and 6198 Å). Contributions from
$OH^+$ and $CO_2^+$ are other important species for remote observation (Jockers 1991).
Wyckoff et al. (1999) looked at the fine structure of some of these emissions in C/
1996 B2 (Hyakutake).

## 5.5.2   Ion Velocities

The ion tails exhibit large scale structure that is observed to move down the tail. Tracking of these features has allowed study of the velocity and acceleration of the ions in the tail. The results have been broadly confirmed by more recent high resolution spectroscopic observations that have determined the Doppler shift of specific lines. Feature tracking of the tail of 1P/Halley during the 1986 apparition gave ion accelerations of around 3–4 m s$^{-2}$ such that ion velocities were approaching 200 km s$^{-1}$ around 45 million kilometres from the nucleus (Celnik and Schmidt-Kaler 1987). However, this is highly variable and probably strongly dependent upon the solar wind conditions and the ion production rates.

## 5.5.3   Disconnection Events

The basic structure of the tail in steady-state can be understood from numerical models (Fig. 5.15). However, solar wind conditions are rarely stable and it is now clear that changes in the solar wind interaction with the comet can lead to spectacular phenomena in the plasma tail. Disconnection events (e.g. Figure 5.15) give the impression that the whole plasma tail is "disconnecting" from the head of the comet and flowing away.

The Niedner-Brandt sequence describes four phases of a disconnection event based upon observation (e.g. Brandt and Chapman 2004; see also Jia et al. 2007). In phase I, the tail becomes narrower than usual, and tail rays can be observed. In phase II, the tail breaks from the nucleus. In phase III, the old tail recedes from the nucleus and a new tail forms. In phase IV, the old tail disappears and the comet resumes its usual appearance.

In the mid-1980s the idea of solar wind magnetic sector boundary crossings provoking disconnection events gained some traction. However, the number of observed tail disconnection events is too high to be explainable solely by interplanetary sector boundary crossings (Jockers 1985). Alternatively, increases in the ambient solar wind pressure might result in plasma instabilities. We now know that changes in solar wind conditions do have consequences. Vourlidas et al. (2007) made the first direct imaging of the interaction of a comet (2P/Encke) with a coronal mass ejection (CME) in the inner heliosphere with high temporal and spatial resolution and observed a clear, correlated, disconnection of the tail. They concluded that the disconnection is driven by magnetic reconnection between the magnetic field entrained in the CME and the interplanetary field draped around the comet.

As noted above, disconnection is often accompanied by the appearance of quasi-linear tail rays. Bundles of these structures may appear either roughly symmetrically to the tail axis or on one side of the main tail only. C/1973 XII Kohoutek was studied by Jockers (1985) and, in this case, one-sided rays were most frequently observed.

**Fig. 5.15** Image of the plasma tail of comet 1P/Halley showing the start of a disconnection event. Enhanced version of data (LSPN 2185) in the International Halley Watch Large Scale Phenomenon Network archive

In the late twentieth century, numerical simulation of disconnection events was challenging because of the computer power necessary. Various specific cases were addressed. Wegmann et al. (1996), for example, compared their model with observations of C/1989 X1 (Austin) which produced multiple tail rays but failed to follow this with a disconnection event. They concluded that a 90° rotation of the interplanetary magnetic field (IMF) could reproduce this scenario although sector boundary crossings with field reversals were not investigated.

Jia et al. (2007) studied disconnection events specifically using a global 3D MHD model with considerably higher spatial resolution. They showed that indeed sector boundary crossings initiated disconnection events supporting the idea as originally promoted in a series of papers by Niedner and Brandt (e.g. Niedner Jr. and Brandt 1978). However, the observational evidence seems to suggest that other large discontinuities in the solar wind can also initiate the process with changes in the magnetic field orientation >90° being necessary. Tail rays are also a natural consequence of tangential discontinuities in the solar wind flow. Jia et al. (2007) showed that if multiple discontinuities occur at intervals of the order of 2 h or greater, strong tail rays result followed by disconnection.

When disconnection events were thought to be exclusively related to interplanetary magnetic field boundary crossings it was asked whether such a crossing has influence on charged dust. Horanyi and Mendis (1987) posed this question noting that because of the reversal of the magnetic field, the convectional electric field in the flowing plasma also changes direction. This leads to the reversal of the electrodynamic acceleration of a charged dust grain as a magnetic sector boundary sweeps by. Modelling suggested that the wavy appearance of the dust tail of C/1965 S1 Ikeya-Seki might have been caused by related effects.

# Chapter 6
# Comet-Like Activity in Related Objects

## 6.1 Active Asteroids

Several objects in asteroidal orbits have been seen to have dust tails and are therefore assumed to be losing dusty material. A recent example of this is the 6 km diameter inner main-belt asteroid, (6478) Gault (Jewitt et al. 2019). In this particular case, material ejection speeds of $0.15 \pm 0.05$ m s$^{-1}$, were derived indicating outflow speeds around 3 orders of magnitude lower than are typically seen at an active comet. These "active asteroids" have low inclinations and orbit between about 2 and 4 AU from the Sun (i.e. $a_s < a_J$) and a Tisserand parameter, $T_J > 3$. Hence, they have dynamical similarities to main-belt asteroids. One can see in Fig. 6.1, the dynamical separation of active asteroids from comets.

The activity is not necessarily evidence of comet-like, sublimation-driven, activity from surface ices. The activity seen in some cases may have arisen in ways that are not strictly comet-like. Jewitt (2012) considered eight different mechanisms leading to observable mass loss. The more probable candidates are listed in Table 6.1.

Combinations of these mechanisms together with other phenomena (such as thermally or impact induced volatile loss) are also conceivable and we have seen evidence for the influence of several of these mechanisms being active on the nucleus of 67P even if, in the case of comets, sublimation is the dominant loss mechanism. In addition to (6478) Gault, asteroidal objects that have shown some form of mass loss include, 345P/LINEAR, 133P/Elst-Pizarro (asteroid number 7968; Figs. 6.1 and 6.2) and 107P/Wilson-Harrington (which has the asteroid designation (4015) Wilson–Harrington).

The loss of material caused by the impact of a small object on a larger asteroidal body may be visible as a short-lived coma or narrow tail. Jewitt (2012) suggested this as a possible mechanism for the production of the thin tail of 354P/LINEAR (also known as P/2010 A2), for example. Impact had previously been suggested as a source for the activity of 133P/(7968) Elst-Pizarro based on its showing a thin,

© Springer Nature Switzerland AG 2020
N. Thomas, *An Introduction to Comets*, Astronomy and Astrophysics Library,
https://doi.org/10.1007/978-3-030-50574-5_6

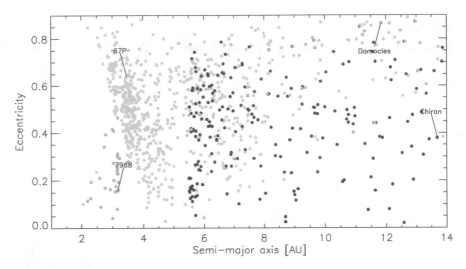

**Fig. 6.1** Semi-major axes and eccentricities of Centaurs (dark blue), comets (yellow-green), active asteroids (cyan) and Damocloids (purple-grey). The most famous objects in each category are marked. Note the separation of the comets from the active asteroids in this diagram and how the Damocloids superpose the longer period comets

**Table 6.1** Plausible mechanisms to explain "active asteroids"

| Mechanism | Description |
|---|---|
| Sublimation | Sublimation with dust dragged away from the surface by gas drag. This is a comet-like mechanism. |
| Impact/collision | Impacts between material within the asteroid belt can occur at velocities of several kilometres per second and hence ejection of material can occur. |
| Rotational instability | Here the centripetal acceleration resulting from rotation becomes close to or equal to the surface gravitational acceleration. |
| Electrostatic forces | Particle and surface charging effects have been proposed to explain phenomena associated with small bodies and it could be responsible for lifting particles to produce faint comae. |
| Insolation weathering | Repeated thermal cycling of surfaces leading to break-up and mass loss through thermal fatigue and thermal shock. This might lead to landslide or collapse although it is hard to envisage this placing material on escape trajectories without some additional force. |
| Radiation pressure sweeping | Here, radiation pressure removes particles directly from the surface. This mechanism must however overcome particle cohesive forces at the surface. |

Extracted from Jewitt (2012)

narrow tail with no appreciable coma suggesting very low particle outflow velocities and little influence of gas drag. However, Hsieh et al. (2004) showed that the activity was recurrent making the impact hypothesis somewhat unlikely.

107P/Wilson-Harrington, (3200) Phaethon, and (2201) Oljato are dynamically asteroidal bodies ($T_J > 3$) in which mass loss has been reported, but they are not in

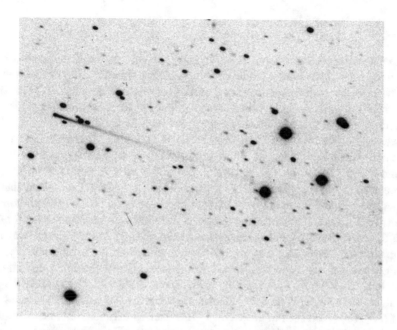

**Fig. 6.2** 133P/Elst-Pizarro observed on 7 August 1996 by E.W. Elst and G. Pizarro. Note the stellar-like head and the long thin tail. Credit: ESO (CC BY 4.0)

the main belt and this has led Jewitt to prefer the term "active asteroids" for this object class rather than the previously used "Main-Belt Comets".

It was inferred that the previously thought to be inactive asteroid (3200) Phaethon was a dead comet based upon IRAS observations of a dust trail visible along its orbital path. Phaethon has now gained a comet designation following clear evidence of aperiodic activity (see e.g. Jewitt 2012). Jewitt had previously suggested radiation pressure sweeping as a contributing mechanism for loss from objects with small perihelia such as Phaethon. A further example is 324P/La Sagra which was originally given an asteroid designation (2010 R2) and is in an asteroid orbit with $T_J = 3.099$. However, it has shown activity consistent with sublimation (Jewitt et al. 2016a).

To summarize, it is arguable whether these objects should be considered to be within a common category or whether they should be treated as individuals. Jewitt (2012), for example, looked at several objects individually to try to assess which of the mechanisms in Table 6.1 might be responsible in specific cases. Activity through sublimation is probable in several cases (e.g. 324P/La Sagra; Jewitt et al. (2016a) and 313P/Gibbs; Jewitt et al. (2015)) and cometary designations assigned. On the other hand, for other objects the situation is less clear.

## 6.2   Activity of Centaurs

As noted earlier, Centaurs are thought to be objects in transition between the scattered disc and the Jupiter family of comets. Approximately 10% of the Centaurs discovered show some evidence of activity (Mazzotta Epifani et al. 2017; Peixinho et al. 2020) through having a diffuse coma and/or tail in optical images. Several of these objects have received comet designations including 174P/Echeclus and 29P/ Schwassmann-Wachmann 1 (e.g. Schambeau et al. 2019). The mass losses from these objects can be substantial. Wong et al. (2019) measured the loss from P/2011 S1 and C/2012 Q1 to be $140 \pm 20$ and $250 \pm 40$ kg s$^{-1}$, respectively which is comparable to the loss from 67P (e.g. Fig. 3.48). One should be aware that the methods to derive the losses from Centaurs are based upon broad-band photometry of the dust comae and are therefore markedly different to the methods used for 67P. With the perihelia of Centaurs being outside the orbit of Jupiter, simple water ice sublimation models cannot explain this activity. However, the exothermic crystallization of amorphous water ice (Sect. 2.9.3.4) remains a plausible model although Jewitt et al. (2017) have found activity in an object, C/2017 K2 (PANSTARRS), for which this explanation is probably inadequate because of its extreme distance to the Sun (23.7 AU) when seen to be active. 174P/Echeclus exhibits outbursts with mass loss rates of the order of 86 kg s$^{-1}$ occurring at 12.9 AU from the Sun (Rousselot 2008). The sporadic and variable activity seen in some Centaurs is probably not consistent with homogeneous objects undergoing crystallization leading Wong et al. (2019) to suggest that outbursts are an indication of nucleus inhomogeneity. On the other hand, super-volatiles such as CO may be significant. Wierzchos et al. (2017) detected CO J(2-1) emission towards 174P/Echeclus with a production rate of around 36 kg s$^{-1}$ at 6.1 AU from the Sun which they noted was 40 times less than seen at 29P/Schwassmann–Wachmann 1 at a similar heliocentric distance (Gunnarsson et al. 2008).

The driving process(es) behind the activity of Centaurs remains unknown at this point and the differences between Centaurs in terms of, for example, CO loss per unit surface area, are significant (see e.g. Wierzchos et al. 2017). On the other hand, outbursts are, in themselves, complex phenomena that may lead to many processes (exposure of the sub-surface, lifting of ice-rich material volatiles into the coma, etc.) that may have stoichastic components that will be difficult to disentangle. But if these objects really are JFC pre-cursors, then understanding of their current activity is clearly of major importance.

## 6.3   Interstellar Visitors

The discovery of the first object with an inbound trajectory placing its origin outside our Solar System (1I/(2017 U1) 'Oumuamua) was a major milestone (Meech et al. 2017). It was not observed to become active during its perihelion passage. However, this was followed by the detection of another object of interstellar origin, 2I/(2019 Q4) Borisov (2019), which was determined to be active with an dust emission rate of approximately 2 kg s$^{-1}$ at 2.7 AU from the Sun (Jewitt and Luu 2019). The existence of activity provides an opportunity to explore the composition of objects originating from other proto-planetary discs but its use is restricted because the exact origins of these objects are likely to remain, in general, poorly known.

It has been speculated that the asteroid (514107) Kaʻepaokaʻawela is of interstellar origin based upon its highly unusual orbit. (514107) Kaʻepaokaʻawela is in a 1:1 resonance with Jupiter but is in a retrograde orbit with the planet. A combination of the inclination and the eccentricity keeps it more than 1 AU away from Jupiter. This makes it an interesting target for exploration although its retrograde orbit would provide a significant challenge to spacecraft dynamicists if a rendezvous with this object is to be achieved.

It is to be expected that gravitational interactions during the formation of solar systems will not merely populate Oort-like clouds around other stars but also lead to ejection from the original system. The same N-body simulation used to produce Fig. 1.12, resulted in 70% of the test particles being ejected from the system with around 15% colliding with the Sun or the giant planets. Although these numbers probably depend upon the exact initial conditions and orbital evolution of the large planets within planetary systems, it is a strong indicator that quite substantial numbers of comet-like objects are thrown out of planetary systems at an early age and so there are probably vast numbers of these objects wandering through interstellar space. Hence, it is not inconceivable that future missions such as ESA's Comet Interceptor could provide a close view of one of these objects within the coming decades.

## 6.4   Comets in Close Proximity to the Sun

As was pointed out in the introduction, Jones et al. (2018) have provided a recent extensive review of comets that come close to or impact the Sun. They proposed a formal definition of different categories that is reproduced in Table 6.2. What follows is a brief précis of the Jones et al. review.

Most sun-grazers and sun-skirters have been detected by space-based observatories and instruments studying the solar corona such as the LASCO instrument

**Table 6.2**   The Jones et al. (2018) classification of comets that come to the Sun

| Category | Perihelion distance [AU] |
|---|---|
| Near-Sun | <0.307 |
| Sun-skirters | 0.016–0.1537 |
| Sun-grazers | 0.0046–0.016 |
| Sun-divers | <0.0046 (1 solar radius) and therefore impacting |

onboard SOHO and the SECCHI instrument onboard STEREO. The detected objects usually belong to one of a series of "families" that have similar orbital characteristics. By far the best known of these families is the Kreutz family which is a group of sun-grazers that includes C/1965 S1 Ikeya-Seki and C/2011 W3 Lovejoy. Other sun-skirting groups include the Meyer, Marsden, and Kracht families. Backward orbital integrations indicate that the Marsden and Kracht families may be related (Ohtsuka et al. 2003).

Families are assumed to arise from the result of the break-up of a parent object with tidal forces at a previous perihelion passage being a logical assumption. Further splitting at subsequent perihelion passages of the fragments is considered likely. Kreutz family comets appear to arrive in clusters that may point to this process. The observed morphologies of Kreutz family objects are diverse which may indicate inhomogeneity of the progenitor.

The absolute brightness of Kreutz family comets appears to peak before they reach perihelion (around 0.005 AU from the Sun) and then declines as they get even closer. This would not be consistent with insolation-driven water ice sublimation but may be caused by sublimation of what are normally considered to be refractories such as olivine and pyroxene as the surface temperature rises (Kimura et al. 2002). Use of Eq. (2.100) results in temperatures above 4000 K at 0.01 AU but of course sublimation will be a strong coolant, the magnitude of which is difficult to quantify, and the point source approximation for the solar input no longer holds at these distances.

The loss of more refractory material can provide additional information on the composition as illustrated by the detection of emission lines from of O, Na, K, Ca, V, Cr, Mn, Fe, Co, Ni, and Cu in C/1965 S1 Ikeya-Seki (Preston 1967; Slaughter 1969). C/2003 K7 was observed at 0.016 AU when grains were rapidly sublimated. The relative abundances of H: C: Si were approximately 1: 0.0035: 0.045 and therefore carbon-poor compared to what we have seen at 67P (cf. Fig. 4.70). On the other hand, analysis of observations of C/2011 W3 Lovejoy has provided abundance ratios by number for H: C: N: O: Si: Fe as 1.0: 0.035: 0.004: 0.5: 0.015: 0.025 respectively which seems to be more in keeping with the 67P results but with a very high iron content. Strength estimates from the break-ups of sun-grazers are of the order of 0.5–10 Pa (e.g. C/2012 S1 ISON; Steckloff et al. 2015).

While sun-grazers are spectacular, it remains a challenge to establish their relationship to other comets and whether we can use the information obtained to constrain the properties of other cometary objects.

# Chapter 7
# The Loss of Comets

Estimates of the mass loss rates caused by sublimation from comets, combined with their sizes and densities, suggest that a typical short period comet can maintain activity for around 1000 revolutions in the inner Solar System before the material is exhausted. As noted above, the timescale for 67P against complete erosion of the nucleus in the current orbit is ~6000 years, for example. This timescale can be increased by airfall deposition of more inert material leading to a slow choking of the activity reducing the mass loss rate. This concept is often assumed to explain the existence of inactive or very weakly active objects in comet-like orbits. Asteroid (3552) Don Quixote (1983 SA), for example, is in an orbit that strongly resembles that of a short-period comet (Bottke et al. 2002) and has the reddened spectrum of a D-type asteroid which is also often taken as an indicator of having a comet-like composition. Mommert et al. (2014) used data from the Spitzer infrared telescope acquired in 2009 to show that the object was emitting dust at a rate of ~1.9 kg s$^{-1}$ (see also Mommert et al. 2020). Emission from gaseous $CO_2$ was also detected.

Visiting objects like (3552) Don Quixote with a spacecraft could provide additional about this hypothesis and indeed, in the past, Don Quixote has been proposed as a target for future space missions. High resolution mapping of the surface could provide an estimate of the length of time that the object has been in this near-inert state. One approach would be to use crater statistics. Teanby (2015) estimated impact rates onto the surface of Mars based partially on HiRISE observations of recent impacts (Daubar et al. 2013). These results are shown in Fig. 7.1 after being adapted to show the area of the surface influenced by impacts over a 1000 year period as a function of the impact size for which the crater diameter is assumed to be a proxy. The turnover at small crater diameters probably arises from the effect of the Martian atmosphere and extrapolation of the more linear part of the curve at larger crater diameters to smaller crater diameters is likely to be more realistic. The integral of this curve gives a value of 6.2% kyear$^{-1}$ which suggests that saturation would occur in around 20,000 years if a Martian impact distribution can be used for this type of estimate.

© Springer Nature Switzerland AG 2020
N. Thomas, *An Introduction to Comets*, Astronomy and Astrophysics Library,
https://doi.org/10.1007/978-3-030-50574-5_7

**Fig. 7.1** The surface area affected by impacts of different sizes derived from the work of Teanby (2015). The results were obtained by studying the surface of Mars and the turnover at small sizes is probably related to the existence of an atmosphere. The integral suggests that around 6.2% of the surface is disturbed by impacts over a period of 1000 years

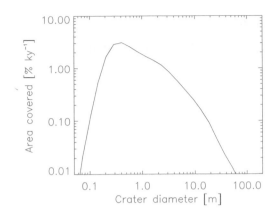

Damocloids are objects with orbits similar to those of Halley-family (Figs. 6.1 and 1.17) and long-period comets but they do not display visible signs of outgassing. The prototype for these objects is (5335) Damocles. Their high eccentricities and inclinations suggest that the Damocloids are the dead or dormant nuclei of long-period comets (Asher et al. 1994; Jewitt 2005).

There are two definitions of Damocloids with one stating that a Damocloid is any point-source object having $T_J \leq 2$. The other definition requires the object to be retrograde or it should be a combination of having a perihelion distance <5.2 AU, a semi-major axis >8.0 AU, and an eccentricity >0.75. The two definitions are roughly, but not precisely, equivalent. There are now over 150 objects that meet these criteria and many are plotted in Fig. 6.1. (5335) Damocles is marked. Note how the Damocloids and longer period comets appear to occupy a similar region in this diagram.

The hypothesis that Damocloids and Halley-type comets are related is strengthened by the detection of outgassing from some Damocloids after their initial detection as inert objects leading to them being re-classified as comets. C/2001 OG108 (LONEOS) is an example (Abell et al. 2003). For this object, with an orbital period of 48.51 years and a radius similar to that of 1P/Halley, we can, by analogy, estimate that its lifetime against complete erosional loss should have been less than a million years. Lightcurves and sizes can be determined to provide further statistics, as with asteroid 2006 BZ8 (Hergenrother 2018), that may place constraints on, for example, tensile strength.

There seems now to be little doubt that short-period and Halley-type comets can become inert and dormant providing one means by which comets are "lost". Further evidence for decrease in activity with time comes from observations of 49P/Arend-Rigaux for which the active area has been inferred to have reduced from ~3% to ~0.2% over six apparitions (Chu et al. 2019).

The measured spin-up of 67P over its perihelion passage has shown that the timescale for this process can be appreciably shorter than the timescale for total erosional loss and that this may lead to splitting of nuclei well before their activity diminishes. There is considerable ground-based evidence for comets splitting. A list

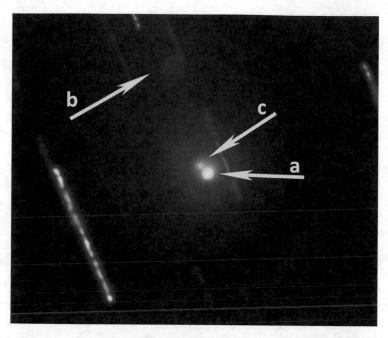

**Fig. 7.2** The separation of fragments of C/2011 J2 (LINEAR) observed on 25 October 2015. The image is ~80 arcseconds on the horizontal dimension. Credit: F.Manzini, V.Oldani, TNG-Telescopio Nazionale Galileo

of comets known to have split is kept by the International Comet Quarterly.[1] A review was provided by Boehnhardt (2004) and an example (C/2011 J2 (LINEAR)) is shown in Fig. 7.2 from Manzini et al. (2016). Other examples include the long-period comet C/1999 S4 (LINEAR) that disintegrated in July 2000 and the short-period comet 332P/Ikeya–Murakami which was well studied using the Hubble Space Telescope (Jewitt et al. 2016b) following a splitting event in January 2016. The exact mechanism for observed splitting in these cases is unknown and indeed dynamical effects arising from asymmetric outgassing and spin-up as indicated by the Rosetta observations, may be only one of several mechanisms (Boehnhardt 2002) that could include internal gas pressure build-up (Smoluchowski 1985), internal stress related to heat penetration (Sekanina and Chodas 2012) or impact phenomena (e.g. Guliyev 2017). For 332P/Ikeya–Murakami, Jewitt et al. (2016b) infer from brightness variations that rotational instability may have played a role while Graykowski and Jewitt (2019) concluded from observations of the splitting of 73P/Schwassmann-Wachmann 3 that, in this case, rotational instability could not be responsible. Steckloff et al. (2015) have suggested nuclei can fragment and disperse through dynamic sublimation pressure, which induces differential stresses within the interior of the nucleus. This was modelled using the sungrazer C/2012 S1 (ISON) as

---

[1]http://www.icq.eps.harvard.edu/ICQsplit.html

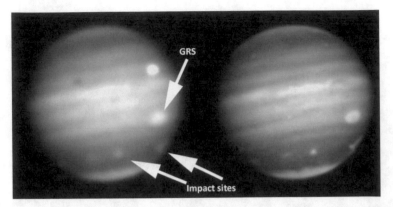

**Fig. 7.3** Observations of Jupiter obtained shortly after the impacts of the fragments of SL9 into the southern hemisphere. The observations were taken two full Jovian rotations apart (23 July 1994 at 03:18 UT and 23:17 UT). The image to the right was obtained in better observational conditions at the 1.5 m ESO telescope on La Silla. Two of the four visible impact sites and the Great Red Spot are marked

an example. Fragments can produce potentially dangerous near-Earth objects. For example, the near-Earth object (NEO) 2004 TG10 may be a fragment of 2P/Encke (Porubčan et al. 2006).

Fracturing and separation of cometary sub-nuclei do not necessarily result in the break-up of a comet unless there are forces that ensure that the fragments reach escape velocity. Re-accumulation of material is possible resulting in a "re-configuration" of the sub-nuclei thereby simply changing the shape and exposing "fresh" surfaces (Jutzi and Benz 2017). This will strongly depend upon the additional forces (either gravitational or non-gravitational arising from the jet action of outgassing) acting on the sub-nuclei. We have at this point no observational evidence of re-accumulation of comets after fragmentation but observations of binary asteroids and even triple asteroid systems have been explained by rotational fission (Margot et al. 2015) suggesting that similar processes can be considered.

The tidal breakup of D/Shoemaker-Levy 9 by Jupiter (Fig. 1.8) followed by collision with the planet's atmosphere (Fig. 7.3) is an explanation solely for this one comet. Comet encounters with Jupiter that significantly perturb the orbit are fairly common. 67P experienced a perturbing encounter in 1959. However, it is necessary that the encounter be close to or within the Roche limit to affect a tidal break-up even in the absence of any tensile strength. Levison et al. (2000) computed impact rates for the giant planets and found that 21% of the objects that hit Jupiter in their simulations were bound to the planet before impact.

Tidal disruption of comets when they come close to the Sun seems to be a plausible mechanism for objects on specific orbits. The Solar and Heliospheric Observatory has observed more than 3000 comets in a field of view that is just a few solar radii in width. Some 85% of these comets were members of the Kreutz group of comets (Jones et al. 2018) which have aphelia at around 170 AU. Almost all of these comets fail to survive the encounter.

**Table 7.1** Crude estimates of the timescales for processes resulting in loss of observable comets

| Process | Approximate timescale |
|---|---|
| Total mass loss through activity | 1000 orbits |
| Spin-up and splitting | <100 orbits |
| Extinction of activity | 10 orbits (e.g. 49P/Arend-Rigeaux) to >70 orbits (e.g. 2P/Encke) |
| Internal pressure buildup/fragmentation disruption | $10^4$–$10^5$ orbits |
| Impact induced disruption | $10^5$–$10^6$ orbits (HTCs > JFCs) |
| Tidal disruption | Unknown (HTCs ≫ JFCs) |

96P/Machholz has been observed by SOHO's LASCO experiment several times. This particular comet does not get as close to the Sun as the Kreutz group and this probably saves it from complete disintegration. Probably the most famous Kreutz comet was C/1965 S1 (Ikeya-Seki) which passed close to the Sun in September 1965 and broke into three pieces shortly before perihelion.

Jutzi et al. (2017) have computed the critical specific impact energies (in [J kg$^{-1}$]) needed to produce a complete disruption of a cometary nucleus by high velocity impact. The specific impact energy of a collision is defined as

$$Q_{col} = \frac{1}{2} \frac{\mu_r v_{col}^2}{(M_N + m_p)} \tag{7.1}$$

where

$$\mu_r = \frac{M_N m_p}{(M_N + m_p)} \tag{7.2}$$

and $M_N$ is the mass of the nucleus, $m_p$ is the mass of the projectile and $v_{col}$ is the collision velocity. The critical specific impact energy for disruption is dependent upon the impact velocity which in the asteroid belt can be 10–15 km s$^{-1}$. Using scaling relations from Jutzi et al. (2017) and Holsapple (1993) and extrapolating the impact rate from Teanby, we get a lifetime against complete disruption by impact in the Earth-Jupiter region of $10^5$–$10^6$ orbits for a 2 km sized nucleus. This is considerably longer than the disruption rate calculated by Belton (2015) from observations of fragmentation events.

To summarize, the timescales for comet loss arising from proposed mechanisms are, in many cases, difficult to quantify. In Table 7.1, there is a crude comparison between estimated timescales. There is also an indication of the relative timescales for HTCs and JFCs arising from the fact that JFCs are close to the ecliptic plane where collisions with asteroidal material is more probable.

# Chapter 8
# Future Investigations of Comets

Given the details revealed by space missions in the past 35 years culminating in the Rosetta mission, it is a valid question to ask whether there is any future for ground-based and Earth-orbiting observations of comets. However, there are several important aspects that space missions cannot answer within financially realistic boundary conditions. The first of these is diversity. We are now aware that comets are by no means identical and that there is enormous diversity in composition. Hence, observing many different objects has a clear goal in relating this diversity to origins.

The evolution of the outgassing with heliocentric distance is a related issue. It is a rule-of-thumb that cometary activity increases from 2.7 AU inwards but this is simply because this is the approximate distance at which the surface reaches the free sublimation temperature of water ice. The relationship between the outgassing of the major species (CO, $CO_2$, and $H_2O$) is more complex than this, as has been clearly demonstrated by, for example, the presence of water ice particles in the inner coma of 103P/Hartley 2 in comparison with 67P or by the evidence for $CO_2$ outgassing from the nightside of the nucleus of 67P which may or may not be a universal phenomenon.

Given the timescales for future missions, it is also interesting to look at the questions that our most recent spacecraft investigations have failed to answer and at questions that have arisen as a consequence of those observations (see also Thomas et al. 2019).

While the observations of nuclei have provided hints on the structure of surface layers and specifically the porosity, the interrelation of volatiles and refractories and their variation with depth at intermediate scales are still uncertain. Many of the issues are connected to

- the density of cometary material with depth
- the refractory to volatile ratio with depth
- the volatile composition with depth

These three quantities would provide us with several pieces of information that are blocking study of cometary activity, namely

© Springer Nature Switzerland AG 2020
N. Thomas, *An Introduction to Comets*, Astronomy and Astrophysics Library,
https://doi.org/10.1007/978-3-030-50574-5_8

- the depth and density of the uppermost inert layer (if it exists, although I would argue that this is almost proven)
- the changing volatile mixing ratio with depth and detection of a sub-surface recondensation layer
- the determination of the positions of species-specific sublimation fronts (should they exist)
- the depth at which internal properties are no longer influenced by the irradiation of the surface

It should be noted that models have been developed that purport to derive these quantities but none of these have actually been measured to confirm and/or constrain the models. It is also notable that coring down to at least 1 m and preferably 3 m was the initial target of the Rosetta mission when it was still a sample return mission in the 1980s. It is perhaps telling that we have not been able to refine this significantly other than to confirm what we already knew.

It is also evident from the work on Rosetta data that our understanding of the microphysics of the surface and sub-surface layers remains rudimentary. This has left two major uncertainties in our understanding of comets.

First, we do not know how the volatile component is associated with the refractory component of cometary material. This is vital to understanding how comets formed and may ultimately provide key information on the solar system formation process. One can envisage ices surrounding or encapsulating the dust component, ices within dust matrices, ices as a component isolated from the refractories, and so on. There is no strong evidence supporting any of these models.

Second, the thermophysical behaviour of the surface and sub-surface layers remain inadequately constrained. We have values for the thermal inertia but the thermal conductivity and heat capacity are unknown, the importance of gas transport and re-condensation are unknown, and the refractory to volatile ratio inside the nucleus remains a subject of fierce debate. Without these, models of cometary evolution are, at best, poorly constrained.

Hence, one of the primary objectives for future missions must be to improve our knowledge of thermophysical parameters since data interpretation and our understanding of cometary activity heavily depends on the output of thermophysical computer simulations of the upper cometary surface layers ($\leq 1$ m). These parameters strongly depend on microphysical details of the cometary surface, such as e.g., the material composition, the morphology of the material (structure, arrangement of the particles, coordination number, void space) and the mixing of the different components (silicates, organics and ices). Besides the thermal modelling aspect, also the macroscopic behaviour of the cometary surface is influenced by the microphysical properties of the material, such as the tensile strength (Gundlach et al. 2018), compressive strength (Lorek et al. 2016; Schräpler et al. 2015), and the thermal conductivity (Gundlach and Blum 2012).

This also illustrates that laboratory work has an important role to play. The properties of ice-refractory mixtures can be studied under idealised conditions to help understand the basic physical processes and to try to establish (possibly empirical) relationships between variables.

It is also apparent from the surface morphology of 67P and the variations in activity (both in magnitude and composition) with position that inhomogeneities in longitude and latitude on the nucleus are present. The variability evident at 9P/ Tempel 1 was perhaps less extreme but nonetheless present. There remains considerable doubt whether this inhomogeneity is primordial or evolutionary. Hence comparisons of areas of different morphology and/or morphological context are highly desirable.

The results obtained by the CONSERT radar on Rosetta demonstrated the unique capability of radars to provide information about the internal structure of a comet nucleus. However, this instrument was operated under unfavorable conditions resulting in considerable uncertainty about the deep interior. A follow-up ground penetrating radar would clearly be an interesting examination of theories on nucleus formation.

The isotopic, elemental, chemical, and chiral composition of non-volatile material, including complex organic material was not established. The elemental composition of ejected dust was found with COSIMA but the composition of complex organics, for example, remains unknown. Clearly, any discussion of comets bringing complex organics to the inner solar system and their participation in the evolution of life requires knowledge of the composition of these organics and it is fully apparent that the optimum way to address composition is by using the power of Earth-based analytical laboratories. Remote-sensing approaches are totally inadequate whereas in situ analysis remains technically challenging and limited by spacecraft resources. It is this aspect that forms the strongest argument for initiating a sample return mission to a comet although the Stardust mission has scratched the surface of this particular topic.

Given the difficulties of executing such investigations, the answers will not be available any time soon. Nevertheless, analytically, we are in a far better position now than 20 years ago with recent decades having seen both progressive increases in performance of instrumentation, huge increases in computer power and ease of use, and breakthroughs such as the introduction of nano-SIMS technology, the spread of con-focal microscopy and sub-micron resolution X-ray microscopes as well as the greater availability of synchrotron-based techniques.

Laboratory analysis (especially of the mineralogy, chemistry and isotopic composition) of solar-system and extra-solar material has been enormously important in furthering our understanding of the origin and evolution of the solar system, as well as shedding light on the nucleosynthetic processes in several types of star. Using samples returned from the Moon, as well as meteorites likely to have come from a variety of parent bodies (including Mars and perhaps the asteroid Vesta), we have been able to determine the age of the Earth and the solar system, the timescales for the dynamic evolution of the solar system, and the timescale over which liquid water existed on early planetesimals. The power of laboratory instrumentation, and the advantage of having samples on Earth, is shown by the way in which lunar samples returned almost 50 years ago are still leading to fresh insights. If we had instead relied solely on in situ analysis of the Moon with 1960s technology, our understanding would be a fraction of what it is today.

The two major issues with returning to Earth are connected to evolution of the sample during the sampling process and the return trip and the effects on the sample during reentry. The latter in particular is a significant problem for attempts to acquire undisturbed cores. The forces due to deceleration on re-entry will undoubtedly compromise the sample's integrity structurally, leading to mixing between products from different depths. It seems apposite to conclude that, for the foreseeable future at least, an ambition to return to Earth, a cored cometary sample is highly challenging.

Rosetta has demonstrated that constraining the spatial distribution of gas and dust sources uniquely is vastly more challenging than originally thought. One should caution that some authors have a relatively low threshold for success in this context. And yet this remains a key question in the efforts to relate surface structure (thermal, mechanical, chemical) to activity and evolution. The observational requirements for future missions needed to break the degeneracy and relate coma gas and dust densities to surface properties are challenging.

The single point density measurements by COPS have shown the importance of this type of measurement. Local gas density measurements at multiple positions within the inner coma would have provided far more rigorous constraints on the gas flow field. In particular, continuous measurements near the sub-solar point (over periods greater than the rotation period at a cadence of $<1$ min$^{-1}$) would have been particularly valuable. When coupled with a local velocity and temperature measurement (and we note here the additional importance of determining the temperature anisotropy as a constraint on the energy distributions), this would have provided us with a far better understanding of outgassing than the rather ambiguous situation we have today.

Remote sensing observations of the gas distribution were obtained by Rosetta but their interpretation remains complex. In a future mission concept, the inversion of density, temperature, and velocity from remote sensing should be traded against local direct multipoint measurements to establish the optimum strategy. The critical aspect is whether the gas flow field can be inverted to produce global maps of source strength at a resolution that is sufficient to relate the production rate to surface morphologies and properties. Given the variation in surface morphology seen on the surface of 67P, the required resolution is potentially $<100$ m and comparable to the mean free path for modest activity cases.

There are several properties of the dust that affect the dust flow field pattern in addition to the source. It is well known that particle size and shape are important. The dust size distribution remains a critical parameter in all studies of the dust flow field and its brightness at all wavelengths. Future missions will need much more thorough instrumentation to address this topic.

The comets that we see today come from three reservoirs: The Oort cloud is the most distant, and the source of long period and dynamically new comets, while short period Jupiter family comets (JFCs) come from the scattered disc of the Kuiper Belt. The modern distinction is based on the comet's orbit (specifically its Tisserand parameter with respect to Jupiter), and splits comets into "ecliptic" comets from the trans-Neptunian region and "nearly isotropic comets" (NICs) from the Oort cloud, the latter including both long period and high inclination short period comets of the Halley type (Levison 1996). The third reservoir is the main asteroid belt, which has recently been recognised to contain a significant population of icy bodies, including the so-called "Main Belt Comets" (MBCs), whose relatively circular orbits stay within the belt (Jewitt et al. 2015; Snodgrass et al. 2017). Cross-comet calibration of the measurements identifying differences in these objects would clearly be worthwhile.

The next mission to a comet is likely to be the European Space Agency's Comet Interceptor, recently selected for launch in 2029, which is designed to make a fast fly-by of a dynamically new comet or, if luck is on one's side, an interstellar visitor. The idea here is that the spacecraft waits at a parking position until a dynamically new comet or an interstellar asteroid is detected on its way to the inner Solar System. The spacecraft leaves its parking position and performs a fast fly-by to study the surface and innermost coma of the object. This mission will address questions of diversity between reservoirs but it is unlikely to constrain the physico-chemical structure of the nucleus and inner coma that we have just discussed.

We can summarize these issues by referring to a table of possible future missions and their goals (Table 8.1) taking in account the fact that spacecraft missions are limited in terms of financial, technical, and programmatic resources. One can see that no one mission can address all of these outstanding questions so that some form of selection or trade-off would be necessary. On the other hand, this shows that there is plenty of scope for designing and implementing our next foray into the world of the comet.

**Table 8.1**  Possible future missions to comets and their primary objectives

| Mission concept | Description | Goals | Comments |
|---|---|---|---|
| The coma swarm | Multiple small satellites or cube-sats orbiting or manoeuvering around the nucleus making local measurements of the gas and dust coma | - Detailed evaluation of the gas and dust dynamics in 3D<br>- Dust size distributions and their variation within the coma<br>- Simultaneous multi-directional remote-sensing of the nucleus to monitor activity | |
| The surface network | Multiple small landed packages placed on the surface of the nucleus. | - Determination of the interior structure through low frequency tomography.<br>- Seismic sensing of the interior.<br>- Local investigation of the surface layer structure at sub-centimetre scales | It is assumed that this concept would not allow a drilling system |
| The surface rover | Single mobile station traversing the cometary nucleus | - Determination of the diversity of cometary material within one object<br>- Local investigation of the surface layer structure at sub-metre scales | It is assumed that this concept would carry a drilling system |
| The new comet fly-by | Single spacecraft waiting in orbit for a new spectacular target | - Determination of the nucleus and inner coma characteristics of an interstellar or pristine Oort cloud comet | This is the concept behind ESA's Comet Interceptor due to launch in 2028. |
| The multi-object fly-by | Single spacecraft making multiple fast fly-bys of cometary and/or comet-like objects | - Characterization of the diversity of comet-like objects<br>- Imaging and detailed measurements of composition | |
| Advanced impact exhumation | Dual spacecraft system with an impactor and a monitoring spacecraft. Improved version of Deep Impact to view in detail the exhumed material through a rendezvous. | - Evaluation of the internal properties of the target<br>- Determination of the layer structure (if present)<br>- Determination of the internal composition | AIDA/Hera concept might be considered |

(continued)

**Table 8.1**  (continued)

| Mission concept | Description | Goals | Comments |
|---|---|---|---|
| Sample return | Spacecraft rendezvous and acquisition of cometary material followed by return to Earth | - Detailed laboratory analysis of cometary material | The CAESAR New Frontiers mission within the NASA programme went through a Phase A study before losing out to Dragonfly in the final selection. |

# Chapter 9
# Final Remarks

Two concepts have driven cometary science for much of the past 80 years. Firstly, if comets really are debris from the formation of our Solar System then studying their chemical and physical structure appears to be the only way to acquire data on the processes occurring in our proto-planetary disc. The diversity now being seen in the structures of other planetary systems suggests that every planetary system is "special"—the initial conditions make an enormous difference to the end result which necessarily limits what we can learn from exoplanetary systems about our own. Consequently, while we may be able to learn about processes from exoplanet astronomy, the knowledge to be gained from our primitive objects has a major role to play in constraining the specific processes and evolution of our own Solar System.

The second reason for studies of comets has been their role in transporting material (principally water and organics) across the Solar System. This is still the subject of debate. Is the high D/H ratio observed at 67P indicating that Earth's water is not of cometary origin or are there other processes at work? At what stage in Earth's evolution did organic material become relevant and did comets contribute to that reservoir? Was there a contribution to the evolution of life?

There remains insufficient information to constrain the theories associated with these concepts. This is in part due to the facts that basic processes are not understood and that details matter. We have only limited understanding of the processes involved in proto-planetesimal formation. We have no understanding of whether life can evolve from cometary material. We remain uncertain about the dynamical history of our own Solar System over the past 4.6 billion years. One could go on.

While acquiring that information is challenging, comets do offer a laboratory for testing hypotheses. It is apparent that to get at the chemical and physical structure without ambiguity, we need to get "up close and personal" with these bodies. In particular, the micro-structure of cometary material may provide us with clues to the processes that were active during Solar System formation. Similarly, the chemical compositions of the surfaces and interiors may provide better understanding of

© Springer Nature Switzerland AG 2020
N. Thomas, *An Introduction to Comets*, Astronomy and Astrophysics Library,
https://doi.org/10.1007/978-3-030-50574-5_9

precisely what comets transported through the Solar System during Earth's formation.

Rosetta was originally described as a mission "towards the origin of the Solar System" and indeed that is certainly what it was. But, both since its inception and following its execution, we have realised that comets will not give up these secrets easily. On the other hand, it is worth remembering that the Rosetta data set is huge. A colleague who is very familiar with one of the major instruments remarked to me that he thought only 2–3% of the data from that instrument had really been explored thoroughly more than 4 years after the completion of the mission. While I am not quite so pessimistic about the data sets I myself am most familiar with, I would agree that analysis of data sets could continue for many years both to extract the maximum return from the investment made but also to prepare for the future. It is clear that students can still now perform very worthwhile studies on the data sets and in particular by linking observations from several instruments to exploit multiple constraints.

One thing, however, has been obvious in compiling this work. The data sets we have acquired, from all missions and observations, have limitations that should not be swept under the carpet. This is, to some extent, a natural consequence of the boundaries imposed when spacecraft missions are constructed and the limits on what can be achieved by Earth-based remote sensing. But degeneracies cannot be ignored and alternative explanations must be explored. In cometary science, there is plenty left to do.

# References

Abell, P. A., Fernandez, Y. R., Pravec, P., French, L. M., Farnham, T. L., Gaffey, M. J., Hardersen, P. S., Kusnirak, P., Sarounova, L., & Sheppard, S. S. (2003). Physical characteristics of asteroid-like comet nucleus C/2001 OG108 (LONEOS). *Lunar and Planetary Science Conference, 1253.*

Acton, C. H. (1996). Ancillary data services of NASA's Navigation and Ancillary Information Facility, *Planetary and Space Science, 44*, 65.

Agarwal, J., A'Hearn, M. F., Vincent, J.-B., Güttler, C., Höfner, S., Sierks, H., Tubiana, C., Barbieri, C., Lamy, P. L., Rodrigo, R., Koschny, D., Rickman, H., Barucci, M. A., Bertaux, J.-L., Bertini, I., Boudreault, S., Cremonese, G., Da Deppo, V., Davidsson, B., Debei, S., De Cecco, M., Deller, J., Fornasier, S., Fulle, M., Gicquel, A., Groussin, O., Gutiérrez, P. J., Hofmann, M., Hviid, S. F., Ip, W.-H., Jorda, L., Keller, H. U., Knollenberg, J., Kramm, J.-R., Kührt, E., Küppers, M., Lara, L. M., Lazzarin, M., Lopez Moreno, J. J., Marzari, F., Naletto, G., Oklay, N., Shi, X., & Thomas, N. (2016). Acceleration of individual, decimetre-sized aggregates in the lower coma of comet 67P/Churyumov-Gerasimenko. *Monthly Notices of the Royal Astronomical Society, 462*, S78.

Agarwal, J., et al. (2017). Evidence of sub-surface energy storage in comet 67P from the outburst of 2016 July 03. *Monthly Notices of the Royal Astronomical Society, 469*, s606.

A'Hearn, M. F. (1982). Spectrophotometry of comets at optical wavelengths. In L. L. Wilkening (Ed.), *Comets* (p. 433). Tucson: University of Arizona Press.

A'Hearn, M. F. (2008). Deep impact and the origin and evolution of cometary nuclei. *Space Science Reviews, 138*, 237–246.

A'Hearn, M. F., & Festou, M. C. (1990). The neutral coma. In W. F. Huebner (Ed.), *Physics and chemistry of comets*. Springer: Berlin.

A'Hearn, M. F., Schleicher, D. G., Millis, R. L., Feldman, P. D., & Thompson, D. T. (1984). Comet Bowell 1980b. *The Astronomical Journal, 89*, 579.

A'Hearn, M. F., Belton, M. J. S., Delamere, A., & Blume, W. H. (2005). Deep impact: A large-scale active experiment on a cometary nucleus. *Space Science Reviews, 117*, 1.

A'Hearn, M. F., Belton, M. J. S., Delamere, W. A., Feaga, L. M., Hampton, D., Kissel, J., Klaasen, K. P., McFadden, L. A., Meech, K. J., Melosh, H. J., Schultz, P. H., Sunshine, J. M., Thomas, P. C., Veverka, J., Wellnitz, D. D., Yeomans, D. K., Besse, S., Bodewits, D., Bowling, T. J., Carcich, B. T., Collins, S. M., Farnham, T. L., Groussin, O., Hermalyn, B., Kelley, M. S., Kelley, M. S., Li, J.-Y., Lindler, D. J., Lisse, C. M., McLaughlin, S. A., Merlin, F., Protopapa, S., Richardson, J. E., & Williams, J. L. (2011). EPOXI at Comet Hartley 2. *Science, 332*, 1396.

A'Hearn, M. F., Feaga, L. M., Keller, H. U., Kawakita, H., Hampton, D. L., Kissel, J., Klaasen, K. P., McFadden, L. A., Meech, K. J., Schultz, P. H., Sunshine, J. M., Thomas, P. C., Veverka,

J., Yeomans, D. K., Besse, S., Bodewits, D., Farnham, T. L., Groussin, O., Kelley, M. S., Lisse, C. M., Merlin, F., Protopapa, S., & Wellnitz, D. D. (2012). Cometary volatiles and the origin of comets. *The Astrophysical Journal, 758*, 29.

A'Hearn, M. F., Krishna Swamy, K. S., Wellnitz, D. D., & Meier, R. (2015). Prompt emission by OH in comet Hyakutake. *The Astronomical Journal, 150*, 5.

Aksnes, K., & Mysen, E. (2011). Nongravitational forces on comets: An extension of the standard model. *The Astronomical Journal, 142*, 81.

Alexander, F. J., & Garcia, A. L. (1997). The direct simulation Monte Carlo Method. *Computers in Physics, 11*, 588.

Altwegg, K. (1996). *Sulfur in comet Halley*. Habilitationsschrift, University of Bern.

Altwegg, K., Balsiger, H., Bar-Nun, A., Berthelier, J. J., Bieler, A., Bochsler, P., Briois, C., Calmonte, U., Combi, M., De Keyser, J., Eberhardt, P., Fiethe, B., Fuselier, S., Gasc, S., Gombosi, T. I., Hansen, K. C., Hässig, M., Jäckel, A., Kopp, E., Korth, A., LeRoy, L., Mall, U., Marty, B., Mousis, O., Neefs, E., Owen, T., Reme, H., Rubin, M., Semon, T., Tzou, C.-Y., Waite, H., & Wurz, P. (2015). 67P/Churyumov-Gerasimenko, a Jupiter family comet with a high D/H ratio. *Science, 347*, 1261952.

Altwegg, K., Balsiger, H., Bar-Nun, A., Berthelier, J.-J., Bieler, A., Bochsler, P., Briois, C., Calmonte, U., Combi, M. R., Cottin, H., De Keyser, J., Dhooghe, F., Fiethe, B., Fuselier, S. A., Gasc, S., Gombosi, T. I., Hansen, K. C., Haessig, M., Jäckel, A., Kopp, E., Korth, A., Le Roy, L., Mall, U., Marty, B., Mousis, O., Owen, T., Reme, H., Rubin, M., Semon, T., Tzou, C.-Y., Waite, J. H., & Wurz, P. (2016). Prebiotic chemicals–amino acid and phosphorus–in the coma of comet 67P/Churyumov-Gerasimenko. *Science Advances, 2*, e1600285.

Altwegg, K., Balsiger, H., Berthelier, J. J., Bieler, A., Calmonte, U., Fuselier, S. A., Goesmann, F., Gasc, S., Gombosi, T. I., Le Roy, L., de Keyser, J., Morse, A., Rubin, M., Schuhmann, M., Taylor, M. G. G. T., Tzou, C.-Y., & Wright, I. (2017). Organics in comet 67P – a first comparative analysis of mass spectra from ROSINA-DFMS, COSAC and Ptolemy. *Monthly Notices of the Royal Astronomical Society, 469*, S130.

Altwegg, K., Balsiger, H., & Fuselier, S. A. (2019). Cometary chemistry and the origin of icy solar system bodies: The view after Rosetta. *Annual Review of Astronomy and Astrophysics, 57*, 113.

Altwegg, K., Balsiger, H., Berthelier, J.-J., Briois, C., Combi, M., Cottin, H., De Keyser, J., Dhooghe, F., Fiethe, B., Fuselier, S. A., Gombosi, T. I., Hänni, N., Rubin, M., Schuhmann, M., Schroeder, I., Sémon, T., & Wampfler, S. (2020). Evidence of ammonium salts in comet 67P as explanation for the nitrogen depletion in cometary comae. *Nature Astronomy, 3*. https://doi.org/10.1038/s41550-019-0991-9.

Amsden, A., Ruppel, H., & Hirt, C. (1980). SALE: A simplified ALE computer program for fluid flow at all speeds. Los Alamos National Laboratories Report, LA-8095:101p. Los Alamos, NM: LANL.

Anisimov, S. I. (1968). Vaporization of metal absorbing laser radiation. *Soviet Physics JETP, 27*(1), 182–183.

Arakawa, S., Tatsuuma, M., Sakatani, N., & Nakamoto, T. (2019). Thermal conductivity and coordination number of compressed dust aggregates. *Icarus, 324*, 8–14.

Asher, D. J., Bailey, M. E., Hahn, G., & Steel, D. I. (1994). Asteroid 5335 Damocles and its implications for cometary dynamics. *Monthly Notices of the Royal Astronomical Society, 267*, 26.

Askari, R., Hejazi, S. H., & Sahimi, M. (2017). Effect of deformation on the thermal conductivity of granular porous media with rough grain surface. *Geophysical Research Letters, 44*, 8285–8293.

Asphaug, E., & Benz, W. (1994). Density of comet Shoemaker-Levy 9 deduced by modelling breakup of the parent 'rubble pile'. *Nature, 370*, 120–124.

Asphaug, E., & Benz, W. (1996). Size, density, and structure of comet shoemaker-levy 9 inferred from the physics of tidal breakup. *Icarus, 121*, 225–248.

Attree, N., et al. (2018). Tensile strength of 67P/Churyumov-Gerasimenko nucleus material from overhangs. *Astronomy and Astrophysics, 611*, A33.

Attree, N., Jorda, L., Groussin, O., Mottola, S., Thomas, N., Brouet, Y., Kührt, E., Knapmeyer, M., Preusker, F., Scholten, F., Knollenberg, J., Hviid, S., Hartogh, P., & Rodrigo, R. (2019). Constraining models of activity on comet 67P/Churyumov-Gerasimenko with Rosetta trajectory, rotation, and water production measurements. *Astronomy and Astrophysics, 630*, A18.

Auger, A.-T., et al. (2018). Meter-scale thermal contraction crack polygons on the nucleus of comet 67P/Churyumov-Gerasimenko. *Icarus, 301*, 173.

Avakyan, S. V., Il'in, R. N., Lavrov, V. M., & Ogurtsov, G. N. (Eds.). (1998). *Collision processes and excitation of UV emission from planetary atmospheric gases: A handbook of cross sections.* Amsterdam: Gordon and Breach.

Bagnold, R. A. (1941). *The physics of blown sand and desert dunes.* New York: Methuen.

Baines, M. J., Williams, I. P., & Asebiomo, A. S. (1965). Resistance to the motion of a small sphere moving through a gas. *Monthly Notices of the Royal Astronomical Society, 130*, 63.

Baker, V. R. (2001). Water and the martian landscape. *Nature, 412*, 228.

Balsiger, H., Altwegg, K., & Geiss, J. (1995). D/H and O-18/O-16 ratio in the hydronium ion and in neutral water from in situ ion measurements in comet Halley. *Journal of Geophysical Research, 100*, 5827–5834.

Bardyn, A., Baklouti, D., Cottin, H., Fray, N., Briois, C., Paquette, J., Stenzel, O., Engrand, C., Fischer, H., Hornung, K., Isnard, R., Langevin, Y., Lehto, H., Le Roy, L., Ligier, N., Merouane, S., Modica, P., Orthous-Daunay, F.-R., Rynö, J., Schulz, R., Silén, J., Thirkell, L., Varmuza, K., Zaprudin, B., Kissel, J., & Hilchenbach, M. (2017). Carbon-rich dust in comet 67P/Churyumov-Gerasimenko measured by COSIMA/Rosetta. *Monthly Notices of the Royal Astronomical Society, 469*, S712.

Baertschi, P. (1976). Absolute $^{18}$O content of standard mean ocean water. *Earth and Planetary Science Letters, 31*, 341.

Barucci, M. A., et al. (2016). Detection of exposed $H_2O$ ice on the nucleus of comet 67P/Churyumov-Gerasimenko as observed by Rosetta OSIRIS and VIRTIS instruments. *Astronomy and Astrophysics, 595*, A102.

Becker, K. H., & Haaks, D. (1973). Measurement of the natural lifetimes and quenching rate constants of OH(2Σ+, v = 0,1) and OD(2Σ+, v = 0,i) Radicals. *Zeitschrift Naturforschung, Teil A, 28*, 249–256.

Belton, M. J. S. (2015). The mass disruption of Jupiter Family comets. *Icarus, 245*, 8.

Belton, M. J. S., Wehinger, P., Wyckoff, S., & Spinrad, H. (1986). A precise spin period for P/Halley. *ESLAB Symposium on the Exploration of Halley's Comet, 250*, 599.

Belton, M. J. S., Julian, W. H., Anderson, A. J., & Mueller, B. E. A. (1991). The spin state and homogeneity of comet Halley's nucleus. *Icarus, 93*, 183.

Belton, M. J. S., et al. (2011). Stardust-NExT, Deep Impact, and the accelerating spin of 9P/Tempel 1. *Icarus, 213*, 345.

Belton, M. J. S., Thomas, P., Li, J.-Y., Williams, J., Carcich, B., A'Hearn, M. F., McLaughlin, S., Farnham, T., McFadden, L., Lisse, C. M., Collins, S., Besse, S., Klaasen, K., Sunshine, J., Meech, K. J., & Lindler, D. (2013). The complex spin state of 103P/Hartley 2: Kinematics and orientation in space. *Icarus, 222*, 595.

Benkhoff, J., & Boice, D. C. (1996). Modeling the thermal properties and the gas flux from a porous, ice-dust body in the orbit of P/Wirtanen. *Planetary and Space Science, 44*, 665–673.

Bensch, F., & Bergin, E. A. (2004). The pure rotational line emission of ortho-water vapor in comets. I. Radiative transfer model. *The Astrophysical Journal, 615*, 531.

Bertini, I., et al. (2017). The scattering phase function of comet 67P/Churyumov-Gerasimenko coma as seen from the Rosetta/OSIRIS instrument. *Monthly Notices of the Royal Astronomical Society, 469*, S404.

Beutler, G. (2005). *Methods of celestial mechanics: Vol. I. Physical, mathematical, and numerical principles.* In cooperation with Leos Mervart and Andreas Verdun. Astronomy and Astrophysics Library. Berlin: Springer. ISBN 3-540-40749-9.

Biele, J., Ulamec, S., Richter, L., Kührt, E., Knollenberg, J., Moehlmann, D., & Philae Team. (2009). The strength of cometary surface material: Relevance of Deep Impact results for Philae landing on a comet, Deep Impact as a world observatory event: Synergies in space, time, and wavelength., *285*.

Bieler, A., Altwegg, K., Balsiger, H., Berthelier, J.-J., Calmonte, U., Combi, M., De Keyser, J., Fiethe, B., Fougere, N., Fuselier, S., Gasc, S., Gombosi, T., Hansen, K., Hässig, M., Huang, Z., Jäckel, A., Jia, X., Le Roy, L., Mall, U. A., Reme, H., Rubin, M., Tenishev, V., Toth, G., Tzou, C.-Y., & Wurz, P. (2015a). Comparison of 3D kinetic and hydrodynamic models to ROSINA-COPS measurements of the neutral coma of 67P/Churyumov-Gerasimenko. *Astronomy and Astrophysics, 583*, A7.

Bieler, A., Altwegg, K., Balsiger, H., Bar-Nun, A., Berthelier, J.-J., Bochsler, P., Briois, C., Calmonte, U., Combi, M., de Keyser, J., van Dishoeck, E. F., Fiethe, B., Fuselier, S. A., Gasc, S., Gombosi, T. I., Hansen, K. C., Hässig, M., Jäckel, A., Kopp, E., Korth, A., Le Roy, L., Mall, U., Maggiolo, R., Marty, B., Mousis, O., Owen, T., Rème, H., Rubin, M., Sémon, T., Tzou, C.-Y., Waite, J. H., Walsh, C., & Wurz, P. (2015b). Abundant molecular oxygen in the coma of comet 67P/Churyumov-Gerasimenko. *Nature, 526*, 678.

Biermann, L. (1951). Kometenschweife und solare Korpuskularstrahlung. *Zeitschrift fur Astrophysik, 29*, 274.

Bird, G. A. (1994). *Molecular gas dynamics and the direct simulation of gas flows*. Oxford: Clarendon Press.

Biver, N., & Bockelée-Morvan, D. (2019). Complex organic molecules in comets from remote-sensing observations at millimeter wavelengths. *ACS Earth and Space Chemistry, 3*, 1550.

Biver, N., Bockelée-Morvan, D., Crovisier, J., Davies, J. K., Matthews, H. E., Wink, J. E., Rauer, H., Colom, P., Dent, W. R. F., Despois, D., Moreno, R., Paubert, G., Jewitt, D., & Senay, M. (1999). Spectroscopic monitoring of comet C/1996 B2 (Hyakutake) with the JCMT and IRAM radio telescopes. *The Astronomical Journal, 118*, 1850.

Biver, N., Bockelée-Morvan, D., Crovisier, J., Colom, P., Henry, F., Moreno, R., Paubert, G., Despois, D., & Lis, D. C. (2002a). Chemical composition diversity among 24 comets observed at radio wavelengths. *Earth Moon and Planets, 90*, 323.

Biver, N., Bockelée-Morvan, D., Colom, P., Crovisier, J., Henry, F., Lellouch, E., Winnberg, A., Johansson, L. E. B., Gunnarsson, M., Rickman, H., Rantakyro, F., Davies, J. K., Dent, W. R. F., Paubert, G., Moreno, R., Wink, J., Despois, D., Benford, D. J., Gardner, M., Lis, D. C., Mehringer, D., Phillips, T. G., & Rauer, H. (2002b). The 1995–2002 long-term monitoring of comet C/1995 O1 (HALE BOPP) at radio wavelength. *Earth Moon and Planets, 90*, 5.

Biver, N., Moreno, R., Bockelée-Morvan, D., Sandqvist, A., Colom, P., Crovisier, J., Lis, D. C., Boissier, J., Debout, V., Paubert, G., Milam, S., Hjalmarson, A., Lundin, S., Karlsson, T., Battelino, M., Frisk, U., Murtagh, D., & Team, O. (2016). Isotopic ratios of H, C, N, O, and S in comets C/2012 F6 (Lemmon) and C/2014 Q2 (Lovejoy). *Astronomy and Astrophysics, 589*, A78.

Biver, N., Bockelée-Morvan, D., Hofstadter, M., Lellouch, E., Choukroun, M., Gulkis, S., Crovisier, J., Schloerb, F. P., Rezac, L., von Allmen, P., Lee, S., Leyrat, C., Ip, W. H., Hartogh, P., Encrenaz, P., Beaudin, G., & Team, M. (2019). Long-term monitoring of the outgassing and composition of comet 67P/Churyumov-Gerasimenko with the Rosetta/MIRO instrument. *Astronomy and Astrophysics, 630*, A19.

Bockelée-Morvan, D. (1987). A model for the excitation of water in comets. *Astronomy and Astrophysics, 181*, 169.

Bockelée-Morvan, D., & Biver, N. (2017). The composition of cometary ices. *Philosophical Transactions of the Royal Society of London Series A, 375*, 20160252.

Bockelée-Morvan, D., & Crovisier, J. (1989). The nature of the 2.8-micron emission feature in cometary spectra. *Astronomy and Astrophysics, 216*, 278.

Bockelée-Morvan, D., J. Crovisier, J., Mumma, M.J. and & Weaver, H. A., (2004a), The composition of cometary volatiles, In M. Festou, H. U. Keller, & H. A. Weaver (Eds.), Comets II, (p. 391). Tucson: University of Arizona Press.

Bockelée-Morvan, D., Biver, N., Colom, P., Crovisier, J., Henry, F., Lecacheux, A., Davies, J. K., Dent, W. R. F., & Weaver, H. A. (2004b). The outgassing and composition of Comet 19P/Borrelly from radio observations. *Icarus, 167*, 113.

Bockelée-Morvan, D., Debout, V., Erard, S., Leyrat, C., Capaccioni, F., Filacchione, G., Fougere, N., Drossart, P., Arnold, G., Combi, M., Schmitt, B., Crovisier, J., de Sanctis, M.-C., Encrenaz, T., Kührt, E., Palomba, E., Taylor, F. W., Tosi, F., Piccioni, G., Fink, U., Tozzi, G., Barucci, A., Biver, N., Capria, M.-T., Combes, M., Ip, W., Blecka, M., Henry, F., Jacquinod, S., Reess, J.-M., Semery, A., & Tiphene, D. (2015). First observations of $H_2O$ and $CO_2$ vapor in comet 67P/Churyumov-Gerasimenko made by VIRTIS onboard Rosetta. *Astronomy and Astrophysics, 583*, A6.

Bockelée-Morvan, D., Crovisier, J., Erard, S., Capaccioni, F., Leyrat, C., Filacchione, G., Drossart, P., Encrenaz, T., Biver, N., de Sanctis, M.-C., Schmitt, B., Kührt, E., Capria, M.-T., Combes, M., Combi, M., Fougere, N., Arnold, G., Fink, U., Ip, W., Migliorini, A., Piccioni, G., & Tozzi, G. (2016). Evolution of $CO_2$, $CH_4$, and OCS abundances relative to $H_2O$ in the coma of comet 67P around perihelion from Rosetta/VIRTIS-H observations. *Monthly Notices of the Royal Astronomical Society, 462*, S170.

Bockelée-Morvan, D., Leyrat, C., Erard, S., Andrieu, F., Capaccioni, F., Filacchione, G., Hasselmann, P. H., Crovisier, J., Drossart, P., Arnold, G., Ciarniello, M., Kappel, D., Longobardo, A., Capria, M.-T., De Sanctis, M. C., Rinaldi, G., & Taylor, F. (2019). VIRTIS-H observations of the dust coma of comet 67P/Churyumov-Gerasimenko: Spectral properties and color temperature variability with phase and elevation. *Astronomy and Astrophysics, 630*, A22.

Bodenheimer, P., Laughlin, G. P., Rozyczka, M., Plewa, T., & Yorke, H. W. (2006). *Numerical methods in astrophysics: An introduction.* Taylor and Francis.

Bodewits, D., Lara, L. M., A'Hearn, M. F., La Forgia, F., Gicquel, A., Kovacs, G., Knollenberg, J., Lazzarin, M., Lin, Z.-Y., Shi, X., Snodgrass, C., Tubiana, C., Sierks, H., Barbieri, C., Lamy, P. L., Rodrigo, R., Koschny, D., Rickman, H., Keller, H. U., Barucci, M. A., Bertaux, J.-L., Bertini, I., Boudreault, S., Cremonese, G., Da Deppo, V., Davidsson, B., Debei, S., De Cecco, M., Fornasier, S., Fulle, M., Groussin, O., Gutiérrez, P. J., Güttler, C., Hviid, S. F., Ip, W.-H., Jorda, L., Kramm, J.-R., Kührt, E., Küppers, M., López-Moreno, J. J., Marzari, F., Naletto, G., Oklay, N., Thomas, N., Toth, I., & Vincent, J.-B. (2016). Changes in the physical environment of the inner coma of 67P/Churyumov-Gerasimenko with decreasing heliocentric distance. *The Astronomical Journal, 152*, 130.

Bodewits, D., Farnham, T. L., Kelley, M. S. P., & Knight, M. M. (2018). A rapid decrease in the rotation rate of comet 41P/Tuttle-Giacobini-Kresak. *Nature, 553*, 186.

Boehnhardt, H. (2002). Comet splitting – observations and model scenarios. *Earth Moon and Planets, 89*, 91.

Boehnhardt, H. (2004). Split comets. In M. Festou, H. U. Keller, & H. A. Weaver (Eds.), *Comets II* (p. 301). Tucson: University of Arizona Press.

Boehnhardt, H., Kaeufl, H. U., Goudfrooij, P., Storm, J., Manfroid, J., & Reinsch, K. (1996). The break-up of periodic comet Schwassmann-Wachmann 3: Image documents from La Silla telescopes. *The Messenger, 84*, 26–29.

Boehnhardt, H., Bibring, J.-P., Apathy, I., Auster, H. U., Ercoli Finzi, A., Goesmann, F., Klingelhöfer, G., Knapmeyer, M., Kofman, W., Krüger, H., Mottola, S., Schmidt, W., Seidensticker, K., Spohn, T., & Wright, I. (2017). The Philae lander mission and science overview. *Philosophical Transactions of the Royal Society of London Series A, 375*, 20160248.

Bogelund, E. G., & Hogerheijde, M. R. (2017). Exploring the volatile composition of comets C/2012 F6 (Lemmon) and C/2012 S1 (ISON) with ALMA. *Astronomy and Astrophysics, 604*, A131.

Bohren, C. F., & Huffman, D. R. (1983). *Absorption and scattering of light by small particles.* New York: Wiley.

Boice, D. C., Soderblom, L. A., Britt, D. T., Brown, R. H., Sandel, B. R., Yelle, R. V., Buratti, B. J., Hicks, M. D., Nelson, R. M., Rayman, M. D., Oberst, J., & Thomas, N. (2002). The Deep Space 1 encounter with comet 19p/Borrelly. *Earth Moon and Planets, 89*, 301.

Bonev, B. P., Mumma, M. J., DiSanti, M. A., Dello Russo, N., Magee-Sauer, K., Ellis, R. S., & Stark, D. P. (2006). A comprehensive study of infrared OH prompt emission in two comets. I. Observations and effective g-Factors. *The Astrophysical Journal, 653*, 774.

Bonev, B. P., Mumma, M. J., Villanueva, G. L., Disanti, M. A., Ellis, R. S., Magee-Sauer, K., & Dello Russo, N. (2007). A search for variation in the $H_2O$ ortho-para ratio and rotational temperature in the inner coma of comet C/2004 Q2 (Machholz). *The Astrophysical Journal, 661*, L97–L100.

Bonev, B. P., Villanueva, G. L., Paganini, L., DiSanti, M. A., Gibb, E. L., Keane, J. V., Meech, K. J., & Mumma, M. J. (2013). Evidence for two modes of water release in Comet 103P/Hartley 2: Distributions of column density, rotational temperature, and ortho-para ratio. *Icarus, 222*, 740–751.

Borchert, T.-H., & Waterman, J. P. (2017). *The book of miracles*. Taschen. ISBN 978-3-8365-6414-4.

Borisov, G. (2019, September 11). Minor Planet Electronic Circular No. 2019-R106.

Bottke, W.F., A. Morbidelli, R. Jedicke, J.-M. Petit, H.F. Levison, P. Michel, and T.S. Metcalfe, (2002), Debiased Orbital and Absolute Magnitude Distribution of the Near-Earth Objects, Icarus, 156, 399.

Botta, O., Martins, Z., & Ehrenfreund, P. (2007). Amino acids in Antarctic CM1 meteorites and their relationship to other carbonaceous chondrites. *Meteoritics and Planetary Science, 42*, 81.

Brandt, J. C., & Chapman, R. D. (2004). *Introduction to comets* (2nd ed.). New York: Cambridge University Press.

Brandt, J. C., Niedner, M. B., & Rahe, J. (1993). *The international Halley watch atlas of large-scale phenomena*. ISBN 1-880768-00-3.

Brasser, R., Duncan, M. J., & Levison, H. F. (2007). Embedded star clusters and the formation of the Oort cloud. II. The effect of the primordial solar nebula. *Icarus, 191*, 413–433.

Brasser, R., & Schwamb, M. E. (2015). Re-assessing the formation of the inner Oort cloud in an embedded star cluster. II. Probing the inner edge. *Monthly Notices of the Royal Astronomical Society, 446*, 3788–3796.

Brin, G. D., & Mendis, D. A. (1979). Dust release and mantle development in comets. *The Astrophysical Journal, 229*, 402–408.

Britt, D. T., & Opeil, C. (2017). The thermal properties of CM carbonaceous chondrites. *American Geophysical Union, Fall Meeting 2017*, abstract #P54B-08.

Britt, D. T., Boice, D. C., Buratti, B. J., Campins, H., Nelson, R. M., Oberst, J., Sandel, B. R., Stern, S. A., Soderblom, L. A., & Thomas, N. (2004). The morphology and surface processes of Comet 19/P Borrelly. *Icarus, 167*, 45–53.

Brouet, Y., Levasseur-Regourd, A. C., Sabouroux, P., Neves, L., Encrenaz, P., Poch, O., Pommerol, A., Thomas, N., Kofman, W., Le Gall, A., Ciarletti, V., Hérique, A., Lethuillier, A., Carrasco, N., & Szopa, C. (2016a). A porosity gradient in 67P/C-G nucleus suggested from CONSERT and SESAME-PP results: An interpretation based on new laboratory permittivity measurements of porous icy analogues. *Monthly Notices of the Royal Astronomical Society, 462*, S89–S98.

Brouet, Y., Neves, L., Sabouroux, P., Levasseur-Regourd, A. C., Poch, O., Encrenaz, P., Pommerol, A., Thomas, N., & Kofman, W. (2016b). Characterization of the permittivity of controlled porous water ice-dust mixtures to support the radar exploration of icy bodies. *Journal of Geophysical Research (Planets), 121*, 2426.

Brown, M. E., Bouchez, A. H., Spinrad, H., & Misch, A. (1998). Sodium velocities and sources in Hale-Bopp. *Icarus, 134*, 228.

Brownlee, D. E. (1977). Interplanetary dust – Possible implications for comets and presolar interstellar grains. In *Protostars and planets: Studies of star formation and of the origin of the solar system* (pp. 134–150). (A79-26776 10-90) Tucson.

Brownlee, D. E., Tsou, P., Anderson, J. D., Hanner, M. S., Newburn, R. L., Sekanina, Z., Clark, B. C., Hörz, F., Zolensky, M. E., Kissel, J., McDonnell, J. A. M., Sandford, S. A., & Tuzzolino, A. J. (2003). Stardust: Comet and interstellar dust sample return mission. *Journal of Geophysical Research (Planets), 108*, 8111.

Brownlee, D. E., Horz, F., Newburn, R. L., Zolensky, M., Duxbury, T. C., Sandford, S., Sekanina, Z., Tsou, P., Hanner, M. S., Clark, B. C., Green, S. F., & Kissel, J. (2004). Surface of young Jupiter Family Comet 81 P/Wild 2: View from the Stardust spacecraft. *Science, 304*, 1764–1769.

Brugger, B., Mousis, O., Morse, A., Marboeuf, U., Jorda, L., Guilbert-Lepoutre, A., Andrews, D., Barber, S., Lamy, P., Luspay-Kuti, A., Mandt, K., Morgan, G., Sheridan, S., Vernazza, P., & Wright, I. P. (2016). Subsurface characterization of 67P/Churyumov-Gerasimenko's Abydos site. *The Astrophysical Journal, 822*, 98.

Brucato, J. R., Castorina, A. C., Palumbo, M. E., Satorre, M. A., & Strazzulla, G. (1997). Ion irradiation and extended CO emission in cometary comae. *Planetary and Space Science, 45*, 835.

Buffa, G., Tarrini, O., Scappini, F., & Cecchi-Pestellini, C. (2000). $H_2O$-$H_2O$ collision rate coefficients. *The Astrophysical Journal Supplement Series, 128*, 597.

Budzien, S. A., Festou, M. C., & Feldman, P. D. (1994). Solar flux variability and the lifetimes of cometary $H_2O$ and OH. *Icarus, 107*, 164.

Burns, J. A., & Safronov, V. S. (1973). Asteroid nutation angles. *Monthly Notices of the Royal Astronomical Society, 165*, 403–411.

Buratti, B. J., Hicks, M. D., Soderblom, L. A., Britt, D., Oberst, J., & Hillier, J. K. (2004). Deep Space 1 photometry of the nucleus of Comet 19P/Borrelly. *Icarus, 167*, 16.

Busarev, V. V. (2016). New reflectance spectra of 40 asteroids: A comparison with the previous results and an interpretation. *Solar System Research, 50*, 13.

Butscher, T., Duvernay, F., Danger, G., Torro, R., Lucas, G., Carissan, Y., Hagebaum-Reignier, D., & Chiavassa, T. (2019). Radical-assisted polymerization in interstellar ice analogues: Formyl radical and polyoxymethylene. *Monthly Notices of the Royal Astronomical Society, 486*, 1953.

Calmonte, U., Altwegg, K., Balsiger, H., Berthelier, J.-J., Bieler, A., De Keyser, J., Fiethe, B., Fuselier, S. A., Gasc, S., Gombosi, T. I., Le Roy, L., Rubin, M., Semon, T., Tzou, C.-Y., & Wampfler, S. F. (2017). Sulphur isotope mass-independent fractionation observed in comet 67P/Churyumov-Gerasimenko by Rosetta/ROSINA. *Monthly Notices of the Royal Astronomical Society, 469*, S787–S803.

Capaccioni, F., et al. (2015). The organic-rich surface of comet 67P/Churyumov-Gerasimenko as seen by VIRTIS/Rosetta. *Science, 347*, aaa0628.

Capanna, C., Gesquière, G., Jorda, L., Lamy, P., & Vibert, D. (2013). Three-dimensional reconstruction using multiresolution photoclinometry by deformation. *Vis Comput., 29*, 825. https://doi.org/10.1007/s00371-013-0821-5.

Capria, M. T., Cremonese, G., Bhardwaj, A., & de Sanctis, M. C. (2005). $O(^1S)$ and $O(^1D)$ emission lines in the spectrum of 153P/2002 C1 (Ikeya-Zhang). *Astronomy and Astrophysics, 442*, 1121.

Carrigy, N. B., Pant, L. M., Mitra, S., & Secanella, M. (2013). Knudsen diffusivity and permeability of PEMFC microporous coated gas diffusion layers for different polytetrafluoroethylene loadings. *Journal of the Electrochemical Society, 160*(2), F81–F89.

Carry, B., Kaasalainen, M., Merline, W. J., Müller, T. G., Jorda, L., Drummond, J. D., Berthier, J., O'Rourke, L., Ďurech, J., Küppers, M., Conrad, A., Tamblyn, P., Dumas, C., Sierks, H., Osiris Team, A'Hearn, M., Angrilli, F., Barbieri, C., Barucci, A., Bertaux, J.-L., Da Deppo, V., Davidsson, B., Debei, S., De Cecco, M., Fornasier, S., Fulle, M., Groussin, O., Gutiérrez, P., Ip, W.-H., Keller, H. U., Koschny, D., Knollenberg, J., Kramm, J. R., Kuehrt, E., Lamy, P., Lara, L. M., Lazzarin, M., López-Moreno, J. J., Marzari, F., Michalik, H., Naletto, G., Rickman, H., Rodrigo, R., Sabau, L., Thomas, N., & Wenzel, K.-P. (2012). Shape modeling technique KOALA validated by ESA Rosetta at (21) Lutetia. *Planetary and Space Science, 66*, 200.

Carter, J. A., Bodewits, D., Read, A. M., & Immler, S. (2012). Simultaneous Swift X-ray and UV views of comet C/2007 N3 (Lulin). *Astronomy and Astrophysics, 541*, A70.

Celnik, W. E., & Schmidt-Kaler, T. (1987). Structure and dynamics of plasma-tail condensations of comet P/Halley 1986 and inferences on the structure and activity of the cometary nucleus. *Astronomy and Astrophysics, 187*, 233.

Ceplecha, Z. (1994). Impacts of meteoroids larger than 1 M into the Earth's atmosphere. *Astronomy and Astrophysics, 286*, 967–970.

Cercignani, C. (1981). Strong evaporation of a polyatomic gas. In *Rarefied gas dynamics* (pp. 305–320). 12th Conference Proceedings.

Cercignani, C. (2000). Rarefied gas dynamics. Cambridge: Cambridge University Press. ISBN 0521650089.

Chamberlain, J. W., & Hunten, D. M. (1987). *Theory of planetary atmospheres. An introduction to their physics and chemistry* (International geophysics series) (Vol. 36). Orlando, FL: Academic Press. ISBN 0-12-167251-4.

Chapovsky, P. L. (2019). Conversion of nuclear spin isomers of water molecules under ultracold conditions of space. *Quantum Electronics, 49*, 473.

Cheng, A. F., Izenberg, N., Chapman, C. R., & Zuber, M. T. (2002). Ponded deposits on asteroid 433 Eros. *Meteoritics and Planetary Science, 37*, 1095.

Chernova, G. P., Kiselev, N. N., & Jockers, K. (1993). Polarimetric characteristics of dust particles as observed in 13 comets – comparisons with asteroids. *Icarus, 103*, 144–158.

Choukroun, M., Keihm, S., Schloerb, F. P., Gulkis, S., Lellouch, E., Leyrat, C., von Allmen, P., Biver, N., Bockelée-Morvan, D., Crovisier, J., Encrenaz, P., Hartogh, P., Hofstadter, M., Ip, W.-H., Jarchow, C., Janssen, M., Lee, S., Rezac, L., Beaudin, G., Gaskell, B., Jorda, L., Keller, H. U., & Sierks, H. (2015). Dark side of comet 67P/Churyumov-Gerasimenko in Aug.-Oct. 2014. MIRO/Rosetta continuum observations of polar night in the southern regions. *Astronomy and Astrophysics, 583*, A28.

Choukroun, M., Altwegg, K., Kührt, E., Biver, N., Bockelée-Morvan, D., Drkazkowska, J., Hérique, A., Hilchenbach, M., Marschall, R., Pätzold, M., Taylor, M. G. G. T., & Thomas, N. (2020). Dust-to-gas and refractory-to-ice mass ratios of Comet 67P/Churyumov-Gerasimenko from Rosetta observations. *Space Science Reviews, 216*, 44.

Christou, C., Dadzie, S. K., Thomas, N., Marschall, R., Hartogh, P., Jorda, L., Kührt, E., Wright, I., & Rodrigo, R. (2018). Gas flow in near surface comet like porous structures: Application to 67P/Churyumov-Gerasimenko. *Planetary and Space Science, 161*, 57.

Christou, C., Dadzie, S. K., Marschall, R., & Thomas, N. (2020). Porosity gradients as a means of driving lateral flows at cometary surfaces. *Planetary and Space Science, 180*, 104752.

Chu, L. E. U., Meech, K. J., Farnham, T. L., Kührt, E., Mottola, S., Keane, J. V., Hellmich, S., Hainaut, O. R., & Kleyna, J. T. (2019) Detailed characterization of low activity Comet 49P/Arend-Rigaux. arXiv e-prints, arXiv:1912.02194.

Chyba, C., & Sagan, C. (1988). Cometary organic matter still a contentious issue. *Nature, 332*, 592.

Claire, M. W., Sheets, J., Cohen, M., Ribas, I., Meadows, V. S., & Catling, D. C. (2012). The evolution of solar flux from 0.1 nm to 160 μm: Quantitative estimates for planetary studies. *The Astrophysical Journal, 757*, 95.

Clauset, A., Shalizi, C. R., & Newman, M. E. J. (2009). Power-law distributions in empirical data. *SIAM Rev., 51*(4), 661–703. https://doi.org/10.1137/070710111.

Clery, D. (2018). Hints of young planets puzzle theorists. *Science, 362*, 1337.

Cochran, A. L. (1985). A re-evaluation of the Haser model scale lengths for comets. *The Astronomical Journal, 90*, 2609.

Cochran, A. L. (2008). Atomic oxygen in the comae of comets. *Icarus, 198*, 181.

Colangeli, L., López-Moreno, J. J., Palumbo, P., Rodriguez, J., Cosi, M., Della Corte, V., Esposito, F., Fulle, M., Herranz, M., Jeronimo, J. M., Lopez-Jimenez, A., Epifani, E. M., Morales, R., Moreno, F., Palomba, E., & Rotundi, A. (2007). The grain impact analyser and dust accumulator (GIADA) experiment for the Rosetta mission: Design, performances and first results. *Space Science Reviews, 128*, 803.

Colwell, J. E., Gulbis, A. A. S., Horányi, M., & Robertson, S. (2005). Dust transport in photoelectron layers and the formation of dust ponds on Eros. *Icarus, 175*, 159.

Combes, M., Moroz, V. I., Crovisier, J., Encrenaz, T., Bibring, J.-P., Grigoriev, A. V., Sanko, N. F., Coron, N., Crifo, J. F., Gispert, R., Bockelée-Morvan, D., Nikolsky, Y. V., Krasnopolsky, V. A., Owen, T., Emerich, C., Lamarre, J. M., & Rocard, F. (1988). The 2.5-12 micron spectrum of Comet Halley from the IKS-VEGA experiment. *Icarus, 76*, 404–436.

Combi, M. R. (1996). Time-dependent gas kinetics in tenuous planetary atmospheres: The cometary coma. *Icarus, 123*, 207.

Combi, M. (2017). SOHO SWAN derived cometary water production rates collection, urn:nasa: pds:soho:swan_derived::1.0. In L. Feaga (Ed.), NASA planetary data system.

Combi, M. R., & Delsemme, A. H. (1980). Neutral cometary atmospheres. I. Average random walk model for dissociation in comets. *Astrophysical Journal, 237*, 633–641.

Combi, M. R., & Feldman, P. D. (1992). IUE observations of H Lyman-α in comet P/Giacobini-Zinner. *Icarus, 97*, 260.

Combi, M. R., Harris, W. M., & Smyth, W. H. (2004). Gas dynamics and kinetics in the cometary coma: Theory and observations. In M. Festou, H. U. Keller, & H. A. Weaver (Eds.), *Comets II* (p. 523). Tucson: University of Arizona Press.

Combi, M. R., Mäkinen, T. T., Bertaux, J.-L., Quémerais, E., Ferron, S., Avery, M., & Wright, C. (2018). Water production activity of nine long-period comets from SOHO/SWAN observations of hydrogen Lyman-alpha: 2013-2016. *Icarus, 300*, 33.

Combi, M., Shou, Y., Fougere, N., Tenishev, V., Altwegg, K., Rubin, M., Bockelée-Morvan, D., Capaccioni, F., Cheng, Y.-C., Fink, U., Gombosi, T., Hansen, K. C., Huang, Z., Marshall, D., & Toth, G. (2020). The surface distributions of the production of the major volatile species, $H_2O$, $CO_2$, CO and $O_2$, from the nucleus of comet 67P/Churyumov-Gerasimenko throughout the Rosetta Mission as measured by the ROSINA double focusing mass spectrometer. *Icarus, 335*, 113421.

Cowley, S. W. H. (1987). ICE observations of comet Giacobini-Zinner. *Philosophical Transactions of the Royal Society of London Series A, 323*, 405.

Cravens, T. E. (1997). *Physics of solar system plasmas*. Cambridge: Cambridge University Press.

Cravens, T. E. (2000). X-ray emission from comets and planets. *Advances in Space Research, 26*, 1443–1451.

Cravens, T. E., & Körösmezey, A. (1986). Vibrational and rotational cooling of electrons by water vapor. *Planetary and Space Science, 34*, 961.

Cravens, T. E., Kozyra, J. U., Nagy, A. F., Gombosi, T. I., & Kurtz, M. (1987). Electron impact ionization in the vicinity of comets. *Journal of Geophysical Research, 92*, 7341.

Cremonese, G., & the European Hale–Bopp Team. (1997). IAU Circular 6631.

Cremonese, G., Boehnhardt, H., Crovisier, J., Rauer, H., Fitzsimmons, A., Fulle, M., Licandro, J., Pollacco, D., Tozzi, G. P., & West, R. M. (1997). Neutral sodium from Comet Hale-Bopp: A third type of tail. *The Astrophysical Journal, 490*, L199–L202.

Cremonese, G., Huebner, W. F., Rauer, H., & Boice, D. C. (2002). Neutral sodium tails in comets. *Advances in Space Research, 29*, 1187–1197.

Crifo, J. F. (1995). A general physicochemical model of the inner coma of active comets. 1: Implications of spatially distributed gas and dust production. *The Astrophysical Journal, 445*, 470–488.

Crifo, J. F., & Rodionov, A. V. (1999). Modelling the circumnuclear coma of comets: Objectives, methods and recent results. *Planetary and Space Science, 47*, 797–826.

Crifo, J. F., Lukianov, G. A., Rodionov, A. V., Khanlarov, G. O., & Zakharov, V. V. (2002a). Comparison between Navier-Stokes and direct Monte-Carlo simulations of the circumnuclear coma. I. Homogeneous, spherical source. *Icarus, 156*, 249–268.

Crifo, J.-F., Rodionov, A. V., Szegö, K., & Fulle, M. (2002b). Challenging a paradigm: Do we need active and inactive areas to account for near-nuclear jet activity? *Earth Moon and Planets, 90*, 227.

Crifo, J.-F., Loukianov, G. A., Rodionov, A. V., & Zakharov, V. V. (2005). Direct Monte Carlo and multifluid modeling of the circumnuclear dust coma. Spherical grain dynamics revisited. *Icarus, 176*, 192–219.

Crovisier, J. (1984). The water molecule in Comets: Fluorescence mechanisms and thermodynamics of the inner coma. *Astronomy and Astrophysics, 135*, 197. (Erratum: Astronomy and Astrophysics, 135, 197).

Crovisier, J., & Encrenaz, T. (1983). Infrared fluorescence of molecules in comets – The general synthetic spectrum. *Astronomy and Astrophysics, 126*, 170.

Crovisier, J., Colom, P., Biver, N., Bockelée-Morvan, D., & Boissier, J. (2013). Observations of the 18-cm OH lines of Comet 103P/Hartley 2 at Nançay in support to the EPOXI and Herschel missions. *Icarus, 222*, 679.

Dankert, C., & Koppenwallner, G. (1984). *14th Symposium on Rarefied Gas Dynamics* (Vol. 1, pp. 477–484).

Danks, A. C., Lambert, D. L., & Arpigny, C. (1974). The C-12/C-13 ratio in comet Kohoutek / 1973f/. *The Astrophysical Journal, 194*, 745–751.

Dartois, E., Augé, B., Boduch, P., Brunetto, R., Chabot, M., Domaracka, A., Ding, J. J., Kamalou, O., Lv, X. Y., Rothard, H., da Silveira, E. F., & Thomas, J. C. (2015). Heavy ion irradiation of crystalline water ice. Cosmic ray amorphisation cross-section and sputtering yield. *Astronomy and Astrophysics, 576*, A125.

Daubar, I. J., McEwen, A. S., Byrne, S., Kennedy, M. R., & Ivanov, B. (2013). The current martian cratering rate. *Icarus, 225*, 506.

Davidsson, B. J. R. (2001). Tidal splitting and rotational breakup of solid biaxial ellipsoids. *Icarus, 149*, 375.

Davidsson, B. J. R. (2008). Comet Knudsen layers. *Space Science Reviews, 138*, 207–223.

Davidsson, B. J. R., & Gutiérrez, P. J. (2004). Estimating the nucleus density of Comet 19P/Borrelly. *Icarus, 168*, 392.

Davidsson, B. J. R., & Skorov, Y. V. (2002a). On the light-absorbing surface layer of cometary nuclei. I. Radiative transfer. *Icarus, 156*, 223.

Davidsson, B. J. R., & Skorov, Y. V. (2002b). On the light-absorbing surface layer of cometary nuclei. II. Thermal modeling. *Icarus, 159*, 239.

Davidson, P. A. (2001). *Introduction to magnetohydrodynamics*. Cambridge: Cambridge University Press.

Davies, J. K., Green, S. F., Stewart, B. C., Meadows, A. J., & Aumann, H. H. (1984). The IRAS fast-moving object search. *Nature, 309*, 315–319.

Davies, J. K., Puxley, P. J., Mumma, M. J., Reuter, D. C., Hoban, S., Weaver, H. A., & Lumsden, S. L. (1993). The infrared (3.2-3.6 micron) spectrum of Comet P/Swift-Tuttle – Detection of methanol and other organics. *Monthly Notices of the Royal Astronomical Society, 265*, 1022.

Debout, V., Bockelée-Morvan, D., & Zakharov, V. (2016). A radiative transfer model to treat infrared molecular excitation in cometary atmospheres. *Icarus, 265*, 110–124.

Decock, A., Jehin, E., Rousselot, P., Hutsemekers, D., Manfroid, J., Raghuram, S., Bhardwaj, A., & Hubert, B. (2015). Forbidden oxygen lines at various nucleocentric distances in comets. *Astronomy and Astrophysics, 573*, A1.

de Gregorio, B. T., Stroud, R. M., Cody, G. D., Nittler, L. R., David Kilcoyne, A. L., & Wirick, S. (2011). Correlated microanalysis of cometary organic grains returned by Stardust. *Meteoritics and Planetary Science, 46*, 1376.

Delamere, P. A. (2006). Hybrid code simulations of the solar wind interaction with Comet 19P/Borrelly. *Journal of Geophysical Research, 111*, A12217. https://doi.org/10.1029/2006JA011859.

Delbó, M., & Harris, A. W. (2002). Physical properties of near-Earth asteroids from thermal infrared observations and thermal modeling. *Meteoritics and Planetary Science, 37*, 1929.

Delbo, M., Libourel, G., Wilkerson, J., Murdoch, N., Michel, P., Ramesh, K. T., Ganino, C., Verati, C., & Marchi, S. (2014). Thermal fatigue as the origin of regolith on small asteroids. *Nature, 508*, 233–236.

Della Corte, V., Rotundi, A., Fulle, M., Gruen, E., Weissman, P., Sordini, R., Ferrari, M., Ivanovski, S., Lucarelli, F., Accolla, M., Zakharov, V., Mazzotta Epifani, E., Lopez-Moreno, J. J., Rodriguez, J., Colangeli, L., Palumbo, P., Bussoletti, E., Crifo, J. F., Esposito, F., Green,

S. F., Lamy, P. L., McDonnell, J. A. M., Mennella, V., Molina, A., Morales, R., Moreno, F., Ortiz, J. L., Palomba, E., Perrin, J. M., Rietmeijer, F. J. M., Rodrigo, R., Zarnecki, J. C., Cosi, M., Giovane, F., Gustafson, B., Herranz, M. L., Jeronimo, J. M., Leese, M. R., Lopez-Jimenez, A. C., & Altobelli, N. (2015). GIADA: Shining a light on the monitoring of the comet dust production from the nucleus of 67P/Churyumov-Gerasimenko. *Astronomy and Astrophysics, 583*, A13.

Della Corte, V., Rotundi, A., Fulle, M., Ivanovski, S., Green, S. F., Rietmeijer, F. J. M., Colangeli, L., Palumbo, P., Sordini, R., Ferrari, M., Accolla, M., Zakharov, V., Epifani, E. M., Weissman, P., Gruen, E., Lopez-Moreno, J. J., Rodriguez, J., Bussoletti, E., Crifo, J. F., Esposito, F., Lamy, P. L., McDonnell, J. A. M., Mennella, V., Molina, A., Morales, R., Moreno, F., Palomba, E., Perrin, J. M., Rodrigo, R., Zarnecki, J. C., Cosi, M., Giovane, F., Gustafson, B., Ortiz, J. L., Jeronimo, J. M., Leese, M. R., Herranz, M., Liuzzi, V., & Lopez-Jimenez, A. C. (2016). 67P/C-G inner coma dust properties from 2.2 au inbound to 2.0 au outbound to the Sun. *Monthly Notices of the Royal Astronomical Society, 462*, S210.

Dello Russo, N., DiSanti, M. A., Magee-Sauer, K., Gibb, E. L., Mumma, M. J., Barber, R. J., & Tennyson, J. (2004). Water production and release in Comet 153P/Ikeya–Zhang (C/2002 C1): Accurate rotational temperature retrievals from hot-band lines near 2.9-μm. *Icarus, 168*, 186.

Dello Russo, N., Mumma, M. J., DiSanti, M. A., Magee-Sauer, K., Gibb, E. L., Bonev, B. P., McLean, I. S., & Xu, L.-H. (2006). A high-resolution infrared spectral survey of Comet C/1999 H1 Lee. *Icarus, 184*, 255.

Dello Russo, N., Vervack, R. J., Weaver, H. A., Biver, N., Bockelée-Morvan, D., Crovisier, J., & Lisse, C. M. (2007). Compositional Homogeneity in the fragmented comet 73P/Schwassmann-Wachmann 3. *Nature, 448*, 172–175.

Dello Russo, N., Vervack, R. J., Lisse, C. M., Weaver, H. A., Kawakita, H., Kobayashi, H., Cochran, A. L., Harris, W. M., McKay, A. J., Biver, N., Bockelée-Morvan, D., & Crovisier, J. (2011). The volatile composition and activity of Comet 103P/Hartley 2 during the EPOXI closest approach. *The Astrophysical Journal, 734*, L8.

Dello Russo, N., Kawakita, H., Bonev, B. P., Vervack, R. J., Gibb, E. L., Shinnaka, Y., Roth, N. X., DiSanti, M. A., & McKay, A. J. (2020). Post-perihelion volatile production and release from Jupiter-family comet 45P/Honda-Mrkos-Pajdušáková. *Icarus, 335*, 113411.

De Niem, D., Kührt, E., Hviid, S., & Davidsson, B. (2018). Low velocity collisions of porous planetesimals in the early solar system. *Icarus, 301*, 196–218.

De Pater, I., & Lissauer, J. J. (2015). *Planetary sciences*. Cambridge: Cambridge University Press. https://doi.org/10.1017/CBO9781316165270.

De Sanctis, M. C., et al. (2015). The diurnal cycle of water ice on comet 67P/Churyumov-Gerasimenko. *Nature, 525*, 500.

De Val-Borro, M., Küppers, M., Hartogh, P., Rezac, L., Biver, N., Bockelée-Morvan, D., Crovisier, J., Jarchow, C., & Villanueva, G. L. (2013). A survey of volatile species in Oort cloud comets C/2001 Q4 (NEAT) and C/2002 T7 (LINEAR) at millimeter wavelengths. *Astronomy and Astrophysics, 559*, A48.

DiSanti, M. A., Bonev, B. P., Villanueva, G. L., & Mumma, M. J. (2013). Highly depleted ethane and mildly depleted methanol in Comet 21P/Giacobini-Zinner: Application of a new empirical $v_2$-band model for $CH_3OH$ near 50 K. *The Astrophysical Journal, 763*, 1.

DiSanti, M. A., Bonev, B. P., Gibb, E. L., Roth, N. X., Dello Russo, N., & Vervack, R. J. (2018). Comet C/2013 V5 (Oukaimeden): Evidence for depleted organic volatiles and compositional heterogeneity as revealed through infrared spectroscopy. *The Astronomical Journal, 156*, 258.

Divine, N. (1981). Numerical models for Halley dust environments. *ESA, SP-174*, 25.

Divine, N., Fechtig, H., Gombosi, T. I., Hanner, M. S., Keller, H. U., Larson, S. M., Mendis, D. A., Newburn, R. L., Reinhard, R., Sekanina, Z., & Yeomans, D. K. (1986). The Comet Halley dust and gas environment. *Space Science Reviews, 43*, 1–104.

Dlugach, J. M., Ivanova, O. V., Mishchenko, M. I., & Afanasiev, V. L. (2018). Retrieval of microphysical characteristics of particles in atmospheres of distant comets from ground-based polarimetry. *Journal of Quantitative Spectroscopy and Radiative Transfer, 205*, 80.

Dollfus, A. (1989). Polarimetry of grains in the coma of P/Halley. II. Interpretation. *Astronomy and Astrophysics, 213*, 469–478.

Dombard, A. J., Barnouin, O. S., Prockter, L. M., & Thomas, P. C. (2010). Boulders and ponds on the Asteroid 433 Eros. *Icarus, 210*, 713.

Dones, L., Weissman, P. R., Levison, H. F., & Duncan, M. J. (2004). Oort cloud formation and dynamics. In M. Festou, H. U. Keller, & H. A. Weaver (Eds.), *Comets II*. Tucson: University of Arizona Press.

Donn, B. (1991). The accumulation and structure of comets. In R. Newburn, M. Neugebauer, & J. Rahe (Eds.), *IAU Colloq. 116: Comets in the post-Halley era* (Vol. 1, p. 335).

Donn, B., Daniels, P. A., & Hughes, D. W. (1985). On the structure of the cometary nucleus. *Bulletin of the American Astronomical Society, 17*, 520.

Draine, B. T. (1988). The discrete-dipole approximation and its application to interstellar graphite grains. *The Astrophysical Journal, 333*, 848.

Draine, B. T., & Flatau, P. J. (2012). User guide for the discrete dipole approximation code DDSCAT 7.2. arXiv e-prints, arXiv:1202.3424.

Drolshagen, E., Ott, T., Koschny, D., Güttler, C., Tubiana, C., Agarwal, J., Sierks, H., Barbieri, C., Lamy, P. I., Rodrigo, R., Rickman, H., A'Hearn, M. F., Barucci, M. A., Bertaux, J.-L., Bertini, I., Cremonese, G., da Deppo, V., Davidsson, B., Debei, S., de Cecco, M., Deller, J., Feller, C., Fornasier, S., Fulle, M., Gicquel, A., Groussin, O., Gutierrez, P. J., Hofmann, M., Hviid, S. F., Ip, W.-H., Jorda, L., Keller, H. U., Knollenberg, J., Kramm, J. R., Kührt, E., Küppers, M., Lara, L. M., Lazzarin, M., Lopez Moreno, J. J., Marzari, F., Naletto, G., Oklay, N., Shi, X., Thomas, N., & Poppe, B. (2017). Distance determination method of dust particles using Rosetta OSIRIS NAC and WAC data. *Planetary and Space Science, 143*, 256–264.

Duffard, R., Pinilla-Alonso, N., Santos-Sanz, P., Vilenius, E., Ortiz, J. L., Mueller, T., Fornasier, S., Lellouch, E., Mommert, M., Pal, A., Kiss, C., Mueller, M., Stansberry, J., Delsanti, A., Peixinho, N., & Trilling, D. (2014). "TNOs are Cool": A survey of the trans-Neptunian region. XI. A Herschel-PACS view of 16 Centaurs. *Astronomy and Astrophysics, 564*, A92.

Dulieu, F., Minissale, M., & Bockelée-Morvan, D. (2017). Production of $O_2$ through dismutation of $H_2O_2$ during water ice desorption: A key to understanding comet $O_2$ abundances. *Astronomy and Astrophysics, 597*, A56.

Duncan, M., Levison, H., & Dones, L. (2004). Dynamical evolution of ecliptic comets. In M. Festou, H. U. Keller, & H. A. Weaver (Eds.), *Comets II*. Tucson: University of Arizona Press.

Durán, O., Andreotti, B., & Claudin, P. (2012). Numerical simulation of turbulent sediment transport, from bed load to saltation. *Physics of Fluids, 24*, 103306.

Duxbury, T. C., Newburn, R. L., & Brownlee, D. E. (2004). Comet 81P/Wild 2 size, shape, and orientation. *Journal of Geophysical Research (Planets), 109*, E12S02.

Eaton, N., Davies, J. K., & Green, S. F. (1984). The anomalous dust tail of comet P/Tempel 2. *Monthly Notices of the Royal Astronomical Society, 211*, 15P.

Eberhardt, P. (1999). Comet Halley's Gas composition and extended sources: Results from the neutral mass spectrometer on Giotto. *Space Science Reviews, 90*, 45.

Eberhardt, P., Krankowsky, D., Schulte, W., Dolder, U., Lammerzahl, P., Berthelier, J. J., Woweries, J., Stubbemann, U., Hodges, R. R., Hoffman, J. H., & Illiano, J. M. (1987). To CO and $N_2$ abundance in Comet p/ Halley. *Astronomy and Astrophysics, 187*, 481.

Eddington, A. S. (1910). The envelopes of Comet Morehouse (1908 c). *Monthly Notices of the Royal Astronomical Society, 70*, 442.

Edgeworth, K. E. (1949). The origin and evolution of the Solar System. *Monthly Notices of the Royal Astronomical Society, 109*, 600.

Ehrenfreund, P., & Charnley, S. B. (2000). Organic molecules in the interstellar medium, comets, and meteorites: A voyage from dark clouds to the early Earth. *Annual Review of Astronomy and Astrophysics, 38*, 427.

Ehrenfreund, P., Gerakines, P. A., Schutte, W. A., van Hemert, M. C., & van Dishoeck, E. F. (1996). Infrared properties of isolated water ice. *Astronomy and Astrophysics, 312*, 263–274.

Eisner, N., Knight, M. M., & Schleicher, D. G. (2017). The rotation and other properties of Comet 49P/Arend-Rigaux, 1984-2012. *The Astronomical Journal, 154*, 196.

Eistrup, C., & Walsh, C. (2019). Formation of cometary $O_2$ ice and related ice species on grain surfaces in the midplane of the pre-solar nebula. *Astronomy and Astrophysics, 621*, A75.

Elbeshausen, D., Wünnemann, K., & Collins, G. S. (2009). Scaling of oblique impacts in frictional targets: Implications for crater size and formation mechanisms. *Icarus, 204*, 716–731. ISSN: 0019-1035.

Elliot, J. L., Kern, S. D., Clancy, K. B., Gulbis, A. A. S., Millis, R. L., Buie, M. W., Wasserman, L. H., Chiang, E. I., Jordan, A. B., Trilling, D. E., & Meech, K. J. (2005). The Deep Ecliptic survey: A search for Kuiper Belt objects and centaurs. II. Dynamical classification, the Kuiper Belt plane, and the core population. *The Astronomical Journal, 129*, 1117.

El-Maarry, M. R., et al. (2015a). Regional surface morphology of comet 67P/Churyumov-Gerasimenko from Rosetta/OSIRIS images. *Astronomy and Astrophysics, 583*, A26.

El-Maarry, M. R., et al. (2015b). Fractures on comet 67P/Churyumov-Gerasimenko observed by Rosetta/OSIRIS. *Geophysical Research Letters, 42*, 5170–5178.

El-Maarry, M. R., Watters, W. A., Yoldi, Z., Pommerol, A., Fischer, D., Eggenberger, U., & Thomas, N. (2015c). Field investigation of dried lakes in western United States as an analogue to desiccation fractures on Mars. *Journal of Geophysical Research (Planets), 120*, 2241.

El-Maarry, M. R., et al. (2016). Regional surface morphology of comet 67P/Churyumov-Gerasimenko from Rosetta/OSIRIS images: The southern hemisphere. *Astronomy and Astrophysics, 593*, A110.

El-Maarry, M. R., et al. (2017a). Regional surface morphology of comet 67P/Churyumov-Gerasimenko from Rosetta/OSIRIS images: The southern hemisphere (Corrigendum). *Astronomy and Astrophysics, 598*, C2.

El-Maarry, M. R., et al. (2017b). Surface changes on comet 67P/Churyumov-Gerasimenko suggest a more active past. *Science, 355*, 1392.

El-Maarry, M. R., Groussin, O., Keller, H. U., Thomas, N., Vincent, J.-B., Mottola, S., Pajola, M., Otto, K., Herny, C., & Krasilnikov, S. (2019). Surface morphology of comets and associated evolutionary processes: A review of Rosetta's observations of 67P/Churyumov-Gerasimenko. *Space Science Reviews, 215*, 36.

Elsila, J. E., Glavin, D. P., & Dworkin, J. P. (2009). Cometary glycine detected in samples returned by Stardust. *Meteoritics and Planetary Science, 44*, 1323.

Emerich, C., Lamarre, J. M., Moroz, V. I., Combes, M., Sanko, N. F., Nikolsky, Y. V., Rocard, F., Gispert, R., Coron, N., Bibring, J. P., Encrenaz, T., & Crovisier, J. (1987). Temperature and size of the nucleus of Comet p/Halley deduced from IKS infrared VEGA-1 measurements. *Astronomy and Astrophysics, 187*, 839.

Engelhardt, I. A. D., Eriksson, A. I., Vigren, E., Vallieres, X., Rubin, M., Gilet, N., & Henri, P. (2018). Cold electrons at comet 67P/Churyumov-Gerasimenko. *Astronomy and Astrophysics, 616*, A51.

Eriksson, A. I., Engelhardt, I. A. D., Andre, M., Bostrom, R., Edberg, N. J. T., Johansson, F. L., Odelstad, E., Vigren, E., Wahlund, J.-E., Henri, P., Lebreton, J.-P., Miloch, W. J., Paulsson, J. J. P., Simon Wedlund, C., Yang, L., Karlsson, T., Jarvinen, R., Broiles, T., Mandt, K., Carr, C. M., Galand, M., Nilsson, H., & Norberg, C. (2017). Cold and warm electrons at comet 67P/Churyumov-Gerasimenko. *Astronomy and Astrophysics, 605*, A15.

Ershkovich, A. I., & Mendis, D. A. (1986). The effects of the interaction between plasma and neutrals on the stability of the cometary ionopause. *The Astrophysical Journal, 302*, 849.

Esmaili, S., Bass, A. D., Cloutier, P., Sanche, L., & Huels, M. A. (2018). Glycine formation in $CO_2$: $CH_4$:$NH_3$ ices induced by 0-70 eV electrons. *Journal of Chemical Physics, 148*, 164702.

Espinasse, S., Coradini, A., Capria, M. T., Capaccioni, F., Orosei, R., Salomone, M., & Federico, C. (1993). Thermal evolution and differentiation of a short-period comet. *Planetary and Space Science, 41*, 409.

Faggi, S., Villanueva, G. L., Mumma, M. J., & Paganini, L. (2018). The volatile composition of Comet C/2017 E4 (Lovejoy) before its disruption, as revealed by high-resolution infrared spectroscopy with iSHELL at the NASA/IRTF. *The Astronomical Journal, 156*, 68.

Fairbairn, M. B. (2005). Planetary photometry: The Lommel-Seeliger law. *Journal of the Royal Astronomical Society of Canada, 99*, 92.

Fan, L. S., & Zhu, C. (2005). *Principles of gas–solid flows.* Cambridge: Cambridge University Press.

Fanale, F. P., & Salvail, J. R. (1984). An idealized short-period comet model – Surface insolation, $H_2O$ flux, dust flux, and mantle evolution. *Icarus, 60*, 476–511.

Farnham, T. L., & Cochran, A. L. (2002). A McDonald observatory study of Comet 19P/Borrelly: Placing the Deep Space 1 observations into a broader context. *Icarus, 160*, 398.

Farnham, T. L., Wellnitz, D. D., Hampton, D. L., Li, J.-Y., Sunshine, J. M., Groussin, O., McFadden, L. A., Crockett, C. J., A'Hearn, M. F., Belton, M. J. S., Schultz, P., & Lisse, C. M. (2007). Dust coma morphology in the Deep Impact images of Comet 9P/Tempel 1. *Icarus, 187*, 26.

Farnham, T. L., Bodewits, D., Li, J.-Y., Veverka, J., Thomas, P., & Belton, M. J. S. (2013). Connections between the jet activity and surface features on Comet 9P/Tempel 1. *Icarus, 222*, 540.

Farquhar, R. (1983). ISEE-3 – A late entry in the great comet chase. *Astronautics Aeronautics, 21*, 50.

Faure, A., & Josselin, E. (2008). Collisional excitation of water in warm astrophysical media. I. Rate coefficients for rovibrationally excited states. *Astronomy and Astrophysics, 492*, 257.

Favre, C., Fedele, D., Semenov, D., Parfenov, S., Codella, C., Ceccarelli, C., Bergin, E. A., Chapillon, E., Testi, L., Hersant, F., Lefloch, B., Fontani, F., Blake, G. A., Cleeves, L. I., Qi, C., Schwarz, K. R., & Taquet, V. (2018). First detection of the simplest organic acid in a protoplanetary disk. *The Astrophysical Journal, 862*, L2.

Feaga, L. M., A'Hearn, M. F., Sunshine, J. M., Groussin, O., & Farnham, T. L. (2007). Asymmetries in the distribution of $H_2O$ and $CO_2$ in the inner coma of Comet 9P/Tempel 1 as observed by Deep Impact. *Icarus, 191*, 134. https://doi.org/10.1016/j.icarus.2007.04.038.

Feaga, L. M., A'Hearn, M. F., Farnham, T. L., Bodewits, D., Sunshine, J. M., Gersch, A. M., Protopapa, S., Yang, B., Drahus, M., & Schleicher, D. G. (2014). Uncorrelated volatile behavior during the 2011 apparition of Comet C/2009 P1 Garradd. *The Astronomical Journal, 147*, 24.

Feldman, P. D., & Brune, W. H. (1976). Carbon production in comet West 1975n. *The Astrophysical Journal, 209*, L45.

Feldman, P. D., Festou, M. C., Tozzi, P., & Weaver, H. A. (1997). The $CO_2/CO$ abundance ratio in 1P/Halley and several other comets observed by IUE and HST. *The Astrophysical Journal, 475*, 829.

Feldman, P. D., Weaver, H. A., & Burgh, E. B. (2002). Far ultraviolet spectroscopic explorer observations of CO and $H_2$ emission in Comet C/2001 A2 (LINEAR). *The Astrophysical Journal, 576*, L91.

Feldman, P. D., Cochran, A. L., & Combi, M. R. (2004). Spectroscopic investigations of fragment species in the coma. In M. Festou, H. U. Keller, & H. A. Weaver (Eds.), *Comets II* (p. 425). Tucson: University of Arizona Press.

Feldman, P. D., A'Hearn, M. F., Bertaux, J.-L., Feaga, L. M., Parker, J. W., Schindhelm, E., Steffl, A. J., Stern, S. A., Weaver, H. A., Sierks, H., & Vincent, J.-B. (2015). Measurements of the near-nucleus coma of comet 67P/Churyumov-Gerasimenko with the Alice far-ultraviolet spectrograph on Rosetta. *Astronomy and Astrophysics, 583*, A8.

Feldman, P. D., Weaver, H. A., A'Hearn, M. F., Combi, M. R., & Dello Russo, N. (2018a). Far-ultraviolet spectroscopy of recent comets with the cosmic origins spectrograph on the Hubble space telescope. *The Astronomical Journal, 155*, 193.

Feldman, P. D., A'Hearn, M. F., Bertaux, J.-L., Feaga, L. M., Keeney, B. A., Knight, M. M., Noonan, J., Parker, J. W., Schindhelm, E., Steffl, A. J., Stern, S. A., Vervack, R. J., & Weaver,

H. A. (2018b). FUV spectral signatures of molecules and the evolution of the gaseous coma of Comet 67P/Churyumov-Gerasimenko. *The Astronomical Journal, 155*, 9.

Feller, C., et al. (2016). Decimetre-scaled spectrophotometric properties of the nucleus of comet 67P/Churyumov-Gerasimenko from OSIRIS observations. *Monthly Notices of the Royal Astronomical Society, 462*, S287–S303.

Fergason, R. L., Christensen, P. R., & Kieffer, H. H. (2006). High-resolution thermal inertia derived from the Thermal Emission Imaging System (THEMIS): Thermal model and applications. *Journal of Geophysical Research (Planets), 111*, E12004.

Fernandez, Y. R. (2009). That's the way the comet crumbles: Splitting Jupiter-family comets. *Planetary and Space Science, 57*, 1218–1227.

Ferrari, S., Penasa, L., La Forgia, F., Massironi, M., Naletto, G., Lazzarin, M., Fornasier, S., Hasselmann, P. H., Lucchetti, A., Pajola, M., Ferri, F., Cambianica, P., Oklay, N., Tubiana, C., Sierks, H., Lamy, P. L., Rodrigo, R., Koschny, D., Davidsson, B., Barucci, M. A., Bertaux, J.-L., Bertini, I., Bodewits, D., Cremonese, G., Da Deppo, V., Debei, S., De Cecco, M., Deller, J., Franceschi, M., Frattin, E., Fulle, M., Groussin, O., Gutiérrez, P. J., Güttler, C., Hviid, S. F., Ip, W.-H., Jorda, L., Keller, H. U., Knollenberg, J., Kührt, E., Küppers, M., Lara, L. M., López-Moreno, J. J., Marzari, F., Shi, X., Simioni, E., Thomas, N., & Vincent, J.-B. (2018). The big lobe of 67P/Churyumov-Gerasimenko comet: Morphological and spectrophotometric evidences of layering as from OSIRIS data. *Monthly Notices of the Royal Astronomical Society, 479*, 1555.

Ferrín, I. (2007). Secular light curve of Comet 28P/Neujmin 1 and of spacecraft target Comets 1P/Halley, 9P/Tempel 1, 19P/Borrelly, 21P/Giacobinni Zinner, 26P/Grigg Skjellerup, 67P/Churyumov Gerasimenko, and 81P/Wild 2. *Icarus, 191*, 22.

Ferrin, I. (2010). Atlas of secular light curves of comets. *Planetary and Space Science, 58*, 365–391.

Festou, M. C. (1981). The density distribution of neutral compounds in cometary atmospheres. I — Models and equations. *Astronomy and Astrophysics, 95*, 69–79.

Filacchione, G., et al. (2016a). Exposed water ice on the nucleus of comet 67P/Churyumov-Gerasimenko. *Nature, 529*, 368.

Filacchione, G., et al. (2016b). Seasonal exposure of carbon dioxide ice on the nucleus of comet 67P/Churyumov-Gerasimenko. *Science, 354*, 1563.

Filacchione, G., Groussin, O., Herny, C., Kappel, D., Mottola, S., Oklay, N., Pommerol, A., Wright, I., Yoldi, Z., Ciarniello, M., Moroz, L., & Raponi, A. (2019). Comet 67P/CG nucleus composition and comparison to other comets. *Space Science Reviews, 215*, 19.

Fillion, J.-H., Bertin, M., Lekic, A., Moudens, A., Philippe, L., & Michaut, X. (2012). Understanding the relationship between gas and ice: Experimental investigations on ortho-para ratios. *EAS Publications Series, 58*, 307–314. https://doi.org/10.1051/eas/1258051.

Fink, U., & Doose, L. (2018). Determination of the coma dust back-scattering of 67P for phase angles from 1.2° to 75°. *Icarus, 309*, 265.

Fink, U., & Rubin, M. (2012). The calculation of Afρ and mass loss rate for comets. *Icarus, 221*, 721.

Finklenburg, S. (2009). *Investigations of the gas dynamics in the near-nucleus region of a comet using a direct simulation Monte Carlo Code*. MSc thesis, Universität Bern.

Finklenburg, S. (2014). *Investigations of the near nucleus gas and dust comae of comets*. PhD thesis, Universität Bern.

Finklenburg, S., Thomas, N., Knollenberg, J., & Kührt, E. (2011). Comparison of DSMC and Euler equations solutions for inhomogeneous sources on comets. *American Institute of Physics Conference Series, 1333*, 1151–1156.

Finklenburg, S., Thomas, N., Su, C. C., & Wu, J.-S. (2014). The spatial distribution of water in the inner coma of Comet 9P/Tempel 1: Comparison between models and observations. *Icarus, 236*, 9.

Finson, M. J., & Probstein, R. F. (1968). A theory of dust comets. I. Model and equations. *The Astrophysical Journal, 154*, 327–352.

Fitzsimmons, A., Williams, I. P., Williams, G. P., & Andrews, P. J. (1990). Narrow-band photometry of Comet P/Halley – OH and $H_2O$ scalelengths and pre-perihelion production rates. *Monthly Notices of the Royal Astronomical Society, 244*, 453–457.

Flynn, G. J., et al. (2006). Elemental compositions of Comet 81P/Wild 2 samples collected by Stardust. *Science, 314*, 1731.

Fornasier, S., et al. (2015). Spectrophotometric properties of the nucleus of comet 67P/Churyumov-Gerasimenko from the OSIRIS instrument onboard the ROSETTA spacecraft. *Astronomy and Astrophysics, 583*, A30.

Fornasier, S., et al. (2016). Rosetta's comet 67P/Churyumov-Gerasimenko sheds its dusty mantle to reveal its icy nature. *Science, 354*, 1566–1570.

Fornasier, S., et al. (2017). The highly active Anhur-Bes regions in the 67P/Churyumov-Gerasimenko comet: Results from OSIRIS/ROSETTA observations. *Monthly Notices of the Royal Astronomical Society, 469*, S93.

Fornasier, S., Feller, C., Hasselmann, P. H., Barucci, M. A., Sunshine, J., Vincent, J.-B., Shi, X., Sierks, H., Naletto, G., Lamy, P. L., Rodrigo, R., Koschny, D., Davidsson, B., Bertaux, J.-L., Bertini, I., Bodewits, D., Cremonese, G., Da Deppo, V., Debei, S., De Cecco, M., Deller, J., Ferrari, S., Fulle, M., Gutierrez, P. J., Güttler, C., Ip, W.-H., Jorda, L., Keller, H. U., Lara, M. L., Lazzarin, M., Lopez Moreno, J. J., Lucchetti, A., Marzari, F., Mottola, S., Pajola, M., Toth, I., & Tubiana, C. (2019). Surface evolution of the Anhur region on comet 67P/Churyumov-Gerasimenko from high-resolution OSIRIS images. *Astronomy and Astrophysics, 630*, A13.

Fortier, A., Alibert, Y., Carron, F., Benz, W., & Dittkrist, K.-M. (2013). Planet formation models: The interplay with the planetesimal disc. *A&A, 549*, A44. https://doi.org/10.1051/0004-6361/201220241.

Fouchard, M., Rickman, H., Froeschlé, C., & Valsecchi, G. B. (2013). Planetary perturbations for Oort Cloud comets. I. Distributions and dynamics. *Icarus, 222*, 20.

Fouchard, M., Rickman, H., Froeschlé, C., & Valsecchi, G. B. (2014a). Planetary perturbations for Oort cloud comets: II. Implications for the origin of observable comets. *Icarus, 231*, 110.

Fouchard, M., Rickman, H., Froeschlé, C., & Valsecchi, G. B. (2014b). Planetary perturbations for Oort cloud comets: III. Evolution of the cloud and production of centaurs and Halley type comets. *Icarus, 231*, 99.

Fouchard, M., Rickman, H., Froeschlé, C., & Valsecchi, G. B. (2017a). On the present shape of the Oort cloud and the flux of "new" comets. *Icarus, 292*, 218–233.

Fouchard, M., Rickman, H., Froeschle, C., & Valsecchi, G. B. (2017b). Distribution of long-period comets: Comparison between simulations and observations. *Astronomy and Astrophysics, 604*, A24.

Fougere, N., Altwegg, K., Berthelier, J.-J., Bieler, A., Bockelée-Morvan, D., Calmonte, U., Capaccioni, F., Combi, M. R., De Keyser, J., Debout, V., Erard, S., Fiethe, B., Filacchione, G., Fink, U., Fuselier, S. A., Gombosi, T. I., Hansen, K. C., Hässig, M., Huang, Z., Le Roy, L., Leyrat, C., Migliorini, A., Piccioni, G., Rinaldi, G., Rubin, M., Shou, Y., Tenishev, V., Toth, G., & Tzou, C.-Y. (2016). Direct Simulation Monte Carlo modelling of the major species in the coma of comet 67P/Churyumov-Gerasimenko. *Monthly Notices of the Royal Astronomical Society, 462*, S156–S169.

Frattin, E., Munoz, O., Moreno, F., Nava, J., Escobar-Cerezo, J., Gomez Martin, J. C., Guirado, D., Cellino, A., Coll, P., Raulin, F., Bertini, I., Cremonese, G., Lazzarin, M., Naletto, G., & La Forgia, F. (2019). Experimental phase function and degree of linear polarization of cometary dust analogues. *Monthly Notices of the Royal Astronomical Society, 484*, 2198.

Freeman, R. L., & Jones, E. M. (1974). CLM-R 137. UKAEA Research Group.

Fulle, M., & Sedmak, G. (1988). Photometrical analysis of the Neck-Line Structure of Comet Bennett 1970II. *Icarus, 74*, 383.

Fulle, M., Cremonese, G., & Böhm, C. (1998). The preperihelion dust environment of C/1995 O1 Hale-Bopp from 13 to 4 AU. *The Astronomical Journal, 116*, 1470.

Fuse, T., Yamamoto, N., Kinoshita, D., Furusawa, H., & Watanabe, J.-I. (2007). Observations of fragments split from nucleus B of Comet 73P/Schwassmann-Wachmann 3 with Subaru telescope. *Publications of the Astronomical Society of Japan, 59*, 381–386.

Galand, M., Heritier, K. L., Odelstad, E., Henri, P., Broiles, T. W., Allen, A. J., Altwegg, K., Beth, A., Burch, J. L., Carr, C. M., Cupido, E., Eriksson, A. I., Glassmeier, K.-H., Johansson, F. L., Lebreton, J.-P., Mandt, K. E., Nilsson, H., Richter, I., Rubin, M., Sagnieres, L. B. M., Schwartz, S. J., Sémon, T., Tzou, C.-Y., Vallieres, X., Vigren, E., & Wurz, P. (2016). Ionospheric plasma of comet 67P probed by Rosetta at 3 au from the Sun. *Monthly Notices of the Royal Astronomical Society, 462*, S331.

Galli, A., Vorburger, A., Wurz, P., Pommerol, A., Cerubini, R., Jost, B., Poch, O., Tulej, M., & Thomas, N. (2018). 0.2 to 10 keV electrons interacting with water ice: Radiolysis, sputtering, and sublimation. *Planetary and Space Science, 155*, 91.

Gan, L., & Cravens, T. E. (1990). Electron energetics in the inner coma of comet Halley. *Journal of Geophysical Research, 95*, 6285.

Garrod, R. T. (2019). Simulations of ice chemistry in cometary nuclei. *The Astrophysical Journal, 884*, 69.

Gasc, S., Altwegg, K., Fiethe, B., Jäckel, A., Korth, A., Le Roy, L., Mall, U., Rème, H., Rubin, M., Hunter Waite, J., & Wurz, P. (2017). Sensitivity and fragmentation calibration of the time-of-flight mass spectrometer RTOF on board ESA's Rosetta mission. *Planetary and Space Science, 135*, 64.

Gautier, D., & Hersant, F. (2005). Formation and composition of planetesimals. *Space Science Reviews, 116*, 25.

Geiss, J., & Gloeckler, G. (1998). Abundances of deuterium and helium-3 in the protosolar cloud. *Space Sci. Rev., 84*, 239–250.

Gerhardt, S. (2004). *Measurements and modeling of the plasma response to electrode biasing in the HSX Stellarator*. PhD thesis, University of Wisconsin-Madison.

Gerig, S.-B., Marschall, R., Thomas, N., Bertini, I., Bodewits, D., Davidsson, B., Fulle, M., Ip, W.-H., Keller, H. U., Küppers, M., Preusker, F., Scholten, F., Su, C. C., Toth, I., Tubiana, C., Wu, J.-S., Sierks, H., Barbieri, C., Lamy, P. L., Rodrigo, R., Koschny, D., Rickman, H., Agarwal, J., Barucci, M. A., Bertaux, J.-L., Cremonese, G., Da Deppo, V., Debei, S., De Cecco, M., Deller, J., Fornasier, S., Groussin, O., Gutierrez, P. J., Güttler, C., Hviid, S. F., Jorda, L., Knollenberg, J., Kramm, J.-R., Kührt, E., Lara, L. M., Lazzarin, M., López Moreno, J. J., Marzari, F., Mottola, S., Naletto, G., Oklay, N., & Vincent, J.-B. (2018). On deviations from free-radial outflow in the inner coma of comet 67P/Churyumov-Gerasimenko. *Icarus, 311*, 1.

Gerig, S.-B., Pinzón-Rodríguez, O., Marschall, R., Wu, J.-S., & Thomas, N. (2020). Dayside-to-nightside dust coma brightness asymmetry and its implications for nightside activity at comet 67P/Churyumov-Gerasimenko. *Icarus, 351*, article id. 113968.

Giacomini, L., et al. (2016). Geologic mapping of the Comet 67P/Churyumov-Gerasimenko's Northern hemisphere. *Monthly Notices of the Royal Astronomical Society, 462*, S352.

Gibb, E. L., Bonev, B. P., Villanueva, G., DiSanti, M. A., Mumma, M. J., Sudholt, E., & Radeva, Y. (2012). Chemical composition of Comet C/2007 N3 (Lulin): Another "atypical" comet. *The Astrophysical Journal, 750*, 102.

Gicquel, A., Bockelée-Morvan, D., Zakharov, V. V., Kelley, M. S., Woodward, C. E., & Wooden, D. H. (2012). Investigation of dust and water ice in comet 9P/Tempel 1 from Spitzer observations of the Deep Impact event. *Astronomy and Astrophysics, 542*, A119.

Glassmeier, K.-H., Boehnhardt, H., Koschny, D., Kührt, E., & Richter, I. (2007). The Rosetta mission: Flying towards the origin of the Solar System. *Space Science Reviews, 128*, 1.

Glavin, D. P., Dworkin, J. P., & Sandford, S. A. (2008). Detection of cometary amines in samples returned by Stardust. *Meteoritics and Planetary Science, 43*, 399.

Goetz, C., Koenders, C., Hansen, K. C., Burch, J., Carr, C., Eriksson, A., Frühauff, D., Güttler, C., Henri, P., Nilsson, H., Richter, I., Rubin, M., Sierks, H., Tsurutani, B., Volwerk, M., & Glassmeier, K. H. (2016a). Structure and evolution of the diamagnetic cavity at comet 67P/Churyumov-Gerasimenko. *Monthly Notices of the Royal Astronomical Society, 462*, S459.

Goetz, C., Koenders, C., Richter, I., Altwegg, K., Burch, J., Carr, C., Cupido, E., Eriksson, A., Güttler, C., Henri, P., Mokashi, P., Nemeth, Z., Nilsson, H., Rubin, M., Sierks, H., Tsurutani, B., Vallat, C., Volwerk, M., & Glassmeier, K.-H. (2016b). First detection of a diamagnetic cavity at comet 67P/Churyumov-Gerasimenko. *Astronomy and Astrophysics, 588*, A24.

Goetz, C., Volwerk, M., Richter, I., & Glassmeier, K.-H. (2017). Evolution of the magnetic field at comet 67P/Churyumov-Gerasimenko. *Monthly Notices of the Royal Astronomical Society, 469*, S268.

Goldberg, B. A., Garneau, G. W., & Lavoie, S. K. (1984). Io's sodium cloud. *Science, 226*, 512.

Gombosi, T. I. (1994). *Gaskinetic theory*. Cambridge: Cambridge University Press.

Gombosi, T. (2014). Physics of cometary magnetospheres. In A. Keiling, C. M. Jackman, & P. A. Delamere (Eds.), *Magnetotails in the Solar System. Geophysical monograph 207*. Wiley.

Gombosi, T. I., & Houpis, H. L. F. (1986). An icy-glue model of cometary nuclei. *Nature, 324*, 43.

Gombosi, T. I., Nagy, A. F., & Cravens, T. E. (1986). Dust and neutral gas modeling of the inner atmospheres of comets. *Reviews of Geophysics, 24*, 667.

Gombosi, T. I., De Zeeuw, D. L., Häberli, R. M., & Powell, K. G. (1996). Three-dimensional multiscale MHD model of cometary plasma environments. *Journal of Geophysical Research, 101*, 15233.

Gomes, R. S., Fernández, J. A., Gallardo, T., & Brunini, A. (2008). The scattered disk: Origins, dynamics, and end states. In M. A. Barucci, H. Boehnhardt, D. P. Cruikshank, & A. Morbidelli (Eds.), *The Solar System beyond Neptune* (p. 259). Tucson: University of Arizona Press.

Gottlieb, C. A., Myers, P. C., & Thaddeus, P. (2003). Precise millimeter-wave laboratory frequencies for CS and $C^{34}S$. *Astrophysical Journal, 580*, 655.

Graykowski, A., & Jewitt, D. (2019). Fragmented Comet 73P/Schwassmann-Wachmann 3. *The Astronomical Journal, 158*, 112.

Grazier, K. R., Horner, J., & Castillo-Rogez, J. C. (2019). The relationship between Centaurs and Jupiter family comets with implications for K-Pg-type Impacts. *Monthly Notices of the Royal Astronomical Society, 490*, 4388.

Greeley, R. (2013). *Introduction to planetary geomorphology*. Cambridge: Cambridge University Press.

Greenwood, J.B., Williams I.D., Smith S.J., & Chutjian A., (2000), Measurement of Charge Exchange and X-Ray Emission Cross Sections for Solar Wind-Comet Interactions, The Astrophysical Journal, 533, L175.

Grimm, S. L., & Stadel, J. G. (2014). The GENGA code: Gravitational encounters in N-body simulations with GPU acceleration. *The Astrophysical Journal, 796*, 23.

Gronkowski, P. (2007). The search for a cometary outbursts mechanism: A comparison of various theories. *Astronomische Nachrichten, 328*, 126–136.

Gronkowski, P., & Wesołowski, M. (2016). A review of cometary outbursts at large heliocentric distances. *Earth Moon Planets, 119*, 23–33.

Groussin, O., Sunshine, J. M., Feaga, L. M., Jorda, L., Thomas, P. C., Li, J.-Y., A'Hearn, M. F., Belton, M. J. S., Besse, S., Carcich, B., Farnham, T. L., Hampton, D., Klaasen, K., Lisse, C., Merlin, F., & Protopapa, S. (2013). The temperature, thermal inertia, roughness and color of the nuclei of Comets 103P/Hartley 2 and 9P/Tempel 1. *Icarus, 222*, 580–594.

Groussin, O., et al. (2015a). Gravitational slopes, geomorphology, and material strengths of the nucleus of comet 67P/Churyumov-Gerasimenko from OSIRIS observations. *Astronomy and Astrophysics, 583*, A32.

Groussin, O., Sierks, H., Barbieri, C., Lamy, P., Rodrigo, R., Koschny, D., Rickman, H., Keller, H. U., A'Hearn, M. F., Auger, A.-T., Barucci, M. A., Bertaux, J.-L., Bertini, I., Besse, S., Cremonese, G., Da Deppo, V., Davidsson, B., Debei, S., De Cecco, M., El-Maarry, M. R., Fornasier, S., Fulle, M., Gutiérrez, P. J., Güttler, C., Hviid, S., Ip, W.-H., Jorda, L., Knollenberg, J., Kovacs, G., Kramm, J. R., Kührt, E., Küppers, M., Lara, L. M., Lazzarin, M., Lopez Moreno, J. J., Lowry, S., Marchi, S., Marzari, F., Massironi, M., Mottola, S., Naletto, G., Oklay, N., Pajola, M., Pommerol, A., Thomas, N., Toth, I., Tubiana, C., & Vincent, J.-B. (2015b).

Temporal morphological changes in the Imhotep region of comet 67P/Churyumov-Gerasimenko. *Astronomy and Astrophysics, 583*, A36.

Groussin, O., Attree, N., Brouet, Y., Ciarletti, V., Davidsson, B., Filacchione, G., Fischer, H.-H., Gundlach, B., Knapmeyer, M., Knollenberg, J., Kokotanekova, R., Kührt, E., Leyrat, C., Marshall, D., Pelivan, I., Skorov, Y., Snodgrass, C., Spohn, T., & Tosi, F. (2019). The thermal, mechanical, structural, and dielectric properties of cometary nuclei after Rosetta. *Space Science Reviews, 215*, 29.

Grove, C. I., Hook, S. J., & Paylor, E. D. (1992). *Laboratory reflectance spectra for 160 minerals 0.4–2.5 micrometers: JPL Publication 92-2*. Pasadena, CA: Jet Propulsion Laboratory.

Grundy, W. M., & Schmitt, B. (1998). The temperature-dependent near-infrared absorption spectrum of hexagonal $H_2O$ ice. *Journal of Geophysical Research, 103*, 25809–25822.

Grün, E. (2007). Solar System dust. In L.-A. McFadden, P. R. Weissman, & T. V. Johnson (Eds.), *Encyclopedia of the Solar System* (2nd ed., p. 621). Academic Press. ISBN 9780120885893.

Grün, E., & Jessberger, E. K. (1990). Dust. In W. F. Huebner (Ed.), *Physics and chemistry of comets*. Berlin: Springer.

Guilbert-Lepoutre, A., & Jewitt, D. (2011). Thermal shadows and compositional structure in comet nuclei. *The Astrophysical Journal, 743*, 31.

Guliyev, A. (2017). Collision with meteoroids as one of possible causes of cometary nucleus splitting. *Planetary and Space Science, 143*, 40.

Gulkis, S., Frerking, M., Crovisier, J., Beaudin, G., Hartogh, P., Encrenaz, P., Koch, T., Kahn, C., Salinas, Y., Nowicki, R., Irigoyen, R., Janssen, M., Stek, P., Hofstadter, M., Allen, M., Backus, C., Kamp, L., Jarchow, C., Steinmetz, E., Deschamps, A., Krieg, J., Gheudin, M., Bockelée-Morvan, D., Biver, N., Encrenaz, T., Despois, D., Ip, W., Lellouch, E., Mann, I., Muhleman, D., Rauer, H., Schloerb, P., & Spilker, T. (2007). MIRO: Microwave instrument for Rosetta Orbiter. *Space Science Reviews, 128*, 561.

Gulkis, S., Allen, M., von Allmen, P., Beaudin, G., Biver, N., Bockelée-Morvan, D., Choukroun, M., Crovisier, J., Davidsson, B. J. R., Encrenaz, P., Encrenaz, T., Frerking, M., Hartogh, P., Hofstadter, M., Ip, W.-H., Janssen, M., Jarchow, C., Keihm, S., Lee, S., Lellouch, E., Leyrat, C., Rezac, L., Schloerb, F. P., & Spilker, T. (2015). Subsurface properties and early activity of comet 67P/Churyumov-Gerasimenko. *Science, 347*, aaa0709.

Gunderson, K., Thomas, N., & Whitby, J. A. (2006). First measurements with the Physikalisches Institut Radiometric Experiment (PHIRE). *Planetary and Space Science, 54*, 1046–1056.

Gundlach, B., & Blum, J. (2012). Outgassing of icy bodies in the Solar System. II: Heat transport in dry, porous surface dust layers. *Icarus, 219*, 618.

Gundlach, B., Schmidt, K. P., Kreuzig, C., Bischoff, D., Rezaei, F., Kothe, S., Blum, J., Grzesik, B., & Stoll, E. (2018). The tensile strength of ice and dust aggregates and its dependence on particle properties. *Monthly Notices of the Royal Astronomical Society, 479*, 1273.

Gunnarsson, M., Rickman, H., Festou, M. C., Winnberg, A., & Tancredi, G. (2002). An extended CO source around Comet 29P/Schwassmann-Wachmann 1. *Icarus, 157*, 309.

Gunnarsson, M., Bockelée-Morvan, D., Biver, N., Crovisier, J., & Rickman, H. (2008). Mapping the carbon monoxide coma of comet 29P/Schwassmann-Wachmann 1. *Astronomy and Astrophysics, 484*, 537.

Gutiérrez, P. J., Jorda, L., Samarasinha, N. H., & Lamy, P. (2005). Outgassing-induced effects in the rotational state of comet 67P/Churyumov-Gerasimenko during the Rosetta mission. *Planetary and Space Science, 53*, 1135.

Gutiérrez, P. J., et al. (2016). Possible interpretation of the precession of comet 67P/Churyumov-Gerasimenko. *Astronomy and Astrophysics, 590*, A46.

Hadamcik, E., Renard, J.-B., Rietmeijer, F. J. M., Levasseur-Regourd, A. C., Hill, H. G. M., Karner, J. M., & Nuth, J. A. (2007). Light scattering by fluffy Mg-Fe-SiO and C mixtures as cometary analogs (PROGRA$^2$ experiment). *Icarus, 190*, 660.

Haddad, G. N., & Samson, J. A. R. (1986). Total absorption and photoionization cross sections of water vapor between 100 and 1000 Å. *Journal of Chemical Physics, 84*, 6623.

Hadraoui, K., Cottin, H., Ivanovski, S. L., Zapf, P., Altwegg, K., Benilan, Y., Biver, N., Della Corte, V., Fray, N., Lasue, J., Merouane, S., Rotundi, A., & Zakharov, V. (2019). Distributed glycine in comet 67P/Churyumov-Gerasimenko. *Astronomy and Astrophysics, 630*, A32.

Hässig, M., Altwegg, K., Balsiger, H., Bar-Nun, A., Berthelier, J. J., Bieler, A., Bochsler, P., Briois, C., Calmonte, U., Combi, M., De Keyser, J., Eberhardt, P., Fiethe, B., Fuselier, S. A., Galand, M., Gasc, S., Gombosi, T. I., Hansen, K. C., Jäckel, A., Keller, H. U., Kopp, E., Korth, A., Kührt, E., Le Roy, L., Mall, U., Marty, B., Mousis, O., Neefs, E., Owen, T., Reme, H., Rubin, M., Sémon, T., Tornow, C., Tzou, C.-Y., Waite, J. H., & Wurz, P. (2015). Time variability and heterogeneity in the coma of 67P/Churyumov-Gerasimenko. *Science, 347*, aaa0276.

Hässig, M., Altwegg, K., Balsiger, H., Berthelier, J. J., Bieler, A., Calmonte, U., Dhooghe, F., Fiethe, B., Fuselier, S. A., Gasc, S., Gombosi, T. I., Le Roy, L., Luspay-Kuti, A., Mandt, K., Rubin, M., Tzou, C.-Y., Wampfler, S. F., & Wurz, P. (2017). Isotopic composition of $CO_2$ in the coma of 67P/Churyumov-Gerasimenko measured with ROSINA/DFMS. *Astronomy and Astrophysics, 605*, A50.

Haerendel, G. (1987). Plasma transport near the magnetic cavity surrounding Comet Halley. *Geophysical Research Letters, 14*, 673.

Halder, P., Chakraborty, A., Deb Roy, P., & Das, H. S. (2014). Java application for superposition T-matrix code to study the optical properties of cosmic dust aggregates. *Computer Physics Communications, 185*(9), 2369–2379.

Hall, K., & Thorn, C. E. (2014). Thermal fatigue and thermal shock in bedrock: An attempt to unravel the geomorphic processes and products. *Geomorphology, 206*, 1–13.

Haltrin, V. I. (2002). One-parameter two-term Henyey-Greenstein phase function for light scattering in seawater. *Applied Optics, 41*, 1022.

Hansen, K. C., Altwegg, K., Berthelier, J.-J., Bieler, A., Biver, N., Bockelée-Morvan, D., Calmonte, U., Capaccioni, F., Combi, M. R., de Keyser, J., Fiethe, B., Fougere, N., Fuselier, S. A., Gasc, S., Gombosi, T. I., Huang, Z., Le Roy, L., Lee, S., Nilsson, H., Rubin, M., Shou, Y., Snodgrass, C., Tenishev, V., Toth, G., Tzou, C.-Y., Wedlund, C. S., & Team, R. (2016). Evolution of water production of 67P/Churyumov-Gerasimenko: An empirical model and a multi-instrument study. *Monthly Notices of the Royal Astronomical Society, 462*, S491–S506.

Hansen, K. C., Bagdonat, T., Motschmann, U., Alexander, C., Combi, M. R., Cravens, T. E., Gombosi, T. I., Jia, Y.-D., & Robertson, I. P. (2007). The plasma environment of Comet 67P/Churyumov-Gerasimenko throughout the Rosetta main mission. *Space Science Reviews, 128*, 133.

Hapke, B. (1981). Bidirectional reflectance spectroscopy. I – Theory. *Journal of Geophysical Research, 86*, 3039.

Hapke, B. (1984). Bidirectional reflectance spectroscopy. 3. Correction for macroscopic roughness. *Icarus, 59*, 41.

Hapke, B. (1986). Bidirectional reflectance spectroscopy. 4. The extinction coefficient and the opposition effect. *Icarus, 67*, 264.

Hapke, B. (2002). Bidirectional reflectance spectroscopy. 5. The coherent backscatter opposition effect and anisotropic scattering. *Icarus, 157*, 523.

Harker, D. E., Woodward, C. E., Kelley, M. S., Sitko, M. L., Wooden, D. H., Lynch, D. K., & Russell, R. W. (2011). Mid-infrared spectrophotometric observations of fragments B and C of Comet 73P/Schwassmann-Wachmann 3. *The Astronomical Journal, 141*, 26.

Harker, D. E., Woodward, C. E., Wooden, D. H., Fisher, R. S., & Trujillo, C. A. (2007). Gemini-N mid-IR observations of the dust properties of the ejecta excavated from Comet 9P/Tempel 1 during Deep Impact. *Icarus, 190*, 432–453.

Harmon, J. K., Nolan, M. C., Ostro, S. J., & Campbell, D. B. (2004). Radar studies of comet nuclei and grain comae. In M. Festou, H. U. Keller, & H. A. Weaver (Eds.), *Comets II* (p. 265). Tucson: University of Arizona Press.

Hartogh, P. (1997). *High-resolution chirp transform spectrometer for middle atmospheric microwave sounding.* Proc. SPIE 3220, Satellite Remote Sensing of Clouds and the Atmosphere II. https://doi.org/10.1117/12.301141.

Hartogh, P., Lis, D. C., Bockelée-Morvan, D., de Val-Borro, M., Biver, N., Küppers, M., Emprechtinger, M., Bergin, E. A., Crovisier, J., Rengel, M., Moreno, R., Szutowicz, S., & Blake, G. A. (2011). Ocean-like water in the Jupiter-family comet 103P/Hartley 2. *Nature, 478*, 218.

Hartzell, C. M., Wang, X., Scheeres, D. J., & Horányi, M. (2013). Experimental demonstration of the role of cohesion in electrostatic dust lofting. *Geophysical Research Letters, 40*, 1038.

Haruyama, J., Yamamoto, T., Mizutani, H., & Greenberg, J. M. (1993). Thermal history of comets during residence in the Oort cloud: Effect of radiogenic heating in combination with the very low thermal conductivity of amorphous ice. *Journal of Geophysical Research, 98*, 15079.

Haser, L. (1957). Liege Inst. Astrophys. Repr. No. 394.

Hasselmann, P. H., Barucci, M. A., Fornasier, S., Bockelée-Morvan, D., Deshapriya, J. D. P., Feller, C., Sunshine, J., Hoang, V., Sierks, H., Naletto, G., Lamy, P. L., Rodrigo, R., Koschny, D., Davidsson, B., Bertaux, J.-L., Bertini, I., Bodewits, D., Cremonese, G., Da Deppo, V., Debei, S., Fulle, M., Gutierrez, P. J., Güttler, C., Deller, J., Ip, W.-H., Keller, H. U., Lara, L. M., De Cecco, M., Lazzarin, M., López-Moreno, J. J. L., Marzari, F., Shi, X., & Tubiana, C. (2019). Pronounced morphological changes in a southern active zone on comet 67P/Churyumov-Gerasimenko. *Astronomy and Astrophysics, 630*, A8.

Helfenstein, P., & Veverka, J. (1987). Photometric properties of lunar terrains derived from Hapke's equation. *Icarus, 72*, 342–357.

Helfenstein, P., & Shepard, M. K. (2011). Testing the Hapke photometric model: Improved inversion and the porosity correction. *Icarus, 215*, 83–100.

Hellmich, R. (1981). The influence of the radiation transfer in cometary dust Halos on the production rates of gas and dust. *Astronomy and Astrophysics, 93*, 341.

Hergenrother, C. W. (2018). Photometry of Damocloid Asteroid 2006 BZ8. *Minor Planet Bulletin, 45*, 64.

Hérique, A., Kofman, W., Beck, P., Bonal, L., Buttarazzi, I., Heggy, E., Lasue, J., Levasseur-Regourd, A. C., Quirico, E., & Zine, S. (2016). Cosmochemical implications of CONSERT permittivity characterization of 67P/CG. *Monthly Notices of the Royal Astronomical Society, 462*, S516.

Hérique, A., Kofman, W., Zine, S., Blum, J., Vincent, J.-B., & Ciarletti, V. (2019). Homogeneity of 67P/Churyumov-Gerasimenko as seen by CONSERT: Implication on composition and formation. *Astronomy and Astrophysics, 630*, A6.

Heritier, K. L., Galand, M., Henri, P., Johansson, F. L., Beth, A., Eriksson, A. I., Vallieres, X., Altwegg, K., Burch, J. L., Carr, C., Ducrot, E., Hajra, R., & Rubin, M. (2018). Plasma source and loss at comet 67P during the Rosetta mission. *Astronomy and Astrophysics, 618*, A77.

Herny, C., Mousis, O., Marschall, R., Thomas, N., Rubin, M., Pinzon, O., & Wright, I.P. (2020). New constraints on the chemical composition and outgassing of 67P/Churyumov-Gerasimenko. *Planetary and Space Science*, submitted.

Hill, J. R., & Mendis, D. A. (1980). On the origin of striae in cometary dust tails. *The Astrophysical Journal, 242*, 395.

Hirabayashi, M., Scheeres, D. J., Chesley, S. R., Marchi, S., McMahon, J. W., Steckloff, J., Mottola, S., Naidu, S. P., & Bowling, T. (2016). Fission and reconfiguration of bilobate comets as revealed by 67P/Churyumov-Gerasimenko. *Nature, 534*, 352–355.

Hirao, K. (1986). The Planet-A Halley encounters. *Nature, 321*, 294.

Ho, T. M., Thomas, N., Boice, D. C., Kollein, C., & Soderblom, L. A. (2003). Comparative study of the dust emission of 19P/Borrelly (Deep space 1) and 1P/Halley. *Advances in Space Research, 31*, 2583.

Ho, T.-M., Thomas, N., Boice, D. C., Combi, M., Soderblom, L. A., & Tenishev, V. (2007). Comparison of the dust distribution in the inner comae of comets 1P/Halley and 19P/Borrelly spacecraft observations. *Planetary and Space Science, 55*, 974–985.

Hoang, M., Altwegg, K., Balsiger, H., Beth, A., Bieler, A., Calmonte, U., Combi, M. R., De Keyser, J., Fiethe, B., Fougere, N., Fuselier, S. A., Galli, A., Garnier, P., Gasc, S., Gombosi, T., Hansen, K. C., Jäckel, A., Korth, A., Lasue, J., Le Roy, L., Mall, U., Reme, H., Rubin, M., Semon, T.,

Toublanc, D., Tzou, C.-Y., Waite, J. H., & Wurz, P. (2017). The heterogeneous coma of comet 67P/Churyumov-Gerasimenko as seen by ROSINA: H2O, CO2, and CO from September 2014 to February 2016. *Astronomy and Astrophysics, 600*, A77.

Hoban, S. (1993). Serendipitous images of methanol in Comet Levy (1990 XX). *Icarus, 104*, 149.

Hoban, S., Samarasinha, N. H., A'Hearn, M. F. A., & Klinglesmith, D. A. (1988). An investigation into periodicities in the morphology of CN jets in comet P/Halley. *Astronomy and Astrophysics, 195*, 331.

Höfner, S., et al. (2017). Thermophysics of fractures on comet 67P/Churyumov-Gerasimenko. *Astronomy and Astrophysics, 608*, A121.

Holsapple, K. A. (1993). The scaling of impact processes in planetary sciences. *Annual Review of Earth and Planetary Sciences, 21*, 333–373.

Holsapple, K. A. (1994). Catastrophic disruptions and cratering of solar system bodies: A review and new results. *Planetary and Space Science, 42*, 1067–1078.

Holsapple, K. A., & Housen, K. R. (2007). A crater and its ejecta: An interpretation of Deep Impact. *Icarus, 187*, 345.

Hoppe, P., Rubin, M., & Altwegg, K. (2018). Presolar isotopic signatures in meteorites and comets: New insights from the Rosetta Mission to Comet 67P/Churyumov–Gerasimenko. *Space Science Reviews, 214*, 106.

Horányi, M., & Mendis, D. A. (1987). The effect of a sector boundary crossing on the cometary dust tail. *Earth Moon Planets, 37*, 71–77.

Hornung, K., Merouane, S., Hilchenbach, M., Langevin, Y., Mellado, E. M., Della Corte, V., Kissel, J., Engrand, C., Schulz, R., Ryno, J., Silen, J., & Team, C. O. S. I. M. A. (2016). A first assessment of the strength of cometary particles collected in-situ by the COSIMA instrument onboard ROSETTA. *Planetary and Space Science, 133*, 63.

Hosek, M. W., Blaauw, R. C., Cooke, W. J., & Suggs, R. M. (2013). Outburst dust production of Comet 29P/Schwassmann-Wachmann 1. *The Astronomical Journal, 145*, 122.

Hosseini, S., West, R., Keller, H. U., Altwegg, K., & Davidsson, B. (2018). *New generation of compact Cometary D/H survey mission.* 42nd COSPAR Scientific Assembly, 42, B1.1-56-18.

Housen, K. R., & Holsapple, K. A. (2011). Ejecta from impact craters. *Icarus, 211*, 856.

Housen, K. R., & Holsapple, K. A. (2012). Craters without ejecta. *Icarus, 219*, 297–306.

Hovenier, J. W., Van Der Mee, C., & Domke, H. (2004). *Transfer of polarized light in planetary atmospheres: Basic concepts and practical methods.* Astrophysics and Space Science Library. Kluwer Academic.

Howell, E. S., Lovell, A. J., Butler, B., & Schloerb, F. P. (2007). Radio OH observations of 9P/Tempel 1 before and after Deep Impact. *Icarus, 187*, 228.

Hsieh, H. H., Jewitt, D. C., & Fernandez, Y. R. (2004). The strange case of 133P/Elst-Pizarro: A comet among the asteroids. *The Astronomical Journal, 127*, 2997–3017.

Hsieh, H. H., & Jewitt, D. (2005). Search for activity in 3200 Phaethon. *The Astrophysical Journal, 624*, 1093.

Huang, J. H. (1971). Effective thermal conductivity of rocks. *Journal of Geophysical Research, 76*, 6420–6427.

Huang, X., & Yung, Y. L. (2004). A common misunderstanding about the Voigt line profile. *Journal of Atmospheric Sciences, 61*, 1630.

Hudson, R. L., & Moore, M. H. (1992). A far-IR study of irradiated amorphous ice: An unreported oscillation between amorphous and crystalline phases. *Journal of Physical Chemistry, 96*, 6500.

Huebner, W. F. (1970). Dust from cometary nuclei. *Astronomy and Astrophysics, 5*, 286–297.

Huebner, W. F. (1987). First polymer in space identified in Comet Halley. *Science, 237*, 628.

Huebner, W. F. (1990). *Physics and chemistry of comets.* Springer.

Huebner, W. F., Keady, J. J., & Lyon, S. P. (1992). Solar photo rates for planetary atmospheres and atmospheric pollutants. *Astrophysics and Space Science, 195*, 1–289.

Huebner, W. F., Benkhoff, J., Capria, M.-T., Coradini, A., De Sanctis, C., Orosei, R., & Prialnik, D. (2006). Heat and gas diffusion in comet nuclei. *ISSI Scientific Report, SR-004.*

Hunten, D. M., Morgan, T. H., & Shemansky, D. E. (1988). The mercury atmosphere. In F. Vilas, C. R. Chapman, & M. S. Matthews (Eds.), *Mercury* (pp. 562–612). Tucson: University of Arizona Press.

Hunten, D. M., Roach, F. E., & Chamberlain, J. W. (1956). A photometric unit for the airglow and aurora. *Journal of Atmospheric and Terrestrial Physics, 8*, 345.

Hyland, M. G., Fitzsimmons, A., & Snodgrass, C. (2019). Near-UV and optical spectroscopy of comets using the ISIS spectrograph on the WHT. *Monthly Notices of the Royal Astronomical Society, 484*, 1347.

Ip, W.-H. (1983). On photochemical heating of cometary comae – The cases of $H_2O$ and CO-rich comets. *The Astrophysical Journal, 264*, 726.

Ip, W.-H., & Axford, W. I. (1987). The formation of a magnetic-field-free cavity at comet Halley. *Nature, 325*, 418.

Itikawa, Y. (2015). Cross sections for electron collisions with carbon monoxide. *Journal of Physical and Chemical Reference Data, 44*, 013105.

Itikawa, Y., & Mason, N. (2005). Cross sections for electron collisions with water molecules. *Journal of Physical and Chemical Reference Data, 34*, 1.

Ivanova, O. V., Dlugach, J. M., Afanasiev, V. L., Reshetnyk, V. M., & Korsun, P. P. (2015). CCD polarimetry of distant comets C/2010 S1 (LINEAR) and C/2010 R1 (LINEAR) at the 6-m telescope of the SAO RAS. *Planetary and Space Science, 118*, 199.

Ivanovski, S. L., Della Corte, V., Rotundi, A., Fulle, M., Fougere, N., Bieler, A., Rubin, M., Ivanovska, S., & Liuzzi, V. (2017a). Dynamics of non-spherical dust in the coma of 67P/Churyumov-Gerasimenko constrained by GIADA and ROSINA data. *Monthly Notices of the Royal Astronomical Society, 469*, S774–S786.

Ivanovski, S. L., Zakharov, V. V., Della Corte, V., Crifo, J.-F., Rotundi, A., & Fulle, M. (2017b). Dynamics of aspherical dust grains in a cometary atmosphere: I. Axially symmetric grains in a spherically symmetric atmosphere. *Icarus, 282*, 333–350.

Jäger, C., Mutschke, H., & Henning, T. (1998). Optical properties of carbonaceous dust analogues. *Astronomy and Astrophysics, 332*, 291.

Jäger, C., Dorschner, J., Mutschke, H., Posch, T., & Henning, T. (2003). Steps toward interstellar silicate mineralogy. VII. Spectral properties and crystallization behaviour of magnesium silicates produced by the sol-gel method. *Astronomy and Astrophysics, 408*, 193.

Jehin, E., Manfroid, J., Hutsemekers, D., Arpigny, C., & Zucconi, J.-M. (2009). Isotopic ratios in comets: Status and perspectives. *Earth Moon and Planets, 105*, 167–180.

Jenniskens, P., & Blake, D. F. (1994). Structural transitions in amorphous water ice and astrophysical implications. *Science, 265*, 753.

Jenniskens, P., Blake, D. F., Wilson, M. A., & Pohorille, A. (1995). High-density amorphous ice, the frost on interstellar grains. *The Astrophysical Journal, 455*, 389.

Jenniskens, P., Blake, D. F., & Kouchi, A. (1998). Amorphous water ice. A Solar System material. *Solar System Ices, 227*, 139.

Jessberger, E. K., Christoforidis, A., & Kissel, J. (1988). Aspects of the major element composition of Halley's dust. *Nature, 332*, 691.

Jewitt, D. (2005). A first look at the damocloids. *The Astronomical Journal, 129*, 530.

Jewitt, D. (2010). Kuiper Belt and comets: An observational persepective. In K. Altwegg, W. Benz, & N. Thomas (Eds.), *Trans-Neptunian objects and comets*. Proceedings of 35th Saas-Fee Conference.

Jewitt, D. (2012). The active asteroids. *The Astronomical Journal, 143*, 66.

Jewitt, D., & Luu, J. (1993). Discovery of the candidate Kuiper belt object 1992 QB1. *Nature, 362*, 730–732.

Jewitt, D. C., & Luu, J. (2004). Crystalline water ice on the Kuiper belt object (50000) Quaoar. *Nature, 432*, 731.

Jewitt, D., & Luu, J. (2019). Initial characterization of interstellar Comet 2I/2019 Q4 (Borisov). *The Astrophysical Journal, 886*, L29.

Jewitt, D., & Sheppard, S. (2004). The nucleus of Comet 48P/Johnson. *The Astronomical Journal, 127*, 1784.

Jewitt, D., Sheppard, S., & Fernández, Y. (2003). 143P/Kowal-Mrkos and the shapes of cometary nuclei. *The Astronomical Journal, 125*, 3366.

Jewitt, D., Li, J., & Agarwal, J. (2013). The dust tail of Asteroid (3200) Phaethon. *The Astrophysical Journal, 771*, L36.

Jewitt, D., Li, J., Agarwal, J., Weaver, H., Mutchler, M., & Larson, S. (2015). Nucleus and mass loss from active Asteroid 313P/Gibbs. *The Astronomical Journal, 150*, 76.

Jewitt, D., Agarwal, J., Weaver, H., Mutchler, M., Li, J., & Larson, S. (2016a). Hubble Space telescope observations of active Asteroid 324P/La Sagra. *The Astronomical Journal, 152*, 77.

Jewitt, D., Mutchler, M., Weaver, H., Hui, M.-T., Agarwal, J., Ishiguro, M., Kleyna, J., Li, J., Meech, K., Micheli, M., Wainscoat, R., & Weryk, R. (2016b). Fragmentation kinematics in Comet 332P/Ikeya-Murakami. *The Astrophysical Journal, 829*, L8.

Jewitt, D., Hui, M.-T., Mutchler, M., Weaver, H., Li, J., & Agarwal, J. (2017). A comet active beyond the crystallization zone. *The Astrophysical Journal, 847*, L19.

Jewitt, D., Kim, Y., Luu, J., Rajagopal, J., Kotulla, R., Ridgway, S., & Liu, W. (2019). Episodically active Asteroid 6478 Gault. *The Astrophysical Journal, 876*, L19.

Jia, P., Andreotti, B., & Claudin, P. (2017). Giant ripples on comet 67P/Churyumov-Gerasimenko sculpted by sunset thermal wind. *Proceedings of the National Academy of Science, 114*, 2509.

Jia, Y.-D., Combi, M. R., Hansen, K. C., & Gombosi, T. I. (2007). A global model of cometary tail disconnection events triggered by solar wind magnetic variations. *Journal of Geophysical Research (Space Physics), 112*, A05223.

Jockers, K. (1985). The ion tail of comet Kohoutek 1973 XII during 17 days of solar wind gusts. *Astronomy and Astrophysics Supplement Series, 62*(791), 1985.

Jockers, K. (1991). Ions in the coma and in the tail of comets – Observations and theory. In *Cometary plasma processes. Geophysical Monograph Series* (Vol. 61, p. 139). Washington, DC: American Geophysical Union.

Johansson, F. L., Odelstad, E., Paulsson, J. J. P., Harang, S. S., Eriksson, A. I., Mannel, T., Vigren, E., Edberg, N. J. T., Miloch, W. J., Simon Wedlund, C., Thiemann, E., Eparvier, F., & Andersson, L. (2017). Rosetta photoelectron emission and solar ultraviolet flux at comet 67P. *Monthly Notices of the Royal Astronomical Society, 469*, S626.

Jones, G. H., Knight, M. M., Battams, K., Boice, D. C., Brown, J., Giordano, S., Raymond, J., Snodgrass, C., Steckloff, J. K., Weissman, P., Fitzsimmons, A., Lisse, C., Opitom, C., Birkett, K. S., Bzowski, M., Decock, A., Mann, I., Ramanjooloo, Y., & McCauley, P. (2018). The science of sungrazers, sunskirters, and other near-sun comets. *Space Science Reviews, 214*, 20.

Jorda, L., Gaskell, R., Capanna, C., Hviid, S., Lamy, P., Durech, J., Faury, G., Groussin, O., Gutierrez, P., Jackman, C., Keihm, S. J., Keller, H. U., Knollenberg, J., Kuehrt, E., Marchi, S., Mottola, S., Palmer, E., Schloerb, F. P., Sierks, H., Vincent, J.-B., A'Hearn, M. F., Barbieri, C., Rodrigo, R., Koschny, D., Rickman, H., Barucci, M. A., Bertaux, J. L., Bertini, I., Cremonese, G., Da Deppo, V., Davidsson, B., Debei, S., De Cecco, M., Fornasier, S., Fulle, M., Guettler, C., Ip, W.-H., Kramm, J. R., Kueppers, M., Lara, L. M., Lazzarin, M., Lopez Moreno, J. J., Marzari, F., Naletto, G., Oklay, N., Thomas, N., Tubiana, C., & Wenzel, K.-P. (2016). The global shape, density and rotation of Comet 67P/Churyumov-Gerasimenko from preperihelion Rosetta/OSI-RIS observations. *Icarus, 277*, 257–278.

Jost, B. (2016). *A laboratory study of spectrophotometric properties of ice-bearing surfaces in the solar system with focus on cometary nuclei*. PhD thesis, University of Bern.

Jost, B., Gundlach, B., Pommerol, A., Oesert, J., Gorb, S. N., Blum, J., & Thomas, N. (2013). Micrometer-sized ice particles for planetary-science experiments – II. Bidirectional reflectance. *Icarus, 225*, 352.

Julian, W. H. (1987). Free precession of the comet Halley nucleus. *Nature, 326*, 57.

Jutzi, M., & Benz, W. (2017). Formation of bi-lobed shapes by sub-catastrophic collisions. A late origin of comet 67P's structure. *Astronomy and Astrophysics, 597*, A62.

Jutzi, M., Thomas, N., Benz, W., El Maarry, M. R., Jorda, L., Kührt, E., & Preusker, F. (2013). The influence of recent major crater impacts on the surrounding surfaces of (21) Lutetia. *Icarus, 226*, 89–100.

Jutzi, M., Benz, W., Toliou, A., Morbidelli, A., & Brasser, R. (2017). How primordial is the structure of comet 67P? Combined collisional and dynamical models suggest a late formation. *Astronomy and Astrophysics, 597*, A61.

Kaasalainen, M. (2001). Interpretation of lightcurves of precessing asteroids. *Astronomy and Astrophysics, 376*, 302–309.

Katayama, D. H., Huffman, R. E., & O'Bryan, C. L. (1973). Absorption and photoionization cross sections for $H_2O$ and $D_2O$ in the vacuum ultraviolet. *Journal of Chemical Physics, 59*, 4309.

Kawakita, H., Dello Russo, N., Vervack, R., Jr., Kobayashi, H., DiSanti, M. A., Opitom, C., Jehin, E., Weaver, H. A., Cochran, A. L., Harris, W. M., Bockelée-Morvan, D., Biver, N., Crovisier, J., McKay, A. J., Manfroid, J., & Gillon, M. (2014). Extremely organic-rich coma of Comet C/2010 G2 (Hill) during its outburst in 2012. *The Astrophysical Journal, 788*, 110.

Kawakita, H., Watanabe, J.-I., Furusho, R., Fuse, T., & Boice, D. C. (2005). Nuclear spin temperature and deuterium-to-hydrogen ratio of methane in Comet C/2001 Q4 (NEAT). *The Astrophysical Journal, 623*, L49–L52.

Keller, H. U. (1990). The nucleus. In W. F. Huebner (Ed.), *Physics and chemistry of comets*. Berlin: Springer.

Keller, H. U., Arpigny, C., Barbieri, C., Bonnet, R. M., Cazes, S., Coradini, M., Cosmovici, C. B., Delamere, W. A., Huebner, W. F., Hughes, D. W., Jamar, C., Malaise, D., Reitsema, H. J., Schmidt, H. U., Schmidt, W. K. H., Seige, P., Whipple, F. L., & Wilhelm, K. (1986). First Halley Multicolour Camera imaging results from Giotto. *Nature, 321*, 320.

Keller, H. U., et al. (1987). Comet P/Halley's nucleus and its activity. *Astronomy and Astrophysics, 187*, 807–823.

Keller, H. U., & Thomas, N. (1989). Evidence for near-surface breezes on Comet P/Halley. *Astronomy and Astrophysics, 226*, L9.

Keller, H. U., Knollenberg, J., & Markiewicz, W. J. (1994). Collimation of cometary dust jets and filaments. *Planetary and Space Science, 42*, 367.

Keller, H. U., Curdt, W., Kramm, J. R., & Thomas, N. (1995). In R. Reinhard, & B. Battrick (Eds.), *Images of the nucleus of comet Halley*. ESA SP-1127.

Keller, H. U., Britt, D., Buratti, B. J., & Thomas, N. (2004). In situ observations of cometary nuclei. In M. Festou, H. U. Keller, & H. A. Weaver (Eds.), *Comets II* (p. 211). Tucson: University of Arizona Press.

Keller, H. U., et al. (2007). OSIRIS. The scientific camera system onboard Rosetta. *Space Science Reviews, 128*, 433.

Keller, H. U., Mottola, S., Davidsson, B., Schröder, S. E., Skorov, Y., Kührt, E., Groussin, O., Pajola, M., Hviid, S. F., Preusker, F., Scholten, F., A'Hearn, M. F., Sierks, H., Barbieri, C., Lamy, P., Rodrigo, R., Koschny, D., Rickman, H., Barucci, M. A., Bertaux, J.-L., Bertini, I., Cremonese, G., Da Deppo, V., Debei, S., De Cecco, M., Fornasier, S., Fulle, M., Gutiérrez, P. J., Ip, W.-H., Jorda, L., Knollenberg, J., Kramm, J. R., Küppers, M., Lara, L. M., Lazzarin, M., Lopez Moreno, J. J., Marzari, F., Michalik, H., Naletto, G., Sabau, L., Thomas, N., Vincent, J.-B., Wenzel, K.-P., Agarwal, J., Güttler, C., Oklay, N., & Tubiana, C. (2015). Insolation, erosion, and morphology of comet 67P/Churyumov-Gerasimenko. *Astronomy and Astrophysics, 583*, A34.

Keller, H. U., Mottola, S., Hviid, S. F., Agarwal, J., Kührt, E., Skorov, Y., Otto, K., Vincent, J.-B., Oklay, N., Schröder, S. E., Davidsson, B., Pajola, M., Shi, X., Bodewits, D., Toth, I., Preusker, F., Scholten, F., Sierks, H., Barbieri, C., Lamy, P., Rodrigo, R., Koschny, D., Rickman, H., A'Hearn, M. F., Barucci, M. A., Bertaux, J.-L., Bertini, I., Cremonese, G., Da Deppo, V., Debei, S., De Cecco, M., Deller, J., Fornasier, S., Fulle, M., Groussin, O., Gutiérrez, P. J., Güttler, C., Hofmann, M., Ip, W.-H., Jorda, L., Knollenberg, J., Kramm, J. R., Küppers, M., Lara, L.-M., Lazzarin, M., Lopez-Moreno, J. J., Marzari, F., Naletto, G., Tubiana, C., & Thomas, N. (2017).

Seasonal mass transfer on the nucleus of comet 67P/Chuyumov-Gerasimenko. *Monthly Notices of the Royal Astronomical Society, 469,* S357.

Kelley, M. S. (2006). *The size, structure, and mineralogy of comet dust.* PhD Thesis, University of Minnesota.

Kelley, M. S., Reach, W. T., & Lien, D. J. (2008). The dust trail of Comet 67P/Churyumov Gerasimenko. *Icarus, 193,* 572–587.

Kieffer, H. H., Christensen, P. R., & Titus, T. N. (2006). CO2 jets formed by sublimation beneath translucent slab ice in Mars' seasonal south polar ice cap. *Nature, 442,* 793–796.

Kimura, H., Mann, I., Biesecker, D. A., & Jessberger, E. K. (2002). Dust grains in the comae and tails of sungrazing comets: Modeling of their mineralogical and morphological properties. *Icarus, 159,* 529–541. https://doi.org/10.1006/icar.2002.6940.

Kitamura, Y. (1986). Axisymmetric dusty gas jet in the inner coma of a comet. *Icarus, 66,* 241–257.

Kivelson, M. G. (1995). In M. G. Kivelson & C. T. Russell (Eds.), *Introduction to space physics.* Cambridge: Cambridge University Press.

Kivelson, M. G., & Russell, C. T. (1995). *Introduction to space physics.* Cambridge: Cambridge University Press.

Klein, P. P. (2012). On the ellipsoid and plane intersection equation. *Applied Mathematics, 3,* 1634–1640.

Klinger, J. (1981). Some consequences of a phase transition of water ice on the heat balance of comet nuclei. *Icarus, 47,* 320.

Knollenberg, J. (1994). *Modellrechnungenzur Staubverteilung in der inneren Koma vom Kometen unter spezieller Berücksichtigung der HMC Daten der Giotto-Mission.* PhD thesis, Universität Goettingen.

Knollenberg, J., Lin, Z. Y., Hviid, S. F., Oklay, N., Vincent, J.-B., Bodewits, D., Mottola, S., Pajola, M., Sierks, H., Barbieri, C., Lamy, P., Rodrigo, R., Koschny, D., Rickman, H., A'Hearn, M. F., Barucci, M. A., Bertaux, J. L., Bertini, I., Cremonese, G., Davidsson, B., Da Deppo, V., Debei, S., De Cecco, M., Fornasier, S., Fulle, M., Groussin, O., Gutierrez, P. J., Ip, W.-H., Jorda, L., Keller, H. U., Kührt, E., Kramm, J. R., Küppers, M., Lara, L. M., Lazzarin, M., Moreno, J. J. L., Marzari, F., Naletto, G., Thomas, N., Güttler, C., Preusker, F., Scholten, F., & Tubiana, C. (2016). A mini outburst from the nightside of comet 67P/Churyumov-Gerasimenko observed by the OSIRIS camera on Rosetta. *Astronomy and Astrophysics, 596,* A89.

Koenders, C., Glassmeier, K.-H., Richter, I., Motschmann, U., & Rubin, M. (2013). Revisiting cometary bow shock positions. *Planetary and Space Science, 87,* 85.

Koenders, C., Glassmeier, K.-H., Richter, I., Ranocha, H., & Motschmann, U. (2015). Dynamical features and spatial structures of the plasma interaction region of 67P/Churyumov-Gerasimenko and the solar wind. *Planetary and Space Science, 105,* 101.

Kok, J. F., Parteli, E. J. R., Michaels, T. I., & Karam, D. B. (2012). The physics of wind-blown sand and dust. *Reports on Progress in Physics, 75,* 106901.

Kokaly, R. F., Clark, R. N., Swayze, G. A., Livo, K. E., Hoefen, T. M., Pearson, N. C., Wise, R. A., Benzel, W. M., Lowers, H. A., Driscoll, R. L., & Klein, A. J. (2017). USGS Spectral Library Version 7: U.S. Geological Survey Data Series 1035, 61.

Kokotanekova, R., Snodgrass, C., Lacerda, P., Green, S. F., Lowry, S. C., Fernández, Y. R., Tubiana, C., Fitzsimmons, A., & Hsieh, H. H. (2017). Rotation of cometary nuclei: New light curves and an update of the ensemble properties of Jupiter-family comets. *Monthly Notices of the Royal Astronomical Society, 471,* 2974.

Kokotanekova, R., Snodgrass, C., Lacerda, P., Green, S. F., Nikolov, P., & Bonev, T. (2018). Implications of the small spin changes measured for large Jupiter-family comet nuclei. *Monthly Notices of the Royal Astronomical Society, 479,* 4665.

Kolokolova, L., Hanner, M. S., Levasseur-Regourd, A.-C., & Gustafson, B. Å. S. (2004). Physical properties of cometary dust from light scattering and thermal emission. In M. Festou, H. U. Keller, & H. A. Weaver (Eds.), *Comets II* (p. 577). Tucson: University of Arizona Press.

Kömle, N. I., Bauer, S. J., & Spohn, T. (Eds.). (1990). *Theoretical modelling of comet simulation experiments.* Verlag der Österreichischen Akademie der Wissenschaften.

Koppenwallner, G., Boettcher, R. D., Detleff, G., & Legge, H. (1986). Rocket exhaust plume flow into space. *ESA ESTEC ESA-SP, 265*, 83.

Kou, J., Wu, F., Lu, H., Xu, Y., & Song, F. (2009). The effective thermal conductivity of porous media based on statistical self-similarity. *Physics Letters A, 374*, 62.

Kouchi, A., Greenberg, J. M., Yamamoto, T., & Mukai, T. (1992). Extremely low thermal conductivity of amorphous ice – relevance to comet evolution. *Astrophysical Journal, 388*, L73.

Kouchi, A., Yamamoto, T., Kozasa, T., Kuroda, T., & Greenberg, J. M. (1994). Conditions for condensation and preservation of amorphous ice and crystallinity of astrophysical ices. *Astronomy and Astrophysics, 290*, 1009–1018.

Kowal, C., Dressler, A., Adams, R., Richstone, D., Boroson, T., Green, R., Marsden, B. G., & Aksnes, K. (1977). Slow-moving object Kowal. *International Astronomical Union Circular, 3134*, 6.

Kozasa, T., Blum, J., & Mukai, T. (1992). Optical properties of dust aggregates. I – Wavelength dependence. *Astronomy and Astrophysics, 263*, 423.

Krankowsky, D., Lammerzahl, P., Herrwerth, I., Woweries, J., Eberhardt, P., Dolder, U., Herrmann, U., Schulte, W., Berthelier, J. J., Illiano, J. M., Hodges, R. R., & Hoffman, J. H. (1986). In situ gas and ion measurements at comet Halley. *Nature, 321*, 326.

Krause, M., & Blum, J. (2004). Growth and form of planetary seedlings: Results from a sounding rocket microgravity aggregation experiment. *Physical Review Letters, 93*, 021103.

Krause, M., Blum, J., Skorov, Y. V., & Trieloff, M. (2011). Thermal conductivity measurements of porous dust aggregates: I. Technique, model and first results. *Icarus, 214*, 286–296.

Krishna Swamy, K. S. (2010). *Physics of comets* (3rd ed.). Singapore: World Scientific.

Kührt, E. K., & Keller, H. U. (1996). On the importance of dust in cometary nuclei. *Earth Moon and Planets, 72*, 79.

Kuiper, G. P. (1951). On the origin of the Solar System. *Proceedings of the National Academy of Science, 37*, 1.

Kurucz, R. L., Furenlid, I., Brault, J., & Testerman, L. (1984). *Solar flux atlas from 296 to 1300 nm.* Sunspot, NM: National Solar Observatory Atlas, National Solar Observatory.

Kurucz, R. L. (2005). New atlases for solar flux, irradiance, central intensity, and limb intensity. *Memorie della Societa Astronomica Italiana Supplementi, 8*, 189.

La Forgia, F., Bodewits, D., A'Hearn, M. F., Protopapa, S., Kelley, M. S. P., Sunshine, J., Feaga, L., & Farnham, T. (2017). Near-UV OH prompt emission in the innermost coma of 103P/Hartley 2. *The Astronomical Journal, 154*, 185.

Lagerros, J. S. V. (1998). Thermal physics of asteroids. IV. Thermal infrared beaming. *Astronomy and Astrophysics, 332*, 1123–1132.

Lammerzahl, P., Krankowsky, D., Hodges, R. R., Stubbemann, U., Woweries, J., Herrwerth, I., Berthelier, J. J., Illiano, J. M., Eberhardt, P., Dolder, U., Schulte, W., & Hoffman, J. H. (1987). Expansion velocity and temperatures of gas and ions measured in the coma of Comet p/Halley. *Astronomy and Astrophysics, 187*, 169.

Lamy, P. L., Malburet, P., Llebaria, A., & Koutchmy, S. (1989). Comet P/Halley at a heliocentric preperihelion distance of 2.6 AU – Jet activity and properties of the dust coma. *Astronomy and Astrophysics, 222*, 316.

Lamy, P., Biesecker, D. A., & Groussin, O. (2003). SOHO/LASCO observation of an outburst of Comet 2P/Encke at its 2000 perihelion passage. *Icarus, 163*, 142–149.

Lamy, P. L., Toth, I., Fernandez, Y. R., & Weaver, H. A. (2004). The sizes, shapes, albedos, and colors of cometary nuclei. In M. Festou, H. U. Keller, & H. A. Weaver (Eds.), *Comets II* (p. 223). Tucson: University of Arizona Press.

Lamy, P. L., Toth, I., Weaver, H. A., A'Hearn, M. F., & Jorda, L. (2009). Properties of the nuclei and comae of 13 ecliptic comets from Hubble Space Telescope snapshot observations. *Astronomy and Astrophysics, 508*, 1045.

Lamy, P. L., Toth, I., Weaver, H. A., A'Hearn, M. F., & Jorda, L. (2011). Properties of the nuclei and comae of 10 ecliptic comets from Hubble Space Telescope multi-orbit observations. *Monthly Notices of the Royal Astronomical Society, 412*, 1573.

Lancaster-Brown, P. (1985). *Halley and his comet* (pp. 76–77). Blandford Press. ISBN 0-7137-1447-6.

Landau, L. D., & Lifshitz, E. M. (1976). *Mechanics* (3rd ed.). Butterworth-Heinemann. ISBN 978-0-7506-2896-9.

Laporta, V., Cassidy, C. M., Tennyson, J., & Celiberto, R. (2012). Electron-impact resonant vibration excitation cross sections and rate coefficients for carbon monoxide. *Plasma Sources Science and Technology, 21*, 045005.

Larson, S. M., & Sekanina, Z. (1984). Coma morphology and dust-emission pattern of periodic Comet Halley. I – High-resolution images taken at Mount Wilson in 1910. *The Astronomical Journal, 89*, 571.

Lebofsky, L. A., Sykes, M. V., Tedesco, E. F., Veeder, G. J., Matson, D. L., Brown, R. H., Gradie, J. C., Feierberg, M. A., & Rudy, R. J. (1986). A refined "standard" thermal model for asteroids based on observations of 1 Ceres and 2 Pallas. *Icarus, 68*, 239.

Lee, S., von Allmen, P., Kamp, L., Gulkis, S., & Davidsson, B. (2011). Non-LTE radiative transfer for sub-millimeter water lines in Comet 67P/Churyumov-Gerasimenko. *Icarus, 215*, 721.

Lefort, A., Russell, P. S., Thomas, N., McEwen, A. S., Dundas, C. M., & Kirk, R. L. (2009). Observations of periglacial landforms in Utopia Planitia with the High Resolution Imaging Science Experiment (HiRISE). *Journal of Geophysical Research (Planets), 114*, E04005.

Le Roy, L., Briani, G., Briois, C., Cottin, H., Fray, N., Thirkell, L., Poulet, G., & Hilchenbach, M. (2012). On the prospective detection of polyoxymethylene in comet 67P/Churyumov-Gerasimenko with the COSIMA instrument onboard Rosetta. *Planetary and Space Science, 65*, 83.

Le Roy, L., Altwegg, K., Balsiger, H., Berthelier, J.-J., Bieler, A., Briois, C., Calmonte, U., Combi, M. R., De Keyser, J., Dhooghe, F., Fiethe, B., Fuselier, S. A., Gasc, S., Gombosi, T. I., Hässig, M., Jäckel, A., Rubin, M., & Tzou, C.-Y. (2015). Inventory of the volatiles on comet 67P/Churyumov-Gerasimenko from Rosetta/ROSINA. *Astronomy and Astrophysics, 583*, A1.

Levasseur-Regourd, A. C., Hadamcik, E., Lasue, J., Renard, J. B., & Worms, J. C. (2004). Physical properties of cometary dust. The Case of 67P/Churyumov-Gerasimenko. In L. Colangeli, E. M. Epifani, & P. Palumbo (Eds.), *The new Rosetta targets. Observations, simulations and instrument performances* (Vol. 311, p. 111).

Levison, H. F. (1996). Comet taxonomy. In T. W. Rettig & J. M. Hahn (Eds.), *Completing the inventory of the solar system. Astronomical Society of the Pacific Conference Proceedings* (Vol. 107, p. 173).

Levison, H. F., & Duncan, M. J. (1997). From the Kuiper Belt to Jupiter-Family comets: The spatial distribution of ecliptic comets. *Icarus, 127*, 13.

Levison, H. F., Duncan, M. J., Zahnle, K., Holman, M., & Dones, L. (2000). Planetary impact rates from ecliptic comets. *Icarus, 143*, 415–420.

Lhotka, C., Bourdin, P., & Narita, Y. (2016). Charged dust grain dynamics subject to solar wind, Poynting-Robertson drag, and the interplanetary magnetic field. *The Astrophysical Journal, 828*, 10.

Li, J.-Y., A'Hearn, M. F., McFadden, L. A., & Belton, M. J. S. (2007). Photometric analysis and disk-resolved thermal modeling of Comet 19P/Borrelly from Deep Space 1 data. *Icarus, 188*, 195.

Li, J.-Y., A'Hearn, M. F., Farnham, T. L., & McFadden, L. A. (2009). Photometric analysis of the nucleus of Comet 81P/Wild 2 from Stardust images. *Icarus, 204*, 209.

Li, J.-Y., Besse, S., A'Hearn, M. F., Belton, M. J. S., Bodewits, D., Farnham, T. L., Klaasen, K. P., Lisse, C. M., Meech, K. J., Sunshine, J. M., & Thomas, P. C. (2013a). Photometric properties of the nucleus of Comet 103P/Hartley 2. *Icarus, 222*, 559.

Li, J.-Y., A'Hearn, M. F., Belton, M. J. S., Farnham, T. L., Klaasen, K. P., Sunshine, J. M., Thomas, P. C., & Veverka, J. (2013b). Photometry of the nucleus of Comet 9P/Tempel 1 from Stardust-NExT flyby and the implications. *Icarus, 222*, 467.

Li, J.-Y., Helfenstein, P., Buratti, B., Takir, D., & Clark, B. E. (2015). Asteroid photometry. In P. Michel, F. E. DeMeo, & W. F. Bottke (Eds.), *Asteroids IV* (pp. 129–150). Tucson: University of Arizona Press. ISBN: 978-0-816-53213-1.

Liao, Y. (2017). *Global explorations of inner neutral gas coma of Comet 67P/Churyumov-Gerasimenko with DSMC approach*. PhD thesis, University of Bern.

Liao, Y., Su, C. C., Marschall, R., Wu, J. S., Rubin, M., Lai, I. L., Ip, W. H., Keller, H. U., Knollenberg, J., Kührt, E., Skorov, Y. V., & Thomas, N. (2016). 3D direct simulation Monte Carlo modelling of the inner gas coma of Comet 67P/Churyumov-Gerasimenko: A parameter study. *Earth Moon and Planets, 117*, 41–64.

Liao, Y., Marschall, R., Su, C. C., Wu, J. S., Lai, I. L., Pinzon, O., & Thomas, N. (2018). Water vapor deposition from the inner gas coma onto the nucleus of Comet 67P/Churyumov-Gerasimenko. *Planetary and Space Science, 157*, 1–9.

Licandro, J., Popescu, M., de Leon, J., Morate, D., Vaduvescu, O., De Pra, M., & Ali-Laoga, V. (2018). The visible and near-infrared spectra of asteroids in cometary orbits. *Astronomy and Astrophysics, 618*, A170.

Lien, D. J. (1990). Dust in comets. I. Thermal properties of homogeneous and heterogeneous grains. *The Astrophysical Journal, 355*, 680.

Lin, Z. Y., Lai, I. L., Su, C. C., Ip, W. H., Lee, J. C., Wu, J. S., Vincent, J. B., La Forgia, F., Sierks, H., Barbieri, C., Lamy, P. L., Rodrigo, R., Koschny, D., Rickman, H., Keller, H. U., Agarwal, J., A'Hearn, M. F., Barucci, M. A., Bertaux, J. L., Bertini, I., Bodewits, D., Cremonese, G., Da Deppo, V., Davidsson, B., Debei, S., De Cecco, M., Fornasier, S., Fulle, M., Groussin, O., Gutierrez, P. J., Guttler, C., Hviid, S. F., Jorda, L., Knollenberg, J., Kovacs, G., Kramm, J. R., Kührt, E., Kuppers, M., Lara, L. M., Lazzarin, M., Lopez-Moreno, J. J., Lowry, S., Marzari, F., Michalik, H., Mottola, S., Naletto, G., Oklay, N., Pajola, M., Rozek, A., Thomas, N., & Tubiana, C. (2016). Observations and analysis of a curved jet in the coma of comet 67P/Churyumov-Gerasimenko. *Astronomy and Astrophysics, 588*, L3.

Lis, D. C., Bockelée-Morvan, D., Güsten, R., Biver, N., Stutzki, J., Delorme, Y., Durán, C., Wiesemeyer, H., & Okada, Y. (2019). Terrestrial deuterium-to-hydrogen ratio in water in hyperactive comets. *Astronomy and Astrophysics, 625*, L5.

Lisse, C. M., Dennerl, K., Englhauser, J., Harden, M., Marshall, F. E., Mumma, M. J., Petre, R., Pye, J. P., Ricketts, M. J., Schmitt, J., Trumper, J., & West, R. G. (1996). Discovery of X-ray and extreme ultraviolet emission from Comet C/Hyakutake 1996 B2. *Science, 274*, 205–209.

Lisse, C. M., Cravens, T. E., & Dennerl, K. (2004). X-ray and extreme ultraviolet emission from comets. In M. Festou, H. U. Keller, & H. A. Weaver (Eds.), *Comets II* (p. 631). Tucson: University of Arizona Press.

Lisse, C. M., VanCleve, J., Adams, A. C., A'Hearn, M. F., Fernandez, Y. R., Farnham, T. L., Armus, L., Grillmair, C. J., Ingalls, J., Belton, M. J. S., Groussin, O., McFadden, L. A., Meech, K. J., Schultz, P. H., Clark, B. C., Feaga, L. M., & Sunshine, J. M. (2006). Spitzer spectral observations of the deep impact ejecta. *Science, 313*, 635–640.

Lisse, C. M., Christian, D. J., Wolk, S. J., Dennerl, K., Bodewits, D., Combi, M. R., Lepri, S. T., Zurbuchen, T. H., Li, J. Y., Dello-Russo, N., Belton, M. J. S., & Knight, M. M. (2013). Chandra ACIS-S imaging spectroscopy of anomalously faint X-ray emission from Comet 103P/Hartley 2 during the EPOXI encounter. *Icarus, 222*, 752.

Litvak, M. M., & Kuiper, E. N. R. (1982). Cometary NH – Ultraviolet and submillimeter emission. *The Astrophysical Journal, 253*, 622.

Loerting, T., Fuentes-Landete, V., Handle, P. H., Seidl, M., Amann-Winkel, K., Gainaru, C., & Böhmer, R. (2015). The glass transition in high-density amorphous ice. *Journal of Non Crystalline Solids, 407*, 423–430.

Lopez-Puertas, M., & Taylor, F. W. (2001). *Non-LTE radiative transfer in the atmosphere*. Singapore: World Scientific.

Lorek, S., Gundlach B., Lacerda P., & Blum J., (2016), Comet formation in collapsing pebble clouds. What cometary bulk density implies for the cloud mass and dust-to-ice ratio, Astronomy and Astrophysics, 587, A128

Lucchetti, A., et al. (2020). The rocky-like behavior of cometary landslides on 67P/Churyumov-Gerasimenko. *Geophysical Research Letters* (in press).

Lummerzheim, D., & Lilensten, J. (1994). Electron transport and energy degradation in the ionosphere: Evaluation of the numerical solution, comparison with laboratory experiments and auroral observations. *Annales Geophysicae, 12*, 1039.

Lupu, R. E., Feldman, P. D., Weaver, H. A., & Tozzi, G.-P. (2007). The fourth positive system of carbon monoxide in the Hubble Space telescope spectra of comets. *The Astrophysical Journal, 670*, 1473.

Luspay-Kuti, A., Hässig, M., Fuselier, S. A., Mandt, K. E., Altwegg, K., Balsiger, H., Gasc, S., Jäckel, A., Le Roy, L., Rubin, M., Tzou, C.-Y., Wurz, P., Mousis, O., Dhooghe, F., Berthelier, J. J., Fiethe, B., Gombosi, T. I., & Mall, U. (2015). Composition-dependent outgassing of comet 67P/Churyumov-Gerasimenko from ROSINA/DFMS. Implications for nucleus heterogeneity? *Astronomy and Astrophysics, 583*, A4.

Luspay-Kuti, A., Mousis, O., Hässig, M., Fuselier, S. A., Lunine, J. I., Marty, B., Mandt, K. E., Wurz, P., & Rubin, M. (2016). The presence of clathrates in comet 67P/Churyumov-Gerasimenko. *Science Advances, 2*, 1501781.

Luu, J. X., & Jewitt, D. C. (1990). Cometary activity in 2060 Chiron. *The Astronomical Journal, 100*, 913.

Lyttleton, R. A. (1951). On the structure of comets and the formation of tails. *Monthly Notices of the Royal Astronomical Society, 111*, 268.

Maeda, K., Wall, M. L., & Carr, L. D. (2015). Hyperfine structure of the hydroxyl free radical (OH) in electric and magnetic fields. *New Journal of Physics, 17*, 045014.

Mahaffy, P. R., Donahue, T. M., Atreya, S. K., Owen, T., & Niemann, H. B. (1998). Galileo probe measurements of D/H and 3He/ 4He in Jupiter's atmosphere. *Space Science Reviews, 84*, 251–263.

Malin, M. C., Caplinger, M. A., & Davis, S. D. (2001). Observational evidence for an active surface reservoir of solid carbon dioxide on Mars. *Science, 294*, 2146.

Maloof, A. C., Kellogg, J. B., & Anders, A. M. (2002). Neoproterozoic sand wedges: Crack formation in frozen soils under diurnal forcing during a snowball Earth. *Earth and Planetary Science Letters, 204*, 1–15.

Mandt, K. E., Eriksson, A., Edberg, N. J. T., Koenders, C., Broiles, T., Fuselier, S. A., Henri, P., Nemeth, Z., Alho, M., Biver, N., Beth, A., Burch, J., Carr, C., Chae, K., Coates, A. J., Cupido, E., Galand, M., Glassmeier, K.-H., Goetz, C., Goldstein, R., Hansen, K. C., Haiducek, J., Kallio, E., Lebreton, J.-P., Luspay-Kuti, A., Mokashi, P., Nilsson, H., Opitz, A., Richter, I., Samara, M., Szego, K., Tzou, C.-Y., Volwerk, M., Wedlund, C. S., & Stenberg Wieser, G. (2016). RPC observation of the development and evolution of plasma interaction boundaries at 67P/Churyumov-Gerasimenko. *Monthly Notices of the Royal Astronomical Society, 462*, S9.

Manfroid, J., Jehin, E., Hutsemékers, D., Cochran, A., Zucconi, J.-M., Arpigny, C., Schulz, R., Stüwe, J. A., & Ilyin, I. (2009). The CN isotopic ratios in comets. *Astronomy and Astrophysics, 503*, 613.

Mannel, T., Bentley, M. S., Schmied, R., Jeszenszky, H., Levasseur-Regourd, A. C., Romstedt, J., & Torkar, K. (2016). Fractal cometary dust – a window into the early Solar system. *Monthly Notices of the Royal Astronomical Society, 462*, S304.

Mannel, T., Bentley, M. S., Boakes, P. D., Jeszenszky, H., Ehrenfreund, P., Engrand, C., Koeberl, C., Levasseur-Regourd, A. C., Romstedt, J., Schmied, R., Torkar, K., & Weber, I. (2019). Dust of comet 67P/Churyumov-Gerasimenko collected by Rosetta/MIDAS: Classification and extension to the nanometer scale. *Astronomy and Astrophysics, 630*, A26.

Manzini, F., Oldani, V., Hirabayashi, M., Behrend, R., Crippa, R., Ochner, P., Pina, J. P. N., Haver, R., Baransky, A., Bryssinck, E., Dan, A., De Queiroz, J., Frappa, E., & Lavayssiere, M. (2016). Comet C/2011 J2 (LINEAR) nucleus splitting: Dynamical and structural analysis. *Planetary and Space Science, 126*, 8–23.

Maquet, L., Colas, F., Jorda, L., & Crovisier, J. (2012). CONGO, model of cometary non-gravitational forces combining astrometric and production rate data. Application to comet 19P/Borrelly. *Astronomy and Astrophysics, 548*, A81.

Marboeuf, U., Schmitt, B., Petit, J.-M., Mousis, O., & Fray, N. (2012). A cometary nucleus model taking into account all phase changes of water ice: Amorphous, crystalline, and clathrate. *Astronomy and Astrophysics, 542*, A82.

Margot, J.-L., Pravec, P., Taylor, P., Carry, B., & Jacobson, S. (2015). Asteroid systems: Binaries, triples, and pairs. In P. Michel, F. E. DeMeo, & W. F. Bottke (Eds.), *Asteroids IV* (pp. 129–150, 355). Tucson: University of Arizona Press. ISBN: 978-0-816-53213-1.

Markel, V. A. (2016). Introduction to the Maxwell Garnett approximation: Tutorial. *Journal of the Optical Society of America A, 33*, 1244.

Markiewicz, W. J., Sablotny, R. M., Keller, H. U., Thomas, N., Titov, D., & Smith, P. H. (1999). Optical properties of the Martian aerosols as derived from Imager for Mars Pathfinder midday sky brightness data. *Journal of Geophysical Research, 104*, 9009.

Marschall, R. (2017). *Inner gas and dust comae of comets*. PhD thesis, University of Bern.

Marschall, R., Su, C. C., Liao, Y., Thomas, N., Altwegg, K., Sierks, H., Ip, W.-H., Keller, H. U., Knollenberg, J., Kührt, E., Lai, I. L., Rubin, M., Skorov, Y., Wu, J. S., Jorda, L., Preusker, F., Scholten, F., Gracia-Berna, A., Gicquel, A., Naletto, G., Shi, X., & Vincent, J.-B. (2016). Modelling observations of the inner gas and dust coma of comet 67P/Churyumov-Gerasimenko using ROSINA/COPS and OSIRIS data: First results. *Astronomy and Astrophysics, 589*, A90.

Marschall, R., Mottola, S., Su, C. C., Liao, Y., Rubin, M., Wu, J. S., Thomas, N., Altwegg, K., Sierks, H., Ip, W.-H., Keller, H. U., Knollenberg, J., Kührt, E., Lai, I. L., Skorov, Y., Jorda, L., Preusker, F., Scholten, F., Vincent, J.-B., Osiris Team, & Rosina Team. (2017). Cliffs versus plains: Can ROSINA/COPS and OSIRIS data of comet 67P/Churyumov-Gerasimenko in autumn 2014 constrain inhomogeneous outgassing? *Astronomy and Astrophysics, 605*, A112.

Marschall, R., Rezac, L., Kappel, D., Su, C. C., Gerig, S.-B., Rubin, M., Pinzón-Rodríguez, O., Marshall, D., Liao, Y., Herny, C., Arnold, G., Christou, C., Dadzie, S. K., Groussin, O., Hartogh, P., Jorda, L., Kührt, E., Mottola, S., Mousis, O., Preusker, F., Scholten, F., Theologou, P., Wu, J.-S., Altwegg, K., Rodrigo, R., & Thomas, N. (2019). A comparison of multiple Rosetta data sets and 3D model calculations of 67P/Churyumov-Gerasimenko coma around equinox (May 2015). *Icarus, 328*, 104.

Marschall, R., Liao, Y., Thomas, N., & Wu, J.-S. (2020). On the limitations gas models have in linking cometary coma data to the surface emission distributions – a Rosetta perspective on data collected on comet 67P. *Icarus, 346*, article id. 113742.

Marsden, B. G., Sekanina, Z., & Yeomans, D. K. (1973). Comets and nongravitational forces, V. *The Astronomical Journal, 78*, 211.

Marshall, D. W., Hartogh, P., Rezac, L., von Allmen, P., Biver, N., Bockelée-Morvan, D., Crovisier, J., Encrenaz, P., Gulkis, S., Hofstadter, M., Ip, W.-H., Jarchow, C., Lee, S., & Lellouch, E. (2017a). Spatially resolved evolution of the local H2O production rates of comet 67P/Churyumov-Gerasimenko from the MIRO instrument on Rosetta. *Astronomy and Astrophysics, 603*, A87.

Marshall, D. W., Hartogh, P., Rezac, L., von Allmen, P., Biver, N., Bockelée-Morvan, D., Crovisier, J., Encrenaz, P., Gulkis, S., Hofstadter, M., Ip, W.-H., Jarchow, C., Lee, S., & Lellouch, E. (2017b). VizieR Online Data Catalog: Local production rates of 67P/CG from MIRO (Marshall+, 2017). *VizieR Online Data Catalog, 360*.

Marshall, D., Groussin, O., Vincent, J.-B., Brouet, Y., Kappel, D., Arnold, G., Capria, M. T., Filacchione, G., Hartogh, P., Hofstadter, M., Ip, W.-H., Jorda, L., Kührt, E., Lellouch, E., Mottola, S., Rezac, L., Rodrigo, R., Rodionov, S., Schloerb, P., & Thomas, N. (2018). Thermal inertia and roughness of the nucleus of comet 67P/Churyumov-Gerasimenko from MIRO and VIRTIS observations. *Astronomy and Astrophysics, 616*, A122.

Massironi, M., et al. (2015). Two independent and primitive envelopes of the bilobate nucleus of comet 67P. *Nature, 526*, 402.

Mastrapa, R. M., Bernstein, M. P., Sandford, S. A., Roush, T. L., Cruikshank, D. P., & Dalle Ore, C. M. (2008). Optical constants of amorphous and crystalline $H_2O$-ice in the near infrared from 1.1 to 2.6 μm. *Icarus, 197*, 307–320.

Mastrapa, R. M. E., Grundy, W. M., & Gudipati, M. S. (2013). Amorphous and crystalline $H_2O$-ice. *Astrophysics and Space Science Library, 356*, 371.

Maté, B., Tanarro, I., Escribano, R., Moreno, M. A., & Herrero, V. J. (2015). Stability of extraterrestrial glycine under energetic particle radiation estimated from 2 keV electron bombardment experiments. *The Astrophysical Journal, 806*, 151.

Matson, D. L., & Brown, R. H. (1989). Solid-state greenhouses and their implications for icy satellites. *Icarus, 77*, 67–81.

Maxwell Garnett, J. C. (1904). Colours in metal glasses and in metallic films. *Philosophical Transactions of the Royal Society of London Series A, 203*, 385.

Mazzotta Epifani, E., Perna, D., Dotto, E., Palumbo, P., Dall'Ora, M., Micheli, M., Ieva, S., & Perozzi, E. (2017). Nucleus of the active Centaur C/2011 P2 (PANSTARRS). *Astronomy and Astrophysics, 597*, A59.

McCoy, R. P., Meier, R. R., Keller, H. U., Opal, C. B., & Carruthers, G. R. (1992). The hydrogen coma of Comet P/Halley observed in Lyman-alpha using sounding rockets. *Astronomy and Astrophysics, 258*, 555–565.

McDonnell, J. A. M., Lamy, P. L., & Pankiewicz, G. S. (1991). Physical properties of cometary dust. *IAU Colloq. 116: Comets in the Post-Halley Era, 167*, 1043.

McEwen, A. S. (1991). Photometric functions for photoclinometry and other applications. *Icarus, 92*, 298.

McKay, A. J., Cochran, A. L., DiSanti, M. A., Dello Russo, N., Weaver, H., Vervack, R. J., Harris, W. M., & Kawakita, H. (2018). Evolution of $H_2O$ production in comet C/2012 S1 (ISON) as inferred from forbidden oxygen and OH emission. *Icarus, 309*, 1–12.

McKenzie, M. L., Cravens, T. E., & Ye, G. (1994). Theoretical calculations of ion acceleration in the vicinity of comet Giacobini-Zinner. *Journal of Geophysical Research, 99*, 6585.

Meador, W. E., & Weaver, W. R. (1975). A photometric function for diffuse reflection, by particulate materials. NASA-TN-D-7903.

Meech, K. J., Hainaut, O. R., & Marsden, B. G. (2004). Comet nucleus size distributions from HST and Keck telescopes. *Icarus, 170*, 463.

Meech, K. J., Weryk, R., Micheli, M., Kleyna, J. T., Hainaut, O. R., Jedicke, R., Wainscoat, R. J., Chambers, K. C., Keane, J. V., Petric, A., Denneau, L., Magnier, E., Berger, T., Huber, M. E., Flewelling, H., Waters, C., Schunova-Lilly, E., & Chastel, S. (2017). A brief visit from a red and extremely elongated interstellar asteroid. *Nature, 552*, 378.

Melosh, H. J., & Schenk, P. (1993). Split comets and the origin of crater chains on Ganymede and Callisto. *Nature, 365*, 731.

Mendis, D. A., & Brin, G. D. (1977). Monochromatic brightness variations of comets.II – Core-mantle model. *Moon, 17*, 359–372.

Mendis, D. A., & Ip, W.-H. (1976). The neutral atmospheres of comets. *Astrophysics and Space Science, 39*, 335.

Mendis, D. A., & Horányi, M. (2013). Dusty plasma effects in comets: Expectations for Rosetta. *Reviews of Geophysics, 51*, 53.

Meng, Z., Yang, P., Kattawar, G. W., Bi, L., Liou, K. N., & Laszlo, I. (2010). Single-scattering properties of tri-axial ellipsoidal mineral dust aerosols: A database for application to radiative transfer calculations. *Journal of Aerosol Science, 41*, 501.

Merouane, S., Stenzel, O., Hilchenbach, M., Schulz, R., Altobelli, N., Fischer, H., Hornung, K., Kissel, J., Langevin, Y., Mellado, E., Ryno, J., & Zaprudin, B. (2017). Evolution of the physical properties of dust and cometary dust activity from 67P/Churyumov-Gerasimenko measured in situ by Rosetta/COSIMA. *Monthly Notices of the Royal Astronomical Society, 469*, S459.

Migliorini, A., Piccioni, G., Capaccioni, F., Filacchione, G., Bockelée-Morvan, D., Erard, S., Leyrat, C., Combi, M. R., Fougere, N., Crovisier, J., Taylor, F. W., De Sanctis, M. C., Capria, M. T., Grassi, D., Rinaldi, G., Tozzi, G. P., & Fink, U. (2016). Water and carbon dioxide distribution in the 67P/Churyumov-Gerasimenko coma from VIRTIS-M infrared observations. *Astronomy and Astrophysics, 589,* A45.

Mishchenko, M. I., Travis, L. D., & Lacis, A. A. (2002). *Scattering, absorption, and emission of light by small particles.* Cambridge: Cambridge University Press.

Mitchell, D. L., Lin, R. P., Anderson, K. A., Carlson, C. W., Curtis, D. W., Korth, A., Reme, H., Sauvaud, J. A., D'Uston, C., & Mendis, D. A. (1989). Complex organic ions in the atmosphere of Comet Halley. *Advances in Space Research, 9,* 35.

Möhlmann, D. (1994). Surface regolith and environment of comets. *Planetary and Space Science, 42,* 933–937.

Möhlmann, D., Seidensticker, K. J., Fischer, H.-H., Faber, C., Flandes, A., Knapmeyer, M., Krüger, H., Roll, R., Scholten, F., Thiel, K., & Arnold, W. (2018). Compressive strength and elastic modulus at Agilkia on comet 67P/Churyumov-Gerasimenko derived from the SESAME/CASSE touchdown signals. *Icarus, 303,* 251–264.

Mohamad, A. A. (2011). *Lattice Boltzmann method.* London: Springer. https://doi.org/10.1007/978-0-85729-455-5_2.

Mommert, M., Hora, J. L., Harris, A. W., Reach, W. T., Emery, J. P., Thomas, C. A., Mueller, M., Cruikshank, D. P., Trilling, D. E., Delbo, M., & Smith, H. A. (2014). The discovery of cometary activity in Near-Earth Asteroid (3552) Don Quixote. *The Astrophysical Journal, 781,* 25.

Mommert, M., Hora, J. L., Trilling, D. E., Biver, N., Wierzchos, K., Harrington Pinto, O., Agarwal, J., Kim, Y., McNeill, A., Womack, M., Knight, M. M., Polishook, D., Moskovitz, N., Kelley, M. S. P., & Smith, H. A. (2020). Recurrent cometary activity in Near-Earth Object (3552) Don Quixote. *The Planetary Science Journal, 1,* 12.

Moore, M. H., & Hudson, R. L. (1992). Far-infrared spectral studies of phase changes in water ice induced by proton irradiation. *The Astrophysical Journal, 401,* 353.

Moosmüller, H., & Arnott, W. P. (2009). Particle optics in the Rayleigh regime. *Journal of the Air & Waste Management Association, 59*(9), 1028–1031.

Moosmüller, H., & Sorensen, C. M. (2018). Small and large particle limits of single scattering albedo for homogeneous, spherical particles. *Journal of Quantitative Spectroscopy and Radiative Transfer, 204,* 250.

Morbidelli, A. (2010). Comets and their reservoirs: Current dynamics and primordial evolution. In K. Altwegg, W. Benz, & N. Thomas (Eds.), *Trans-Neptunian objects and comets.* Proceedings of 35th Saas-Fee Conference.

Morbidelli, A., & Brown, M. E. (2004). The Kuiper Belt and the primordial evolution of the solar system. In M. Festou, H. U. Keller, & H. A. Weaver (Eds.), *Comets II* (p. 175). Tucson: University of Arizona Press.

Mousis, O., Guilbert-Lepoutre, A., Brugger, B., Jorda, L., Kargel, J. S., Bouquet, A., Auger, A.-T., Lamy, P., Vernazza, P., Thomas, N., & Sierks, H. (2015). Pits formation from volatile outgassing on 67P/Churyumov-Gerasimenko. *The Astrophysical Journal, 814,* L5.

Mousis, O., Ronnet, T., Brugger, B., Ozgurel, O., Pauzat, F., Ellinger, Y., Maggiolo, R., Wurz, P., Vernazza, P., Lunine, J. I., Luspay-Kuti, A., Mandt, K. E., Altwegg, K., Bieler, A., Markovits, A., & Rubin, M. (2016). Origin of molecular oxygen in Comet 67P/Churyumov-Gerasimenko. *The Astrophysical Journal, 823,* L41.

Mukai, T., Ishimoto, H., Kozasa, T., Blum, J., & Greenberg, J. M. (1992). Radiation pressure forces of fluffy porous grains. *Astronomy and Astrophysics, 262,* 315.

Mullen, P. D., Cumbee, R. S., Lyons, D., Gu, L., Kaastra, J., Shelton, R. L., & Stancil, P. C. (2017). Line ratios for solar wind charge exchange with comets. *The Astrophysical Journal, 844,* 7.

Mumma, M. J., Krasnopolsky, V. A., & Abbott, M. J. (1997). Soft X-rays from four comets observed with EUVE. *The Astrophysical Journal, 491,* L125–L128.

Muñoz, O., Moreno, F., Gómez-Martín, J. C., Vargas-Martín, F., Guirado, D., Ramos, J. L., Bustamante, I., Bertini, I., Frattin, E., Markannen, J., Tubiana, C., Fulle, M., Güttler, C., Sierks, H., Rotundi, A., Corte, V. D., Ivanovski, S., Zakharov, V. V., Bockelée-Morvan, D., Blum, J., Merouane, S., Levasseur-Regourd, A. C., Kolokolova, L., Jardiel, T., & Caballero, A. C. (2020).

Experimental phase function and degree of linear polarization curves of millimeter-sized cosmic dust analogs. *The Astrophysical Journal Supplement Series, 247*, 19.

Murray, C. D., & Dermott, S. F. (1999). *Solar System dynamics*. Cambridge: Cambridge University Press.

Najib, D., Nagy, A. F., Tóth, G., & Ma, Y. (2011). Three-dimensional, multifluid, high spatial resolution MHD model studies of the solar wind interaction with Mars. *Journal of Geophysical Research, 116*, A05204. https://doi.org/10.1029/2010JA016272.

Narahari Rao, K. (1949). Structure of the Cameron bands of carbon monoxide. *The Astrophysical Journal, 110*, 304.

Neishtadt, A. I., Scheeres, D. J., Sidorenko, V. V., & Vasiliev, A. A. (2002). Evolution of comet nucleus rotation. *Icarus, 157*, 205–218.

Nelson, R. M., Smythe, W. D., Hapke, B. W., & Hale, A. S. (2002). *Planetary and Space Science, 50*, 849–856.

Nemeth, Z. (2020). The dynamics of the magnetic-field-free cavity around comets. *The Astrophysical Journal, 891*, 112.

Neubauer, F. M., Glassmeier, K. H., Pohl, M., Raeder, J., Acuna, M. H., Burlaga, L. F., Ness, N. F., Musmann, G., Mariani, F., Wallis, M. K., Ungstrup, E., & Schmidt, H. U. (1986). First results from the Giotto magnetometer experiment at comet Halley. *Nature, 321*, 352.

Neugebauer, M., Goldstein, B. E., Goldstein, R., Lazarus, A. J., Altwegg, K., & Balsiger, H. (1987). The pick-up of cometary protons by the solar wind. *Astronomy and Astrophysics, 187*, 21–24.

Newburn, R. L., & Yeomans, D. K. (1982). Halley's Comet. *Annual Review of Earth and Planetary Sciences, 10*, 297.

Niedner, M. B., Jr., & Brandt, J. C. (1978). Interplanetary gas: XXIII. Plasma tail disconnection events in comets: Evidence for magnetic field line reconnection at interplanetary sector boundaries? *Astrophysical Journal, 223*, 655.

Nishioka, K. (1998). Finite lifetime fragment model 2 for synchronic band formation in dust tails of comets. *Icarus, 134*(1), 24–34.

Nordheim, T. A., Jones, G. H., Halekas, J. S., Roussos, E., & Coates, A. J. (2015). Surface charging and electrostatic dust acceleration at the nucleus of comet 67P during periods of low activity. *Planetary and Space Science, 119*, 24. https://doi.org/10.1016/j.pss.2015.08.008.

Oba, Y., Takano, Y., Naraoka, H., Kouchi, A., & Watanabe, N. (2017). Deuterium fractionation upon the formation of hexamethylenetetramines through photochemical reactions of interstellar ice analogs containing deuterated methanol isotopologues. *The Astrophysical Journal, 849*, 122.

O'Dell, C. R., Robinson, R. R., Krishna Swamy, K. S., McCarthy, P. J., & Spinrad, H. (1988). C 2 in Comet Halley: Evidence for its being third generation and resolution of the vibrational population discrepancy. *The Astrophysical Journal, 334*, 476.

Odelstad, E., Eriksson, A. I., Johansson, F. L., Vigren, E., Henri, P., Gilet, N., Heritier, K. L., Vallieres, X., Rubin, M., & Andre, M. (2018). Ion velocity and electron temperature inside and around the diamagnetic cavity of Comet 67P. *Journal of Geophysical Research (Space Physics), 123*, 5870.

Ohishi, M., Suzuki, T., Hirota, T., Saito, M., & Kaifu, N. (2019). Detection of a new methylamine ($CH_3NH_2$) source: Candidate for future glycine surveys. *Publications of the Astronomical Society of Japan, 71*, 86.

Ohtsuka, K., Nakano, S., & Yoshikawa, M. (2003). On the association among periodic comet 96P/Machholz, Arietids, the Marsden comet group, and the Kracht comet group. *Publications of the Astronomical Society of Japan, 55*, 321–324.

O'Leary, M. H. (1988). Carbon isotopes in photosynthesis. *BioScience, 38*, 328–336.

Oort, J. H. (1950). The structure of the cloud of comets surrounding the Solar System and a hypothesis concerning its origin. *Bulletin of the Astronomical Institutes of the Netherlands, 11*, 91.

Ootsubo, T., Kawakita, H., Hamada, S., Kobayashi, H., Yamaguchi, M., Usui, F., Nakagawa, T., Ueno, M., Ishiguro, M., Sekiguchi, T., Watanabe, J., Sakon, I., Shimonishi, T., & Onaka,

T. (2012). AKARI near-infrared spectroscopic survey for $CO_2$ in 18 comets. *The Astrophysical Journal, 752*, 15.

Osterbrock, D. E. (1989). *Astrophysics of gaseous nebulae and active galactic nuclei*. University Science Books. ISBN 0-935702-22-9.

Ott, T., et al. (2017). Dust mass distribution around comet 67P/Churyumov-Gerasimenko determined via parallax measurements using Rosetta's OSIRIS cameras. *Monthly Notices of the Royal Astronomical Society, 469*, S276–S284.

Pachucki, K., & Komasa, J. (2008). Ortho-para transition in molecular hydrogen. *Physical Review A, 77*, 030501.

Pähtz, T., & Durán, O. (2016). How common are aeolian processes on planetary bodies with very thin atmospheres? *EGU General Assembly Conference Abstracts, 18*, EPSC2016-2280.

Pätzold, M., Andert, T., Hahn, M., Asmar, S. W., Barriot, J.-P., Bird, M. K., Häusler, B., Peter, K., Tellmann, S., Grün, E., Weissman, P. R., Sierks, H., Jorda, L., Gaskell, R., Preusker, F., & Scholten, F. (2016). A homogeneous nucleus for comet 67P/Churyumov-Gerasimenko from its gravity field. *Nature, 530*, 63–65.

Pätzold, M., Andert, T. P., Hahn, M., Barriot, J.-P., Asmar, S. W., Häusler, B., Bird, M. K., Tellmann, S., Oschlisniok, J., & Peter, K. (2019). The nucleus of comet 67P/Churyumov-Gerasimenko – Part I: The global view – nucleus mass, mass-loss, porosity, and implications. *Monthly Notices of the Royal Astronomical Society, 483*, 2337–2346.

Paganini, L., Mumma, M. J., Villanueva, G. L., Keane, J. V., Blake, G. A., Bonev, B. P., DiSanti, M. A., Gibb, E. L., & Meech, K. J. (2014a). C/2013 R1 (Lovejoy) at IR wavelengths and the variability of CO abundances among Oort Cloud comets. *The Astrophysical Journal, 791*, 122.

Paganini, L., DiSanti, M. A., Mumma, M. J., Villanueva, G. L., Bonev, B. P., Keane, J. V., Gibb, E. L., Boehnhardt, H., & Meech, K. J. (2014b). The unexpectedly bright Comet C/2012 F6 (Lemmon) unveiled at near-infrared wavelengths. *The Astronomical Journal, 147*, 15.

Pajola, M., et al. (2017). The pristine interior of comet 67P revealed by the combined Aswan outburst and cliff collapse. *Nature Astronomy, 1*, 0092.

Palchik, V. (1999). Influence of porosity and elastic modulus on uniaxial compressive strength in soft brittle porous sandstones. *Rock Mechanics and Rock Engineering, 32*(4), 303–309.

Paquette, J. A., Hornung, K., Stenzel, O. J., Ryno, J., Silen, J., Kissel, J., & Hilchenbach, M. (2017). The 34S/32S isotopic ratio measured in the dust of comet 67P/Churyumov–Gerasimenko by Rosetta/COSIMA. *Monthly Notices of the Royal Astronomical Society, 469*, S230–S237.

Peixinho, N., Thirouin, A., Tegler, S. C., Di Sisto, R., Delsanti, A., Guilbert-Lepoutre, A., & Bauer, J. G. (2020). From Centaurs to Comets – 40 years. In D. Prialnik, M. A. Barucci, & L. Young (Eds.). ISBN: 9780128164907 *The Trans-Neptunian Solar System* (p. 307). Elsevier.

Penzias, A. A., Solomon, P. M., Wilson, R. W., & Jefferts, K. B. (1971). Interstellar carbon monosulfide. *The Astrophysical Journal, 168*, L53.

Pérez, S., Casassus, S., Baruteau, C., Dong, R., Hales, A., & Cieza, L. (2019). Dust unveils the formation of a mini-Neptune planet in a protoplanetary ring. *The Astronomical Journal, 158*, 15.

Petrovic, J. J. (2003). Mechanical properties of ice and snow. *Journal of Materials Science, 38*, 1. https://doi.org/10.1023/A:1021134128038.

Pike, R. J. (1988). Geomorphology of impact craters on Mercury. In *Mercury* (p. 165). Tucson: University of Arizona Press.

Piquette, M., & Horányi, M. (2017). The effect of asymmetric surface topography on dust dynamics on airless bodies. *Icarus, 291*, 65–74.

Pirali, O., Boudon, V., Carrasco, N., & Dartois, E. (2014). Rotationally resolved IR spectroscopy of hexamethylenetetramine (HMT) $C_6N_4H_{12}$. *Astronomy and Astrophysics, 561*, A109.

Poch, O., Pommerol, A., Jost, B., Carrasco, N., Szopa, C., & Thomas, N. (2016a). Sublimation of ice-tholins mixtures: A morphological and spectro-photometric study. *Icarus, 266*, 288.

Poch, O., Pommerol, A., Jost, B., Carrasco, N., Szopa, C., & Thomas, N. (2016b). Sublimation of water ice mixed with silicates and tholins: Evolution of surface texture and reflectance spectra, with implications for comets. *Icarus, 267*, 154–173.

Poch, O., Istiqomah, I., Quirico, E., Beck, P., Schmitt, B., Theulé, P., Faure, A., Hily-Blant, P., Bonal, L., Raponi, A., Ciarniello, M., Rousseau, B., Potin, S., Brissaud, O., Flandinet, L., Filacchione, G., Pommerol, A., Thomas, N., Kappel, D., Mennella, V., Moroz, L., Vinogradoff, V., Arnold, G., Erard, S., Bockelée-Morvan, D., Leyrat, C., Capaccioni, F., De Sanctis, M. C., Longobardo, A., Mancarella, F., Palomba, E., & Tosi, F. (2020). Ammonium salts are a reservoir of nitrogen on a cometary nucleus and possibly on some asteroids. *Science, 367,* eaaw7462. https://doi.org/10.1126/science.aaw7462.

Pommerol, A., et al. (2015). OSIRIS observations of meter-sized exposures of $H_2O$ ice at the surface of 67P/Churyumov-Gerasimenko and interpretation using laboratory experiments. *Astronomy and Astrophysics, 583,* A25.

Pommerol, A., Jost, B., Poch, O., Yoldi, Z., Brouet, Y., Gracia-Berná, A., Cerubini, R., Galli, A., Wurz, P., Gundlach, B., Blum, J., Carrasco, N., Szopa, C., & Thomas, N. (2019). Experimenting with mixtures of water ice and dust as analogues for icy planetary material. Recipes from the Ice Laboratory at the University of Bern. *Space Science Reviews, 215,* 37.

Pontoppidan, K. M., Salyk, C., Bergin, E. A., Brittain, S., Marty, B., Mousis, O., & Oberg, K. I. (2014). Volatiles in protoplanetary disks. In H. Beuther, R. S. Klessen, C. P. Dullemond, & T. Henning (Eds.), *Protostars and Planets VI* (p. 363). Tucson: University of Arizona Press.

Poppe, A. R., Piquette, M., Likhanskii, A., & Horányi, M. (2012). The effect of surface topography on the lunar photoelectron sheath and electrostatic dust transport. *Icarus, 221,* 135.

Porubčan, V., Kornoš, L., & Williams, I. P. (2006). The Taurid complex meteor showers and asteroids. *Contributions of the Astronomical Observatory Skalnate Pleso, 36,* 103.

Preston, G. W. (1967). The spectrum of Ikeya-Seki (1965f). *Astrophysical Journal, 147,* 718–742.

Preusker, F., Scholten, F., Matz, K.-D., Roatsch, T., Willner, K., Hviid, S. F., Knollenberg, J., Jorda, L., Gutiérrez, P. J., Kührt, E., Mottola, S., A'Hearn, M. F., Thomas, N., Sierks, H., Barbieri, C., Lamy, P., Rodrigo, R., Koschny, D., Rickman, H., Keller, H. U., Agarwal, J., Barucci, M. A., Bertaux, J.-L., Bertini, I., Cremonese, G., Da Deppo, V., Davidsson, B., Debei, S., De Cecco, M., Fornasier, S., Fulle, M., Groussin, O., Güttler, C., Ip, W.-H., Kramm, J. R., Küppers, M., Lara, L. M., Lazzarin, M., Lopez Moreno, J. J., Marzari, F., Michalik, H., Naletto, G., Oklay, N., Tubiana, C., & Vincent, J.-B. (2015). Shape model, reference system definition, and cartographic mapping standards for comet 67P/Churyumov-Gerasimenko – Stereo-photogrammetric analysis of Rosetta/OSIRIS image data. *Astronomy and Astrophysics, 583,* A33.

Preusker, F., Scholten, F., Matz, K.-D., Roatsch, T., Hviid, S. F., Mottola, S., Knollenberg, J., Kührt, E., Pajola, M., Oklay, N., Vincent, J.-B., Davidsson, B., A'Hearn, M. F., Agarwal, J., Barbieri, C., Barucci, M. A., Bertaux, J.-L., Bertini, I., Cremonese, G., Da Deppo, V., Debei, S., De Cecco, M., Fornasier, S., Fulle, M., Groussin, O., Gutiérrez, P. J., Güttler, C., Ip, W.-H., Jorda, L., Keller, H. U., Koschny, D., Kramm, J. R., Küppers, M., Lamy, P., Lara, L. M., Lazzarin, M., Lopez Moreno, J. J., Marzari, F., Massironi, M., Naletto, G., Rickman, H., Rodrigo, R., Sierks, H., Thomas, N., & Tubiana, C. (2017). The global meter-level shape model of comet 67P/Churyumov-Gerasimenko. *Astronomy and Astrophysics, 607,* L1.

Prialnik, D. (1991). A model of gas flow through comet nuclei. In N. I. Kömle, S. J. Bauer, & T. Spohn (Eds.), *Theoretical modelling of comet simulation experiments* (p. 1). Verlag der Österreichischen Akademie der Wissenschaften.

Price, M. C., Kearsley, A. T., Burchell, M. J., Hörz, F., Borg, J., Bridges, J. C., Cole, M. J., Floss, C., Graham, G., Green, S. F., Hoppe, P., Leroux, H., Marhas, K. K., Park, N., Stroud, R., Stadermann, F. J., Telisch, N., & Wozniakiewicz, P. J. (2010). Comet 81P/Wild 2: The size distribution of finer (sub-10 μm) dust collected by the Stardust spacecraft. *Meteoritics and Planetary Science, 45,* 1409.

Price, O., Jones, G. H., Morrill, J., Owens, M., Battams, K., Morgan, H., Druckmuller, M., & Deiries, S. (2019). Fine-scale structure in cometary dust tails. I: Analysis of striae in Comet C/2006 P1 (McNaught) through temporal mapping. *Icarus, 319,* 540.

Protopapa, S., Sunshine, J. M., Feaga, L. M., Kelley, M. S. P., A'Hearn, M. F., Farnham, T. L., Groussin, O., Besse, S., Merlin, F., & Li, J.-Y. (2014). Water ice and dust in the innermost coma of comet 103P/Hartley 2. *Icarus, 238*, 191–204.

Purcell, E. M., & Pennypacker, C. R. (1973). Scattering and absorption of light by nonspherical dielectric grains. *The Astrophysical Journal, 186*, 705.

Rauer, H. (2010). Comets. In K. Altwegg, W. Benz, & N. Thomas (Eds.), *Trans-Neptunian objects and comets*. Proceedings of 35th Saas-Fee Conference.

Reinhard, R. (1986). The Giotto encounter with comet Halley. *Nature, 321*, 313.

Reitsema, H. J., Delamere, W. A., Williams, A. R., Boice, D. C., Huebner, W. F., & Whipple, F. L. (1989). Dust distribution in the inner coma of comet Halley: Comparison with models. *Icarus, 81*, 31.

Richardson, J. E., Melosh, H. J., Lisse, C. M., & Carcich, B. (2007). A ballistics analysis of the Deep Impact ejecta plume: Determining Comet Tempel 1's gravity, mass, and density. *Icarus, 190*, 357.

Richter, K., & Keller, H. U. (1995). On the stability of dust particle orbits around cometary nuclei. *Icarus, 114*, 355–371.

Rickman, H. (2017). *Origin and evolution of comets: Ten years after the Nice model and one year after Rosetta*. Singapore: World Scientific.

Rigby, A., Cruz, F., Albertazzi, B., Bamford, R., Bell, A. R., Cross, J. E., Fraschetti, F., Graham, P., Hara, Y., Kozlowski, P. M., Kuramitsu, Y., Lamb, D. Q., Lebedev, S., Marques, J. R., Miniati, F., Morita, T., Oliver, M., Reville, B., Sakawa, Y., Sarkar, S., Spindloe, C., Trines, R., Tzeferacos, P., Silva, L. O., Bingham, R., Koenig, M., & Gregori, G, (2018). Electron acceleration by wave turbulence in a magnetized plasma. *Nature Physics, 14*, 475.

Robbins, S. J., Riggs, J. D., Weaver, B. P., Bierhaus, E. B., Chapman, C. R., Kirchoff, M. R., Singer, K. N., & Gaddis, L. R. (2018). Revised recommended methods for analyzing crater size-frequency distributions. *Meteoritics and Planetary Science, 53*, 891.

Roberts, J. H., Kahn, E. G., Barnouin, O. S., Ernst, C. M., Prockter, L. M., & Gaskell, R. W. (2014). Origin and flatness of ponds on asteroid 433 Eros. *Meteoritics and Planetary Science, 49*, 1735.

Rodgers, S. D., & Charnley, S. B. (2002). A model of the chemistry in cometary comae: Deuterated molecules. *Monthly Notices of the Royal Astronomical Society, 330*, 660.

Rodgers, S. D., Charnley, S. B., Huebner, W. F., & Boice, D. C. (2004). Physical processes and chemical reactions in cometary comae. In M. Festou, H. U. Keller, & H. A. Weaver (Eds.), *Comets II* (pp. 505–522). Tucson: University of Arizona Press.

Rodionov, A. V., Crifo, J.-F., Szego, K., Lagerros, J., & Fulle, M. (2002). An advanced physical model of cometary activity. *Planetary and Space Science, 50*, 983–1024.

Roll, R., Witte, L., & Arnold, W. (2016). Rosetta lander Philae—soil strength analysis. *Icarus, 280*, 359–365.

Rosenbush, V. K., Ivanova, O. V., Kiselev, N. N., Kolokolova, L. O., & Afanasiev, V. L. (2017). Spatial variations of brightness, colour and polarization of dust in comet 67P/Churyumov-Gerasimenko. *Monthly Notices of the Royal Astronomical Society, 469*, S475.

Roth, N. X., Gibb, E. L., Bonev, B. P., DiSanti, M. A., Mumma, M. J., Villanueva, G. L., & Paganini, L. (2017). The composition of Comet C/2012 K1 (PanSTARRS) and the distribution of primary volatile abundances among comets. *The Astronomical Journal, 153*, 168.

Rothman, L. S., Jacquemart, D., Barbe, A., Chris Benner, D., Birk, M., Brown, L. R., Carleer, M. R., Chackerian, C., Jr., Chance, K., Coudert, L. H., Dana, V., Malathy Devi, V., Flaud, J.-M., Gamache, R. R., Goldman, A., Hartmann, J.-M., Jucks, K. W., Maki, A. G., Mandin, J.-Y., Massie, S. T., Orphal, J., Perrin, A., Rinsland, C. P., Smith, M. A. H., Tennyson, J., Tolchenov, R. N., Toth, R. A., Vander Auwera, J., Varanasi, P., & Wagner, G. (2005). The HITRAN 2004 molecular spectroscopic database. *JQSRT, 96*, 139–204.

Rotundi, A., et al. (2015). Dust measurements in the coma of comet 67P/Churyumov-Gerasimenko inbound to the Sun. *Science, 347*, aaa3905.

Rousselot, P. (2008). 174P/Echeclus: A strange case of outburst. *Astronomy and Astrophysics, 480*, 543.

Rozitis, B., & Green, S. F. (2011). Directional characteristics of thermal-infrared beaming from atmosphereless planetary surfaces – a new thermophysical model. *Monthly Notices of the Royal Astronomical Society, 415*, 2042.

Rubin, M., Hansen, K. C., Combi, M. R., Daldorff, L. K. S., Gombosi, T. I., & Tenishev, V. M. (2012). Kelvin-Helmholtz instabilities at the magnetic cavity boundary of comet 67P/Churyumov-Gerasimenko. *Journal of Geophysical Research (Space Physics), 117*, A06227.

Rubin, M., Fougere, N., Altwegg, K., Combi, M. R., Le Roy, L., Tenishev, V. M., & Thomas, N. (2014a). Mass transport around comets and its impact on the seasonal differences in water production rates. *The Astrophysical Journal, 788*, 168.

Rubin, M., Koenders, C., Altwegg, K., Combi, M. R., Glassmeier, K.-H., Gombosi, T. I., Hansen, K. C., Motschmann, U., Richter, I., Tenishev, V. M., & Tóth, G. (2014b). Plasma environment of a weak comet – Predictions for Comet 67P/Churyumov-Gerasimenko from multifluid-MHD and Hybrid models. *Icarus, 242*, 38.

Rubin, M., Combi, M. R., Daldorff, L. K. S., Gombosi, T. I., Hansen, K. C., Shou, Y., Tenishev, V. M., Tóth, G., van der Holst, B., & Altwegg, K. (2014c). Comet 1P/Halley multifluid MHD model for the Giotto fly-by. *The Astrophysical Journal, 781*, 86.

Rubin, M., Altwegg, K., van Dishoeck, E. F., & Schwehm, G. (2015). Molecular oxygen in Oort Cloud Comet 1P/Halley. *The Astrophysical Journal, 815*, L11.

Rubin, M., Altwegg, K., Balsiger, H., Berthelier, J.-J., Bieler, A., Calmonte, U., Combi, M., De Keyser, J., Engrand, C., Fiethe, B., Fuselier, S. A., Gasc, S., Gombosi, T. I., Hansen, K. C., Haessig, M., Le Roy, L., Mezger, K., Tzou, C.-Y., Wampfler, S. F., & Wurz, P. (2017). Evidence for depletion of heavy silicon isotopes at comet 67P/Churyumov-Gerasimenko. *Astronomy and Astrophysics, 601*, A123.

Rubin, M., Altwegg, K., Balsiger, H., Berthelier, J.-J., Combi, M. R., De Keyser, J., Drozdovskaya, M., Fiethe, B., Fuselier, S. A., Gasc, S., Gombosi, T. I., Hänni, N., Hansen, K. C., Mall, U., Rème, H., Schroeder, I. R. H. G., Schuhmann, M., Sémon, T., Waite, J. H., Wampfler, S. F., & Wurz, P. (2019). Elemental and molecular abundances in comet 67P/Churyumov-Gerasimenko. *Monthly Notices of the Royal Astronomical Society, 489*, 594.

Russell, H. N. (1916). On the albedo of the planets and their satellites. *The Astrophysical Journal, 43*, 173–196.

Russell, H. W. (1935). Principles of heat flow in porous insulators. *Journal of the American Ceramic Society, 18*, 1.

Rybicki, G. B. (1984). Escape probability methods. In W. Kalkofen (Ed.), *Methods of radiative transfer* (p. 21). Cambridge: Cambridge University Press.

Sagan, C., & Khare, B. N. (1979). Tholins: Organic chemistry of interstellar grains and gas. *Nature, 277*, 102.

Sagdeev, R. Z., Blamont, J., Galeev, A. A., Moroz, V. I., Shapiro, V. D., Shevchenko, V. I., & Szego, K. (1986a). Vega spacecraft encounters with comet Halley. *Nature, 321*, 259.

Sagdeev, R. Z., Avanesov, G. A., Ziman, Y. L., Moroz, V. I., Tarnopolsky, V. I., Zhukov, B. S., Shamis, V. A., Smith, B., & Toth, I. (1986b). *Tv experiment on the VEGA mission: Photometry of the nucleus and the inner coma.* ESLAB Symposium on the Exploration of Halley's Comet, ESA SP-250, 317.

Salo, H. (1988). Monte Carlo modeling of the net effects of coma scattering and thermal reradiation on the energy input to cometary nucleus. *Icarus, 76*, 253.

Samarasinha, N. H., & Belton, M. J. S. (1995). Long-term evolution of rotational stress and nongravitational effects for Halley-like cometary nuclei. *Icarus, 116*, 340–358.

Samarasinha, N. H., Mueller, B. E. A., Belton, M. J. S., & Jorda, L. (2004). Rotation of cometary nuclei. In M. Festou, H. U. Keller, & H. A. Weaver (Eds.), *Comets II* (p. 281). Tucson: University of Arizona Press.

Samarasinha, N. H., & Mueller, B. E. A. (2015). Component periods of non-principal-axis rotation and their manifestations in the lightcurves of asteroids and bare cometary nuclei. *Icarus, 248*, 347.

Sandford, S. A., Alexander, C. M. O. D., Araki, T., Bajt, S., Baratta, G. A., Borg, J., Bradley, J. P., Brownlee, D. E., Brucato, J. R., Burchell, M. J., Busemann, H., Butterworth, A., Clemett, S. J., Cody, G., Colangeli, L., Cooper, G., D'Hendecourt, L., Djouadi, Z., Dworkin, J. P., Ferrini, G., Fleckenstein, H., Flynn, G. J., Franchi, I. A., Fries, M., Gilles, M. K., Glavin, D. P., Gounelle, M., Grossemy, F., Jacobsen, C., Keller, L. P., Kilcoyne, A. L. D., Leitner, J., Matrajt, G., Meibom, A., Mennella, V., Mostefaoui, S., Nittler, L. R., Palumbo, M. E., Papanastassiou, D. A., Robert, F., Rotundi, A., Snead, C. J., Spencer, M. K., Stadermann, F. J., Steele, A., Stephan, T., Tsou, P., Tyliszczak, T., Westphal, A. J., Wirick, S., Wopenka, B., Yabuta, H., Zare, R. N., & Zolensky, M. E. (2006). Organics captured from comet 81p/Wild 2 by the Stardust spacecraft. *Science, 314*(5806), 1720–1724.

Sarid, G., Volk, K., Steckloff, J. K., Harris, W., Womack, M., & Woodney, L. M. (2019). 29P/ Schwassmann-Wachmann 1, A Centaur in the Gateway to the Jupiter-family Comets. *The Astrophysical Journal, 883*, L25.

Sarugaku, Y., Ishiguro, M., Ueno, M., Usui, F., & Watanabe, J. (2010). Outburst of Comet 217P/ LINEAR. *The Astrophysical Journal, 724*, L118–L121.

Scanlon, T. J., Roohi, E., White, C., Darbandi, M., & Reese, J. M. (2010). An open source, parallel DSMC code for rarefied gas flows in arbitrary geometries. *Computers & Fluids, 39*, 2078–2089.

Schambeau, C. A., Fernández, Y. R., Samarasinha, N. H., Woodney, L. M., & Kundu, A. (2019). Analysis of HST WFPC2 observations of Centaur 29P/Schwassmann–Wachmann 1 while in outburst to place constraints on the nucleus' rotation state. *Astronomical Journal, 158*, 259.

Scheeres, D. J., Hartzell, C. M., Sanchez, P., & Swift, M. (2010). Scaling forces to asteroid surfaces: The role of cohesion. *Icarus, 210*, 968.

Schleicher, D. G. (2010). The fluorescence efficiencies of the CN violet bands in comets. *The Astronomical Journal, 140*, 973.

Schleicher, D. G., & Bair, A. N. (2011). The composition of the interior of Comet 73P/ Schwassmann-Wachmann 3: Results from narrowband photometry of multiple components. *The Astronomical Journal, 141*, 177.

Schleicher, D. G., & A'Hearn, M. F. (1988). The fluorescence of cometary OH. *The Astrophysical Journal, 331*, 1058–1077.

Schleicher, D. G., & Farnham, T. L. (2004). Photometry and imaging of the coma with narrowband filters. In M. Festou, H. U. Keller, & H. A. Weaver (Eds.), *Comets II* (p. 449). Tucson: University of Arizona Press.

Schleicher, D. G., Woodney, L. M., & Millis, R. L. (2003). Comet 19P/Borrelly at multiple apparitions: Seasonal variations in gas production and dust morphology. *Icarus, 162*, 415.

Schleicher, D. G., Knight, M. M., Eisner, N. L., & Thirouin, A. (2019). Gas jet morphology and the very rapidly increasing rotation period of Comet 41P/Tuttle-Giacobini-Kresák. *The Astronomical Journal, 157*, 108.

Schloerb, F. P., Keihm, S., von Allmen, P., Choukroun, M., Lellouch, E., Leyrat, C., Beaudin, G., Biver, N., Bockelée-Morvan, D., Crovisier, J., Encrenaz, P., Gaskell, R., Gulkis, S., Hartogh, P., Hofstadter, M., Ip, W.-H., Janssen, M., Jarchow, C., Jorda, L., Keller, H. U., Lee, S., Rezac, L., & Sierks, H. (2015). MIRO observations of subsurface temperatures of the nucleus of 67P/ Churyumov-Gerasimenko. *Astronomy and Astrophysics, 583*, A29.

Schlosser, W., Schulz, R., & Koczet, P. (1986). *The cyan shells of Comet P/Halley*. ESLAB Symposium on the Exploration of Halley's Comet, ESA SP-250, 495.

Schmidt, H. U., Wegmann, R., Huebner, W. F., & Boice, D. C. (1988). Cometary gas and plasma flow with detailed chemistry. *Computer Physics Communications, 49*, 17–59.

Schmidt-Voigt, M. (1989). Time-dependent MHD simulations for cometary plasmas. *Astronomy and Astrophysics, 210*, 433.

Schöier, F. L., van der Tak, F. F. S., van Dishoeck, E. F., & Black, J. H. (2005). An atomic and molecular database for analysis of submillimetre line observations. *Astronomy and Astrophysics, 432*, 369.

Schräpler, R., Blum, J., von Borstel, I., & Güttler, C., (2015),The stratification of regolith on celestial objects,Icarus,257,33

Schroeder I. R. H. G., Altwegg, K., Balsiger, H., Berthelier, J.-J., De Keyser, J., Fiethe, B., Fuselier, S. A., Gasc, S., Gombosi, T. I., Rubin, M., Semon, T., Tzou, C.-Y., Wampfler, S. F., & Wurz, P. (2018). The 16O/18O ratio in water in the coma of Comet 67P/Churyumov-Gerasimenko measured with the Rosetta/ROSINA double-focusing mass spectrometer. arXiv e-prints, arXiv:1809.03798.

Schuhmann, M., Altwegg, K., Balsiger, H., Berthelier, J.-J., De Keyser, J., Fiethe, B., Fuselier, S. A., Gasc, S., Gombosi, T. I., Hänni, N., Rubin, M., Tzou, C.-Y., & Wampfler, S. F. (2019). Aliphatic and aromatic hydrocarbons in comet 67P/Churyumov-Gerasimenko seen by ROSINA. *Astronomy and Astrophysics, 630*, A31.

Schulz, R., & Schlosser, W. (1989). CN-shell structures and dynamics of the nucleus of Comet P/Halley. *Astronomy and Astrophysics, 214*, 375–385.

Schultz, P. H., Eberhardy, C. A., Ernst, C. M., A'Hearn, M. F., Sunshine, J. M., & Lisse, C. M. (2007). The Deep Impact oblique impact cratering experiment. *Icarus, 191*, 84–122.

Schultz, P. H., Hermalyn, B., & Veverka, J. (2013). The Deep Impact crater on 9P/Tempel-1 from Stardust-NExT. *Icarus, 222*, 502–515.

Sears, D. W. G., & Dodd, R. T. (1988). Overview and classification of meteorites. In J. F. Kerridge & M. S. Matthews (Eds.), *Meteorites and the early solar system* (pp. 3–31). Tucson: The University of Arizona Press.

Sears, D. W. G., Tornabene, L. L., Osinski, G. R., Hughes, S. S., & Heldmann, J. L. (2015). Formation of the "ponds" on asteroid (433) Eros by fluidization. *Planetary and Space Science, 117*, 106.

Seiferlin, K. (1991). The thermal conductivity of porous ice with application to KOSI sample material: A review. In N. I. Kömle, S. J. Bauer, & T. Spohn (Eds.), *Theoretical modelling of comet simulation experiments* (p. 49). Verlag der Österreichischen Akademie der Wissenschaften.

Séjourné, A., Costard, F., Gargani, J., Soare, R. J., Fedorov, A., & Marmo, C. (2011). Scalloped depressions and small-sized polygons in western Utopia Planitia, Mars: A new formation hypothesis. *Planetary and Space Science, 59*, 412.

Sekanina, Z. (2017). Major outburst and splitting of long-period Comet C/2015 ER61 (Pan-STARRS). arXiv e-prints, arXiv:1712.03197.

Sekanina, Z., & Chodas, P. W. (2012). Comet C/2011 W3 (Lovejoy): Orbit determination, outbursts, disintegration of nucleus, dust-tail morphology, and relationship to new cluster of bright sungrazers. *The Astrophysical Journal, 757*, 127.

Sen, A. K., Botet, R., Vilaplana, R., Choudhury, N. R., & Gupta, R. (2017). The effect of porosity of dust particles on polarization and color with special reference to comets. *Journal of Quantitative Spectroscopy and Radiative Transfer, 198*, 164.

Shahid-Saless, B., & Yeomans, D. K. (1994). Relativistic effects on the motion of asteroids and comets. *The Astronomical Journal, 107*, 1885–1889.

Shao, Y., & Lu, H. (2000). A simple expression for wind erosion threshold friction velocity. *Journal of Geophysical Research, 105*, 22.

Shepard, M. K. (2017). *Introduction to planetary photometry*. Cambridge: Cambridge University Press.

Shepard, M. K., & Helfenstein, P. (2007). A test of the Hapke photometric model. *Journal of Geophysical Research (Planets), 112*, E03001.

Shepard, M. K., & Helfenstein, P. (2011). A laboratory study of the bidirectional reflectance from particulate samples. *Icarus, 215*, 526–533.

Shi, X., Hu, X., Sierks, H., Güttler, C., A'Hearn, M., Blum, J., El-Maarry, M. R., Kührt, E., Mottola, S., Pajola, M., Oklay, N., Fornasier, S., Tubiana, C., Keller, H. U., Vincent, J.-B., Bodewits, D., Höfner, S., Lin, Z.-Y., Gicquel, A., Hofmann, M., Barbieri, C., Lamy, P. L., Rodrigo, R., Koschny, D., Rickman, H., Barucci, M. A., Bertaux, J.-L., Bertini, I., Cremonese, G., Da Deppo, V., Davidsson, B., Debei, S., De Cecco, M., Fulle, M., Groussin, O., Gutierrez, P. J., Hviid, S. F., Ip, W.-H., Jorda, L., Knollenberg, J., Kovacs, G., Kramm, J.-R., Küppers, M., Lara, L. M., Lazzarin, M., Lopez-Moreno, J. J., Marzari, F., Naletto, G., & Thomas, N. (2016). Sunset jets observed on comet 67P/Churyumov-Gerasimenko sustained by subsurface thermal lag. *Astronomy and Astrophysics, 586*, A7.

Shinnaka, Y., Kawakita, H., Kobayashi, H., Jehin, E., Manfroid, J., Hutsemekers, D., & Arpigny, C. (2011). Ortho-to-para abundance ratio (OPR) of ammonia in 15 comets: OPRs of ammonia versus 14N/15N ratios in CN. *The Astrophysical Journal, 729*, 81.

Shinnaka, Y., Kawakita, H., Jehin, E., Decock, A., Hutsemekers, D., Manfroid, J., & Arai, A. (2016). Nitrogen isotopic ratios of NH2 in comets: Implication for 15N-fractionation in cometary ammonia. *Monthly Notices of the Royal Astronomical Society, 462*, S195–S209.

Shkuratov, Y., Ovcharenko, A., Zubko, E., Miloslavskaya, O., Muinonen, K., Piironen, J., Nelson, R., Smythe, W., Rosenbush, V., & Helfenstein, P. (2002). *Icarus, 159*, 396–416.

Sierks, H., et al. (2015). On the nucleus structure and activity of comet 67P/Churyumov-Gerasimenko. *Science, 347*, aaa1044.

Silsbee, K., & Draine, B. T. (2016). Radiation pressure on fluffy submicron-sized grains. *The Astrophysical Journal, 818*, 133.

Singh, P. D., de Almeida, A. A., & Huebner, W. F. (1992). Dust release rates and dust-to-gas mass ratios of eight comets. *The Astronomical Journal, 104*, 848.

Sirono, S. (2017). Heating of porous icy dust aggregates. *The Astrophysical Journal, 842*, 11.

Skorov, Y. V., & Rickman, H. (1999). Gas flow and dust acceleration in a cometary Knudsen layer. *Planetary and Space Science, 47*, 935.

Skorov, Y. V., Markelov, G. N., & Keller, H. U. (2006). Direct statistical simulation of the near-surface layers of a cometary atmosphere.II: A nonspherical nucleus. *Solar System Research, 40*, 219–229.

Skorov, Y. V., van Lieshout, R., Blum, J., & Keller, H. U. (2011). Activity of comets: Gas transport in the near-surface porous layers of a cometary nucleus. *Icarus, 212*, 867–876.

Skorov, Y. V., Rezac, L., Hartogh, P., & Keller, H. U. (2017). Is near-surface ice the driver of dust activity on 67P/Churyumov-Gerasimenko. *Astronomy and Astrophysics, 600*, A142.

Slaughter, C. D. (1969). The emission spectrum of comet Ikeya-Seki 1965-f at perihelion passage. *Astronomical Journal, 74*, 929–943.

Slavin, J. A., Goldberg, B. A., Smith, E. J., McComas, D. J., Bame, S. J., Strauss, M. A., & Spinrad, H. (1986). The structure of a cometary Type I tail: Ground-based and ice observations of P/Giacobini-Zinner. *Geophysical Research Letters, 13*, 1085.

Smith, D., Arnaud, A., Julie, B., Benoit, N.-A., Pennec, F., Tessier-Doyen, N., Otsu, K., Matsubara, H., Elser, P., & Gonzenbach, U. (2013). Thermal conductivity of porous materials. *Journal of Materials Research, 28*. https://doi.org/10.1557/jmr.2013.179.

Smoluchowski, R. (1985). Amorphous and porous ices in cometary nuclei. *NATO Advanced Science Institutes (ASI) Series C, 156*, 397.

Snios, B., Kharchenko, V., Lisse, C. M., Wolk, S. J., Dennerl, K., & Combi, M. R. (2016). Chandra observations of Comets C/2012 S1 (ISON) and C/2011 L4 (PanSTARRS). *The Astrophysical Journal, 818*, 199.

Snios, B., Lichtman, J., & Kharchenko, V. (2018). The presence of dust and ice scattering in X-ray emissions from Comets. *The Astrophysical Journal, 852*, 138.

Snodgrass, C., Agarwal, J., Combi, M., Fitzsimmons, A., Guilbert-Lepoutre, A., Hsieh, H. H., Hui, M.-T., Jehin, E., Kelley, M. S. P., Knight, M. M., Opitom, C., Orosei, R., de Val-Borro, M., & Yang, B. (2017). The main belt comets and ice in the Solar System. *Astronomy and Astrophysics Review, 25*, 5.

Sone, Y., & Sugimoto, H. (1993). Kinetic theory analysis of steady evaporating flows from a spherical condensed phase into a vacuum. *Physics of Fluids A, 5*, 1491–1511.

Spencer, J. R., Lebofsky, L. A., & Sykes, M. V. (1989). Systematic biases in radiometric diameter determinations. *Icarus, 78*, 337.

Spohn, T., Seiferlin, K., & Benkhoff, J. (1989). Thermal conductivities and diffusivities of porous ice samples at low pressures and temperatures and possible modes of heat transfer in near surface layers of comets. In *Physics and mechanics of cometary materials*. ESA SP-302,77.

Spohn, T., Knollenberg, J., Ball, A. J., Banaszkiewicz, M., Benkhoff, J., Grott, M., Grygorczuk, J., Hüttig, C., Hagermann, A., Kargl, G., Kaufmann, E., Kömle, N., Kührt, E., Kossacki, K. J., Marczewski, W., Pelivan, I., Schrödter, R., & Seiferlin, K. (2015). Thermal and mechanical properties of the near-surface layers of comet 67P/Churyumov-Gerasimenko. *Science, 349*, 2.464.

Squyres, S. W., McKay, C. P., & Reynolds, R. T. (1985). Temperatures within comet nuclei. *Journal of Geophysical Research, 90*, 12,381.

Stansberry, J., Grundy, W., Brown, M., Cruikshank, D., Spencer, J., Trilling, D., & Margot, J.-L. (2008). Physical properties of Kuiper Belt and Centaur objects: Constraints from the Spitzer Space telescope. In M. A. Barucci, H. Boehnhardt, D. P. Cruikshank, & A. Morbidelli (Eds.), *The Solar System beyond Neptune* (p. 161). Tucson: University of Arizona Press.

Steckloff, J. K., & Samarasinha, N. H. (2018). The sublimative torques of Jupiter Family Comets and mass wasting events on their nuclei. *Icarus, 312*, 172.

Steckloff, J. K., Johnson, B. C., Bowling, T., Jay Melosh, H., Minton, D., Lisse, C. M., & Battams, K. (2015). Dynamic sublimation pressure and the catastrophic breakup of Comet ISON. *Icarus, 258*, 430.

Steckloff, J. K., Graves, K., Hirabayashi, M., Melosh, H. J., & Richardson, J. E. (2016). Rotationally induced surface slope-instabilities and the activation of $CO_2$ activity on comet 103P/Hartley 2. *Icarus, 272*, 6.

Steiner, G. (1990). Two considerations concerning the free molecular flow of gases in porous ices. *Astronomy and Astrophysics, 240*, 533–536.

Steiner, G., Kömle, N. I., & Kührt, E. (1991). Thermal modelling of comet simulation experiments. In N. I. Kömle, S. J. Bauer, & T. Spohn (Eds.), *Theoretical modelling of comet simulation experiments* (p. 11). Verlag der Österreichischen Akademie der Wissenschaften.

Sterken, C., & Manfroid, J. (1992). *Astronomical photometry. A guide*. Astrophysics and Space Science Library. Berlin: Springer. ISBN 0-7923-1653-3.

Sterken, V. J., Altobelli, N., Kempf, S., Schwehm, G., Srama, R., & Grün, E. (2012). The flow of interstellar dust into the solar system. *Astronomy and Astrophysics, 538*, A102.

Stern, S. A., et al. (2019). Initial results from the New Horizons exploration of 2014 $MU_{69}$, a small Kuiper Belt object. *Science, 364*, aaw9771.

Su, C. C. (2013). PhD thesis, National Chiao Tung University, Taiwan.

Sunshine, J. M., A'Hearn, M. F., Schultz, P. H., Groussin, O., Feaga, L., & Deep Impact Science Team. (2005). The spatial and temporal distribution of ice excavated by the Deep Impact experiment on the Comet 9/P Tempel 1. *Bulletin of the American Astronomical Society, 37*, 189.09.

Sunshine, J. M., A'Hearn, M. F., Groussin, O., Li, J.-Y., Belton, M. J. S., Delamere, W. A., Kissel, J., Klaasen, K. P., McFadden, L. A., Meech, K. J., Melosh, H. J., Schultz, P. H., Thomas, P. C., Veverka, J., Yeomans, D. K., Busko, I. C., Desnoyer, M., Farnham, T. L., Feaga, L. M., Hampton, D. L., Lindler, D. J., Lisse, C. M., & Wellnitz, D. D. (2006). Exposed water ice deposits on the surface of Comet 9P/Tempel 1. *Science, 311*, 1453.

Sunshine, J. M., Thomas, N., El-Maarry, M. R., & Farnham, T. L. (2016). Evidence for geologic processes on comets. *Journal of Geophysical Research (Planets), 121*, 2194.

Sykes, M. V., Lebofsky, L. A., Hunten, D. M., & Low, F. (1986). The discovery of dust trails in the orbits of periodic comets. *Science, 232*, 1115–1117.

Sykes, M. V., Grün, E., Reach, W. T., & Jenniskens, P. (2004). The interplanetary dust complex and comets. In M. Festou, H. U. Keller, & H. A. Weaver (Eds.), *Comets II* (p. 677). Tucson: University of Arizona Press.

Taff, L. G. (1985). *Celestial mechanics: A computational guide for the practitioner*. Wiley-Interscience.

Tancredi, G., Rickman, H., & Greenberg, J. M. (1994). Thermochemistry of cometary nuclei 1: The Jupiter family case. *Astronomy and Astrophysics, 286*, 659.

Taquet, V., Furuya, K., Walsh, C., & van Dishoeck, E. F. (2016). A primordial origin for molecular oxygen in comets: A chemical kinetics study of the formation and survival of O2 ice from clouds to discs. *Monthly Notices of the Royal Astronomical Society, 462*, S99.

Teanby, N. A. (2015). Predicted detection rates of regional-scale meteorite impacts on Mars with the InSight short-period seismometer. *Icarus, 256*, 49.

Tenishev, V., Combi, M. R., & Rubin, M. (2011). Numerical simulation of dust in a cometary coma: Application to Comet 67P/Churyumov-Gerasimenko. *The Astrophysical Journal, 732*, 104.

Tenishev, V., Fougere, N., Borovikov, D., Combi, M. R., Bieler, A., Hansen, K. C., Gombosi, T. I., Migliorini, A., Capaccioni, F., Rinaldi, G., Filacchione, G., Kolokolova, L., & Fink, U. (2016). Analysis of the dust jet imaged by Rosetta VIRTIS-M in the coma of comet 67P/Churyumov-Gerasimenko on 2015 April 12. *Monthly Notices of the Royal Astronomical Society, 462*, S370.

Tennyson, J., Bernath, P. F., Brown, L. R., Campargue, A., Carleer, M. R., Csaszar, A. G., Gamache, R. R., Hodges, J. T., Jenouvrier, A., Naumenko, O. V., Polyansky, O. L., Rothman, L. S., Toth, R. A., Vandaele, A. C., Zobov, N. F., Daumont, L., Fazliev, A. Z., Furtenbacher, T., Gordon, I. E., Mikhailenko, S. N., & Shirin, S. V. (2009). IUPAC critical evaluation of the rotational–vibrational spectra of water vapor. Part I—Energy levels and transition wavenumbers for $H_2{}^{17}O$ and $H_2{}^{18}O$. *Journal of Quantitative Spectroscopy and Radiative Transfer, 110*, 573.

Tennyson, J., Bernath, P. F., Brown, L. R., Campargue, A., Csaszar, A. G., Daumont, L., Gamache, R. R., Hodges, J. T., Naumenko, O. V., Polyansky, O. L., Rothman, L. S., Toth, R. A., Vandaele, A. C., Zobov, N. F., Fally, S., Fazliev, A. Z., Furtenbacher, T., Gordon, I. E., Hu, S.-M., Mikhailenko, S. N., & Voronin, B. A. (2010). IUPAC critical evaluation of the rotational–vibrational spectra of water vapor. Part II: Energy levels and transition wavenumbers for $HD{}^{16}O$, $HD{}^{17}O$, and $HD{}^{18}O$. *Journal of Quantitative Spectroscopy and Radiative Transfer, 111*, 2160.

Tennyson, J., Bernath, P. F., Brown, L. R., Campargue, A., Csaszar, A. G., Daumont, L., Gamache, R. R., Hodges, J. T., Naumenko, O. V., Polyansky, O. L., Rothman, L. S., Vandaele, A. C., Zobov, N. F., Al Derzi, A. R., Fabri, C., Fazliev, A. Z., Furtenbacher, T., Gordon, I. E., Lodi, L., & Mizus, I. I. (2013). IUPAC critical evaluation of the rotational–vibrational spectra of water vapor, Part III: Energy levels and transition wavenumbers for $H_2{}^{16}O$. *Journal of Quantitative Spectroscopy and Radiative Transfer, 117*, 29.

Tennyson, J., Bernath, P. F., Brown, L. R., Campargue, A., Csaszar, A. G., Daumont, L., Gamache, R. R., Hodges, J. T., Naumenko, O. V., Polyansky, O. L., Rothman, L. S., Vandaele, A. C., Zobov, N. F., Denes, N., Fazliev, A. Z., Furtenbacher, T., Gordon, I. E., Hu, S.-M., Szidarovszky, T., & Vasilenko, I. A. (2014). IUPAC critical evaluation of the rotational–vibrational spectra of water vapor. Part IV. Energy levels and transition wavenumbers for $D_2{}^{16}O$, $D_2{}^{17}O$, and $D_2{}^{18}O$. *Journal of Quantitative Spectroscopy and Radiative Transfer, 142*, 93.

Tennyson, J., Rahimi, S., Hill, C., Tse, L., Vibhakar, A., Akello-Egwel, D., Brown, D. B., Dzarasova, A., Hamilton, J. R., Jaksch, D., Mohr, S., Wren-Little, K., Bruckmeier, J., Agarwal, A., Bartschat, K., Bogaerts, A., Booth, J.-P., Goeckner, M. J., Hassouni, K., Itikawa, Y., Braams, B. J., Krishnakumar, E., Laricchiuta, A., Mason, N. J., Pandey, S., Petrovic, Z. L., Pu, Y.-K., Ranjan, A., Rauf, S., Schulze, J., Turner, M. M., Ventzek, P., Whitehead, J. C., & Yoon, J.-S. (2017). QDB: A new database of plasma chemistries and reactions. *Plasma Sources Science Technology, 26*, 055014.

Terai, T., Itoh, Y., Oasa, Y., Furusho, R., & Watanabe, J. (2016). Photometric measurements of $H_2O$ ice crystallinity on trans-Neptunian objects. *The Astrophysical Journal, 827*, 65.

Thiel, K., Koelzer, G., Kochan, H., Ratke, L., Gruen, E., & Koehl, H. (1989). Dynamics of crust formation and dust emission of comet nucleus analogues under isolation. In *Physics and mechanics of cometary materials*. ESA SP-302, 221.

Thomas, N. (1992). Optical observations of Io's neutral clouds and plasma torus. *Surveys in Geophysics, 13*, 91.

Thomas, N., & Keller, H. U. (1987). Fine dust structures in the emission of comet P/Halley observed by the Halley Multicolour Camera on board Giotto. *Astronomy and Astrophysics, 187*, 843.

Thomas, N., & Keller, H. U. (1989). The colour of Comet P/Halley's nucleus and dust. *Astronomy and Astrophysics, 213*, 487–494.

Thomas, N., & Keller, H. U. (1990a). Photometric calibration of the Halley multicolor camera. *Applied Optics, 29*, 1503.

Thomas, N., & Keller, H. U. (1990b). Interpretation of the inner coma observations of comet P/Halley by the Halley Multicolour Camera. *Annales Geophysicae, 8*, 147.

Thomas, N., & Keller, H. U. (1991). Comet P/Halley's dust production rate at Giotto encounter derived from Halley Multicolour Camera observations. *Astronomy and Astrophysics, 249*, 258–268.

Thomas, N., Markiewicz, W. J., Sablotny, R. M., Wuttke, M. W., Keller, H. U., Johnson, J. R., Reid, R. J., & Smith, P. H. (1999). The color of the Martian sky and its influence on the illumination of the Martian surface. *Journal of Geophysical Research, 104*, 8795.

Thomas, N., Alexander, C., & Keller, H. U. (2008). Loss of the surface layers of Comet Nuclei. *Space Science Reviews, 138*, 165–177.

Thomas, N., Barbieri, C., Keller, H. U., Lamy, P., Rickman, H., Rodrigo, R., Sierks, H., Wenzel, K. P., Cremonese, G., Jorda, L., Küppers, M., Marchi, S., Marzari, F., Massironi, M., Preusker, F., Scholten, F., Stephan, K., Barucci, M. A., Besse, S., El-Maarry, M. R., Fornasier, S., Groussin, O., Hviid, S. F., Koschny, D., Kührt, E., Martellato, E., Moissl, R., Snodgrass, C., Tubiana, C., & Vincent, J.-B. (2012). The geomorphology of (21) Lutetia: Results from the OSIRIS imaging system onboard ESA's Rosetta spacecraft. *Planetary and Space Science, 66*, 96–124.

Thomas, N., et al. (2015a). The morphological diversity of comet 67P/Churyumov-Gerasimenko. *Science, 347*, 0440.

Thomas, N., et al. (2015b). Redistribution of particles across the nucleus of comet 67P/Churyumov-Gerasimenko. *Astronomy and Astrophysics, 583*, A17.

Thomas, N., El Maarry, M. R., Theologou, P., Preusker, F., Scholten, F., Jorda, L., Hviid, S. F., Marschall, R., Kührt, E., Naletto, G., Sierks, H., Lamy, P. L., Rodrigo, R., Koschny, D., Davidsson, B., Barucci, M. A., Bertaux, J. L., Bertini, I., Bodewits, D., Cremonese, G., Da Deppo, V., Debei, S., De Cecco, M., Fornasier, S., Fulle, M., Groussin, O., Gutièrrez, P. J., Güttler, C., Ip, W. H., Keller, H. U., Knollenberg, J., Lara, L. M., Lazzarin, M., Lòpez-Moreno, J. J., Marzari, F., Tubiana, C., & Vincent, J. B. (2018). Regional unit definition for the nucleus of comet 67P/Churyumov-Gerasimenko on the SHAP7 model. *Planetary and Space Science, 164*, 19.

Thomas, N., Ulamec, S., Kührt, E., Ciarletti, V., Gundlach, B., Yoldi, Z., Schwehm, G., Snodgrass, C., & Green, S. F. (2019). Towards new comet missions. *Space Science Reviews, 215*, 47.

Thomas, P. C., Veverka, J., Belton, M. J. S., Hidy, A., A'Hearn, M. F., Farnham, T. L., Groussin, O., Li, J.-Y., McFadden, L. A., Sunshine, J., Wellnitz, D., Lisse, C., Schultz, P., Meech, K. J., & Delamere, W. A. (2007). The shape, topography, and geology of Tempel 1 from Deep Impact observations. *Icarus, 187*, 4.

Thomas, P. C., A'Hearn, M., Belton, M. J. S., Brownlee, D., Carcich, B., Hermalyn, B., Klaasen, K., Sackett, S., Schultz, P. H., Veverka, J., Bhaskaran, S., Bodewits, D., Chesley, S., Clark, B., Farnham, T., Groussin, O., Harris, A., Kissel, J., Li, J.-Y., Meech, K., Melosh, J., Quick, A., Richardson, J., Sunshine, J., & Wellnitz, D. (2013a). The nucleus of Comet 9P/Tempel 1: Shape and geology from two flybys. *Icarus, 222*, 453–466.

Thomas, P. C., A'Hearn, M. F., Veverka, J., Belton, M. J. S., Kissel, J., Klaasen, K. P., McFadden, L. A., Melosh, H. J., Schultz, P. H., Besse, S., Carcich, B. T., Farnham, T. L., Groussin, O., Hermalyn, B., Li, J.-Y., Lindler, D. J., Lisse, C. M., Meech, K., & Richardson, J. E. (2013b). Shape, density, and geology of the nucleus of Comet 103P/Hartley 2. *Icarus, 222*, 550–558.

Tiscareno, M. S., & Malhotra, R. (2003). The dynamics of known Centaurs. *The Astronomical Journal, 126*, 3122.

Tomasko, M. G. (1976). Photometry and polarimetry of Jupiter. In *Jupiter: Studies of the interior, atmosphere, magnetosphere and satellites* (p. 486). Proceedings of the Colloquium, Tucson, AZ, May 19–21, 1975, (A77-12001 02-91). Tucson: University of Arizona Press.

Tucker, W. H. (1975). *Radiative processes in astrophysics*. MIT Press.

Turcotte, D. L., & Schubert, G. (2002). *Geodynamics* (2nd ed., p. 472). Cambridge: Cambridge University Press.

Urquhart, M. L., & Jakosky, B. M. (1996). Constraints on the solid-state greenhouse effect on the icy Galilean satellites. *Journal of Geophysical Research, 101*, 21169–21176.

Vaisberg, O. L., Zastenker, G., Smirnov, V., Khazanov, B., Omelchenko, A., Fedorov, A., & Zakharov, D. (1987). Spatial distribution of heavy ions in Comet p/ Halley's Coma. *Astronomy and Astrophysics, 187*, 183.

Verigin, M. I., Gringauz, K. I., Richter, A. K., Gombosi, T. I., Remizov, A. P., Szego, K., Apathy, I., Szemerey, I., Tatrallyay, M., & Lezhen, L. A. (1987). Plasma properties from the upstream regions to the cometopause of Comet p/ Halley – VEGA observations. *Astronomy and Astrophysics, 187*, 121.

Veverka, J., Farquhar, B., Robinson, M., Thomas, P., Murchie, S., Harch, A., Antreasian, P. G., Chesley, S. R., Miller, J. K., Owen, W. M., Williams, B. G., Yeomans, D., Dunham, D., Heyler, G., Holdridge, M., Nelson, R. L., Whittenburg, K. E., Ray, J. C., Carcich, B., Cheng, A., Chapman, C., Bell, J. F., Bell, M., Bussey, B., Clark, B., Domingue, D., Gaffey, M. J., Hawkins, E., Izenberg, N., Joseph, J., Kirk, R., Lucey, P., Malin, M., McFadden, L., Merline, W. J., Peterson, C., Prockter, L., Warren, J., & Wellnitz, D. (2001). The landing of the NEAR-Shoemaker spacecraft on asteroid 433 Eros. *Nature, 413*, 390.

Veverka, J., Klaasen, K., A'Hearn, M., Belton, M., Brownlee, D., Chesley, S., Clark, B., Economou, T., Farquhar, R., Green, S. F., Groussin, O., Harris, A., Kissel, J., Li, J.-Y., Meech, K., Melosh, J., Richardson, J., Schultz, P., Silen, J., Sunshine, J., Thomas, P., Bhaskaran, S., Bodewits, D., Carcich, B., Cheuvront, A., Farnham, T., Sackett, S., Wellnitz, D., & Wolf, A. (2013). Return to Comet Tempel 1: Overview of Stardust-NExT results. *Icarus, 222*, 424.

Vigren, E., & Galand, M. (2013). Predictions of ion production rates and ion number densities within the diamagnetic cavity of Comet 67P/Churyumov-Gerasimenko at Perihelion. *The Astrophysical Journal, 772*, 33.

Vigren, E., Edberg, N. J. T., Eriksson, A. I., Galand, M., Henri, P., Johansson, F. L., Odelstad, E., Rubin, M., & Valliéres, X. (2019). The evolution of the electron number density in the coma of Comet 67P at the location of Rosetta from 2015 November through 2016 March. *The Astrophysical Journal, 881*, 6.

Villanueva, G. L., Mumma, M. J., Bonev, B. P., Novak, R. E., Barber, R. J., & Disanti, M. A. (2012). Water in planetary and cometary atmospheres: $H_2O/HDO$ transmittance and fluorescence models. *Journal of Quantitative Spectroscopy and Radiative Transfer, 113*, 202–220.

Villanueva, G. L., Smith, M. D., Protopapa, S., Faggi, S., & Mandell, A. M. (2018). Planetary Spectrum Generator: An accurate online radiative transfer suite for atmospheres, comets, small bodies and exoplanets. *Journal of Quantitative Spectroscopy and Radiative Transfer, 217*, 86–104.

Vincent, J.-B., et al. (2015). Large heterogeneities in comet 67P as revealed by active pits from sinkhole collapse. *Nature, 523*, 63–66.

Vincent, J.-B., et al. (2016a). Are fractured cliffs the source of cometary dust jets? Insights from OSIRIS/Rosetta at 67P/Churyumov-Gerasimenko. *Astronomy and Astrophysics, 587*, A14.

Vincent, J.-B., et al. (2016b). Summer fireworks on comet 67P. *Monthly Notices of the Royal Astronomical Society, 462*, S184.

Vinogradoff, V., Duvernay, F., Fray, N., Bouilloud, M., Chiavassa, T., & Cottin, H. (2015). Carbon dioxide influence on the thermal formation of complex organic molecules in interstellar ice analogs. *The Astrophysical Journal, 809*, L18.

Virkar, Y., & Clauset, A. (2014). Power-law distributions in binned empirical data. *The Annals of Applied Statistics, 8*(1), 89–119. https://doi.org/10.1214/13-AOAS710.

Volwerk, M., Goetz, C., Richter, I., Delva, M., Ostaszewski, K., Schwingenschuh, K., & Glassmeier, K.-H. (2018). A tail like no other. The RPC-MAG view of Rosetta's tail excursion at comet 67P/Churyumov-Gerasimenko. *Astronomy and Astrophysics, 614*, A10.

von Rosenvinge, T. T., Brandt, J. C., & Farquhar, R. W. (1986). The international cometary explorer mission to Comet Giacobini-Zinner. *Science, 232*, 353.

Vourlidas, A., Davis, C. J., Eyles, C. J., Crothers, S. R., Harrison, R. A., Howard, R. A., Moses, J. D., & Socker, D. G. (2007). First direct observation of the interaction between a comet and a coronal mass ejection leading to a complete plasma tail disconnection. *The Astrophysical Journal, 668*, L79.

Wallis, M. K. (1974). Expansion velocities of cometary gas. *Astrophysics and Space Science, 30*, 343–346.

Wang, A., Marashdeh, Q., Teixeira, F. L., & Fan, L.-S. (2015). *Industrial tomography* (pp. 529–549). Woodhead Publishing Series in Electronic and Optical Materials.

Wang, F., & Li, X. (2017). The stagnant thermal conductivity of porous media predicted by the random walk theory. *International Journal of Heat and Mass Transfer, 107*, 520–533.

Wang, Z., Chen, X., Gao, F., Zhang, S., Zheng, X.-W., Ip, W.-H., Wang, N., Liu, X., Zuo, X.-T., Gou, W., & Chang, S.-Q. (2017). Observations of the hydroxyl radical in C/2013 US10 (Catalina) at 18 cm wavelength. *The Astronomical Journal, 154*, 249.

Warren, S. G., & Brandt, R. E. (2008). Optical constants of ice from the ultraviolet to the microwave: A revised compilation. *Journal of Geophysical Research, 113*, D14220. https://doi.org/10.1029/2007JD009744.

Watanabe, J.-I., Honda, M., Ishiguro, M., Ootsubo, T., Sarugaku, Y., Kadono, T., Sakon, I., Fuse, T., Takato, N., & Furusho, R. (2009). Subaru/COMICS mid-infrared observation of the near-nucleus region of Comet 17P/Holmes at the early phase of an outburst. *Publications of the Astronomical Society of Japan, 61*, 679.

Weaver, H. A., Feldman, P. D., McPhate, J. B., A'Hearn, M. F., Arpigny, C., & Smith, T. E. (1994). Detection of CO Cameron band emission in Comet P/Hartley 2 (1991 XV) with the Hubble Space telescope. *The Astrophysical Journal, 422*, 374.

Weaver, H. A., Feldman, P. D., A'Hearn, M. F., Dello Russo, N., & Stern, S. A. (2011). The carbon monoxide abundance in Comet 103P/Hartley 2 during the EPOXI flyby. *The Astrophysical Journal, 734*, L5.

Wegmann, R., Schmidt, H. U., & Bonev, T. (1996). The three-dimensional structure of the plasma tail of comet Austin 1990. *Astronomy and Astrophysics, 306*, 638.

Weissman, P. R. (1986). Are cometary nuclei primordial rubble piles? *Nature, 320*, 242.

Werner, R. A., & Scheeres, D. J. (1997). Exterior gravitation of a polyhedron derived and compared with harmonic and mascon gravitation representations of Asteroid 4769 Castalia. *Celestial Mechanics and Dynamical Astronomy, 65*, 313–344.

Westphal, A. J., Fakra, S. C., Gainsforth, Z., Marcus, M. A., Ogliore, R. C., & Butterworth, A. L. (2009). Mixing fraction of inner solar system material in comet 81P/Wild 2. *The Astrophysical Journal, 694*, 18–28.

Westphal, A. J., Bridges, J. C., Brownlee, D. E., Butterworth, A. L., de Gregorio, B. T., Dominguez, G., Flynn, G. J., Gainsforth, Z., Ishii, H. A., Joswiak, D., Nittler, L. R., Ogliore, R. C., Palma, R., Pepin, R. O., Stephan, T., & Zolensky, M. E. (2017). The future of Stardust science. *Meteoritics and Planetary Science, 52*, 1859–1898.

Wilhelm, K. (1987). Rotation and precession of comet Halley. *Nature, 327*, 27.

Wilhelm, K., Cosmovici, C. B., Delamere, W. A., Huebner, W. F., Keller, H. U., Reitsema, H., Schmidt, H. U., & Whipple, F. L. (1986). A three-dimensional model of the nucleus of Comet Halley. *ESLAB Symposium on the Exploration of Halley's Comet, 250*, 367.

Whipple, F. L. (1950). A comet model. I. The acceleration of Comet Encke. *The Astrophysical Journal, 111*, 375–394.

Whipple, F. L. (1951). A comet model. II. Physical relations for comets and meteors. *The Astrophysical Journal, 113*, 464.

White, C., Borg, M. K., Scanlon, T. J., Longshaw, S. M., John, B., Emerson, D. R., & Reese, J. M. (2018). dsmcFoam+: An OpenFOAM based direct simulation Monte Carlo solver. *Computer Physics Communications, 224*, 22–43. ISSN 0010-4655.

Wierzchos, K., Womack, M., & Sarid, G. (2017). Carbon monoxide in the distantly active Centaur (60558) 174P/Echeclus at 6 AU. *The Astronomical Journal, 153*, 230.

Williams, I. P., & Wu, Z. (1993). The Geminid meteor stream and asteroid 3200 Phaethon. *Monthly Notices of the Royal Astronomical Society, 262*, 231.

Williams, K. E., McKay, C. P., & Heldmann, J. L. (2015). Modeling the effects of martian surface frost on ice table depth. *Icarus, 261*, 58.

Wong, I., Mishra, A., & Brown, M. E. (2019). Photometry of active Centaurs: Colors of dormant active Centaur Nuclei. *The Astronomical Journal, 157*, 225.

Wright, I. P., Barber, S. J., Morgan, G. H., Morse, A. D., Sheridan, S., Andrews, D. J., Maynard, J., Yau, D., Evans, S. T., Leese, M. R., Zarnecki, J. C., Kent, B. J., Waltham, N. R., Whalley, M. S., Heys, S., Drummond, D. L., Edeson, R. L., Sawyer, E. C., Turner, R. F., & Pillinger, C. T. (2007). Ptolemy an instrument to measure stable isotopic ratios of key volatiles on a cometary nucleus. *Space Science Reviews, 128*, 363–381.

Wright, I. P., Sheridan, S., Barber, S. J., Morgan, G. H., Andrews, D. J., & Morse, A. D. (2015). CHO-bearing organic compounds at the surface of 67P/Churyumov-Gerasimenko revealed by Ptolemy. *Science, 349*, 2.673.

Wu, J.-S., & Lian, Y. (2003). Parallel three-dimensional direct simulation Monte Carlo method and its applications. *Computers & Fluids, 32*, 1133.

Wu, J.-S., & Tseng, K.-C. (2005). Parallel DSMC method using dynamic domain decomposition. *International Journal for Numerical Methods in Engineering, 63*, 37.

Wu, J.-S., Tseng, K.-C., & Wu, F.-Y. (2004). Parallel three-dimensional DSMC method using mesh refinement and variable time-step scheme. *Computer Physics Communications, 162*, 166.

Wurm, G., & Blum, J. (1998). Experiments on preplanetary dust aggregation. *Icarus, 132*, 125.

Wyckoff, S., Heyd, R. S., & Fox, R. (1999). Unidentified molecular bands in the plasma tail of Comet Hyakutake (C/1996 B2). *The Astrophysical Journal, 512*, L73.

Xie, X., & Mumma, M. J. (1992). The effect of electron collisions on rotational populations of cometary water. *The Astrophysical Journal, 386*, 720.

Yamada, T., Rezac, L., Larsson, R., Hartogh, P., Yoshida, N., & Kasai, Y. (2018). Solving non-LTE problems in rotational transitions using the Gauss-Seidel method and its implementation in the Atmospheric Radiative Transfer Simulator. *Astronomy and Astrophysics, 619*, A181.

Yao, Y., & Giapis, K. P. (2017). Dynamic molecular oxygen production in cometary comae. *Nature Communications, 8*, 15298.

Yeomans, D. K., & Chodas, P. W. (1989). An asymmetric outgassing model for cometary nongravitational accelerations. *The Astronomical Journal, 98*, 1083–1093.

Yoldi, Z., Pommerol, A., Poch, O., Jost, B., & Thomas, N. (2018). Laboratory reflectance measurements of ice and dust mixtures. Application to permanently shaded regions on the Moon and Mercury. *Lunar and Planetary Science Conference, 49*, 2207.

Zakharov, V., Bockelée-Morvan, D., Biver, N., Crovisier, J., & Lecacheux, A. (2007). Radiative transfer simulation of water rotational excitation in comets. Comparison of the Monte Carlo and escape probability methods. *Astronomy and Astrophysics, 473*, 303.

Zakharov, V. V., Ivanovski, S. L., Crifo, J.-F., Della Corte, V., Rotundi, A., & Fulle, M. (2018). Asymptotics for spherical particle motion in a spherically expanding flow. *Icarus, 312*, 121.

Zubko, E., Videen, G., Hines, D. C., & Shkuratov, Y. (2016). The positive-polarization of cometary comae. *Planetary and Space Science, 123*, 63.

# Index

**0-9, and Symbols**

Λ-doubling, 256
1I/(2017 U1) 'Oumuamua, 430
2I/(2019 Q4) Borisov, 431
2P/Encke, 5, 176, 372, 423, 436
(4) Vesta, 68
9P/Tempel 1, xxxv, 5, 34–37, 45, 46, 74, 84, 113, 115, 118, 121, 158, 159, 169, 232, 269–271, 377, 387–389, 441
14P/Wolf, 47
17P/Holmes, 176
19P/Borrelly, xxxv, 8, 34, 109, 113, 190, 268, 339, 351, 422
20D/Westphal, 8
(21) Lutetia, 43, 68, 119, 120
21P/Giacobini-Zinner, xxxiv, xxxv, 208, 417, 421
22P/Kopff, 269
29P/Schwassmann-Wachmann 1, 21, 184, 430
41P/Tuttle–Giacobini–Kresák, 46, 272
45P/HMP (Honda–Mrkos–Pajdušáková), 208
48P/Johnson, 27
49P/Arend–Rigaux, 46
55P/Tempel–Tuttle, 12
67P Regions
    Aker, 90, 116, 117, 132, 151
    Anhur, 86, 88–90, 116, 174, 271
    Anubis, 86, 87, 118, 158, 160
    Anuket, 91, 131, 138
    Apis, 86, 89, 118
    Ash, 86–90, 92, 123, 124, 127, 144, 145, 149, 333
    Aten, 88, 89, 91, 177
    Atum, 86, 89, 118, 271
    Babi, 88–90, 132

    Bastet, 92, 93, 116
    Bes, 87, 89, 90, 139, 142, 143, 159, 171, 271, 360
    Geb, 73, 86, 89, 271
    Hapi, 90–92, 109, 110, 116, 117, 126, 128, 138–140, 145, 158, 160, 163, 165, 167, 171, 239, 241, 242, 270, 341
    Hathor, 92, 138
    Hatmehit, 92, 93, 127
    Imhotep, 87–89, 123, 124, 139–141, 149, 152, 153, 156–159, 270
    Khepry, 87, 89, 90, 139, 140, 151, 152, 171
    Khonsu, 86, 118, 270
    Ma'at, 92, 124, 127, 144, 145, 150, 166, 333
    Maftet, 91, 93, 127, 166, 168
    Neith, 91, 270
    Nut, 91, 166
    Serqet, 91, 147
    Seth, 88, 89, 126, 127, 130, 139, 140, 149
    Sobek, 91, 143, 158, 270
    Wosret, 91–93, 133, 135, 136, 152, 270
73P/Schwassmann-Wachmann 3, 71, 72, 435
81P/Wild 2, xxxv, 5, 35, 42, 113, 121, 125, 386, 387
96P/Machholz, 437
103P/Hartley 2, xxxv, 35, 37, 40, 51, 70, 84, 109, 113, 144, 183, 184, 197, 208, 233, 268, 269, 327, 335, 359, 406, 439
107P/Wilson-Harrington, 427, 428
133P/(7968) Elst-Pizarro, 427
143P/Kowal-Mrkos, 27, 47
153P/Ikeya-Zhang, 9, 267, 268, 407
162P/Siding Spring, 47
174P/Echeclus, 430
217P/LINEAR, 176

© Springer Nature Switzerland AG 2020
N. Thomas, *An Introduction to Comets*, Astronomy and Astrophysics Library,
https://doi.org/10.1007/978-3-030-50574-5

313P/Gibbs, 429
324P/La Sagra, 429
332P/Ikeya–Murakami, 435
345P/LINEAR, 427
354P/LINEAR, 427
(433) Eros, 119, 151
1992 QB$_1$, 19
2000 CR$_{105}$, 18
2006 BZ8, 434
2008 AP1$_{29}$, 106
2012 VP113, 18
2013 FT$_{28}$, 18
2016 LN8, 21
(2060) Chiron, 17, 18
(2201) Oljato, 428
(3200) Phaethon, 428
(3552) Don Quixote, 15, 433
(4015) Wilson–Harrington, 427
(5145) Pholus, 22
(5335) Damocles, 15, 434
(5370) Taranis, 15
(6478) Gault, 427
(8405) Asbolus, 22
(10199) Chariklo, 22
(10370) Hylonome, 22
(15809) 1994 JS, 18
(42355) Typhon, 106
(50000) Quaoar, 18, 19, 106
(90377) Sedna, 18
(90482) Orcus, 18, 106
(95626) 2002 GZ32, 22
(119979) 2002 WC$_{19}$, 18
(136108) Haumea, 106
(136199) Eris, 18
(136472) Makemake, 18
(145486) 2005 UJ438, 22
(248835) 2006 SX368, 22
(281371) 2008 FC76, 22
(486958) Arrokoth, 70
(514107) Ka'epaoka'awela, 431


**A**

Absorption cross-section, 185, 187, 188, 203,
    204, 420
Absorption efficiency, 291, 376
Active asteroids, 12
Airfall, 76, 124, 144, 145, 150, 154, 333, 359,
    363, 368, 433
Amorphous to crystalline transition, 101,
    104–107
Anharmonicity, 193, 195, 200
Anti-tail, 371
Aswan, 127, 129, 130
Augsburg Book of Miracles, 3

**B**

Ballistic Particle-Cluster Aggregation
    (BPCA), 296
Bayeux Tapistry, 3
Bernoulli effect, 326
Brightness temperature, 31–33, 84, 108, 187
Bulk density, 38, 39, 46, 71, 96, 297, 311, 312,
    324


**C**

C/1965 S1 Ikeya-Seki, 425, 432
C/1973 XII Kohoutek, 423
C/1989 X1 (Austin), 424
C/1995 O1 (Hale-Bopp), xxix, 2, 3, 272
C/1996 B2 (Hyakutake), xxix, 3, 266, 268, 406,
    422
C/1999 H1 Lee, 208
C/1999 S4 (LINEAR), 407, 435
C/2000 WM1 (LINEAR), 266, 268
C/2001 A2 (LINEAR), 267
C/2001 Q4 (NEAT), 268
C/2004 Q2 (Machholz), 266
C/2007 N3 (Lulin), 208, 406
C/2009 P1 (Garradd), 268
C/2010 G2 (Hill), 176
C/2011 J2 (LINEAR), 435
C/2011 L4 (PanSTARRS), 407
C/2011 W3 (Lovejoy), 176
C/2012 F6 (Lemmon), 208
C/2012 K1 (PanSTARRS), 208
C/2012 Q1, 430
C/2012 S1 (ISON), 268, 407, 435
C/2013 R1 (Lovejoy), 208
C/2013 US10 (Catalina), 257
C/2013 V5 (Oukaimeden), 278
C/2014 Q2 (Lovejoy), 268, 278
C/2015 ER$_{61}$ (PAN-STARRS), 176
C/2017 E4 (Lovejoy), 208
C/2017 K2 (PANSTARRS), 106, 430
Cameron bands, 267
Carbonaceous chondrites, 134
Catenae, 12
Centaurs, 17, 18, 20–22, 428, 430
Centripetal acceleration, 54, 131, 132, 324, 428
Chamberlain model, 365
Charge-exchange, 251, 402, 407, 418
Chemical compounds
    acetylene ($C_2H_2$), 208, 253, 278
    ammonia ($NH_3$), 193, 206, 208, 209, 274,
        278, 280, 391, 392
    ammonium salts, 209, 275
    carbonyl group, 390
    cyanogen (CN), 199, 251, 253–256,
        263–265, 269, 271, 272, 275, 278, 279

ethane ($C_2H_6$), 208, 278
ethanol ($C_2H_5OH$), 278
fayalite, 389
formaldehyde ($H_2CO$), 206, 208, 268, 272, 278
forsterite, 389
glycine ($NH_2$-$CH_2$-$COOH$), 209, 386, 391
hexamethylenetetramine (HMT), 279, 392
hydroxyl (OH), 256, 265, 390
methane ($CH_4$), 208, 278, 280, 391
methanol ($CH_3OH$), 183, 193, 208, 209, 272, 278
octoanoic acid, 390
polycyclic aromatic hydrocarbons, 386
polyoxymethylene, 209
pyroxene, 134, 387, 432
tholin, 327
2-hexanone, 390
Circular polarization, 285
Cis-Neptunian objects, 17
Clathrates, 101, 272
Coherent backscatter opposition effect (CBOE), 59
Cohesive forces, 154, 155, 162–164, 166, 325–327, 428
Collisional excitation rate, 203
Collision strength, 203
Cometopause, 414
Complex refractive index, 62, 292, 336
Compressive strength, 72–74, 122, 123, 127, 440
Contact surface, 203, 269, 415, 418
Coronal mass ejection, 423
Coulomb collisions, 399
Cubewanos, 19, 20

**D**
Damping timescale, 52
Dayside to nightside brightness asymmetry, 357
Desiccation polygons, 136
Diamagnetic cavity, 415
Differential scattering cross-section, 288
Directional-hemispherical albedo, 28, 54, 64
Discrete dipole approximation (DDA), 299
Doppler broadening, 187, 188
Doppler width, 187, 188
Drag coefficient, 298, 316, 322, 333
D/Shoemaker-Levy 9, 12, 71, 72, 436
Dune pit, 168

Dunes, 115, 168
Dust size distribution, 148, 318, 320, 322, 333, 338, 344, 380, 393, 394, 442
Dust trails, 372
Dynamic (ram) pressure, 401, 413

**E**
Eccentric anomaly, 6
Effective active fraction, 25, 93, 110, 238
Ejecta blanket, 123
Electron cooling, 203, 420
Electron impact ionization, 251, 407, 418, 419
Electrostatic levitation, 152
Enki Catana, 12
Euler angles, 49, 51
Excess energy, 258, 259, 419
Extinction cross-section, 186, 291
Extinction efficiency, 291

**F**
Fireballs, 72
Flat-bottomed depressions, 121
Flat floor, 113
Fluidization, 152, 154–157, 166
Fluid threshold, 162, 164
Fountain model, 313
Fourth Positive system, 267
Fractionation, 76, 98, 99, 101–104, 205, 233, 276, 279
Fraunhofer absorptions, 263
Free sublimation temperature, 73, 76, 79, 83, 98, 100, 111, 126, 223, 246, 272, 380, 439
Frozen-in magnetic field, 399, 414, 416

**G**
Galactic tides, 14, 16
Ganymede, 12, 13
Geminid meteor stream, 372
Geometric albedo, 27, 29, 34, 54, 58, 63, 65, 294
g factor, 200
GHOSST database, 106
Giant molecular clouds (GMCs), 14
Got bands, 200
Greenstein effect, 265
Gyrofrequency, 410, 411
Gyroradius, 400, 410, 411, 422

**H**
Henyey-Greenstein (H-G) functions, 59
Hill sphere, 360, 368, 383
HL Tau, xxix, xxx
Hot bands, 196, 201, 206
Hyperactive comets, 279

**I**
Icy-glue, 69, 70
Impact cratering, 119, 125
Infrared beaming parameter, 110
Infrared emissivity, 31
Inter-mixture, 327
Internal temperature, 99, 102–104, 246, 280
Interplanetary dust particles, 298
Intra-mixture, 327
Ion-electron recombination, 409
Ion pile-up, 409
Isotopologues, 190–192, 197, 205

**J**
Jacobi constant, 11
Johnson-Morgan magnitude system, 3, 28

**K**
Kelvin-Helmholtz instability (KHI), 418
Kirchhoff's laws of spectroscopy, 190
Knudsen layer, 212, 220, 221, 224–234, 323
KOmeten-Simulation (KOSI), xxxiii, 74, 98
Kracht family, 432
Kreutz family, 311, 432

**L**
Lambertian surface, 28, 57, 58, 64
Landslide deposits, 144
Largest liftable mass, 100, 325
Larmor radius, 410
LASCO, 383, 431, 437
Laws and equations
    Ampere's law, 399
    Barker's equation, 7
    Boltzmann equation, 214, 224–228
    Born-Oppenheimer approximation, 185
    Bruggeman's theory, 304
    Clausius-Clapeyron equation, 77, 81
    Collisionless Boltzmann Equation, 226, 335
    Euler equations, 214, 216, 218–220, 228, 257
    Faraday's law, 399

    Fick's law, 169
    Haser model, 251–258
    Hertz-Knudsen equation, 78, 79
    incompressible Euler momentum, equation,217
    Kepler's equation for parabolic motion, 7
    Kepler's third law, 6
    Kirchhoff's law, 32, 186, 377
    Lommel-Seeliger law, 58
    Lunar-Lambert law, 64
    Maxwell Garnett theory, 304
    Minnaert function, 63, 64
    Navier-Stokes equations, 214, 218, 219, 417
    Ockham's razor, 355
    Planck's radiation law, 31, 187
    radiative transfer equation, 185, 187, 198
    Rayleigh-Jeans law, 31, 187
    Sobolev's method, 198
    Sutherland's law, 324
Linear mixing model, 175
Linear polarization, 287
Local thermal equilibrium (LTE), xxxiv, 185, 211
Long Axis Mode (LAM), 49
Lyman-α, 258

**M**
Magnetic induction, 410
Magnetic reconnection, 423
Magnetic sector boundary crossings, 423
Magnetohydrodynamics (MHD), 415
Marsden family, 432
Massive ice, 173
Maxwell-Boltzmann distribution, 78, 211
Mean anomaly, 6
Mean free path, 167, 185, 210, 211, 214, 216, 221, 228–230, 240, 241, 244, 261, 323, 335, 420, 442
Mesas, 113
Meyer family, 432
MiARD project, 175
Micrometeoroid impact, 153
Momentum transfer coefficient, 25
MSPCD technique, 175, 176
Müller matrix, 286

**N**
Natural broadening, 187
Neck-line structure, 370
Newton's false-root method, 6
Niedner-Brandt sequence, 423

Non-gravitational forces (NGFs), 5, 14, 23, 24, 38, 153
Non-principal-axis rotational states, 47
North Polar Crater Cluster (NPCC), 120, 121
Numbers
    Knudsen number, 98, 214–216, 219, 228, 233, 244
    Knudsen penetration number, 228–232, 240, 241, 345
    Mach number, 217, 220, 221, 230, 231, 346, 412
    Reynolds number, 322
Numerical tools
    DSMC, 212, 226, 227, 233–235, 238, 242, 269, 334, 345
    GENGA, 15
    HITRAN, 198
    hydrocode, 121
    iSALE, 121
    OpenFOAM, 227
    Parallelized Direct Simulation Monte Carlo Code, 227
    Quantemol database, 201
    SMILE, 227
    smoothed particle hydrodynamics, 70
    Splatalogue database, 256
    Web Vectorial Model, 258

**O**

Oort cloud, 14–18, 21, 105, 276, 442, 444
Optical depth, 142, 186, 191, 201, 203–205, 260, 267, 302–304, 308, 323, 334, 347, 354, 367, 376, 404, 405, 420
Orbital eccentricity, 6, 17

**P**

P/2004 A1, 12
P/2008 CL94 Lemmon, 21
P/2010 TO20 Linear-Grauer, 21
P/2010 V1 (Ikeya-Murakami), 176
P/2011 S1, 12, 430
Particle fragmentation, 352, 355
Pebble accretion, 130
Permeability, 244, 306, 401
Philae experiments
    APXS, viii
    CIVA, viii
    COSAC, vii
    MUPUS, vii, 74, 98
    PTOLEMY, vii, 209
    ROLIS, vii

    ROMAP, vii
    SD2, vii
    SESAME, vii
Philae landing sites
    Abydos, 74, 99
    Agilkia, 73
Photo-dissociation, 212, 216, 251
Photoemission, 420
Pitch-angle scattering, 411
*Pit halo*, 113
Pits, 90–92, 126–130, 137, 150, 159, 168, 169
Planck function, 29, 32, 191, 197, 206, 312, 378, 380, 419
Planetary migration, 277
Planetary Spectrum Generator (PSG), 201, 202
Plasma turbulence, 413
Polarization, 263, 283, 285
Ponded deposits, 89, 90, 151–153
Pore area fractal dimension, 98
Pore ice, 173
Porosity, 38, 40, 74, 95–98, 100, 101, 122, 123, 127, 153, 155, 169, 192, 243, 244, 246, 296, 297, 306, 325, 382, 439
Poynting-Robertson effect, 368
Prefix, 8
Pressure broadening, 187
Prompt emission, 265

**R**

Radiance, 30, 32, 55, 57, 64, 65, 139, 185–187, 282, 283, 294, 302–304, 308, 316, 381
Radius of gyration, 297
Rayleigh (unit), 204
Rayleigh scattering, 282
Reactive torque, 25, 44
Re-condensation, 100, 440
Reflectance coefficient, 57
Reflectance factor, 57, 61, 63, 65, 66, 68, 308
Reptation, 162, 164, 166
Resonant fluorescence, 199, 200, 256, 263, 268, 274
Restricted three-body problem, 11
Rigid rotator model, 189
Ring-masking, 339
Ripple, 162, 165
Rocket effect, 368
Rosetta experiments
    ALICE, viii, 268, 269
    CONSERT, viii, 71, 384, 392, 441
    COSIMA, viii, 279, 319, 384, 441
    COSISCOPE, 320
    GIADA, viii, 319, 333

Rosetta experiments (*cont.*)
  MIDAS, viii, 298
  MIRO, vii, 85, 104, 135, 189–191, 193,
    195, 223
  OSIRIS, vii, 43, 44, 63, 66, 68, 116, 117,
    120, 131, 139, 141, 142, 145, 157, 241,
    268, 269, 309, 343, 350, 351, 359, 366,
    369, 376
  ROSINA, viii, 4, 26, 206, 209, 232,
    239–241, 247, 268, 270, 271, 273, 278,
    279
  ROSINA/COPS, viii, 4, 239, 241, 247, 442
  RPC, viii
  RSI, viii, 127, 365
  VIRTIS, vii, 85, 205, 206, 270, 378–380,
    389
Rubble pile, 69, 71, 72

**S**
Saffman lift force, 166, 326
Scattered Disc Objects (SDOs), 20
Scattering cross-section, 185, 294, 335, 353,
    379
SECCHI, 432
Seismic shaking, 152, 153
Self-heating, 76, 109, 110, 137
Self-shadowing, 109, 110, 126
Shadow-hiding opposition effect, 59
Short Axis Mode (SAM), 49
Single scattering albedo, 58, 62, 291, 376
Sintering, 101, 123, 150
Size-frequency distribution (SFD), 316
Size parameter, 288, 300, 311, 338, 376
Slipface, 165
Sodium tail, 3, 274, 275
Solar radiation pressure, 1, 146, 274, 311–313,
    330, 357, 368
Solar wind, xxxvi, 2, 261, 377, 383, 399–402,
    406–408, 411–416, 418, 422–424
Solid-state greenhouse effect, 96
Spacecraft
  Chandra, 407
  Deep Impact, xxxv, 34, 39, 45, 51, 74, 83,
    113, 116, 123, 159, 169, 232, 266, 269,
    377, 387, 444
  Deep Space 1, xxxiii, xxxv, 34, 114, 340,
    351, 357, 422
  EPOXI, xxxv, 51, 116, 335, 359, 360
  EUVE, 406
  Far Ultraviolet Spectroscopic Explorer, 267,
    268

  Giotto, xxxii, xxxv, 8, 34, 65, 66, 69, 70, 81,
    184, 209, 278, 313, 318, 334, 339–341,
    351, 353, 354, 375, 376, 392, 409, 413,
    415
  Herschel, 276, 277
  HST, 268
  Hubble Space Telescope, 12, 27, 43, 45,
    267, 268, 435
  IRAS, 372, 429
  ISEE-3, xxxiv
  NEAR Shoemaker, 119, 120
  New Horizons, 70
  ROSAT, 406
  Sakigake, xxxv
  SOHO, 258, 383, 432, 437
  Spitzer, 43, 372, 387–389, 433
  Stardust, xxxiii, xxxv, 45, 74, 125, 298,
    385–387, 389
  Suisei, xxxv
  Vega, xxxv, 34, 51, 107, 206, 341, 350, 384
Spacecraft potential, 420
Specific heat capacity, 81, 85, 100, 218
Spectral gradient, 68, 375
SPICE kernels, 46
Splash, 157, 164, 166
Standard thermal model (STM), 110
Stefan's problem, 100
Stereo photogrammetry (SPG), 35, 36
Stokes parameters, 283, 288
Stoss, 165, 166
Striations, 383
Sublimation coefficient, 79, 81
Sublimation front, 95, 98, 100, 101, 249
Sunward spike, 371
Surface gravitational potential, 40
Swan band, 276
Swings effect, 264
Swiss cheese, 126

**T**
Tail rays, 423, 424
Talus, 88, 89, 91, 92, 138–140, 142, 158, 171
Tectonics, 135
Tensile strength, 38, 46, 54, 70–72, 122, 127,
    132, 434, 436, 440
Thermal diffusivity, 81
Thermal emissivity, 80, 108
Thermal fatigue, 134, 428
Thermal inertia, 5, 23, 29, 31, 81–85, 94, 96,
    98, 134, 135, 233, 241, 242, 249, 358,
    370, 440

Thermal parameter, 83, 370
Thermal shock, 134, 428
Thermal skin depth, 82, 95, 102, 135, 248
Threshold friction velocity, 162
Tisserand parameter, 10–12, 17, 20, 21, 427, 442
T-matrix, 295, 316
Topographic focussing, 344
Tortuosity, 98, 100, 244, 325
Trans-Neptunian Objects, xxxiii, 17
True anomaly, 6, 7, 24, 396
Truncated Pareto distribution, 321
TW Hydrae, 209

**V**
van der Waals force, 326

Vectorial model, 258
Velocity distribution function (VDF), 80, 185, 211, 220, 323
Vienna Canyon Diablo Troilite (VCDT), 279
Vienna Standard Mean Ocean Water, 279
Voids, 71, 95, 96, 127, 304
Voigt profile, 189
Volume absorption, 96

**X**
X-ray emission, 407

**Z**
Zodiacal light, 375

Printed in the United States
by Baker & Taylor Publisher Services